COLORED SHEEP AND WOOL

Exploring Their Beauty and Function

The Proceedings of
The World Congress on Coloured Sheep
U.S.A. 1989

Edited by Kent Erskine

Black Sheep Press
Ashland, Oregon U.S.A.

For Sachiye Jones, who brought life to the World Congress dream,
For Dona Zimmerman, without whose help this book would not have been completed,
For all the authors, whose participation in the World Congress made it a special success,
And our friends on the World Congress Committee, who worked so hard to nurture the seeds of communication, which grew and flowered into a world-wide gathering of natural colored wool enthusiasts.

Library of Congress Cataloging-in-Publication Data

World Congress on Coloured Sheep (1989:
Eugene, OR)
Colored Sheep and Wool: Exploring Their Beauty and Function
1. Sheep breeds—Congresses. 2. Sheep—Color—Congresses. 3. Wool—Congresses.
I. Erskine, Kent. II. Title.
SF371.2.W67 1989 636.3'145 89-7012
ISBN 0-9608736-2-7

For information and to order copies, contact:

Black Sheep Press
1690 Butler Creek Road
Ashland, OR 97520 U.S.A.

Design and Typesetting: Bookends Typesetting, Ashland, Oregon
Illustrator: Carol Watkins, Mt. Horeb, Wisconsin
Layout: Crystal Castle Graphics, Ashland, Oregon
Printing: IPCO Printing, Ashland, Oregon

CONTENTS

NATURAL COLORED SHEEP FROM AROUND THE WORLD

BREEDING, CONSERVATION AND GENETICS OF NATURAL COLORED SHEEP

SHEEP HEALTH AND MANAGEMENT

UNDERSTANDING AND PREPARING WOOL

PROCESSING AND USING NATURAL COLORED WOOL

MARKETING NATURAL COLORED WOOL PRODUCTS

INTRODUCTION

To natural colored sheep we owe our very lives. At the dawn of history our survival in a harsh environment depended to a large degree on the first domestication of the wild colored sheep. The success of our species is attributable in part to our woolly friends, whose rich fleece covering has provided warmth, shelter and a medium for artistic expression in developing civilizations for thousands of years.

Natural colored sheep have earned my deep love and respect, for even today they mean life itself to those of us who tend them. They are creatures who sustain us, and we find ourselves in turn sustaining them and honoring their existence. There are certainly ways in this modern economy to turn a greater profit than sheep raising, and yet we find ourselves repeating the cycle of birth and production year after year. When asked why, most would answer, "Simply because we want to do this, and can't imagine a world without sheep grazing on grassy pastures."

At the World Congress on Coloured Sheep, held every five years in a different location, producers and consumers of natural colored wool share their knowledge and rekindle their interest in sheep. Eugene, Oregon, U.S.A., hosts the World Congress in 1989. The papers selected for inclusion in this book were submitted by the speakers of this Congress, who came to share their knowledge and insights on all the aspects of natural colored sheep and wool. We on the organizing committee of the World Congress feel that the information is so valuable as to warrant being made available to everyone via this book. We trust that we have accurately recorded the proceedings of the World Congress, and have provided an inspirational text about much of the work being done on raising natural colored sheep and using colored wool all over the world.

The papers have been edited for clarity and content. However, each article expresses the information in local idiom, and in some cases they were translated to English from the author's language. We have attempted to provide consistency of terminology, and realize that each author has expressed his or her knowledge in the clearest way possible.

I would particularly like to thank a few people, without whose help this book, and the World Congress, would not have taken the form it has. Norio Inaba, our good friend from Iwate, Japan, placed the Congress project squarely on its feet at the beginning by granting a substantial donation to the fledgling organization. Carol Watkins of Wisconsin, U.S.A., whose artwork graces the pages of this book, did not hesitate when it came to happily illustrating our published material. Dr. Melinda Burrill of California proofread these articles for consistency and accuracy, and helped me to integrate the new terminology with the old. And Dona Zimmerman gave me endless hours of her already overbooked time, in advice and artistic direction. There are too many other people to mention whose efforts were vital to the success of this project, and who gave much of themselves to overcome tight deadlines and maintain a high level of quality in their work. Thank you all.

Above all else, I honor the spirit of Sachiye Jones, who brought home from the 1984 New Zealand Congress the first embers of the plan for a World Congress on Coloured Sheep in the U.S.A. She did not live to see its eventual completion. To her, and to all who came before us, this book and the World Congress on Coloured Sheep is dedicated.

— Kent Erskine
Editor

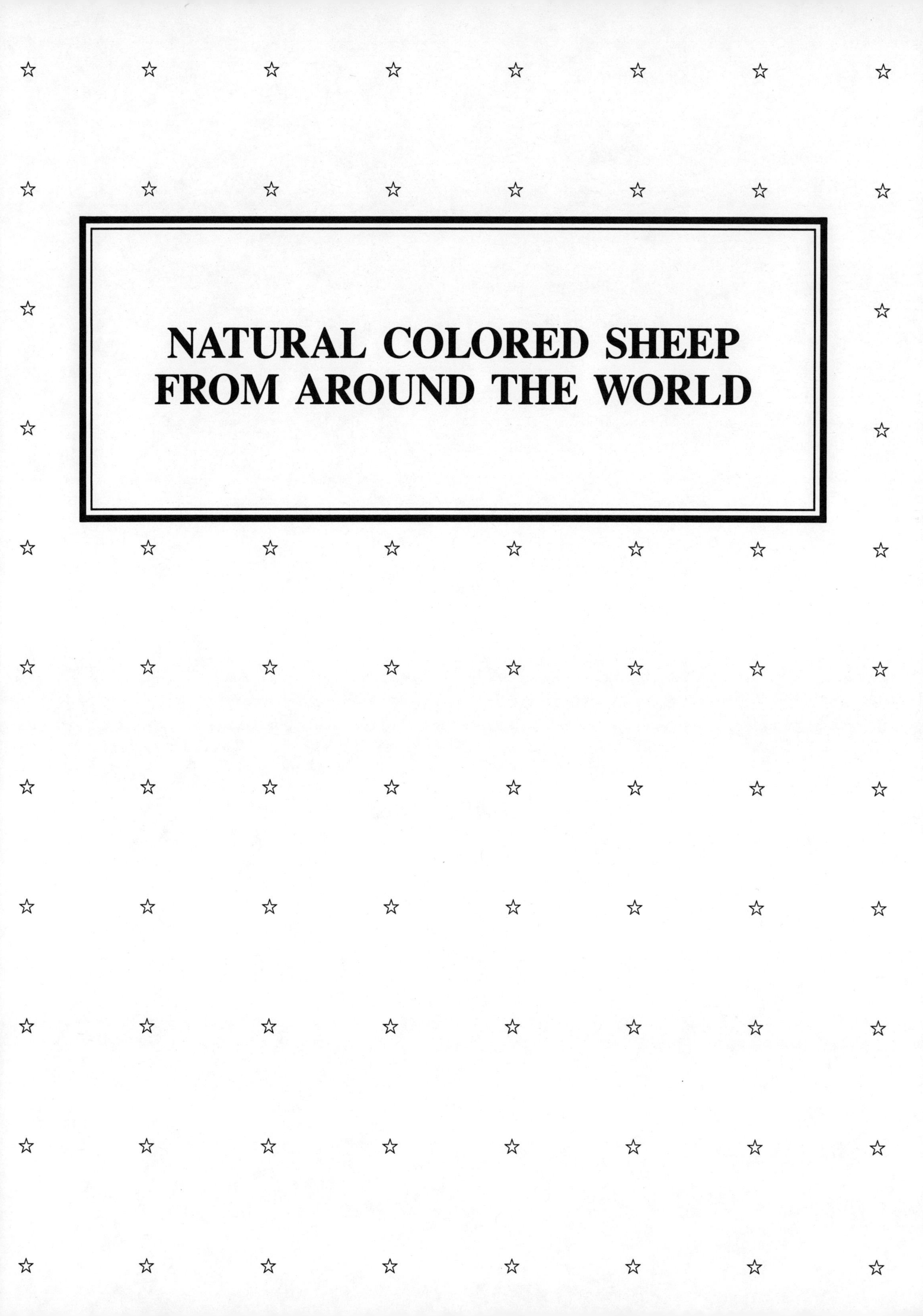

NATURAL COLORED SHEEP FROM AROUND THE WORLD

Bucking the Trend with Coloured Wool

Michael L. Ryder
4 Osprey Close, Southhampton, SO1 8EX U.K.

For thousands of years people have been selectively breeding sheep for white wool. The present interest in coloured wool is therefore a reversal of this trend with its aim of re-creating colours that were common in the past. Even before sheep with a fleece had been developed, mutant white animals are likely to have had a novelty value, but the real stimulus to breed for white wool came only after the development of dyes that allowed wool to be stained with a greater range of colours than found among the natural colours (Ryder, 1983).

Colour in the Wild and Its Function

As a zoologist my instinctive approach to colour is to go back to wild animals to see what range of colours exists in these and to look at the function of colour in the wild. The colours of the coat of wild mammals range from white through grey to black, and through brown to yellow and red. Domestication did not create any new colours and so sheep have the same range as in the wild ancestor. Colour has two opposing functions in wild animals. On the one hand it provides a means of communication with other animals, giving a warning to other species, or indicating mating desires to members of the opposite sex of the same species. On the other hand colour functions in camouflage and concealment, and these are important in the adaptation of an animal to its environment. Although, since most mammals are colour-blind, colour itself is less important than its shade or tone, and the pattern of the colour.

Concealing patterns can be either disruptive or cryptic. A disruptive pattern is provided by the zebra's stripes and these function best at a distance. Cryptic patterns resemble the background and function best at close quarters. They often change with the season, becoming paler in winter. A cryptic pattern is brought about by the super-imposition of spots or stripes on agouti-banded hair. In agouti banding the outer hairs (but not the underwool) have a black tip. This pointed and tapering part of the outer hair is a few millimetres in length, and below this is a brown band up to a centimetre long. This band is shorter and paler in the hairs of the winter coat. The rest of the hair is grey, and this is paler in the winter than in the summer hair. The complicated coat patterns of mammals therefore involve much variation in the colour and shade of individual hairs.

The term "agouti" is derived from a mammal with the same name and is in turn used as the name of the A colour gene series, which I shall be describing in detail. Wild sheep have agouti banding, which is associated with the pale belly possessed by wild sheep. My descriptions refer to the European mouflon. There are variations in other varieties of wild sheep such as the bighorn, which I will not go into. Although a pale belly may have originated through the lack of a need to have protection against ultra violet radiation on the underside of the body (such protection being another function of pigment, particularly in the skin of hairless areas) it has the concealing effect in hoofed animals of neutralising the dark shadow on the ventral surface.

Most of the upper part of the body appears brown in wild sheep, and here as well as agouti hairs there are others that are all brown. The saddle patch of the mouflon has hairs that are completely white. There are also hairs that are all black, and these are found in the black band that separates the pale belly from the brown upper parts, as well as in the black mane. The raising of the mane provides an example of the way in which the conflict between concealment and display can be resolved by the raising of the outer hairs.

Since black objects radiate more heat than white ones, it has been thought that arctic animals with a white coat keep warmer by losing less heat. In fact this radiation difference does not apply to animals, and so the function of a white coat in polar regions must be entirely camouflage.

The Evolutionary Loss of Colour Since Domestication

Sheep were domesticated about 9000 BC and the coat of the first domestic sheep (of the Neolithic period) would have been the same as that of the wild ancestor, i.e., it had coarse outer kemps, which obscured fine underwool.

It took several thousand years for a fleece to develop. We know this because the Mouflon sheep of Corsica and Sardinia are now thought to be not truly wild sheep but feral descendants of domestic animals introduced by farming settlers about 6000 BC (Poplin, 1979). Since these have a coat

Figure 1 White, self-colour (black or brown) and piebald "hair" sheep in ancient Egypt about 2500 BC (from Ryder, 1983, p. 105).

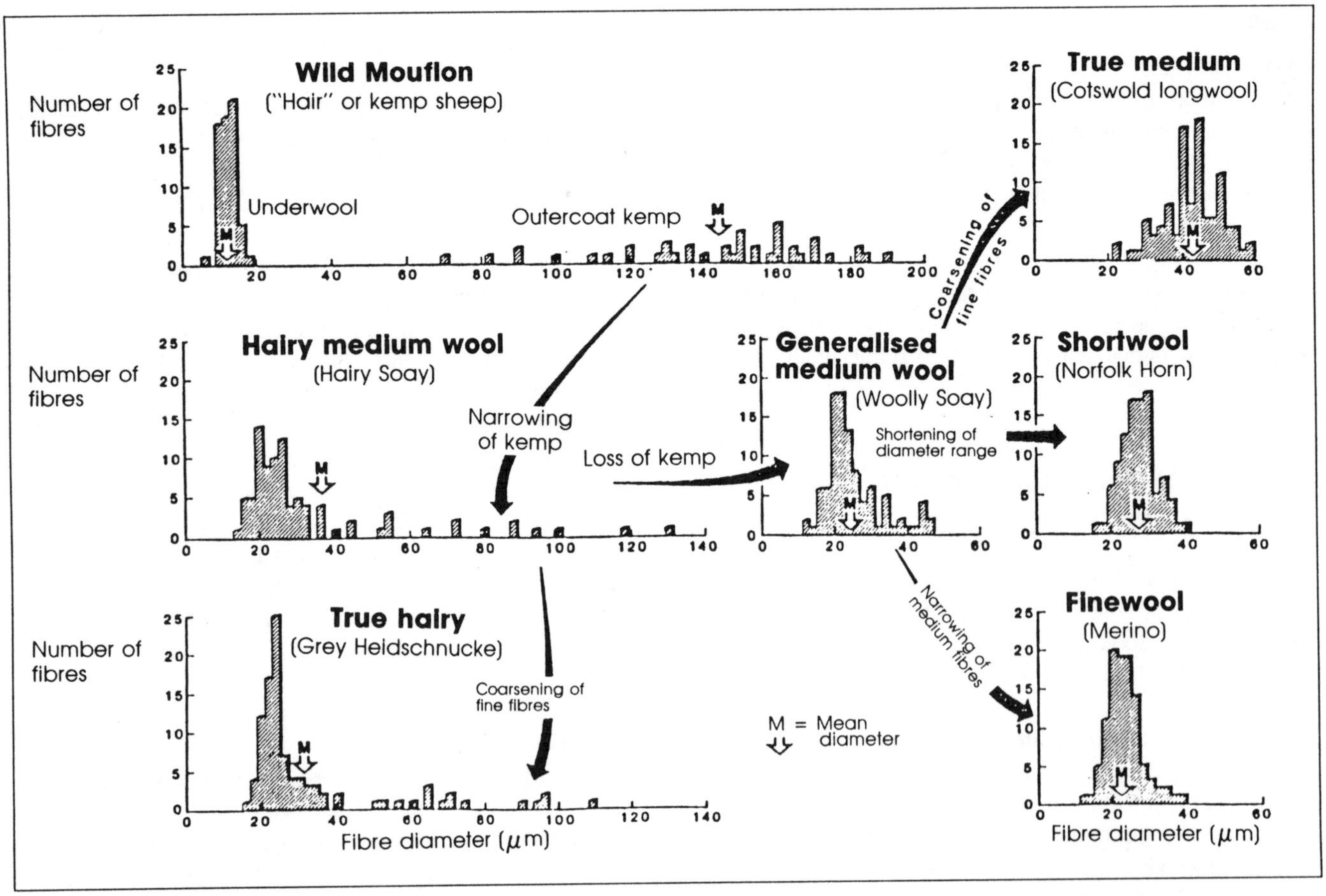

Wild Mouflon
("Hair" or kemp sheep)
Number of fibres
Underwool
Outercoat kemp
M
True medium
(Cotswold longwool)
Coarsening of fine fibres
Hairy medium wool
(Hairy Soay)
Narrowing of kemp
Loss of kemp
Generalised medium wool
(Woolly Soay)
Shortening of diameter range
Shortwool
(Norfolk Horn)
True hairy
(Grey Heidschnucke)
Coarsening of fine fibres
Narrowing of medium fibres
Finewool
(Merino)
M = Mean diameter
Fibre diameter (μm)

Figure 2 Changes in fibre diameter distribution during the evolution of modern fleece types by a progressive narrowing of the outer coat kemps (from Ryder 1983, pp. 46 and 131).

that is apparently no different in structure from that of the bighorn wild sheep of North America (Ryder, 1958), which was never domesticated, it appears that no change took place during the 3000 to 4000 years that elapsed between domestication and the introduction of sheep into these Mediterranean Islands.

The Neolithic type of coat survives little changed in the kempy, so-called "hair" breeds of tropical Africa and India. These have black and white animals in addition to the brown of the wild ancestor. Indeed black or brown, white and piebald "hair" sheep were depicted very early in ancient Egypt (Figure 1). The existence of white "hair" sheep before the development of a fleece indicates that the initial loss of colour did not have a textile stimulus. Black, and more rarely white, colour variants occur in wild ruminants and domestication allowed such mutants not only to survive, but eventually to multiply, when human selective breeding began to be applied. I detailed the way in which different fleece types evolved in the Proceedings of the Second Congress in New Zealand (Ryder, 1984a) and Figure 2 is reminder of the evolutionary changes (see also Ryder, 1987).

Sheep with a fleece first appear during the Bronze Age, which began in Europe about 2000 BC, and the predominant colour in wool cloth remains from this period is brown. The measurement of wool fibres in such textiles supports the conclusion from skeletal remains that the sheep of the Bronze Age were like the Soay breed (Figure 3). Soay sheep survive on the St. Kilda Islands off northwest Scotland. Here they live in a feral condition, i.e., they are domestic animals that have gone wild. But some have been taken to the mainland and many people in Britain keep small flocks of Soay sheep for their brown wool, which has two main shades.

The Soay, like the wild ancestor, has a white belly and the upper parts can be either dark brown (in sheep carrying the black gene) or light brown (in sheep carrying the brown gene). The self-colour gene produces a coloured belly when sheep with the black gene become all black, while animals with the brown gene become completely brown (see details of genotypes below). Black sheep are not uncommon in the Soay breed, but white sheep are rare, and grey animals unknown. Piebald Soay sheep are common, however.

White wool did not become common in cloth remains until the Iron Age, which began about 1000 BC in Europe. Now with the full development of textile crafts there is a real stimulus to breed white sheep, not just as a novelty, but to produce wool that can be dyed. This was when grey wool first apppeared and this colour predominated (Ryder, 1989), being produced in sheep by a mixture of black and white fibres. Despite the increase in number of white fleeces, naturally-coloured wools continued to be used in the weaving of patterns through the Roman period and the Middle Ages until recent times. There is the so-called "Falkirk tartan" of Roman Scotland (Ryder, 1983, p. 753), in which coloured and white wool was woven together to produce a check (see Figure 4). And seventeenth-century bog burials from Co. Derry, Ireland had wools of different shades of brown, as well as possible grey, woven together (Ryder, 1983, p. 518).

The native sheep of the Orkney and Shetland islands to the north of Scotland, appear to be survivors from the Iron Age. These breeds have brown, black, grey and white

Figure 3 Light and dark brown Soay sheep with Mouflon pattern (wild type) white belly (from Ryder, 1983, p. 35).

anmals, as well as piebald, but sheep with the wild-type white belly of the Soay are virtually absent. Similar breeds remain in pockets throughout Europe, to form the European vari-coloured type. A common feature of these sheep is the predominance of grey animals, and natural grey cloth used to be widespread, from the *grautucher* of Switzerland to the hodden grey of the English Lake District. Shetland knitwear continues to embody natural black, brown, and grey in Fair Isle patterns on a white background.

The very popular four-horned and piebald Jacob breed could have had a similar origin to the other four-horned breeds of Britain despite numerous attempts to seek exotic origins. But it probably contains other influence since the fleece is coarser and the tail is no longer short. The Black Welsh Mountain breed, too, may be a survivor of an earlier type since unlike the ordinary Welsh Mountain breed its colour is dominant. There are specific Welsh names for the different colour patterns, the white belly being called *torwen* and white sheep with a black belly (badger-face) *torddu*.

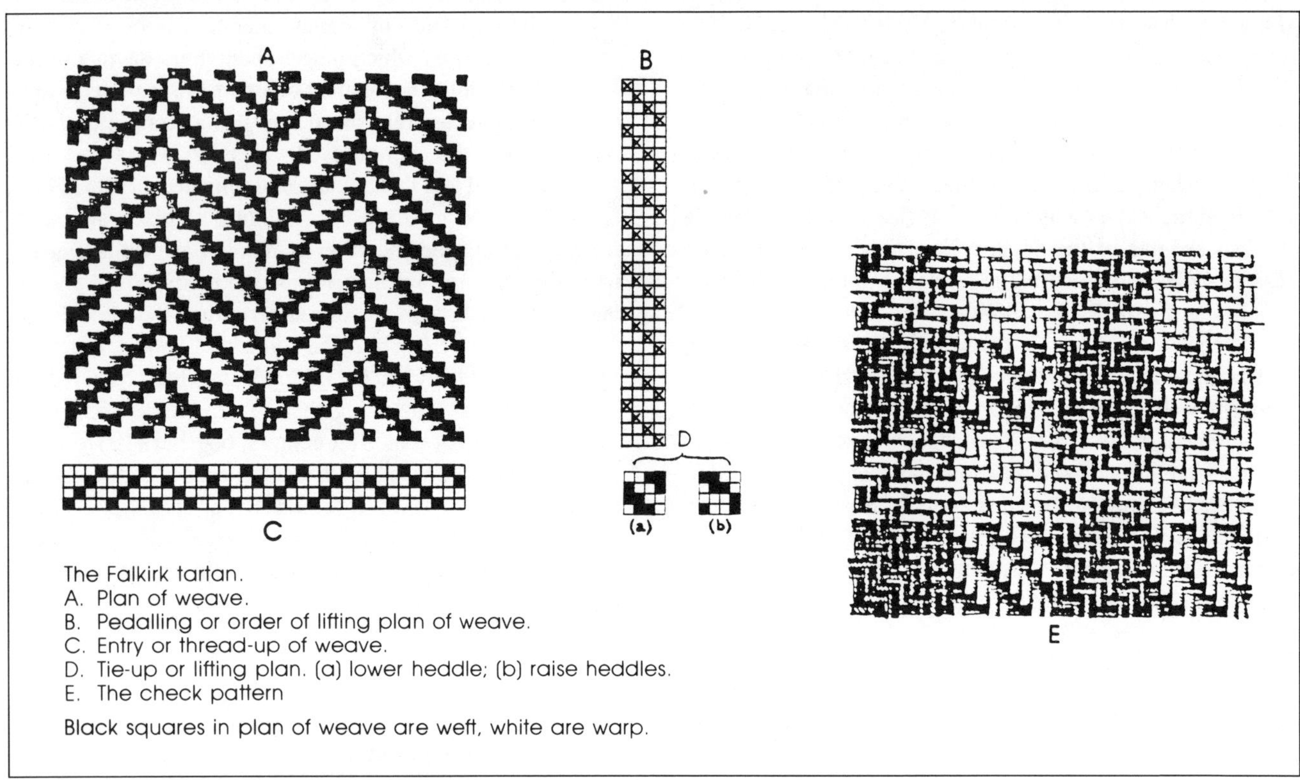

The Falkirk tartan.
A. Plan of weave.
B. Pedalling or order of lifting plan of weave.
C. Entry or thread-up of weave.
D. Tie-up or lifting plan. (a) lower heddle; (b) raise heddles.
E. The check pattern
Black squares in plan of weave are weft, white are warp.

Figure 4 Roman check woven from naturally-coloured wool.

The Colours in Surviving Primitive Breeds of Europe

Perhaps the best-known group of European vari-coloured sheep is the Northern short-tail, of which I have already mentioned the Orkney and Shetland breeds. Each Scandinavian country has or had its native breed with coloured individuals from Iceland and Faroe through the Spaelsau of Norway and the Goth of Sweden to the Finnish Landrace and the Romanov of Russia. The last two are noted for large litters, and the Finnish Landrace is now mostly white. Others in Britain are the four-horned breeds—the black four-horned breed that used to be called the St. Kilda and has been re-named the Hebridean to avoid confusion with the Soay, and the brown four-horned breed, the Loghtan of the Isle of Man (see Figure 5).

Spain has the Guirra breed and France has the Ushant of the island with the same name, as well as the Corsican, which has been shown by Lauvergne and Adalsteinsson (1976) to have similar colour genotypes to Icelandic sheep. This indicates a common ancestry, and supports the idea of a European vari-coloured type as discussed here.

A group of breeds, known as Heath Breeds, formerly extended across the north European Plain from the Netherlands to Poland, and indeed into Russia if one regards the Romanov breed as a link between the Northern Short-tail and the Heath Breeds. The three Dutch Heath breeds are the Drenthe, the Veluwe and the Kempen. These have the full range of colours with many piebald sheep, but the tail is long. The Heidschnucke of Germany has a short tail and there are two varieties—one with horns, a black face and a grey fleece and one without horns with a white fleece. The Wrzosowka of Poland has the full range of colours including the different shades of grey found in the Swedish Gotland breed, as well as various colour patterns.

In the Alps, Switzerland had the Bundner breed with brown as well as grey sheep, and the Steinschaf remains in Austria with the full colour range and the same genes as in Icelandic and Corsican sheep. The Della Roccia is the Italian example, and Italy south of the Alps has the grey or brown Massa breed and the black Carapella.

The breeds of South East Europe are divided into hairy, Zackel sheep and finer-woolled, but more primitive Ruda sheep. The Zackel type starts in southern Poland with the Polish Mountain breed, which has a range of colours and occurs in Hungary as the Racka. This has corkscrew horns and only black and white varieties. The Zackel is represented in Romania by the Turcana breed, which has coloured varieties. There is a greater range of colours in the Ruda type of Romania, the Tsigai breed. A Zackel breed that is found further south is the Karakachan of Greece as well as Bulgaria, which has the full colour range. The Ruda type of Bulgaria is the Karnobat, which is unusual in having a brown fleece with a black face (Ryder, 1984b). Similar breeds in Greece are the Drama and the Chalkidiki.

The Caucasus has the fat-tailed Karachaev breed, which has black and grey sheep. This breed raises the interesting question of whether fleece variation, including colour, had evolved before the development of a fat tail, or whether fleece variation developed separately in different broad types of sheep, i.e., thin-tailed and fat-tailed. Moving out of Europe, there is a similar breed on the Altai Steppe of Siberia with brown as well as black and grey sheep.

The Evolution of Different Colour Genotypes

Coat colour provides a good indication of the complexity of genetics. The different colours are clearly visible, and appear to be controlled by a relatively few major genes, yet it is often difficult to distinguish which of several theories provides the true explanation for a series of breeding observations. One reason for this is that in the past different people described the same colour in different ways; also, what are actually different colours are often given the same name by different people.

Another trouble with fleece colour genetics is that there are few coloured breeds to study. Research on the subject was reviewed by Ryder and Stephenson (1968), but the real breakthrough was made by Adalsteinsson (1970) using thousands of observations on coloured Icelandic sheep. The literature was reviewed again by Ryder (1980). It was the worldwide demand for naturally-coloured wools by handicraft workers that created an interest in breeding sheep with coloured wool, which in turn led to the international conferences of which this is the third. The Australian conference in 1979, and the one in New Zealand in 1984, provided the basis for the first international discussions on the subject by scientists, and particularly on the terminology of fleece colour. At the 1984 conference the "Committee of the Genetic Nomenclature of Sheep and Goats" (COGNOSAG) was set up and Lauvergne (1984) has suggested the use of comparative colour scales to define and standardise shade.

Going back to basics again, most research on colour genetics has been carried out with mice, and at least seven gene loci on different chromosomes are recognised. It is probable that these gene loci are present in most mammals and their importance certainly varies between species (Searle, 1968). The seven loci (positions at which series of colour genes can occur) and their importance in sheep, are shown in Table 1.

TABLE 1 The Importance of Different Colour Gene Loci in Sheep

Agouti	A	Very important—determine pattern.
Brown	B	Very important—determine colour.
Albino	C	Very rare.
Dilute	D	Unimportant (clumps pigment granules).
Extension	E	Known, makes colour dominant.
Pink eye	P	Not known.
Spotting	S	Moderately important.

Gene loci A, B and S are the most important ones in European sheep. Locus B is concerned with the colour type, and two genes are known, black Bl or B and brown B2 or little b. Black is dominant to brown, therefore brown sheep are homozygous recessives (bb) carrying two little b genes. Both black and brown belong to the black (eumelanin) type of pigment; the change to the tan (red-yellow, phaeomelanin) type of pigment is governed by genes at the A locus.

Since black is dominant to brown it is impossible to have black and brown fibres in the same fleece. Where this is apparently seen, the "brown" fibres are not genetically brown but tan. I was confused by goats seemingly having a mixture of black and brown hairs before I realised that the "brown" hairs are in fact of tan colour. Two species as closely related as sheep and goats would be expected to have similar colour inheritance, but goats apparently have no brown gene. "Brown" breeds of goats are therefore really tan and in fact genetically white.

Such subtle distinctions stress the importance of strict terminology. With older descriptions of light brown sheep one is never sure whether they are genetically brown or tan. I will later mention mixtures of brown and white fibres, and roan, which is uncommon in sheep. The difficulty becomes even worse with sheep descriptions from antiquity. I was "thrown" for several years when a classicist drew my attention to the reference to "red" sheep. I even investigated the possibility that the eating of a dye plant might cause the fleece to become red. I put alizarin, the dye of the madder plant, into the drinking water of white mice, but the coat did not change. The answer was all the time "staring me in the face" in such common English names as "red" deer. And indeed there was a medieval cloth colour called "russet", but there is no certainty that the wool was naturally-coloured.

Returning to the colour genes, there are at least eight genes at the A locus for colour pattern. These show their effect on a brown as well as a black background. In other words they act only in combination with the genes of the B series. Pigment inhibition, which produces white animals, is dominant over the production of eumelanin pigment. The top dominant gene of the A series, A1 or Awh (dominant white) inhibits all eumelanin production, that is black and brown colour, but it does not inhibit phaeomelanin production. This explains why white sheep can have tan fibres. The bottom recessive gene of the series, A5 or little *a*, allows full expression of black and brown; this is the self colour gene giving a coloured belly.

The remaining common genes of the A series are A2 or Ag (grey) and A4 or Aw (Mouflon or wild pattern giving white belly). Grey is dominant to wild pattern and both of these genes again show their effect on a brown as well as a black background. In Soay sheep therefore the A4 gene gives a white belly, and with the black gene gives dark brown upper parts, while with the brown gene it gives light brown upper parts.

Soay, Orkney and Shetland sheep were the first breeds outside Iceland to which the theory of Adalsteinsson was applied (Ryder, Land and Ditchburn, 1974) and some of the apparent genotypes are shown in Table 2.

TABLE 2 Colour Genotypes in Soay, Orkney and Shetland Sheep

Dark brown wild-pattern (white belly)	AwAwBB Awa BB	AwAwBb Awa Bb
Light brown wild pattern	AwAwbb	Awa bb
Black (self-colour)	aaBB (homozygous)	aaBb (heterozygous)
Brown	aabb (homozygous recessive)	

Dark brown Soays are genetically black and Table 2 shows the four possible genotypes. There are four because the two genes can be either homozygous (e.g. BB) or heterozygous (e.g. Bb). This also explains why dark Soays are more common than light ones. With the light brown, wild pattern

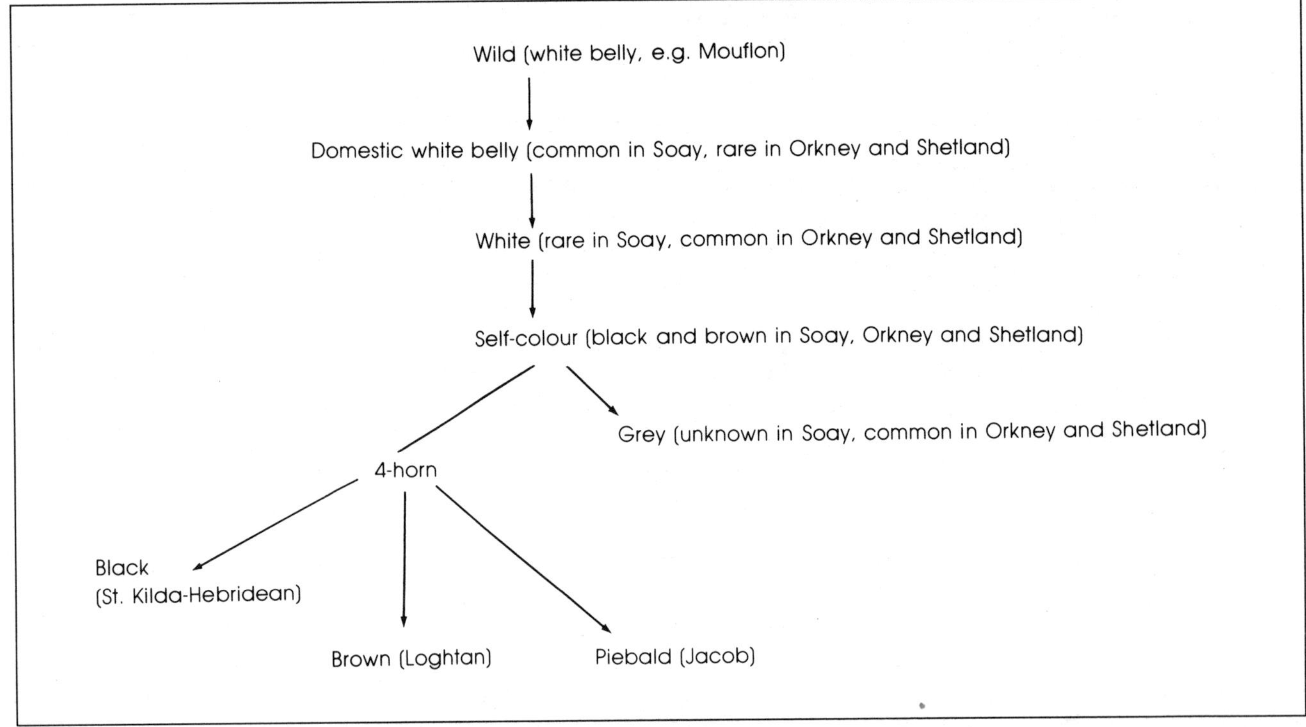

Figure 5 Possible relationships of colour types and four-horned breeds.

Soays there are only two genotypes because each is homozygous for brown (bb). There are also two genotypes for black (aaBB and aaBb). Heterozygous (aaBb) black sheep appear no different from homozygous (aaBB) animals and the only way to distinguish them is from the colour of their offspring: whereas homozygous blacks breed true, if heterozygous blacks are mated together, one brown lamb is produced for every three black ones, i.e.—

black (aaBb) X black (aaBb)
= black (aaBB) + 2 black (aaBb) + brown (aabb)

Only one of the three black lambs will breed true, the other two are heterozygotes like their parents. It was the appearance of a brown lamb in a flock of black Shetland sheep I was working with that caused me to begin studying sheep colour genetics. Also, if a heterozygous black (aaBb) sheep is mated with a brown one (aabb), half the lambs produced will be black (aaBb)—all heterozygotes—and half brown (aabb), i.e.—

black (aaBb) X brown (aabb) =
black (aaBb) + brown (aabb)

Brown is therefore the simplest colour with only one genotype (aabb).

Turning now to Orkney and Shetland sheep, the brown (moorit) and black genotypes are apparently identical to those in the Soay. But in addition to self-colour (*a*) and wild-pattern (Aw) there are white (Awh) and grey (Ag). White is the top dominant gene of the A series and self-colour the bottom recessive giving the following series: Awh, Ag, Aw (rare outwith the Soay and wild sheep), and *a*. Other less common genes of the series have been omitted from this simple summary.

The grey gene on a brown background gives a mixture of brown and white fibres termed brown-grey, or more picturesquely by the Norwegians: "brown skimlet". There are two genotypes for "brown skimlet" (AgAgbb and Agabb) and four for true grey ("black skimlet") produced by the Ag gene on a black background, light grey and dark grey having different genotypes (Table 3). Brown-grey is not in fact roan, which is rare in sheep, but more common in cattle. Another similar mixture of fibres is tan (phaeomelanin) and white found only in white sheep (below). These similarities make it very difficult, if not impossible, to assign to a particular genotype a mixture of brown and white fibres, even when viewed under the microscope. There are many genotypes for white if all the genes in the A series are included in the various possible combinations. Table 3 shows the nine white genotypes including only the Awh, Ag and *a* genes.

It remains to consider piebald or spotted sheep. White spotting is produced by genes at the S locus; there are two genes here, Sl or big S producing unbroken colour in pigmented sheep, and S2 or little *s*, which produces broken colour, i.e., white patches. All black and brown sheep must therefore carry the big S gene, which is dominant to little *s*. Piebald sheep are therefore homozygous recessives (*ss*) and they do in fact breed true. As an aside it can be remarked that piebald sheep are therefore coloured animals with white markings, and not white sheep with black markings, which is what they appear to be.

Loss of Colour During Fleece Evolution by Mutation

The major change in colour occurred between the Bronze Age Soay sheep and the Iron Age vari-coloured type and mutations in Icelandic sheep observed by Adalsteinsson suggest one way in which this change came about. He has observed the wild pattern (white belly) gene (Aw) to mutate to white (Awh). White sheep therefore appeared first rather than last. This could explain the rarity of sheep with a white belly subsequent to the Bronze Age type. Then white (Awh) has been observed to mutate to self-colour (*a*) giving all brown or all black animals. This explains the occurrence of some white wool in Bronze Age textiles, and also the brown and black sheep in the Soay breed. Finally, self-colour (*a*) has been observed to mutate to grey (Ag), which causes a mixture of coloured and white fibres. The fact that this is the last mutation in the series explains the lack of grey in the Soay, and its predominance in the vari-coloured (Iron Age) type. These changes are summarised in Figure 5.

Selective breeding for white sheep could have been carried out in at least two ways. First, by the selection of all white mutants, i.e., animals carrying the Awh gene, which is dominant over colour genes. Secondly, by the selection of piebald sheep with greater and greater areas of white wool

TABLE 3 Colour Genotypes in Orkney and Shetland Sheep

Brown-grey (brown skimlet)		AgAgbb	Agabb
Black-grey (black skimlet)	light	AgAgBB	AgAgBb
	dark	Aga BB	Aga Bb
White	AwhAwhBB	AwhAwhBb	AwhAwhbb
	AwhAg BB	AwhAg Bb	AwhAg bb
	Awh a BB	Awh a Bb	Awh a bb

in each generation. The great variation in the extent of white markings in piebald Soays indicates the feasibility of this approach, but it would have resulted in recessive white sheep, and breeds of this type apparently do not exist. Adalsteinsson suggests that because coloured sheep in modern breeds frequently have white markings, selection probably took place for dominant white unconsciously in combination with the spotting gene. Since the white wool of piebald sheep is often finer than the black, the second approach would also have led to finer fleeces (Ryder and Adalsteinsson, 1987).

Selection for white sheep in the past is unlikely to have affected the colour genes. Black, brown and grey are therefore almost certainly still present in modern white breeds, but are hidden owing to the inhibiting action of dominant white and also of spotting genes. Breeders of coloured sheep obviously want to reveal these genes by crossing with coloured rams. Any coloured lambs obtained in the first cross will unwittingly tell us something about breed relationships, and so open up a whole new area of research into breed origins and sheep history.

Adalsteinsson and Ryder (1984) showed how the colour of lambs from matings of a coloured ram with white ewes can reveal which colour genes the mother was carrying. Half of the lambs produced by each ewe will be white, but the other half will have the colour of the genes that were hidden in the mother. If the ewe is carrying black and the ram is black the colours can be black, grey or black with a white belly. If the ewe is carrying brown and a brown ram is used, the colours become brown, brown-grey, and brown with a white belly. There is also badger-face, which I have not mentioned before, with white upper parts and a coloured belly.

Brown rams in fact reveal more since they can distinguish homozygous (BB) ewes from heterozygous (Bb) ewes. Heterozygous ewes produce two white lambs for every two coloured ones, one being black and the other brown as in the above series. If a spotted brown ram is used, even more information is revealed—any spotted lambs indicate that the ewe was carrying the spotting gene. Although piebald sheep breed true, different coloured offspring can be obtained by crossing with other breeds, at least when using the spotted Jacob breed, which is dominant black, i.e., going back to Table 1, it has the dominant allele of the E gene. Black lambs have resulted from crosses with the Dorset Horn, black and white lambs, roughly in the proportion three to one, have resulted from crosses with Scottish Blackface sheep, white offspring have been obtained from Cheviot crosses, and piebald lambs from matings with the Polled Dorset breed.

The Distribution of Pigment within the Fibre

Keratin, the horn substance of wool, is colourless. Natural colour occurs in wool in granules. In this it differs from dyed colours and discoloration, which are diffused through the fibre. Whereas phaeomelanin pigment granules are spherical, eumelanin granules are ovoid. Those in human hair are 0.7 microns long by 0.3 microns wide. Since one micron equals one thousandth of a mm, they are less than a thousandth of a mm across. Hairs and coarser fibres are usually more densely pigmented than finer fibres, and this is most marked in grey fleeces.

Cross sections of fibres viewed under the microscope allow one to see the distribution of pigment granules within the fibre. In most coloured wool fibres of medium or coarse diameter the distribution of granules is fairly even within the cortex, which forms the bulk of the fibre. Crimped, that is, wavy, wool fibres frequently have pigment on one side of the fibre only—inside the curve, or there may be a greater concentration on that side as seen in brown (moorit) wool from the Shetland breed, and the underwool of the wild ancestor. The outer cuticle of wool also usually contains pigment, which is variable in the central medulla, where in some species it is lacking.

There are variations in the hairs of other animals which provide features used in the identification of the species. In human hair the pigment is concentrated on the outside of the cortex under the cuticle. In horse hair there is a ring of pigment in the middle of the cortex, while in camel hair and pig bristles, the pigment is concentrated towards the inside around the medulla. Alpaca, camel, dog and pig have no pigment in the cuticle so that the hairs appear to have a colourless rind.

There are also changes in pigmentation along the length of fibres, as in the agouti banding already mentioned. In self-coloured sheep the tip of a fibre that is growing to replace one that has shed has the densest pigmentation, while the base is paler and pigment production often ceases altogether before the fibre stops growing and enters the rest period that precedes a moult. But at the time of harvesting, the tips of the wool staples of black sheep have often faded to dark brown as a result of a year's exposure to sunlight. As an abnormality, a deficiency of copper in the diet can cause black wool to grow with white bands.

The Relation of Pigment Granules to Colour and Shade

The granules responsible for colour are produced by pigment cells in the wool follicle bulb. The melanin within the granules is an indole pigment formed by the oxidation of the colourless amino acid tyrosine. Most of the colour variation in sheep is not related to the difference between the more common brown-black eumelanin and the less common yellow-red phaeomelanin. The only colour produced by phaeomelanin in sheep is the uncommon "tan" of individual fibres within a fleece.

The difference between brown and black in sheep must be brought about by a difference in the size, and/or concentration of the granules, or perhaps even in the intensity of pigment within the granule, as in human hair. Compared with other mammals, research is lacking on this aspect of colour in sheep. Grey in sheep, on the other hand, is produced by a mixture of eumelanin-coloured and white fibres. True grey (black and white) is much more common than brown-grey (brown and white) in sheep. The grey of the underwool of wild sheep, however, is apparently brought about by one of the other factors such as a reduced concentration of granules, and not by a mixture of black and white fibres.

Quantification of Fibre Pigmentation

When I first began to record pigment in microscopic preparations of wool, my only quantification was into: little, moderate, and much, pigmentation from an overall impression of the number of fibres with pigment and the intensity of pigmentation within fibres. I now count the percentage of pigmented fibres, so that a sample with no coloured fibres is clearly white, while one with 100 per cent pigmented fibres could be black or brown wool, and those with a mixture of pigmented and white fibres are from grey fleeces. These show much variation from less than 5 per cent pigmented fibres, which would be difficult to detect with the naked eye, and this is named "White grey" by Swedish breeders of the grey Gotland breed. They recognise four more grades: Light grey (13% pigmented fibres), Medium grey (23% pigmented), Dark grey (33%), and Black grey (52%). In my experience these percentages are low, particularly that of the black-grey, which I think has over 90 per cent black fibres. There is also the question that biological characteristics of this kind show continuous variation, with no clear cut demarcations.

The Biosynthesis and Chemistry of Pigment

Melanin granules are formed by cells called melanocytes, which can actually move within the skin. These cells do not enter the fibre; the pigment granules are transferred from the melanocytes to the cells of the fibre during its growth. Eumelanin is produced by the oxidation of the essential amino

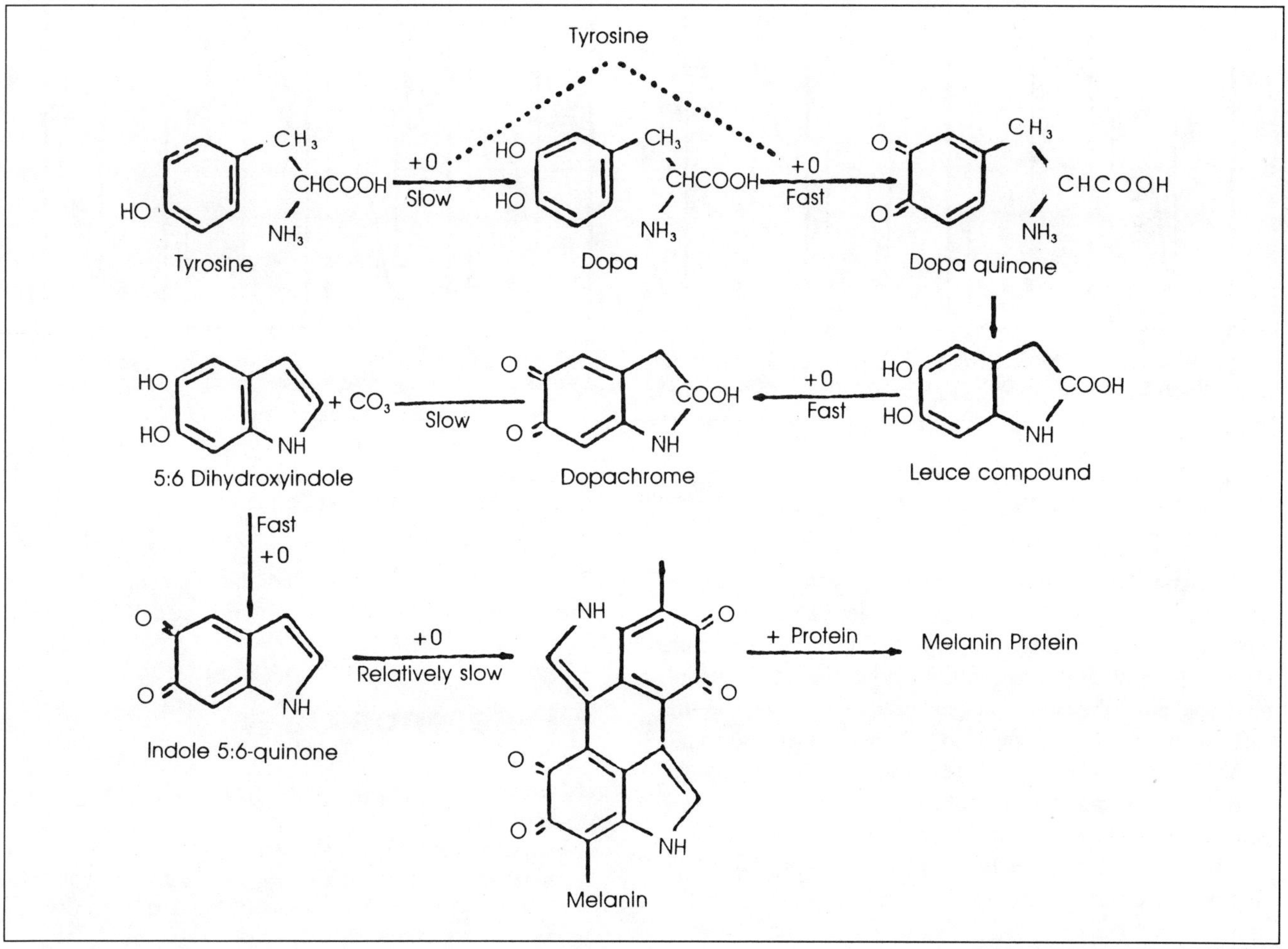

Figure 6 Biochemical steps in the formation of melanin.

acid tyrosine, which is also found in keratin, with the help of the enzyme tyrosinase (Searle, 1968). The stages involved in this transformation are, first, the conversion of tyrosine to 3, 4-dihydroxyphenylalanine ("DOPA") by the catalytic action of tyrosinase, a copper-containing oxidase or phenolase (see Figure 6). The DOPA is then oxidised to DOPA quinone, which undergoes a number of further oxidative changes and polymerization before becoming melanin itself.

Eumelanin is therefore an insoluble, brown indolequinone polymer of high molecular weight, which is always attached to a protein. It contains about 63% carbon, 5% hydrogen and 5 to 10% of nitrogen as well as 0.1 to 4% sulphur, with smaller amounts of chlorine. The variation in the amount of nitrogen and sulphur occurs between species and has not been explained. Eumelanin can be obtained from hair in the form of a black powder by treatment with boiling hydrochloric acid, but considerable purification is necessary to remove the accompanying protein.

compare the ESR of melanin from black hair with melanin synthesised from DOPA, and similar results were obtained. Then a difference was found that makes it possible to distinguish between eumelanin and phaeomelanin. Black human hair (containing eumelanin) had a single peak in the ESR spectrum, while red human hair (containing phaeomelanin) produced an ESR spectrum with more than one peak. The method was tested on wool from Karakul lambs and single peaks were obtained from black, grey and dark brown wool, while multiple peaks were obtained from several lighter shades of brown, indicating that these were caused by phaeomelanin. In another investigation with European sheep, single peaks were obtained with black and brown Icelandic wool as well as with light and dark brown Soay wool. This confirmed not only that black and brown, but also that light as well as dark brown in the Soay are produced by eumelanin (see Figure 7). Tan Icelandic wool, as well as red kemp from the Welsh Mountain breed, produced two

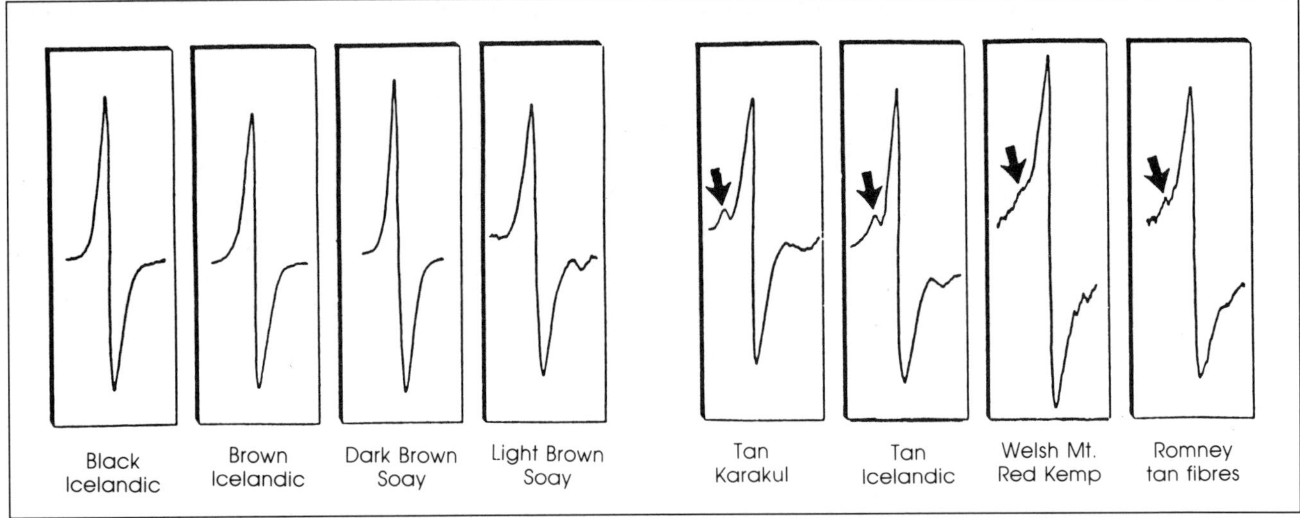

Figure 7 Electron Spin Resonance spectra showing the subsidiary peaks produced by yellow-red phaeomelanin in contrast to the single peak of brown-black eumelanin (from Vsevolodov, Adalsteinsson and Ryder, 1987).

Phaeomelanin, unlike eumelanin, is soluble in dilute alkali, and so whether or not such treatment removes pigment from wool is a crude way of distinguishing tan from brown pigment. Less is known about the synthesis of phaeomelanin. Tyrosine again seems to be necessary in the early stages, but there appears to be a switch mechanism involving the amino acid tryptophan, which causes the pigment to end up as phaeomelanin, which has more sulphur than eumelanin (Prota and Thompson, 1976).

The insolubility of melanins makes them impossible to analyse by the usual methods. A new method known as Electron Spin Resonance (ESR) is therefore of particular value. The method produces spectra in which the number of peaks varies with the kind of pigment. The first approach was to

ESR peaks, so confirming that the pigment of these is phaeomelanin (Vsevolodov, Adalsteinsson and Ryder, 1987).

References

Adalsteinsson, S. (1970). Colour inheritance in Icelandic sheep. *J. Agric. Res., Iceland* 2, 3–135.

Adalsteinsson, S., and Ryder, M. L. (1984). "Detecting hidden coloured genes in white breeds," pp. 160–162 in H. T. Blair (ed.), *Coloured sheep and their products.* (Proc. 2nd World Congress on Coloured Sheep, New Zealand, 1984).

Lauvergne, J. J. (1984). "Physical measurements of the colour of melanin-pigmented wool," pp. 242–253 in H. T. Blair (ed.),

Coloured sheep and their products. (Proc. 2nd World Congress on Coloured Sheep, New Zealand, 1984).

Lauvergne, J. J., and Adalsteinsson, S. (1976). Genes pour la couleur de la toison de la brebis *Corse. Ann. Genet. Sel. anim.* 8 (2), 153–172.

Poplin, F. (1979). Origin of the corsican Mouflon in a new palaeolontological perspective by feralising. *Am. Genet. Sel. Anim.* 11, 133–143.

Prota, G., and Thompson, R. H. (1976). Melanin pigmentation in mammals. *Endeavour* 25, 32–37.

Ryder, M. L. (1958). Follicle arrangement in skin from wild sheep and primitive domestic sheep. *Nature* 182, 781–783.

Ryder, M. L. (1980). Fleece colour in sheep and its inheritance. *Animal Breeding Abstracts* 48, 305–324.

Ryder, M. L. (1983). *Sheep and Man.* Duckworth, London.

Ryder, M. L. (1984a). The historical development of fleece type and loss of colour. Pp. 212–221 in H. T. Blair (ed.), *Coloured sheep and their products.* (Proc. World Congress on Coloured Sheep, New Zealand, 1984).

Ryder, M. L. (1984b). The coloured sheep of Europe. Pp. 56–69 in H. T. Blair (ed.), *Coloured sheep and their products.* (Proc. 2nd World Congress on Coloured Sheep, New Zealand, 1984).

Ryder, M. L. (1987). The evolution of the fleece. *Scientific American* 255 (1), 112–119.

Ryder, M. L. (1989). Skin, and wool-textile remains from Hallstatt, Austria. *Oxford J. Archaeol.* In press.

Ryder, M. L., and Adalsteinsson, S. (1987). A note on differences in coat structure between the black and white areas of piebald lambs. *J. agric. Sci. Cambridge.* 108, 379–382.

Ryder, M. L., Land, R. B., and Ditchburn, R. (1974). Colour inheritance in Soay, Orkney and Shetland sheep. *J. Zoology, London.* 173, 477–485.

Searle, A. G. (1968). *Comparative genetics of coat colour in mammals.* Logos/Academic Press, London.

Vsevolodov, E. B., Adalsteinsson, S., and Ryder, M. L. (1987). Electron Spin Resonance spectrometrical study of the melanins in the wool of some North European sheep in relation to their colour inheritance. *J. Heredity.* 78, 120–122.

The Navajo Sheep Project: A Blueprint for Genetic Conservation

Dr. Lyle G. McNeal
Professor, Animal, Dairy & Veterinary Science Department
Utah State University
Logan, Utah 84322-4815 USA

Perhaps the title of this paper and presentation could be a little misleading, since I do not plan to provide in this discussion actual blueprints or a formal plan, because a blueprint implies, forethought, discrete planning, a fairly structured road map and generally an organized effort to accomplish a preplanned project. In reality, the Navajo Sheep Project has succeeded largely without any blueprint or orchestrated effort through a day-by-day, visceral feel for the work at hand. Like the Navajo sheep, where the fittest survives in their indigenous hostile environment, so has the author survived his hostile professional environment in order to direct a cause that may someday serve as a blueprint for others to follow. Akin to our early explorers, and contemporary heroes who have accomplished "first ever" feats, the best made plans and blueprints for success cannot cover all of the unexpected. So has been my effort to initiate and pave the way for other farm animal conservation programs and projects in the U.S.

Situation and Concern

Thousands of local races or breeds of domesticated animals evolved as Man carried his agricultural pursuits to the far corners of the world. Each had diverse features and qualities that suited the needs and environment of its area, and this process of increasing diversity continued for thousands of years. But in the present century, as breeders learned genetics, reproductive physiology and other animal sciences, the trend was reversed. Animal breeders began concentrating on fewer and fewer breeds. So-called "superior breeds" began replacing the local animals from which these breeds had often been derived. Just like last-year's-model farm machinery, hundreds of breeds have come to be considered obsolete.

There is danger in this, for agriculture is not static. As bio-technology, marketing, consumer preference, fuel prices, national policies and scientific knowledge change, so agriculture changes too.

Many breeds of domestic farm animals are obsolete because during most of this century, livestock breeders have concentrated only on an animal's gross production traits. For example, Holsteins are now the predominant dairy animal in the U.S., constituting over 80% of the national dairy herd. Jerseys, Milking Shorthorns, Guernseys, and Ayrshires compose the balance. Although the U.S. beef cattle industry has seen a great influx of exotics and other temperate zone breeds of cattle, the Hereford breed still comprises a majority of the beef cow herd. One of the fastest growing breeds of sheep in the U.S. that originated in England, the Suffolk, has in recent years found itself in a genetic cesspool of problems. With no immediate solutions in sight, Suffolks have suffered significant setbacks as a popular breed. The Suffolk was developed originally by crossing the Southdown with the Norfolk Horn, whose last remnants died in the mid-1970's. The Suffolk breeders have no way to retrace the steps to re-develop and correct the insidious genetic disease Spider Lamb, that faces the breed. Should the national dairy herd of Holstein cattle face similar health consequences, such as confronted the Suffolk sheep breed, milk production could be dramatically curtailed and cause food shortage of dairy products for the human population. It is an awareness of these potentially hazardous consequences that brought the author to an interest in the genetic conservation of endangered farm or domestic animals in the U.S.

Animal genetic resources are of immense importance to humankind, since they affect the domestic livestock and fowl which provide food, fiber, and draft and associated products for human welfare throughout the world. There are aesthetic and cultural benefits from conservation of breeds of livestock, just as for conservation of buildings, machinery, art, books, and other items of historical, social or commercial importance. Benefits from livestock improvement accrue to the consumer rather than to the breeder or producer, and so a national viewpoint is taken. Any benefits from conserving an endangered breed of livestock will depend on the total value of the national (or international) market, on the costs of conservation, on the proportion of genes used in future commercial production, on the proportion of gain in economic efficiency over current livestock, on the number of years until commercial use, and on the period of years of use. However, the time involved in dissemination or substitution, to achieve the required breed mix, may reduce the benefits obtained.

The Origins of Domestic Animal Conservation

Historically, in the United States the preservation and conservation of endangered species of wild flora and fauna began after World War II. However, the interest and value of preserving animals and plants that have served the human population didn't really surface until the early 1970's. The earliest activities of farm animal preservation that I became familiar with is the work of Mr. Jim Henson, and the Rare Breed Survival Trust in the U.K. The work accomplished by Mr. Henson has certainly become a benchmark to follow for other individuals and groups interested in the survival of rare breeds.

My early interest and exposure to the possibility of some domestic farm animals disappearing from the gene pool scene started in 1971, while I was a member of the Animal Science faculty at California State Polytechnic University at San Luis Obispo, California. Leading my students on one of our annual field trips, we discovered a livestock producer raising two almost-extinct genotypes of farm animals; the mule-footed pig, and the "old-type" Navajo sheep. Although his animal numbers were small, my inner interest was triggered toward the Navajo sheep. Since my practical and professional training was centered in sheep and wool production, it was very natural to store that impression for later retrieval. In 1977 I learned that this individual intended to get rid of his "old-type" Navajo sheep flock. I personally felt that was a shame and thought that perhaps this person would consider donating a small number of breeding ewes and two differently related rams for me to work with, while I was at Cal Poly. Arrangements were made for me to select six ewes and two four-horned rams from the flock and that started my active involvement with genetic conservation. It wasn't until sometime after I returned to San Luis Obispo with the new flock, that I began to worry about the reality of future breeding plans etc. However, it was that rocky and uncertain beginning that started the Navajo Sheep Project. With scrounged pasture ground and borrowed equipment the humble Navajo flock received some mixed signals from my colleagues. All during the years of my professional training, these sheep, the "old-type" Navajo were referred to as "scrubs", and an "unimproved" breed type. Over the last decade I have continued to receive considerable professional negative signals on my efforts to save and preserve these unique animals. Although, in this presentation it may appear that the blueprint was formally laid out, I share the practical aspects of the genetic conservation of domestic animals, only through hindsight and "hands-on" personal experiences.

About the same time that my work with the "old-type" Navajo sheep began in California, the American Minor Breeds Conservancy (AMBC) was formed to alert U.S. citizens about the importance of preserving minor farm animal breeds. AMBC however, was located in New England and the concept was so new that no assistance was available. Both the Navajo Sheep Project and AMBC had attainable goals in genetic conservation, but neither had significant funds because the concept of rare breeds of farm animal preservation hadn't caught on yet. It was too new of an idea. It was still in vogue to save rare earthworms, water beetles, and shrubs, but not regular farm animals. Both the NSP and AMBC struggled in those formative years holding the movement together, mainly with the zeal of an evangelist and the guts and fortitude of a combat soldier.

The Navajo Sheep Project has succeeded because of a network of people throughout the United States and beyond, who strategically contributed their financial means, and vocal support.

In early 1979 I was offered a position on the faculty in the Animal, Dairy and Veterinary Sciences Department at Utah State University, Logan, Utah. My assignment was to serve as an on campus sheep and wool specialist to augment established expertise and reputation in the sheep and wool sciences. I was given primary responsibility to develop an attractive and well balanced undergraduate program in the sheep and wool area, incorporating industry internships with the university, new courses for students to receive "hands-on" training and the development of a strong liason between the university and the sheep industry in Utah. One of the conditions of my coming to USU was that I could have my small

Navajo sheep flock to continue my work and interest. So in mid-summer 1979 the Navajo flock was moved from San Luis Obispo, California to Logan, Utah.

The flock has grown from six to one-hundred and forty-five head of mature breeding ewes. During this ten year period the growing flock had to relocate six times. During the past decade the USU Navajo Sheep Project played a major role in the formation of a Navajo-Churro registry made up of breeders and non-breeders alike. AMBC also provided major assistance in the formation of the American Navajo-Churro Sheep Association.

What Is the Navajo-Churro Sheep and Why Save It?

Actually, the "old-type" Navajo sheep is really not of Navajo origin, it is of Spanish lineage. It just happens that the last people in the United States who capitalized on this

Navajo Ewe 93k, 1949 U.S.D.A. Southwestern Sheep Breeding Laboratory, Fort Wingate, New Mexico (from collection of Dr. Lyle G. McNeal).

genotype of sheep were the Navajo flock growers and weavers. In 1538, the Spanish conquistador Hernando Cortez brought the first Spanish Merino sheep, as well as Churro sheep, to his hacienda at Cuernavaca, near Mexico City. He later distributed them among the missions of Mexico.

In 1540, Coronado brought large herds of cattle and sheep from Mexico into what is now the United States. These herds helped supply his expedition searching for the mythical Seven Cities of Cibola and were the first animals of their kind to enter into what is now the states of New Mexico and Arizona. Two years later when he departed, disappointed and broken in health, he left a number of sheep at Pecos Pueblo in northern New Mexico with Fray Luis de Escalera, who remained behind to teach the Holy Faith to the Indians. No later accounts are found concerning these sheep and it is believed that they perished with the zealous Escalera.

Half a century later in 1598, the colonizer Juan de Onate, came into the Rio Grande Valley from Mexico. He brought with him large flocks of Churro (scrub type) sheep and good Merinos whose descendants remain to the present day in some of the more remote areas of the Southwest. On the virgin ranges of present day Arizona, New Mexico, and Utah, these sheep thrived, surpassing all other classes of livestock.

It was from these early importations that the colonies of Hispanics, the pueblo indians and the nomadic Navajo obtained their seedstock. Meat, milk and textiles were the important products from these scrub sheep. There was no significant genetic artificial selection program to perpetuate these sheep, only natural selection and "survival of the fittest." The economy of the Southwestern U.S. was based on the products from these lowly four-legged wool producing ruminants. As a result both the traditional Navajo and Hispanic peoples of the Southwest U.S. today still have strong, cultural, social, religious, and economic ties to the sheep and wool industries.

Perhaps in another perspective, one can imply that the metropolitan areas of Albuquerque, New Mexico, Scottsdale-Phoenix-Sun City, Arizona, and the accompanying civilization and development of the Southwestern U.S., owe their roots to the first civilizing influence of the Churro sheep. It was this animal that provided the raw resources of survival and a basis for the beginning of a capitalistic economy for the native peoples of this region. For the record, this animal can be labeled the oldest breed of sheep in the U.S. The so-called "improved" breeds from the United Kingdom didn't start coming into the United States until two-hundred years after the first Churro came to the Southwest.

Many of the "improved" breeds were brought to the southwest by the U.S. government and Indian traders. The Navajo-Churro carpet type wool was not suited for the eastern textile mills. The lambs produced were not as large and meaty as those sheep raised in improved pastures and grassland areas. But most of the "improved" sheep brought to the harsh climate and range conditions, could not survive and the fine wooled crossbreeds began to destroy the fiber quality of the historic Navajo and Hispanic weaving.

In 1934 the U.S.D.A. established the Southwestern Sheep Breeding Laboratory at Fort Wingate, New Mexico to help determine which breed or genotype of sheep was best suited for that region. A fine representative flock of "original old-type" Navajo sheep was assembled from Navajo flocks around the Reservation. Over 30 years of extensive research determined that the "old-type" Navajo (Navajo-Churro) was the most suitable sheep for the ecosystem of the southwest, and its fiber, best for the hand processed Navajo textiles. But the station was closed, the research terminated, and materials and documentation destroyed. The research flock of "old-type" Navajo sheep were sold; some were butchered, some went to flocks in remote portions of the Reservation, and a few rams were taken to the U.S.D.A. station at Clay Center, Nebraska for crossbreeding experiments. It is this early work at Fort Wingate that the NSP used to help develop its "blueprint" for the redevelopment of the Navajo-Churro.

When I started the NSP in California, reliable estimates of the "old-type" Navajo sheep were numbered at less than 500 head. Estimates now have Navajo-Churro national flock numbering more than 1,100 breeding animals.

Why save the Navajo-Churro? This question has been asked of me on many occasions, usually in an intimidating manner. My first response is commitment. I am personally motivated to do things that I have dedicated and committed myself to, such as family, religion, profession, etc. I have invested much, beyond recognition in this project to see the Navajo-Churro sheep saved into perpetuity. Beyond my personal commitment, is my desire to do something, while I am on this earth "to pay for my rent." The Navajo Sheep Project is the vehicle, in my own opinion to help pay that rent. Another major motivating force to see the Navajo Sheep Project succeed, resulted from my developmental years when I witnessed gross acts of racial prejudice against the Native Americans in many parts of our country. I saw many dehumanizing acts and attitudes against our Indian brothers and sisters, which caused deep lesions of guilt in my conscience. Broken treaties, and forked tongue oratories have dominated our dealings with our red skinned native inhabitants, and I wanted to be part of the restitution process. I found a way with the little, lowly, nondescript, unimproved, Navajo-Churro sheep. Over the last several years I have learned to love and respect my friends, the Navajo and Hispanic growers. I have an ever greater desire to incorporate this rare breed rescue project into one of providing a means for self-reliance, economic prosperity, and an increase in self-esteem for our Native Americans.

The Blueprint

Before presenting the "blueprint" for genetic conservation one must examine the cause which places a breed on the list of endangered domestic animals. In the case of the Navajo-Churro, the sheep was used as a target for destruction to subjugate the Navajo to the Anglo government. This was reminiscent of the almost complete annihilation of the American Bison that the Plains Indian tribes relied on so heavily for their survival. The Navajo sheep unfortunately was caught in the cultural crossfire of the white and red peoples of the southwest region. This crossfire continues even today; but the bullets and arrows have been replaced with rhetoric of those in leadership and influential positions on both sides.

The success of any genetic conservation project depends on a valid objective. It must be more than just preserving a living organism for its own sake. I am a pragmatist, and my original goal was not just to save the Navajo for posterity, but to return the sheep to its rightful heirs. Since the sheep had virtually disappeared from the daily scene of the Navajo and Hispanic cultures for almost two generations, a process of familiarization and education had to take place before the sheep could be safely returned to them. Following this process of education, it was hoped the people would have a desire

Hubbell's Trading Post National Historic Site, 1984 and Utah State University Navajo Sheep Project Field Staff:
Standing:
Left—William A. Varga, Extension Horticulturalist, USU.
Right—Bill Malone, Resident Trader, Hubbells Trading Post & Advisory Board Member, Navajo Sheep Project.
Kneeling:
Left—Dr. Lyle G. McNeal, Professor & Director Navajo Sheep Project.
Center—Reservation Bred Navajo-Churro ewe lamb.
Right—James D. Keyes, Agricultural Extension Agent, San Juan County, USU, & Advisory Board Member, Navajo Sheep Project.

to be involved again in raising the animals and integrating them back into their lifestyle. The NSP staff has always stressed voluntary involvement and individual choice had to be paramount in accomplishing the objectives of this delicate process. Initially, greatest interest was exhibited by the oldest living members of each of these two cultures. Some individuals shed tears of joy and amazement upon seeing one of these remnants of the past for the first time in many years. The practical aspects of the NSP program mission has kept the focus of genetic conservation at the center of our efforts.

The day-to-day management of the central nucleus flock at Logan is shared by my students majoring in Animal Science. The NSP has brought to my students an awareness and need for the genetic conservation and preservation of our domestic animals, regardless of specie. Although my thrust has been the Navajo-Churro sheep, the American Minor Breeds Conservancy with its Rare Breeds Rescue Project has brought an awareness of the multitude of many other minor breeds of domestic animals that are endangered and in need of someone shepherding their rescue. Once genes are lost from the face of the earth, no man, not even genetic engineering can replace the germ plasma for an exact genotype. On the other hand, where breeds or species are left in small populations, they must be maintained in a pure state. Crossbreeding must not be tolerated and the utmost in breeding integrity must exist if the genotype is to be maintained into perpetuity.

Genetic conservation is not a game for the amateur or hobbyist. A thorough understanding of animal breeding is a prerequisite. Concepts of gene frequency, selection differential, generation interval, inbreeding, outbreeding, cytogenetics, population genetics must be the tools of the conservationist. Nutrition, reproductive physiology, history, marketing, public relations, fund raising, are also talents that the conservationist must carry in his/her haversack.

Conclusions

The Navajo Sheep Project at Utah State University has led the way and developed a blueprint for the genetic conservation of endangered domestic animals in the United States. The focus in recent years for the preservation of minor breeds of farm animals has surfaced. Although the NSP has primarily dealt with the preservation and develop ment of one genotype of one specie, it has established itself as a pioneer for other similar endangered animals to be preserved for future generations. Genetic diversity in agricultural plant and animal systems is essential to help insure that the human race will have a reliable source of food and fiber for future generations. For that justification alone the Navajo Sheep Project is proud to share that responsibility and to succeed with its blueprint for genetic conservation.

Breeds of Coloured Sheep in Great Britain and Their Potential Value

Lawrence Alderson
Colonsay, Hampton Lovett,
Droitwich, Worcs WR9 OLZ, England

Great Britain enjoys the benefit of possessing several long-established breeds of coloured sheep, and this long history of purebreeding, based on specific colours, ensures that breeders can predict the results of matings of their sheep with greater certainty. More recently, this base of breeding stock has been augmented as coloured sheep from white breeds have been retained for breeding. Some of these have been included in special sections within the parent Flock Books, while others have been established as new breeds.

The coloured sheep in Great Britain are found mainly among the primitive or hill breeds, and twenty years ago many of them did not possess a Breed Society. A registration programme was established for these breeds in 1974 with the publication of the Combined Flock Book by Rosenberg and Alderson through Countrywide Livestock Ltd. From that time proper pedigree details have been recorded, but even beforehand purebreeding was practised by most of the flockmasters who owned these sheep. However, from time to time, lambs are born which indicate that some breeders were not averse to a degree of deliberate crossbreeding, or that their fence maintenance was not proof against ram libido!

The coloured breeds of sheep found in Great Britain are detailed in Table 1, but as time goes by it is possible that more coloured breeds will be created from the coloured animals resulting from recessive colour genes in white flocks. Recessive black is found in many longwool breeds (eg. Wensleydale, Leicester), imported breeds (eg. Finnish Landrace) and new breeds (eg. British Milksheep). The recessive badger-face pattern is found in breeds such as the Ryeland and Cheviot that were derived from the old dun-faced type of sheep that previously inhabited the British Isles.

In addition, there are breeds which are born with coloured wool, but which usually lose the colour within the first year of life. These include the Portland, which is a foxy-red colour at birth, and the Suffolk and Norfolk Horn, which are often almost black at birth.

In Great Britain wool is not a major item of output in a sheep enterprise, and normally accounts for less than 10 per cent of the total output. Thus it is important, even within specialist wool-producing flocks, to ensure that other commercial characteristics are given due attention and subjected to improvement programmes. This process has been applied to almost all breeds of coloured sheep in Great Britain.

Soay

The Soay is a small, primitive sheep, and probably is the most typical descendant of the wild Moufflon. All present-day stock are derived from the feral flock on the Isle of Soay in the St. Kilda group of islands beyond the Outer Hebrides in the Atlantic Ocean. A flock of 20 rams, 44 ewes and 43 lambs were moved to the island of Hirta in the same group in 1932 when the human population was evacuated, and the mainland flocks have been derived from the sheep on Hirta.

TABLE 1 Breeds of Coloured Sheep in Great Britain

Breed	*Registration Organisation*	*Colours*	*Notes*
Primitive (short-tail) breeds:			
Soay	Combined Flock Book	Fawn, chocolate, moufflon pattern	Also feral population
Castlemilk Moorit	Combined Flock Book and Breed Society	Fawn, moufflon pattern	
North Ronaldsay	Combined Flock Book and Sheep Court	Various	Also semi-feral population
Shetland	Breed Society	Various	
Gotland	Breed Society	Grey	
Hill breeds:			
Hebridean	Combined Flock Book	Black, dark-brown	
Manx Loghtan	Combined Flock Book	Moorit	Also on the Isle of Man
Jacob	Breed Society	Spotted	
Black Welsh Mtn.	Breed Society	Black	
Torddu	Breed Society	Badger-face pattern	Variant of Welsh Mountain
Torgwen	Breed Society	Badger-face pattern	Variant of Welsh Mountain
Herdwick	Breed Society	Fading black	

A mature ewe weighs about 25 kg. and both sexes on the mainland are horned. The breed is noted for its resistance to footrot and fly-strike and, as an easy-management breed, it has been used to reclaim the waste tips from the English China Clay workings in Cornwall and the gravel pits in the Thames Valley. It produces a lean carcase, as most fat in the body is stored internally.

Soay sheep are found in various colours. A few animals are black, and a few on Hirta have white markings, but the majority are various shades of brown. Chocolate is dominant to fawn, and the moufflon pattern is dominant to self-colour. The quality of the wool is very variable, because of the proportion of strong hairs, but the underwool is soft and fine. The staple length is 5–10 cm. and the Bradford Count is 44–50. The wool is not popular with hand-spinners and, because the fleece is shed naturally, it is often not easy to gather. Soay sheep which are fawn self-colour have a higher mortality and lower growth rate than other Soays.

Castlemilk Moorit

This is a relatively modern breed, which was created by the Buchanan-Jardine family in southern Scotland in the early years of the present century. It is very similar in appearance to a larger fawn Soay with the moufflon pattern, but the wool is rather more consistent in quality. It was created by crossing various breeds, including the Soay and Manx Loghtan. The numbers at present are too small for the breed to be of any significance as a source of coloured wool.

North Ronaldsay

The two main flocks of the North Ronaldsay breed are found in the Orkney Islands; one on its island of origin of the same name, and the other on the island of Linga Holm, which is owned by the Rare Breeds Survival Trust. A few sheep are found on the mainland.

The sheep are about the same size as the Soay, and similarly have a short tail, fine bones, and a concave face profile. The rams are horned, but most ewes are polled. The most notable quality of these sheep is their ability to exist on an exclusive diet of seaweed, and as a result their physiology has been modified to adapt them to tolerate unusual concentrations of copper, iodine and urea. Within its special environment the North Ronaldsay is productive. Like several primitive, northern short-tail breeds it is fairly prolific. The flock on Linga Holm rears a lamb crop in excess of 150 per cent each year, but under grassland systems on the mainland the prolificacy drops to little more than 100 per cent.

The wool is found in a wide range of colours. White is dominant, and moorit is recessive, but between these extremes there is a series of greys, roans and black. The wool generally is of good quality, with a Bradford Count of 50–56, and it can be spun and knitted into speciality shawls and scarves. However, grey wool is often coarser, due to the stronger black fibres, and the wool of sheep that graze on the foreshore usually is heavily contaminated with sand and organic matter. This reduces its value, but nevertheless an embryonic cottage industry has been established in the islands to use the wool.

Shetland

The Shetland is closely related to the North Ronaldsay, but it has been subjected to more selection and improvement programmes. Consequently, it is a larger animal with mature ewes weighing on average 37 kg. In comparative trials it has shown itself to be more productive under some circumstances than the popular Scottish Blackface, and its numbers on the mainland have increased dramatically under the Combined Flock Book registration scheme. However, on its native islands, many of the sheep are not pure Shetlands, and the coloured varieties are declining.

Traditionally, the famous Shetland Islands woollen industry was based on the wool of the true Shetland breed. Shetland shawls were so fine that they could be drawn through a wedding ring, and the multi-coloured Fair Isle sweaters were knitted with undyed wool in traditional patterns that had been handed down from generation to generation. The emphasis placed by the British Wool Marketing Board on white wool, and the lower prices obtained for coloured wool, undermined this industry. The numbers of coloured Shetland sheep fell to a low level, and high quality wool on the islands no longer receives the premium necessary to encourage purebreeding. However, the trend has been reversed on the mainland where the majority of Shetland sheep are coloured.

The wool is particularly soft and fine. It has a Bradford Count of 56–60, and a staple length of 5–15 cm. The colours are similar to those found in North Ronaldsay sheep, and several are known by local dialect names, which indicate the close affinities of the Shetland Islands to Norway. For example, dark grey is known as shaela from the Old Norse word 'hela', meaning hoar frost, and moorit is derived from 'moraudr', meaning peat-brown. Selection for quality of wool has reduced the incidence of coarse black fibres in the grey fleeces.

Gotland

The Gotland is an improved Swedish breed that was imported in 1972 by Mr. and Mrs. W. F. Macdonald. It has a grey (blue) fleece that is the result of a long-term, intensive selection programme, and it is used in the manufacture of speciality sheepskin coats. The wool is lustrous and curly, and of a uniform grey colour and quality throughout the fleece. Here again, the coarse black fibres have been eliminated by selection.

The Gotland is larger than the Shetland, and it is relatively prolific, but it still has a 'primitive' conformation and, under British conditions, it suffers badly from footrot. Like many of the breeds related to the Northern short-tail group, it has a lower metabolic rate during the autumn and winter which makes it difficult for them to lay down flesh at these times of the year.

Hebridean

The Hebridean originated on the islands off the West coast of Scotland, but for many years it has been found in the parks surrounding stately homes and country houses on the mainland. Its aesthetic appeal was enhanced because it is a multi-horned breed, but more recently its commercial qualities have been appreciated more fully. When mated to rams of a meat breed, it can produce 20 per cent more weight of lamb per hectare than the popular crossbred ewes in Great Britain. As a purebred mature ewes weigh on average 36 kg., and the lambs produce a carcase with a lean content of 58.16 per cent, which is significantly higher than almost all other breeds.

The wool is black or dark-brown. The black wool is stronger and more lustrous, and tends to become silvery with age. The colour is determined by a recessive black gene, but some crossing in the past with Black Welsh Mountain and/or Jacob sheep has confused the position. As a result, sometimes white lambs are born, while a good proportion of lambs have a white spot on the poll. These are evidence of the dominant black and spotting genes.

The wool has a Bradford Count of 48–52, and a staple length of 10–15 cm. A variety of goods are made from the wool by private enterprise, but the business suffers through the spoilage of much of the wool by poor handling procedures on the farm and in transit. The knitted goods are of high quality.

Manx Loghtan

This breed is related closely to the Hebridean, being multi-horned and with a half-length tail, but it is moorit in colour and a little larger in size. Mature ewes weigh on average 40 kg. It originated on the Isle of Man, and its name is derived from the Manx words 'lugh' (mouse) and 'dhoan' (brown).

The wool is suitable for both knitted garments and worsted materials and, despite the short supply of wool, a few people have much-prized suits made from Manx Loghtan wool. When the lambs are born, the wool is chocolate brown, but it fades to light brown as it is exposed and bleached. The Bradford Count is 44–52, and the staple length is 8–10 cm. The typical fleece is tight, with a short staple, but handspinners prefer a longer staple, and some breeders have started selecting for this characteristic. Some fleeces contain a significant amount of kemp, but this has been much reduced by selection. As with the Hebridean, white marks in the fleeces sometimes occur, but they are undesirable and moorit was established as the correct colour eighty to ninety years ago.

Jacob

The Jacob is the third multi-horned breed in Britain. However, its origin is more likely to be found in Spain than on the western isles of Britain. It is a larger sheep than the Hebridean and Manx Loghtan, and it has a longer tail. It is likely that the breed has been changed in character to some degree by crossing with the Dorset Horn in the period before the Breed Society was formed. It has increased numerically in the last twenty years, so that it has now become a well-recognised breed in Great Britain.

The wool of the Jacob is comparable in quality to that of the Hebridean and Manx Loghtan, but is spotted in colour. The Jacob also carries the dominant black gene, so that crossbred lambs usually are black, except when the Jacob is mated to white, pink-nosed breeds such as the Dorset Horn, when the lambs are generally spotted. The wool of the black spots is not crimped, and thus stands out above the level of the remainder of the fleece. The mixture of black and white wool in the fleece can be a disadvantage, unless a blended effect is required.

The Jacob breed and its wool products have been vigorously promoted by an active Breed Society, and special arrangements have been reached with the British Wool Marketing Board, in the same way that the Rare Breeds Survival Trust has negotiated exemptions for several other coloured breeds from the Board's monopoly on marketing.

Black Welsh Mountain

The growth of the Black Welsh Mountain as a commercial breed in recent years is comparable to that of the Jacob, largely as a result of the work of Black Sheep Ltd. Although the Black Welsh Mountain is a variant of the Welsh Mountain type, it has not arisen as a result of retaining 'recessive black' sheep in white flocks. The black colour of the breed is determined by a dominant gene, and thus it can be used effectively to produce black wool of different qualities by crossbreeding.

By the standard measures of quality (Bradford Count 44–52; staple length 8–10 cm) the wool would seem to be similar to that of the Manx Loghtan, but in practice it is stronger, and is better used in the manufacture of items such as heavy duty sweaters and worsted, rather than softer knitting wools. However, Black Sheep Ltd. market a wide range of goods, including ties, gloves and socks.

Torddu and Torgwen

These are recent variants of the Welsh Mountain breed, with the badger-face (black belly) and reverse badger-face patterns respectively. They are numerically very small, and their contribution to the coloured wool industry is negligible at present.

Herdwick

The Herdwick is found mainly in the Lake District in North-West England, although some 'hobby' flocks have been established elsewhere. It is adapted to tolerate the severe winters and high rainfall of its native area, and probably is Britain's hardiest breed of sheep. It is likely that it was developed from an admixture of Scandinavian sheep and the black-faced hill sheep of northern Britain.

The wool of the Herdwick is coarse and harsh, with a Bradford Count of 28–32, and contains a considerable amount of kemp. It is a carpet wool but, as a result of a major publicity and marketing campaign launched by the National Trust, its value in the manufacture of distinctive tweeds has been realised. Although only a local breed, it exists in sufficient numbers to support such a marketing campaign. At birth the fleece is black, but changes progressively to grey. The rate of this colour change varies from individual to individual, but in the majority of sheep the grey colour appears during the first year of life.

Longwools

The increased interest in coloured wool, which has developed during the last decade, has been evident in the greater number of coloured sheep from white flocks that are now retained for breeding. These are found mainly in the longwool breeds. The Wensleydale is of particular importance, because of the unique quality of its wool. It possesses a distinctive characteristic known as central checking, which inhibits the growth of coarse fibres found in most longwools. As a result, the wool is of uniform quality throughout the fleece, and even when crossed with hairy breeds the crossbred progeny have no kemp in their fleece. The wool has a high lustre, a Bradford Count of 44–48, and a staple length of about 30 cm.

The Wensleydale is a large animal with adult rams weighing up to 135 kg., and both sexes are polled. The large size and high growth rate of the breed have given it some value as a sire of lambs for the production of heavyweight carcases with a high yield of lean meat. About 15–20 per cent of the lambs are black, and this is associated with the blue pigmentation of the skin which is a characteristic of the breed.

Black lambs are also born in the Leicester Longwool breed, and a section for coloured sheep was opened in the Flock Book in 1986. The black colour of the Leicester should more properly be described as a type of grey. The Bradford Count is 40–46, and the staple length is 25–30 cm. Leicesters are strong, robust sheep with good longevity. As purebreds they achieve a lamb crop of 160 per cent, but the crossbred daughters by British Milksheep rams produce large lamb crops in excess of 200 per cent, with a high milk yield and a heavy fleece of quality lustre wool. Black Leicester sheep have a superior carcase quality with a higher content of lean meat, and this characteristic is being investigated further.

References

Adalsteinsson, S. Colour Inheritance in Icelandic Sheep. J. Agr. Res. Icel. II 3-135 (1970).

Alderson, G. L. H. The Observer's Book of Farm Animals. Frederick Warne, London (1976).

Alderson., G. L. H. Comparative Performance Standards for some Minority Breeds of Sheep. The Ark Vol. IV No 7 (1977).

Alderson, G. L. H. Rare Breeds. Shire Publications, Aylesbury (1984).

Burns, M. Wensleydale Sheep—the Kemp Killers. The Ark Vol. V No 2 (1978).

Elwes, H. J. A Guide to the Primitive Breeds of Sheep and their Crosses. Edinburgh (1913).

Jewell, P. A., et al. (eds.) Island Survivors: the Ecology of the Soay Sheep of St. Kilda. Athlone Press, London (1974).

Werner, A. R. An Enquiry into the Origin of Piebald or Jacob Sheep. Countrywide Livestock Ltd., Droitwich (1975).

The Transformation of a Gene Pool in the Process of the Tajik Breed Development

G. A. Aliev
Academy of Science of Tajik SSR,
Lenin av. 33, Dushanbe 734742 USSR

In the 1920's and 30's aboriginal sheep with low wool production were widely crossed with rams of fine-wooled and long-wooled English breeds in large regions of Russia, the Ukraine, Kazakhstan, and the Republics of Central Asia.[3,4,5] The experience of many years has shown that English sheep can't adapt to dry, hot climates and long distance driving to summer mountain pastures.

22 months ewe (weight 115 kg).

At the beginning of the 1940's it became obvious that the gene pool of fat-rumped sheep created through hundreds of years of natural and artificial selection had to be kept. It was also clear that this gene pool had to be changed by introduction of genes determining the improvement of wool characteristics. So the aim was to develop sheep of a completely new type. These sheep were intended to have a complex of different characteristics: good adaptation for zone of breeding, high body weight, capability to accumulate a large amount of fat in the rump, and high carpet wool production. By 1963 the work had been successfully finished. The state commission had accepted a new meat-fat-wool breed—the Tajik sheep. This paper is devoted to the ways of Tajik breed development: overcoming numerous difficulties of the animal's development from conflicting characteristics; and to the modern trend in sheep breeding—combination of meat, fat and wool in animals of one breed.

A schematic plan of Tajik breed development is shown in Figure 1. Five thousand Hissar ewes with a desirable size of body, shape of fat rump and constitution were chosen to cross with Saraja rams brought from Turkmenistan. Different variants of crosses were used. Animals of different amounts of Hissar and Saraja blood were produced. The basic group of sheep consisted of hybrids of different generations which were ½ Saraja and ½ Hissar. Among these animals the main

A group of Tajik ewes of different lines.

Tajik ewes and their lambs in the pasture.

A group of Tajik rams. Live weight 156 kg.

Tan Tajik lambs of a new line.

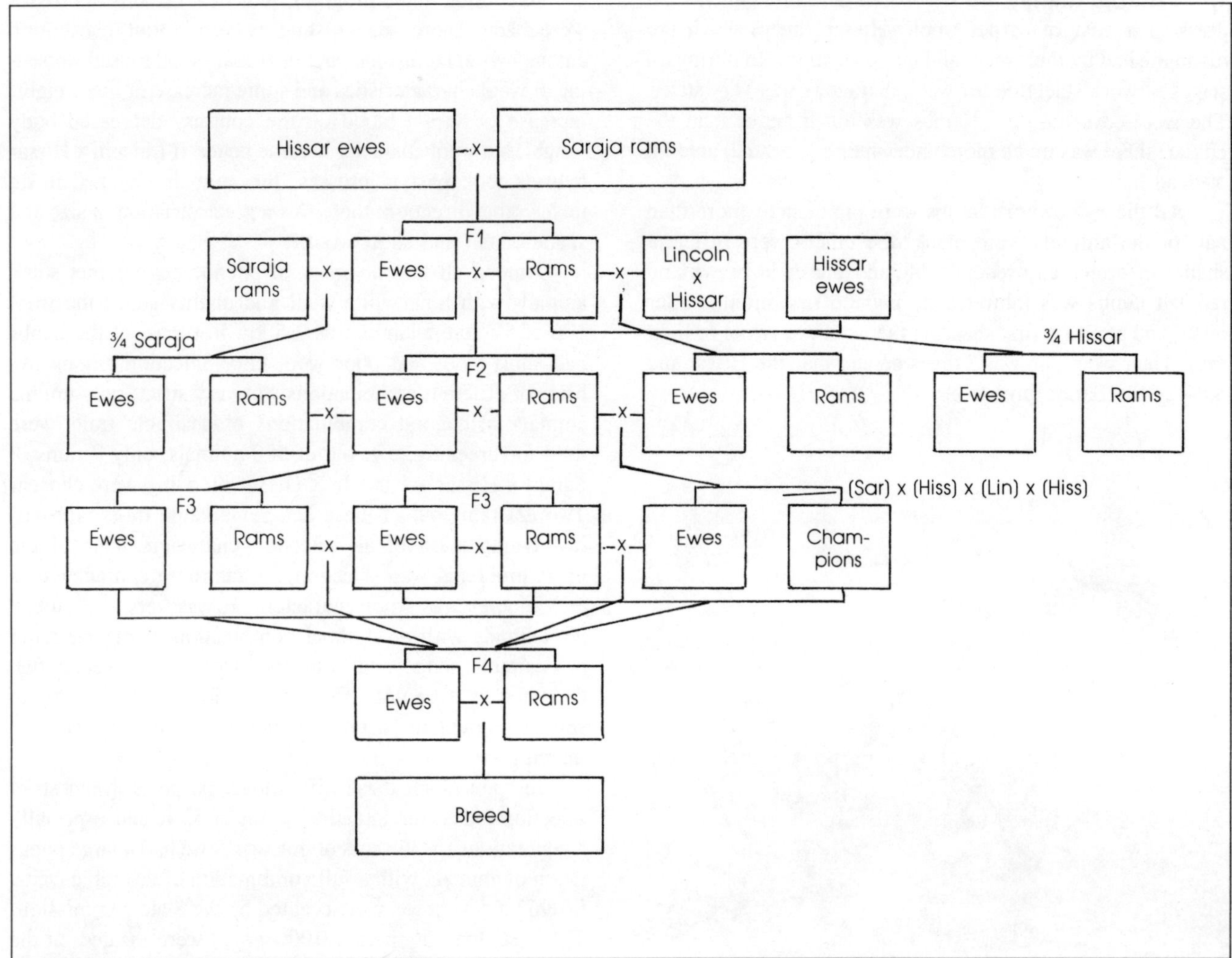

Figure 1 Scheme of Tajik Breed Development.

part of selection had been carried out. In addition, Lincoln x Hissar ewes were used (B. N. Vasin worked with these animals).

The process of breed development had been broken into 3 main stages.

1. Obtaining animals of desirable type by crossing Saraja rams with Hissar ewes and Saraja x Hissar rams with Lincoln x Hissar ewes.
2. Consolidation of animals of desirable type by crossing and special individual selection.
3. Subsequent breeding and improvement of sheep with fat rump and carpet wool.

It is necessary to underline that at all stages of work artificial insemination and strict selection were used. It gave us a chance to develop the breed during a comparatively short period. Animals of desirable type had to have the following characteristics: strong constitution, high live weight, comparatively long muscular neck, wide back and rump, deep and wide breast, strong legs with good firm hoofs, big or middle width elevated fat rump, white wool.

The sheep were bred in seasonal pastures: in summer and autumn in the mountains, in winter and spring in the valleys and hills. In cold rainy weather and on winter nights lactating ewes with newborn lambs were kept in a special building. In winter and early spring they were supplied with additional food. Food abundancy was the main reason for the Tajik breed's successful development.

The first cross of Saraja rams with Hissar ewes were made in 1947. F_1 lambs were very similar to Hissar in the characters of growth and development, live weight and constitution. In all these traits they exceed Saraja sheep greatly. This is understandable since Hissar sheep are the biggest sheep in the world but there are many dead hairs in their wool-covering so that their wool can't be used in industry. A high rate of growth of F_1 lambs was determined both by maternal effect and heterosis.

The main strategy in a new breed development was "to dress in a coat of carpet wool" Hissar sheep which are distinguished by their meat and fat production. So during all stages of work selection for wool characters was very strict. The wool-covering of F_1 lambs was much better than the Hissar: there was much more under-hair and a small amount of dead hair.

All the F_1 newborn lambs were pigmented: more than half of the animals were black, the others were different shades of brown and reddish. Melanogenesis in brown and reddish lambs was inhibited during the first months after birth, and after the first shearing the wool was white or light grey. Their wool clip was 3 times greater than the Hissar and 60% of the Saraja breed.

The ram—breed champion (live weight 177 kg).

The wool structure of F_1 lambs (investigation of 23 samples of wool) was the following: underhair—51.88% (diameter 17.0 microns), medium hair—15.87% (33.16 microns), hair 21.32% (46.72 microns), dead hair—0.86%.

Thus there was no intermediate inheritance. Lambs were close to the maternal breed—Hissar in live weight, fat rump size and exterior, but in a structure of wool covering close to the Saraja breed. Vasin got the same results by crossing Lincoln rams with Hissar ewes.[5]

Apparently a small number of recessive genes controls the presence of dead hairs. The general wooled type also dominates. Thus animals of the first generation had a good combination of desirable traits: high live weight, good characteristics of fat rum and wool characters. So the first results inspired us for future work. But certainly animals of F_1 were very far from ideal: first of all, they didn't satisfy high demands for wool characteristics: structure, thickness, lustre, silky traits, character of suint, etc.

According to the program, four new variants of crosses were made. There was a distinct pattern in trait distribution among hybrid lambs: increase of Hissar blood caused worsening of wool characteristics and some increase of live weight; increase of Saraja blood on the contrary decreased body weight, but wool character became better. If Lincoln x Hissar hybrids took part in crosses, the animals changed in an undesirable direction: there was a great variation in size and shape of tail and short wool.

Among all this diversity of phenotypes we met some animals with long white wool and quality suint, the wool clip of 5-months lambs was 2.5 kg. But most of the lambs had short wool and poor wool clip. Selection among hybrids of different combinations was very strict: only unique animals with good combinations of desirable traits were used for breeding, e.g. out of 500 animals, only 7 rams of Saraja x Hissar x Lincoln x Hissar breeding were chosen. Two (2) rams out of these had remarkable traits, an optimal combination of all selection characters. One of the great problems was selection for fat rump characteristics and another was wool characters. It was very difficult to get animals with an optimal combination of big fat rump of desirable shape and noble wool of carpet type. Often their hooves were not strong enough for durable driving to summer mountain pastures, which was rather difficult for the animals.

In spite of all these difficulties, purposeful and strict selection led to the situation when in 3, 4, and especially 5 generations (in 16 years of our work) we had a large population of animals with a full combination of desirable traits. In 1963 a new breed was accepted by the State commission. The best stock of sheep (10,000 ewes) were located on the experimental farm of the Tajik Institute of Animal Breeding "Dagana-Kiik." Sheep of this population had remarkable productive traits, strong constitution and good health.

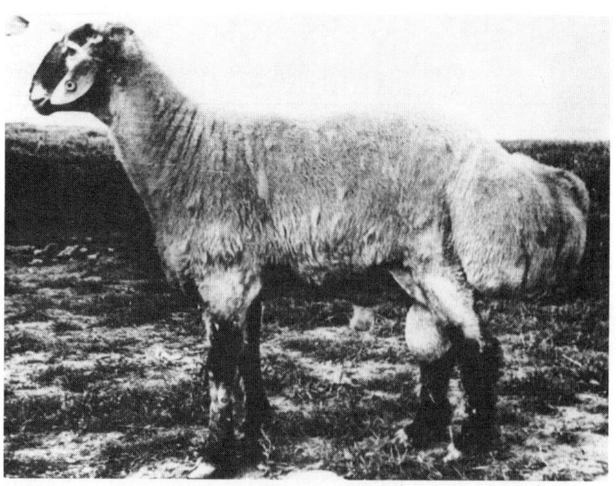

The ram of highest meat and fat production.

In 1984 we began to study Tajik sheepskin characters. Three years of investigation had shown that we can get four types of high quality production from Tajik breed: wool, meat, fat and excellent lambskins.

At different stages of work seven (7) lines of animals had been developed by special selection of prominent rams for ewes with specific traits. Nowadays, we have five (5) basic lines.

We should underline that in 1963 when the breed had been developed, it had an optimal combination of wool, meat and fat production, was very well adapted to the continental climate of Tajikistan and to durable driving to mountain pastures at a distance of 300–400km. Ethologic and physiologic investigations had shown that Tajik sheep didn't lag behind the aboriginal Hissar breed in all characteristics.

10 months ewe (live weight 58 kg).

Conclusion

When the work over the Tajik breed development began, most of the great scientists didn't believe it would succeed. At that time it was generally accepted that to combine in an animal high development of conflicting traits was impossible. The task of getting animals with prominent development of both live weight, fat rump and wool production seemed theoretically impossible. Such a point of view is understandable since fat rump sheep are usually coarse-wooled like wild sheep.

However the Tajik breed development once again proves the unlimited possibilities of artificial selection, when scientists have a great variety of specific gene pools of ancient breeds from which to choose.

So to a great extent we managed to integrate in a new breed a complex of genes, providing high development of meat and fat traits (characteristics of Hissar) with a group of genes for carpet wool characters (characteristics of Saraja).

These results were extremely important from the methodologic point of view too. Tajik breed development has stimulated the organization of similar work in other Soviet Asiatic and Caucasian republics[1,2] i.e. a new trend in sheep breeding began to develop.

The ewe—breed champion and her lambs (live weight 143 kg).

1 month lamb (live weight 14 kg).

These sheep are a live realization of scientist's ideas. We have managed to develop animals of almost ideal type—with excellent exterior, prominent live weight (150–170 kg) and fat rump, high carpet wool clip. Presence of such splendid animals (Figures 2–11) proves that there are no biological barriers for future progress in breed improvement. Purposeful selection under optimized, environmental factors would lead to the creation of a great population of these unique animal champions. Probably in the future new methods of biotechnology could lead to an increase in selection effectiveness.

Literature

[1]Aliev G. A. Tajikskaya myaso-salno-sherstyanaya poroda ovets. Dushanbe: Irfon, 1967, 346 c.

[2]Botbaev I. M. Alaiskaya poroda ovetz i eye selektsiya. Frunze: Kyrghistan, 1982, 184 s.

[3]Filyansky K. D. Ob ispolzovanii pustynnyh i visokogornyh pastbitsh.—Sovetskaya zootecnhiya, 1949, 7.

[4]Filyansky K. D. Reservy povysheniya proizvodstva odnorodnoi shersti.—Sovetskaya zootechniya, 1949, 8.

[5]Vasin B. N. Evolutsiya sherstnogo pokrova ovets. Novosibirsk: Nauka, 1969, 283 s.

Some Unique Features of Karakul Fat-Tail Sheep

Maurice Shelton
Texas Agricultural Experiment Station
7887 North Highway 87
San Angelo, Texas 76901 USA

Fat-tail sheep constitute one of the more prevalent types found in the world, and they almost certainly make the greatest contribution to mankind. Most of these are triple purpose animals producing meat, milk and fiber. Most fat-tail sheep tend to have certain traits in common which, in addition to accumulation of fat in the tail, include the production of carpet wool, adaptation to arid regions or more specifically arid regions with great variability in temperature and feed supply, a higher rate of milk production than most breeds and generally low fecundity with good lamb survival. There are many breeds or genotypes of fat-tail sheep, but perhaps the two most widespread, or at least best known, are the Awassi and the Karakul. The latter is thought of as a fur type with the lambs being sacrificed at a very young age to produce Persian lamb skins. However, in practice many lambs are not sacrificed at birth, but are utilized at later ages for meat. The only breed of fat-tail sheep known to be present in the U.S. is the Karakul. These are often used to produce wool for the handicraft trade because of their color and some unique characteristics of the fleece. The genetics of color in the breed are reasonably well worked out and will not be reviewed at this point. However, there are a number of characteristics of this type of sheep which should be of interest.

The Texas Agricultural Experiment Station established a small flock of Karakul sheep approximately 10 years ago. One of the reasons for doing this was to assist in the preservation of a genetic resource which appeared to be in danger of being lost in this country. The second reason had to do with using this breed as an example of the larger fat-tail population of the world in a series of studies contributing to the U.S. AID Small Ruminant Collaborative Research Support Program. The flock was managed under Texas range conditions with winter or very early spring lambing. A series of studies have been conducted and the results will be reviewed and summarized in this paper. Many of these studies have been previously reported in more detail.

Results and Discussion

Fleece Quality

A bulk lot of 865 pounds of Karakul wool was evaluated for a number of characteristics with the following results by Lupton and Shelton.[4]

Average grease fleece weight, 2.86 kg.
Yield, 58.1%
Residual grease (following scouring), 0.6%
Vegetable matter, 1.3%
Fiber diameter, 29.2 microns
Standard deviation in diameter, 11.1 microns

The color ranged from white to black. There were significant differences in fiber diameter between colors, but it is not known what significance to attach to this. It may in fact represent purity of the type. Black colored wools were of the coarsest type. Variability within the fleeces was high (standard deviation of 11.1 microns) compared with typical values of less than 5.0 microns for finewool type sheep. A substantial portion of the fleeces were medullated, but it proved difficult to provide a value for the percent of medullation because the pigmentation prevented evaluation of medullation in the usual manner.

Reproductive Efficiency

Karakul ewes have been maintained on the same property as other breeds (primarily the Rambouillet) or crosses, but they were not always managed as one group. In general, Karakuls were easier keepers (required less supplemental feed), but tended to be more difficult to handle. Most of the ewes cycled beginning in mid-summer, but some difficulty was encountered concerning servicing ability of the ram due to the fat tail. The lambing rate tended to be low (i.e. 1.17 lambs per ewe lambing compared to 1.43 for Rambouillet) or below all other types maintained in experimental flocks. Lamb survival was the highest of the various breeds, even when adjusted for type of birth. Milk production data were not collected, but early lamb growth rates indicate they were the best milking ewes available on the station.

Effect of Docking on Reproduction

It is possible to dock fat-tail lambs, but it must be done very early as fat accumulates in the tail at a rapid rate. In the experimental flock, one-half of the lambs were docked and one-half left undocked. The ewe lambs were subsequently followed through a productive cycle. In this study, a higher percentage of the docked ewes lambed (92.9 vs. 78.9%); lambed earlier (7.8 days) and had a higher lambing rate (1.23 vs. 1.12). The first two traits apparently reflect the difficulty the rams experienced in mating the undocked ewes. The explanation for the docked ewes having a higher lambing rate is not apparent. In any case producers who raise these sheep would be encouraged to dock their ewe lambs.

Grazing Behavior of Karakul vs. Other Types of Sheep

Since these types of sheep evolved in an arid environment, often with very sparse feed conditions it appeared reasonable that they might exhibit different grazing behavior. The primary point of interest would be their willingness to consume a wider variety of plant material such as shrubs. As a part of other studies, the grazing behavior of Karakul sheep was compared with the Rambouillet and the Barbados Blackbelly (a hair sheep). There was no significant difference between the Karakul and Rambouillet as intake patterns of the two were almost exact duplicates.[7] In some cases the Barbados Blackbelly did browse more extensively than the other two breeds. The explanation for this is not apparent, but should hold interest.

The Eating Quality of the Meat of Fat-Tail Sheep

There is a widespread belief in many parts of the world that the meat of fat-tail sheep is superior to other types from the standpoint of its organoleptic properties (taste and smell). In view of the generally low level of (and frequent aversion to) lamb consumption in the U.S., this feature held considerable interest. Several studies were conducted which were directed at identifying or elucidating this point. Two studies were conducted in which sensory panel ratings were obtained from loin chops and leg steaks of different breeds of sheep in which the Karakul was included. The results did not identify a significant or consistent difference favoring the Karakul breed. Since taste is generally considered to be attributable to the fat component of the meat, the fatty acid content (C14 to C18) of the fat from the various breeds was determined. In one of the studies, the Karakul had a significantly lower level of C18 fatty acids and a higher level of unsaturated fatty acids. This was not corroborated in a second set of data. The general conclusions from these studies were that the alleged difference did not exist or that the approach used did not identify a difference. In discussing the meat of these type of sheep, people often refer to odor or the lack of odor. It has been recently suggested that the source of odor in meat is not found in the fatty acids of the C14 to C18 range, but in more volatile, shorter chain acids.[8] This possibility needs to be investigated. The potential of altering the acceptability of sheep meat in the U.S. market through genetic means should hold considerable interest.

Carcass Traits of Fat-Tail Compared to Other Breeds

Several studies have been made which involved comparison of carcass traits of Karakul and other breeds or types. One set of data is shown in Table 1. These data were derived from wether lambs fed to produce U.S. Choice grade lamb carcasses. These data have been reported in more detail by Edwards, et al.[1] The crossbred animals in this tabulation represent Rambouillet x Suffolk crosses. The overwhelming conclusion from these data is that under comparable feeding conditions and at comparable slaughter weights the Karakul is fatter. This resulted in a carcass fat trim of 15.4% as compared to 6.6% and 6.7% respectively for the Rambouillet and crossbred groups. This fat trim was largely associated with the tail as identified as trim from the leg and loin in Table 1. However, the Karakul was fatter overall as evidenced by a significantly higher dressing percentage and a greater fat thickness over the 12th rib. The Karakul also had a markedly lower yield of trimmed wholesale cuts due to the necessary fat trim. Other studies have supported similar

TABLE 1 Influence of Breed of Lamb on Selected Carcass Traits

	Breed Group			
Carcass trait	*Rambouillet*	*Crossbred*	*Karakul*	*SD*
No. observations	20	16	20	
USDA yield grade*	2.7^c	2.4^c	3.0^d	.63
USDA quality grade*	11.1	11.5	10.9	1.05
Dressing percentage	52.6^c	53.8^c	56.1^d	2.94
Carcass fat trim (%)	6.6^c	6.7^c	15.4^d	2.21
Trimmed wholesale cuts (%)	78.9^c	80.2^c	72.7^d	2.65
Fat trim from leg (%)	8.7^c	8.1^c	32.1^d	3.69
Fat trim from loin (%)	9.8^c	9.0^c	21.0^d	5.08
Ribeye area, 12 rib (cm²)	14.3	14.7	14.2	1.91
Leg conformation score*	11.7^c	12.7^d	10.5^e	.92
Fat thickness, 12 rib (mm)	3.2^c	3.1^c	4.4^d	1.40

[c,d,e] Means on the same line concerning all breeds with different superscripts are different (P 0.05). The same applies to other tables in this report.

*These are coded values with the lower values for yield grades representing leaner carcasses. Higher values are more desirable in case of leg conformation or quality grade.

conclusions. O'Donavan et al.[5] reported a carcass fat percentage of 33.3 for Iranian fat-tail sheep. This high level of body fat is perhaps the major factor contributing to survival of these sheep under arid conditions or periods of nutritional stress, but it represents a very serious waste in terms of carcass value in those countries which do not have a ready market for carcasses of this type.

The Influence of Docking on Growth and Carcass Traits

The previous section indicated a considerable amount of waste in the fat trim from the Karakul lambs. Most, though not all, of this was that associated with the tail fat or fat tail. Thus a logical question is what would docking do to the accumulation of fat which must be trimmed to make the carcasses acceptable in non-traditional markets.

As mentioned earlier, one-half of the lambs born in the experimental flock were docked with the remainder left undocked. Some of the males (castrated) were processed through a research laboratory along with comparable Rambouillet lambs as controls. Some of these data are shown in Tables 2 and 3 and in Figure 1.

Death losses and the relative growth rates of the docked vs undocked lambs are shown in Table 2. Differences between years, sex and type of birth were statistically significant. Differences in growth rate between lambs which were docked and those that were not docked were not significant, but when differences existed, they favored the docked lambs. Other researchers have reported slower growth rates for docked animals. In this study, docked lambs had a higher death loss although this difference was not statistically significant. Further research on this point seems warranted.

As shown in Figure 1, docking markedly reduced the tail or dock fat trim from 6.5% to 1.7% of the carcass. This reduced the overall fat-trim in the carcass by 5%. In this study, docking did not increase the thickness of the external fat cover and only a marginal increase in kidney and pelvic fat content. In a study conducted by Joubert and Ueckermann,[3] the docked animals showed a slight, but non-significant, increase in kidney and pelvic fat, but no increase in external fat cover. Other studies have shown an increase in subcutaneous and internal (kidney/pelvic) fat deposition, but such an increase was not sufficient to compensate for the total reduction of fat in the tail. Sefidbakht and Ghorban[6] reported 16% separable fat of docked animals as compared to 27% for non-docked controls. These data suggest a reduced total fat deposition in docked animals. This should be advantageous to carcass value in all cases except potential markets outlets where the price received for fat is equal to that of red meat. Other work has shown a tendency for increased intramuscular fat deposition of the loin in docked animals (i. e., 4.5% for docked vs 4.1% for undocked animals).

TABLE 2 Least Square Means for Weaning Weight (kg) and Lamb Survival (%) by Treatment Group

Year	*Type of birth*	*Sex*	*Type of Tail*	*Wean wt.*	*Lamb survival (%)*
1981 — 29.4^b					
1982 — 30.9^b	Single — 34.7^a	Male — 34.8^a	Docked	33.4	84.3
1983 — 25.0^c					
1984 — 39.6^a	Twin — 28.1^b	Female — 31.5^b	Undocked	33.0	90.7
1985 — 39.4^a					

TABLE 3 Comparison of Rambouillet and Docked vs Non-Docked Karakuls with Respect to Certain Carcass Traits

		Karakul	
Trait	*Rambouillet*	*Docked*	*Non-Docked*
Dressing percentage	54.03[a]	56.73[b]	57.13[b]
Cooler shrink (%)	6.23	6.94	7.35
Fat thickness, 12 rib (mm)	3.81	6.10	6.85
Ribeye area (cm^2)	14.19	14.45	13.09
Total fat (%)	9.00[a]	15.53[b]	19.76[c]
Rack fat (%)	0.68	1.14	1.11
Dock fat (%)	0.25[a]	1.68[b]	6.49[c]
Kidney/pelvic fat (%)	4.44	4.34	4.00

Epstein[2] reports a total body fat content of undocked Awassi lambs at 14.0%, compared to 12.4% for docked lambs. These values are of interest in that this researcher reports much lower fat percentages. This could potentially relate to the breed used, but much more likely is explained by lower slaughter weights (approximately 27 kg as compared to 50+ kg in the Texas study). Observations indicate that fat stores increase, both in actual terms and as a percentage as the animal matures. This is true for all types of sheep, but appears to be more marked for the fat-tail types.

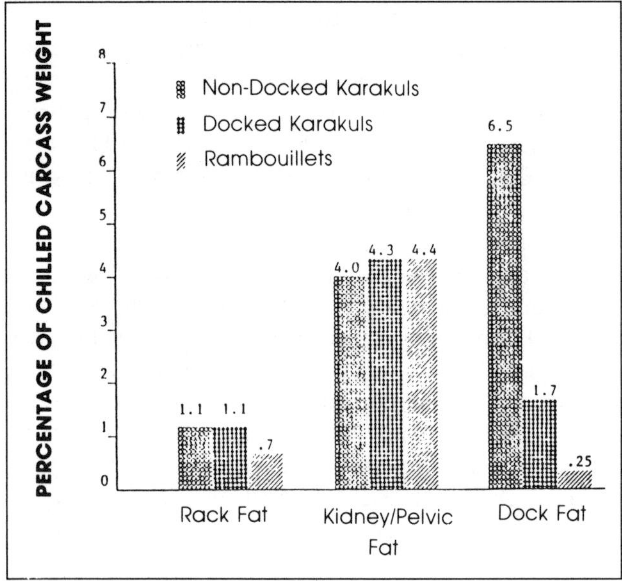

Figure 1 Comparison of Percentages of Rack Fat, Kidney/Pelvic Fat and Dock Fat Among Non-Docked and Docked Karakuls and Rambouillet Sheep.

The Influence of Crossbreeding on Growth and Carcass Traits

Producers of fat-tail (in this case Karakul) sheep, either for the unique pelt or fiber types or because of their unique adaptation to arid environments, have the option of crossing the ewes to rams of other breeds to possibly improve the carcass value of the resulting lambs. Except for the case of Karakuls maintained for pelt production, this could be done with a large portion of the ewe flock while breeding a smaller portion of the ewe flock to generate replacements.

In one study, four different types of lambs were generated by breeding Rambouillet, Suffolk or Karakul rams to Rambouillet or Karakul ewes. The four types of lambs are outlined below:

Rambouillet x Rambouillet
Rambouillet x Suffolk
Rambouillet x Karakul
Karakul x Karakul

It can be seen that the two crossbred types of lambs were generated from Rambouillet ewes as the number of available Karakul ewes were few. Within the four groups, approximately one-half of the males were castrated with one-half left intact, resulting in 12 breed x sex groups. Within each of these groups, one-half of the lambs were docked and one-half were left undocked resulting in 24 breed x sex x docking subgroups.

All of the lambs were grown to approximately 45 kgs and were slaughtered through a commercial abattoir. Data collected include birth weight, pre- and post-weaning gains and a variety of carcass data. The latter included dressing percent, carcass and leg length and several measures of degree of fatness. In addition, the tail was removed and weighed for a total tail weight as well as trimmable fat from the tail and dock region.

Lambs out of Karakul ewes had heavier birth weight and better preweaning gains. This almost certainly traces to a higher level of milk production. Lamb survival also favored Karakul lambs. Sex and type of birth (single vs twins) effects generally followed the expected trends. Irrespective of the tail fat, Karakul and Karakul cross lambs were fatter in all measures of fatness (fat thickness over loin, kidney and pelvic fat as well as condition scores). The pure Karakul lambs had greater leg and carcass length and both pure Karakul and one-half Karakul lambs had lower leg conformation scores. A major interest in the study was the influence of docking and crossbreeding on tail fat trim. The undocked female Karakul lambs had an average of fat trim of 1151.1 grams or 4.5% of the carcass. For the docked crossbred female this value was

318.5 grams or 1.3% of the carcass. As expected, the male lambs had slightly less fat than females.

Thus the combination of the practices of crossbreeding and docking largely eliminated the problem of waste fat trim. In the case of wether lambs the crossbred lambs had only 174 grams more fat than the Suffolk cross lambs.

What is the Significance or Explanation for the Accumulation of Fat in the Tail?

A phylogenetic explanation for the accumulation of fat in the tail must hold theoretical, as well as practical, interest. Most fat-tail sheep are located in arid regions which are subject to extended periods of feed shortage. In addition, such sheep tend to be concentrated in areas subject to great variation in environmental temperature between night and day or between seasons. Few fat-tail sheep are found in humid or tropical climates, although fat rumped sheep may be present in these areas. This is generally explained in that animals in tropical environments do not need large deposits of fat, and climatic conditions do not encourage deposition of fat (i.e., total feed intake is reduced under conditions of high temperature stress). It has also been observed that fat-tail animals suffer in humid environments because the skin covered by the tail flap does not dry out—resulting in unsanitary and sometimes necrotic conditions.

If one seeks an explanation for the presence of the fat-tail from people in the areas where these sheep are produced, an almost invariable answer is that the fat tail is necessary for the animal to survive extended dry periods. However, this is a valid answer only if the accumulation of fat in the tail is physiologically different to fat deposited at some other place in or on the body.

In the simplest terms, an accumulation of fat in the tail must result from natural or artificial (that imposed by man) selection. There is actually a tendency for accumulation of fat in the posterior regions in many breeds or species. For instance, there are several breeds of sheep characterized by deposition of fat around the rump. Thus, fat-tail sheep differ from other types of sheep only in the degree and specific location of their fat deposition. If man has actually selected for the fat tail it may be hypothesized that: (a) such animals were thought to be more adaptable or productive, (b) the meat of this type was preferred, and (c) the fat of the sheep was needed for cooking, seasoning or for use in preservation of other food products. This use might be comparable to that made of pork fat by farm families in the U.S. in earlier years. A quotation cited earlier that the tail fat was "the butter of Central Asia, and ideal for cooking purposes" emphasizes this point. There is an important distinction in that these sheep evolved and were used in the Middle East over thousands of years. In an arid or desert environment, animal fat (from sheep) could be collected much more easily if it was concentrated at one place in or on the body. The author is of the opinion that this is the most likely explanation for selection/propagation of sheep with this unique accumulation of fat. With the widespread availability and use of vegetable oils and changes in dietary habits, the fat tail or tail fat is no longer in great demand and in many markets it is removed from the carcass before delivery to retail outlets. In some cases, it has almost no value or is used for industrial purposes at very low prices. Thus, heavy fat accumulation constitutes considerable loss or waste for up to one-third of the world's sheep population. Many carcasses in market channels have the fat tail removed before shipment, suggesting some resistance to the excess fat on the part of the consuming public.

On the other hand, if natural selection is the primary mechanism for the accumulation of fat in the tail, an explanation for the adaptive advantages must hold interest. Other potential sites for fat deposition are internal (kidney and pelvic region), subcutaneous, intermuscular and intramuscular. It could be theorized that the presence of large amounts of internal fat interfere with the ability of the animal to consume large amounts of feed when it is available, or that subcutaneous fat would interfere with heat dissipation at times of heat stress. Both of these could possibly be true at some time or some place, but are these factors adequate to overcome the interference of the tail with reproduction under natural mating conditions? A series of investigations was carried out to compare Karakul with other sheep in terms of their response to heat stress, and to compare docked vs undocked animals. It might be expected that docking results in an increase in subcutaneous or internal fat deposition and thus these comparisons should provide some information on the question of tail fat deposition as a factor in dealing with heat stress.

Karakul sheep were less affected by temperature stress than Rambouillets, but were more stressed than meat-type (non-fiber producing) sheep or goats. In general, there was little difference between docked and undocked animals, and differences that did exist tended to favor the docked animals. These studies did not support the theory that the accumulation of fat in the tail is beneficial to the animal in dealing with heat stress, and strongly indicated the reverse—that docking is beneficial. The existence of breed differences in respect to heat stress between Karakul and Rambouillet perhaps has an alternative explanation.

Another suggested or potential advantage for accumulation of fat in the tail is that reduced internal fat storage would permit greater feed intake if intake became a critical factor. This would be more likely to be critical only for the pregnant ewe, particularly for those carrying twins. Also, docking the ewe could theoretically increase the internal (kidney/pelvic) fat deposition, but this has not been proven to be the case. An experiment was conducted in which ewes of three breeds (Rambouillet, Karakul and Finnish Landrace), in late pregnancy, were provided *ad libitum* feeding of a mixed and pelleted ration. The Karakul group contained docked and undocked ewes. The ewes were kept on feed for a period of time after lambing to coincide with the early stage of lactation. The results of these studies have not been clear cut. Two differences were noted. The Karakul ewes carrying twin fetuses ate significantly less feed in late gestation than those carrying singles when this was expressed as a function of body weight. This was not true with the other breeds. This may be explained by the fact that Karakul lambs were larger, which resulted in more crowding or limitation in capacity. When expressed as a function of body weight, the Karakul ewes ate more total feed than the other breeds. In another independent study Karakul ewes ate more feed than finewool or hair type sheep. This was true in actual terms and also when expressed in terms of body weight. Docked Karakul ewes tended to eat less than non-docked ewes, but these differences were not statistically significant.

These series of studies suggest to the writer that fat-tail sheep are adapted to arid regions because of greater total body fat reserves, as established in carcass studies. Also, that a higher level of feed intake at times when forage is available facilitates greater fat deposition. These studies do not confirm or deny that the presence of the fat in the tail facilitates greater total feed intake or total fat deposition. Any differences which were observed are very small, and in these studies, non-significant. However, small differences may take on added significance when considered over thousands of years of evolutionary development.

Literature Cited

[1]Edwards, R. L., D. D. Crenwelge, J. W. Savell, M. Shelton and G. C. Smith. 1982. Cutability and Palatability of Rambouillet Blackface-Crossbred and Karakul Lambs. Int. Goat and Sheep Res. 2:77–80.

[2]Epstein, H. 1961. The Development of Body Composition of Docked and Undocked Fat-tail Awassi Lambs. Empire J. of Experimental Agriculture 29:110–118.

[3]Joubert, D. M. and L. Ueckermann. 1971. A Note on the Effect of Docking on Fat Deposition in Fat-tailed Sheep. Animal Prod. 13:191–192.

[4]Lupton, C. J. and M. Shelton. 1986. Some Fleece and Fiber Characteristics of Texas Karakul Sheep. Texas Agri. Expt. Sta. PR–4400.

[5]O'Donovan, P. B., M. B. Ghadaki, R. D. Behesti, B. A. Saleh and D. H. L. Rollison. 1973. Performance and Carcass Composition of Docked and Control Fat-Tailed Kellakai Lambs. Anim. Prod. 16:67–76.

[6]Sefidbakht, N. and K. Ghorban. 1972. Changes Arising from Docking of Fat-Tailed Sheep in Feedlot Performance. Iran J. Agric. Res. 1:72–77.

[7]Warren, L. E., D. N. Ueckert and J. M. Shelton. 1984. Comparative Diets of Rambouillet, Barbado and Karakul Sheep and Spanish and Angora Goats. J. of Range Manag. 37:172–179.

[8]Wong, E., C. B. Johnson and L. N. Nixon. 1975. The Contribution of 4 Methyloctanoic (hircinoic) acid to Mutton and Goat Meat Flavor. N.2. J. of Ag. Res. 18:261–6.

Barbados Blackbelly Sheep in the Caribbean

Stephan Wildeus
Agricultural Experiment Station
University of the Virgin Islands
RR. 2, Box 10,000, Kingshill
St. Croix, U.S. Virgin Islands 00850 USA

Barbados Blackbelly sheep, and hair sheep in general, are part of the same species *Ovis aries* as their more familiar woolly counterparts, although they appear to resemble phenotypically a goat (*Capra hircus*) more than a sheep upon cursory examination. A crude means of separating the two is the erect tail of the goat opposed to the hanging tail of the sheep. In the mixed flocks of hair sheep and goats that are common in the Caribbean, behavioral differences between the two species (diet selection, ease of handling, etc.) are also evident.

Characteristics of the Breed

Apart from the apparent differences in coat characteristics between hair (Barbados Blackbelly) and wool sheep, the former also tend to be smaller in mature size and faster maturing. However, any differences between the two types are confounded by tropical and temperate environmental factors, such as temperature, humidity, feed supply, photoperiod and parasite burden under which they developed. Studies of both types of sheep in the same environment indicate, however, that there are no significant differences in their digestive[1,2] and reproductive[3,4] physiology.

The differences in coat characteristics between the two types are primarily the result of variations in the ratio of secondary to primary skin follicles. Primary follicles are larger, contain sweat glands and erector muscles, and produce kemp and heterotypic (hair) fibers, while the secondary follicles produce unmedullated wool fibers. In hair sheep, the ratio of secondary to primary follicles ranges from 2/1 to 4/1 with a low density of primary follicles.[5] Wool is sporadically present on hair sheep (possibly as the result of earlier crosses with wool sheep) and usually takes the form of a cotted fleece along the dorsal, posterior region of the animal.

Barbados Blackbelly display a very distinct color pattern (Figure 1) as implied by their name. They have been described by Rastogi et al.[6] as varying from light to dark reddish-brown, except for conspicuous black underparts, including chin, throat, breast, belly, axillary and inguinal regions, and inside of legs. The inside of the ear is black and there is a black stripe on the face above and anterior to each eye and to the tip of the muzzle ("badger face"). The black areas are sharply delimited, except in the areas of longer hair. The brown color is paler along the face, neck and flank. Males have a pronounced throat tuft (mane).

Origin of the Breed

Barbados Blackbelly sheep, like most Caribbean hair sheep, are expected to have originated in western Africa.[7,8] At the time of colonization of Barbados in 1624 only feral pigs were present on the island and the subsequent development of a sheep population would have been the result of importation of stock from Africa and/or Europe. In early reports dating to 1657, the presence of both wool and hair sheep on Barbados has been described,[8] but wool sheep were no longer found in significant numbers by 1750,[10] possibly a result of lack of adaptation to the tropical environment.

Figure 1 Barbados Blackbelly sheep.

The western African stock would have come to Barbados during the early periods of the slave trade (1624–1657). Two types of hair sheep are found in this part of Africa, one being a larger, long-legged, lop-eared type found in the drier northern region and the smaller, compact, horizontal-eared type in the more humid southern region.[11,12] Both types have thin tails, are generally horned and display a variety of color patterns. Barbados Blackbelly would have been derived from the latter type and the African breed most closely resembling them are the Fouta Djallon found in Cameroon.[7] However, these sheep do not approach the size and prolificacy of the Blackbelly sheep in the Caribbean.

It is not entirely clear how these West African hair sheep evolved into the Barbados Blackbelly sheep. The fact that Barbados Blackbelly sheep are polled, as are most other Caribbean hair sheep, is most likely the result of pre-selection against horns in sheep transported across the ocean in restricted quarters. The increase in size over their African ancestors may be attributed to the transfer into a less stressful climatic environment; differences in mature size of this type of sheep have been observed between different climatic regions in West Africa.[12] However, the effects of crossbreeding with wool breeds of larger size can not be ruled out, particularly since both types of sheep were present on the island during the development of the breed.[9]

A high level of prolificacy is the most outstanding characteristic of the Barbados Blackbelly breed[13] and exceeds those of the native West African sheep. There has been some speculation as to the nature and origin of this prolificacy. Mason[7] pointed to the possibility of some genetic variation in the fertility of the West African base stock, but also acknowledges that these West African sheep give little indication for the prolificacy found in the Barbados Blackbelly. A strong case is made by Combs[8] for the early crossing (1624–1657) of the West African sheep with large and prolific, but poorly adapted, European wool breeds, most likely of Dutch origin. This would support the theory that the Barbados Blackbelly as a breed were developed as such on the island of Barbados and were not transferred there from northeastern Brazil as has also been suggested.[14,15]

Distribution of the Breed

As a result of their high prolificacy there has been considerable interest in these sheep for breeding stock, though a ban on exports was imposed by the government of Barbados in the 1970's. Within the Caribbean these sheep were exported to St. Lucia in 1902 and later to Antigua,[6] and can now be found in varying numbers ranging from Jamaica and the Bahamas to Trinidad and Tobago, Guyana and the Netherlands Antilles. In many of these locations the original Barbados Blackbelly stock has been exposed to crossbreeding with both local wool and hair sheep, and the total number of pure Barbados Blackbelly sheep remains relatively small.

Exports to South and Central America include Venezuela, Mexico and Panama. Barbados Blackbelly exported to Venezuela in 1961 were used to establish purebred lines in an experimental flock at a research station in Maracay.[16] Breeding stock was also exported to North America, although animals destined for Canada displayed blue tongue titers and were destroyed upon arrival.[17] Barbados Blackbelly for the United States (four ewes and one ram) were imported by the U.S. Department of Agriculture in 1904[18] and were initially quarantined in Bethesda, Maryland.[19] All Barbados Blackbelly sheep found in the U.S. would most likely be descendants of this early import. As a result of cross breeding the number of Blackbelly type sheep in the U.S. increased to as much as 200,000 to 300,000 in the early 1970's.[20] However, since then numbers have declined and the Barbados Blackbelly find their most common application in research and as crosses with the Mouflon for game purposes.

Performance of the Breed

When evaluating the performance of Barbados Blackbelly sheep it is important to consider the environmental conditions under which the breed was developed and the management systems under which the sheep were kept. Livestock breeds developed to perform under tropical conditions will generally trade high growth and reproductive rates for the reduced productivity associated with an increased level of adaptation to conditions of heat stress, seasonal malnutrition and increased disease susceptibility. In order to ensure survival under these harsh environmental conditions, the sheep must be able to utilize lower quality tropical forages and cope with a high level of gastrointestinal parasitism. In contrast, the lambing interval of tropical hair sheep is lower than that of temperate wool sheep in the absence of a photoperiod-induced seasonal anestrus.

The traditional husbandry system for the Barbados Blackbelly sheep on Barbados and throughout the Caribbean is a small flock of 5 to 15 animals, maintained under a moderate to low level of management. Animals are usually grazed during the day either tethered along roadsides or on pasture, and are penned overnight to control predation and larceny. Supplementation is practiced only during the dry season. Rams are run with the flock throughout the year and it is not customary to wean or castrate lambs. Treatment for disease, particularly internal parasites, is usually done on demand and not as routine flock health management. Animals are removed from the flock as needed for slaughter or sale.

TABLE 1 Birth, Weaning and Mature Ewe Weight of Barbados Blackbelly (BB) and Their Crosses at Various Locations

Breed type	*Location*	*Birth weight[1] (kg)*	*Weaning weight[1] (kg)*	*Weaning age (weeks)*	*Mature weight[1] (kg)*	*Reference*
BB pure	Barbados	3.1 (280)	9.9 (112)	12	31.9 (121)	23
BB pure	Tobago	3.1 (29)	13.9 (27)	6–12	44.9 (98)	27
BB pure	Venezuela	2.5 (282)	12.1 (228)	13		16
BB pure	Guyana	2.3 (69)	13.3 (33)	17	32.7 (33)	26
BB grade	Barbados	3.0 (279)				25
BB grade	Tobago	2.8 (29)	13.8 (26)	12	32.2 (49)	27
BB grade	U.S. (MS)				45.3 (-)	24
BB grade	Virgin I.	2.8 (19)	7.3 (18)	9	36.8 (7)	22
BB cross[2]	Guyana	2.7 (24)	12.7 (20)	17		26
BB cross[3]	U.S. (NC)	3.9 (29)			49.2 (10)	21
BB cross[3]	U.S. (MS)	3.2 (112)	15.6 (109)	8.5		24
BB cross[4]	Guyana	2.3 (24)	6.9 (24)	8.5		29

[1]number in parenthesis indicate number of observations
[2]Creole x BB crosses
[3]Dorset x BB crosses
[4]average of two feeding regimes

Growth and Mature Size

Mature size of Barbados Blackbelly sheep is comparable to other breeds of Caribbean hair sheep. Mature weight of ewes ranges from 30 to 45 kg, dependent on location and management (Table 1) and can average 50 kg in Barbados Blackbelly x Dorset crosses.[21] Barbados Blackbelly rams in the Caribbean reach a mature weight of 50 to 70 kg,[13,22] up to 90 kg,[23] and 60–68 kg in the southern U.S.[24]

Birth weights of lambs vary according to litter size, sex of lamb and age of dam, but generally range from 2.3 to 3.1 kg (Table 1). Bradford et al.[25] reported a range of 3.52 kg for single lambs to 2.54 kg for quadruplets on Barbados, with twins and triplets weighing 2.88 and 2.76 kg, respectively. A similar range for birth weights of different types was reported by Patterson.[23] The birth weights of lambs from young (< 19 mo), mature (19–35 mo) and older ewes (> 35 mo) on Barbados was 3.45, 3.90 and 3.13 kg, respectively.[23] Birth weights reported from other locations in the Caribbean were generally lower (3.0 kg).[22,23,27]

Weaning weights (Table 1) and pre-weaning growth are also affected by litter size, sex and age of dam. Under tropical conditions, gain from birth to weaning in Barbados Blackbelly lambs ranged from 90[16,26] to 127 g/day,[28] with higher rates of daily gain for singles (99 g), compared to twins (87 g) and triplets (75 g). These growth rates, however, were higher than those of West African and Criollo sheep maintained under the same conditions.[16] Pre-weaning supplementation of Barbados Blackbelly lambs is not commonly practiced, but will result in increased weaning weights.[29]

Post weaning growth rates for Barbados Blackbelly on pasture range from 69 to 84 g/day, dependent on birth type,[16] but increased to 150 g/day under feedlot conditions.[21] Barbados Blackbelly crosses with Dorset[21] and Rambouillet[30] achieved post weaning growth rates of 200 and 176 g/day, respectively, but did not approach the growth rates of contemporary wool breed crosses. Hence Barbados Blackbelly achieve slaughter weight at a relative late age under Caribbean conditions and have a low dressing percentage of 40 to 45%.[22,23] Their performance improves in the feedlot, but does not approach that of Barbados Blackbelly x Dorset and Barbados Blackbelly x Rambouillet crosses (Table 2) or temperate wool breeds.

Reproductive Performance

The most outstanding characteristic of the Barbados Blackbelly sheep is their prolificacy.[13] The average litter size under diverse management systems and at different locations ranges from 1.35 to 2.00 (Table 3). The frequency of multiple birth is high and the incidence of ewes giving birth to triplets and quadruplets has been reported to approach 25% in some flocks.[23,31] These multiple birth litters are greatly affected by lamb mortality, which may reach up to 43%.[26] Litter size has a marked effect on lamb survival in the first few weeks after birth, mortality ranging from 4.6 to 6.5% in single and twin lambs and increasing to 32.5 and 40.0% in quadruplets and quintuplets.[25] The influence of litter size on lamb survival is less pronounced in older lambs[16,23] and was 4% lower post weaning than prior to weaning at 90 days.[16] Sex of lamb has no apparent effect on lamb mortality.[23,25]

Barbados Blackbelly ewes display estrus and ovulate throughout the year under Caribbean conditions and no seasonal peaks in lambing distribution and litter size have been observed under continuous mating.[23,25] The lambing interval in flocks under these conditions ranged from 8.3 to 8.5 months (Table 3), but can be shortened by a month with improved management.[31,32] The lambing interval was found to be shorter (213 vs. 234 days) following single vs. twin birth, but was shorter again following triplet (216) and quintuplet (206) births,[25] suggesting that the post partum interval in Barbados Blackbelly ewes is not directly related to lactational stress. This is supported by McPherson[33] who observed only a 10% increase in milk production of Blackbelly ewes nursing

TABLE 2 Carcass Characteristics of Barbados Blackbelly (BB) and Their Crosses at Different Locations

	Barbados	*Barbados*	*Trinidad*	*U.S. (MS)*	*U.S. (TX)*
N	23	30	24	18	20
Breed	BB pure	BB pure	BB cross¹	BB cross²	BB cross³
Management	pasture	feedlot	feedlot	feedlot	feedlot
Age (days)	450		277	220	
Live wt. (kg)	27.0	49.6	40.1	42.3	39.5
Carcass wt. (kg)	11.5	24.0	17.4	22.4	22.0
Dressing %	43.0	51.8⁴	43.5	53.1	55.7
Rib eye (cm²)		13.4	19.2	13.5	11.9
HKP (%)		3.2	1.6	5.5	6.2
Backfat (cm)		2.8		6.1	
Reference	23	34	35	24	30

¹BB x West African (hair sheep) cross
²BB x Suffolk cross
³BB x Rambouillet cross
⁴calculated on shrinkage corrected live weight (7%)

TABLE 3 Lambing Performance of Barbados Blackbelly Sheep (BB) and Their Crosses at Various Locations

Breed	Location	Litter Size[1]	Birth Type (%)				Lambing Interval (mo)	Reference
			Single	Twin	Triplet	4+		
BB pure	Barbados	1.99 (1031)	28	48	.21	3	8.5	23
BB pure	Tobago	1.35 (68)	68	29	3	—		27
BB pure	Barbados	1.86 (489)	35	47	16	2	8.3	25
BB pure	Venezuela	1.45 (195)	61	33	6	—		16
BB pure	Guyana	1.68 (145)	42	48	10	—		15
BB pure	Jamaica	1.39 (365)	63	34	3	—	8.3	31
BB pure	Guyana	1.66 (41)	39	56	5	—		26
BB grade[2]	Jamaica	2.05 (63)	27	48	19	6	7.5	31
BB grade	U.S. (NC)	1.56 (36)						36
BB grade	U.S. (MS)	1.70 (66)	20	75	5	—		24
BB grade	U.S. (CA)	1.81 (272)	31	58	11	<1	7.3	32
BB grade	Virgin Islands	1.77 (13)	31	62	8	—	8.9	22

[1]number in parenthesis indicate number of observations
[2]mature ewes only (2+ litters)

twin over those nursing single lambs. No seasonal effects on the lambing interval were observed.[25]

Conclusion

Information on the performance of Barbados Blackbelly sheep in the Caribbean is limited and there is a need to define the production potential ofthe breed under different husbandry systems. Data available indicate a wide range in performance and suggest that the productivity of the breed can be increased by means of selection and improved management. However, the traditional small holder systems are not well suited for performance testing and changes in the productivity of the breed may be slow. Considerable information on the Barbados Blackbelly (though on a limited genetic cross section) has been gathered in the U.S. for their use under harsh production conditions and in accelerated lambing programs. Data indicate that the breed and their F_1 crosses have only limited application for market lamb production, but may be of use in improving lambing performance.

References

[1]Mann, D. L., L. Goode and K. R. Pond. 1987. Voluntary intake, gain digestibility, rate of passage and gastrointestinal tract fill in tropical and temperate breeds of sheep. J. Anim. Sci. 64:880.

[2]Quick, T. C. and B. A. Dehority. 1986. A comparative study of feeding behavior and digestive function in dairy goats, wool sheep and hair sheep. J. Anim. Sci. 63:1516.

[3]Evans, R. C. 1987. Selected parameters of reproduction in Rambouillet and St. Croix ewes. M. S. Thesis, Utah State University, Logan.

[4]Panter, K. E. and W. C. Foote. 1978. Reproduction and endocrine function in the St. Croix ewe. Proc. West. Sec. Am. Soc. Anim. Sci. 29:290.

[5]Burns, M. 1967. The Katsina wool project. I. The coat and skin histology of some northern Nigerian hair sheep and their Merino crosses. Trop. Agric. (Trinidad) 44:173.

[6]Rastogi, R. K., H. E. Williams and F. G. Youssef. 1980. Barbados Blackbelly sheep. In: I. L. Mason, Prolific Tropical Sheep, pp. 5–28, FAO Animal Production and Health Paper 17. FAO, Rome.

[7]Mason, I. L. 1980. The origin of the American Hairsheep. In: Prolific Tropical Sheep, pp. 59–64, FAO Animal Production and Health Paper 17. FAO, Rome.

[8]Combs, W. 1983. A history of the Barbados Blackbelly sheep. In: H.A. Fitzhugh and G. E. Bradford (ed.) Hair Sheep of Western Africa and the Americas.—A Genetic Resource for the Tropics, pp. 179–197, Westview Press, CO.

[9]Ligon, R. 1657. A true and exact history of the island of Barbados. (2nd ed., 1673) Reprinted 1976 by Frank Cass, London.

[10]Hughes, Rev. G. 1750. The natural history of Barbados. Reprint ed. 1972. Arno Press, New York.

[11]Epstein, H. 1971. The Origin of Domestic Animals of Africa. Vol. II. Africana Publishing Comp., New York, London and Munich.

[12]Bradford, G. E. and H. A. Fitzhugh. 1983. Hair sheep: A general description. In: H. A. Fitzhugh and G. E. Bradford (ed.) Hair Sheep of Western Africa and the Americas.—A Genetic Resource for the Tropics, pp. 3–22, Westview Press, CO.

[13]Maule, J. P. 1977. Barbados Blackbelly sheep. World Anim. Rev. 24:19.

[14]Devendra, C. 1972. Barbados Blackbelly sheep of the Caribbean. Trop. Agric. (Trinidad) 49:23.

[15]Devendra, C. 1977. Sheep of the West Indies. World Rev. Anim. Prod. 13:31.

[16]Bodisco, B., C. M. Duque and S. Valle. 1973. Comportamiento productivo de ovinos tropicales en el periodo 1968–1972. Agronomia Tropicales 23:517.

[17]Williams, H. E. 1975. Animal health limitations within the region. Seminar on the Utilization of Local Ingredients in Animal Feeding Stuffs. Jamaica.

[18]Rommel, G. M. 1904. Barbados Blackbelly sheep. The Breeders Gazette 46:845.

[19]Patterson, H. C. 1976. The Barbados Blackbelly sheep. Ministry of Agriculture, Science and Technology, Bridgetown, Barbados.

[20]Shelton, M. 1983. The Barbados Blackbelly ("Barbado") breed in Texas. In H. A. Fitzhugh and G. E. Bradford (ed.) Hair Sheep of Western Africa and the Americas.—A Genetic Resource for the Tropics, pp. 289-291, Westview Press, CO.

[21]Goode. L., T. A. Yazwinski, D. J. Moucol, A. C. Linnerud, G. W. Morgan and D. F. Tugman. 1983. Research with Barbados Blackbelly sheep in North Carolina. In: H. A. Fitzhugh and G. E. Bradford (ed.) Hair Sheep of Western Africa and the Americas.—A Genetic Resource for the Tropics, pp. 257-274. Westview Press, CO.

[22]Wildeus, S. unpublished data.

[23]Patterson, H. C. 1983. Barbados Blackbelly and crossbred sheep performance in an experimental flock in Barbados. In: H. A. Fitzhugh and G. E. Bradford (ed.) Hair Sheep of Western Africa and the Americas.—A Genetic Resource for the Tropics, pp. 151-162, Westview Press, CO.

[24]Boyd, L. H. 1983. Barbados Blackbelly sheep in Mississippi. In: H. A. Fitzhugh and G. E. Bradford (ed.) Hair Sheep of Western Africa and the Americas.—A Genetic Resource for the Tropics, pp. 299-304, Westview Press, CO.

[25]Bradford, G. E., H. A. Fitzhugh and A. Dowding. 1983. Reproduction and birth weight of Barbados Blackbelly sheep in the Golden Grove flock, Barbados. In: H. A. Fitzhugh and G. E. Bradford (ed.) Hair Sheep of Western Africa and the Americas.—A Genetic Resource for the Tropics, pp. 163-170, Westview Press, CO.

[26]Nurse, G., N. Cumberbatch and P. McKenzie. 1983. Performance of Barbados Blackbelly sheep and their crosses at the Ebini Station, Guyana. In: H.A. Fitzhugh and G. E. Bradford (ed.) Hair Sheep of Western Africa and the Americas.—A Genetic Resource for the Tropics, pp. 119-123. Westview Press, CO.

[27]Rastogi, R. K., K. A. E. Archibald and M. J. Keens-Dumas. 1983. Sheep production in Tobago with special reference to the Blenheim Sheep Station. In: H. A. Fitzhugh and G. E. Bradford (ed.) Hair Sheep of Western Africa and the Americas.—A Genetic Resource for the Tropics, pp. 141-150, Westview Press, CO.

[28]Rastogi, R., F. G. Yousef, M. J. Keens-Dumas and D. Davis. 1979. Note on early growth rates of lambs of some tropical breeds. Tropical Agriculture (Trinidad) 56:259.

[29]McPherson, V. O. M., J. Seaton and F. Wilson. 1982. Studies on the performance of sheep in Guyana. I. The effects of supplementing ewes during late pregnancy and their lambs before weaning. IV. Regional Livestock Meeting, Georgetown, Guyana.

[30]Shelton, M. 1983. Crossbreeding with the "Barbado" for market lamb and wool production in the United States. In: H. A. Fitzhugh and G. E. Bradford (ed.) Hair Sheep of Western Africa and the Americas.—A Genetic Resource for the Tropics, pp. 293-297, Westview Press, CO.

[31]Bradford, G. E., A. J. Muschette, V. Lyttle and D. Miller. 1983. A note on performance of Barbados Blackbelly sheep in Jamaica. In: H. A. Fitzhugh and G. E. Bradford (ed.) Hair Sheep of Western Africa and the Americas.—A Genetic Resource for the Tropics, pp. 177-178, Westview Press, CO.

[32]Levine, J. M. and G. M. Spurlock. 1983. Barbados Blackbelly sheep in California. In: H. A. Fitzhugh and G. E. Bradford (ed.) Hair Sheep of Western Africa and the Americas.—A Genetic Resource for the Tropics, pp. 305-311. Westview Press, CO.

[33]McPherson, V. O. M. 1984. Studies on the performance of sheep in Guyana. IV. Milk production of Barbados Blackbelly ewes, V. Regional Livestock Meeting, Nassau, Bahamas.

[34]Swartz, H. A., M. Alexander, M. Hunte, M. Ellersieck, J. Vaughn and A. Devonish. 1987. Carcass characteristics of Blackbelly lambs raised in confinement in Barbados. J. Anim. Sci. 65(Suppl. 1):287.

[35]Lallo, C. H. O., F. A. Neckles and G. W. Garcia. 1984. Intensive lamb finishing on sugarcane based rations at the Sugarcane Feed Centre and the implications for improving the value and contribution of hair sheep in the Caribbean. V. Regional Livestock Meeting, Nassau, Bahamas.

[36]Goode, L. and D. Tugman, 1975. Performance of crossbred Barbados Blackbelly and Finish Landrace ewes. Mimeo., North Carolina State University, Raleigh.

Coloured Sheep and Wool of Canada

Joan Wootton
Bo-Peep's Place
Box 3361, Langley
British Columbia, V3A 4R7 Canada

Canada is a vast country, second only in size to Russia. While the sheep population at one time numbered over three million (the peak being in 1932) there was a steady decline, until by 1960 there were only half that number. By 1986 The Canadian Bureau of Statistics reported 700,935, and while there is some evidence that numbers may be increasing slightly, no further figures are available at the time of writing this article. The largest numbers of sheep are found in the province of Ontario in Eastern Canada, and the province of Alberta in Western Canada. While it is known that there are natural coloured sheep in all ten provinces of Canada, it is not possible to estimate the numbers with any degree of accuracy. Among the four coloured breeds that are recognized as purebred and registerable with Canadian National Livestock Records (CNLR) only a small percentage of each is registered, and while these figures give a hint as to the popularity, they do not represent the total numbers of each. There are no really large flocks of coloured sheep in Canada, and it is not possible to locate all the individual breeders who keep perhaps half a dozen, or even just one or two.

In Canada we have not escaped the usual "growing pains" associated with acceptance of black sheep. A pure white sheep that produced little black lambs would, until recently, be hastily culled, because white would be the standard for the flock.

The author, her dog and sheep at home in British Columbia.

It is encouraging to report that progress is gradually being made in Canada in recognition of natural coloured sheep. While black lambs have been seen for many years in market classes in our exhibitions, the first known classes for breeding stock were held in 1984 at the annual two-day sheep and wool fair put on by the Lower Mainland Sheep Producers' Association in Matsqui, British Columbia. There were classes for coloured yearlings and lambs at the World Sheep and Wool Congress held in Edmonton, Alberta, in 1986. A further break-through came in 1988, which with combined efforts of commercial and purebred sheep breeders, and officials of the Pacific National Exhibition in Vancouver classes were held for coloured ewe and ram lambs, and ewe and ram yearlings.

Purebred Natural Coloured Sheep of Canada

At the present time four natural coloured breeds of sheep are recognized as purebred and registerable with CNLR, jointly with the Canadian Sheep Breeders' Association. These are the Icelandic, Jacob, Karakul and Romanov. A brief description of each follows.

Icelandic: These are rather small, colourful sheep that have been known for centuries to include the "moorit"—a colour described as chocolate, cinnamon or beige, prized by handcrafters. Stefania Dignum and her husband at Parham, Ontario, imported twelve in 1985. These were accepted for registration by CNLR in 1986, and forty-six were recorded in 1987. At this time of writing CNLR list of new registrations for each breed in 1988 are not yet available.

Jacob: Jacob sheep have been known in Canada for many years, but the first registrations (numbering two) were recorded in 1988, according to Mr. J. A. Kittle, General Manager of CNLR. Mr. T. I. Hughes, President of the Ontario Humane Society at Markham, Ontario, reports he has a flock of fourteen Jacobs. There are crossbred Jacobs in many areas in Canada. Their wool, being "patchwork" of white, browns and/or black, is much in demand for handcrafting, since the variation of colours can be utilized without the need for dyeing. Different strains of the Jacob have two, four or six horns, and there is a hornless variety.

Karakul: According to Mr. J. A. Kittle of the CNLR, the first registrations of Karakul sheep in Canada took place in the 1930's. From then until the end of 1987, a total of fourteen hundred and sixty-six have been recorded. Their numbers are declining, so that only two new registrations were recorded in 1987, none in 1986, one in 1985 and five in 1984. Notable for the day-old lambskins that are required in the manufacture of Persian Lamb coats, and coat trim, there are not sufficient numbers of Karakuls in Western Canada to

Jacob Sheep in Ontario.

develop this market. Thus they have been bred here mainly because of the consistently jet black coloured lambs. Crossbred with other breeds they insure a supply of natural coloured wool for handcrafts.

Romanov: The coloured Russian Romanov (known as the "Romanovskaya" in Russia) came to Canada in 1980. To Dr. J. A. Vesely, Senior Research Scientist at the Animal breeding station at Lethbridge, Alberta, goes the credit for relaying the interesting story that led up to this importation:

In the late 1960's French General Charles de Gaulle visited Russia and obtained an agricultural and cultural exchange agreement between these two countries. One of the benefits resulting for France was its license to import Romanov sheep. Since that time several European countries purchased this breed from France, and in October of 1980 Canada brought in a small flock of fourteen ewes and two rams from that source. Extensive research was carried out at the Lennoxville, Quebec, research station, and in 1982 eight rams were transferred to the Lethbridge research station, where further studies were carried out by Dr. Vesely.

These studies in cross-breeding the Romanov with Canadian breeds of sheep proved that claims were justified for the Romanov's inherent prolificacy, its ability to breed at seven or eight months of age, and its production of twins or triplets as yearlings. A further favourable characteristic is that they are polyestrous, breeding most months of the year, so that lambings every eight months are entirely possible. Mature Romanov ewes often have "litters" of four or five lambs. While some reports state the lambs are very active at birth, getting up and nursing quickly without assistance, others report stillbirths are fairly common in multiple births, and

Purebred Romanov ewe and lambs at Doug and Pat Gardiner's ranch.

mortality rates proportionate to the quality of flock management. This includes insuring pregnant ewes receive proper nutrition, and get lots of exercise, and careful monitoring of the actual birthing. Purebred Romanov lambs have the short tail of other Asian breeds of sheep, hence require no docking. The colourful, short fleece of the mature Romanov is pale grey, interspersed with long black coarse hairs, especially about the neck area. The lambs are born black, with white markings on face, legs and other parts of the body. The black tends to fade to greys as the sheep age, retaining the white markings. Crossbred lambs are often multicoloured, with patches of browns, beiges and red tinges, but again, all tend to fade out.

The fleece of the Romanov is considered to have little commercial value in Canada, but it is suitable for such items as horse saddle blankets, and some weavers like its unusual qualities for use in wall hangings. Depending on the breed of sheep the Romanov is crossed with, quality wool for handcrafts could result.

Eligible for registration in Canada, six hundred and twenty-one were recorded in 1986, and in 1987 there were three hundred and eighty-eight.

Other Natural Coloured Sheep of Canada and Their Breeders

Besides the four breeds of coloured sheep accepted for registration as purebreds in Canada, a number of other breeds produce black lambs, with varying frequency. Since the original wild sheep, ancestors of our domesticated breeds, were multicoloured, it is understandable that the odd black lamb will appear occasionally, particularly in the black-faced breeds. There are Black Welsh Mountain sheep, and crossbreds from them, in Canada, but according to Mr. J. A. Kittle of the CNLR, they have not been accepted for registration, and it is difficult to obtain information about them.

A request for information about coloured sheep that appeared in The Western Producer, a weekly farm magazine published in Saskatoon, Saskatchewan, resulted in a response from Jean Flett of Fort Qu'Appelle. She told of her little flock of Shetland sheep, which originated from an importation of four animals in 1948. This has been kept as a closed flock, with no crossbreeding. Jean likes the varied colours of the fleece, especially the browns, for handcrafting.

A late addition to the list of breeders of coloured sheep is Mr. and Mrs. Gurnot Zillig, of Scotch Village, Nova Scotia, who have a flock of over three hundred breeding ewes, about one quarter of which are coloured. Apparently Edith Zillig is a superior craftswoman, and fashions rugs, wall hangings, cushions, etc. with her own wool. She designs patterns that are very original and artistic. Another of her projects is that of tanning her own sheep and lamb skins.

We have Dr. R. D. Crawford of the University of Saskatchewan to thank for reporting another flock of Shetland sheep, owned by Colonel Dailley of Cambridge, Ontario. Dr. Crawford also reported there were Barbados Black-Belly sheep at the Joywind Farm Rare Breeds Conservancy at Marmora, Ontario.

It has been reported that the Dorset occasionally produces spotted, or piebald lambs, brown and white.

Two reports were received of the possibility that the Soay breed may appear in Canada, but it was not possible to confirm this information.

Mr. John Lee of Leeland Farms, Highgate, Ontario, very kindly contributed details of this historic sheep farm, which began breeding purebred Lincolns well over one hundred years ago, in the early 1870's. His father, and grandfather, imported Lincolns from England, and from these coloured lambs appeared fairly frequently. Interest in natural coloured wool for handcrafts created a ready market for breeding stock, and several coloured Lincoln rams were sold to British Columbia, as well as to other areas of Canada and the United States. No doubt many of the crossbred Lincolns seen today originated with Leeland Farms purebred Lincolns. Mr. Lee reports other breeders who raised coloured Lincolns, such as the Walker Brothers in Ontario, whose flock was established in 1831. Also Gardhouse & Sons, whose flock history goes back to the 1830's. Mr. Lee took over management of Leeland Farms when he returned from service with Canadian Naval Forces in 1945. He continues to participate in shows and sales of his Lincolns in both Canada and the United States. Judging by his correspondence it appears Mr. Lee is not only a shepherd of repute, but a student of classical literature as well.

Coloured Romney sheep are seen frequently in Canada, and there are a number of serious breeders who are striving to promote their acceptance by CNLR as purebred, and registerable. However, at the time of writing this paper, the matter is still under discussion. Several importations have been made, mainly of rams, from the United States, where coloured Romneys are accepted for registration by the Romney Society there.

According to Flora Baartz of Lasqueti Island, B.C. there are coloured feral sheep there, ancestry unknown, and there may be feral sheep on other islands off the coast of B.C. It is a singular fact that when an unusual, or very different type of sheep, is advertised for sale, and the question asked as to where it came from, the answer invariably is, "from an island up the coast"; however, the island appears to be unnamed, or mythical, in most cases.

Commercial Processors of Natural Coloured Wool in Canada

An important part of the coloured sheep and wool story in Canada is played by commercial processors of raw fleece. One of the best-known in western Canada is Custom Woolen Mills Ltd. at Carstairs, Alberta. Established in 1978 by Mr. and Mrs. Bill Purves-Smith, they purchased antique machinery from two small mills that were closing down. Some of these machines date back to 1860, and a spinning mule was manufactured in 1910. Combining old with the new machinery over the years, complete service is offered, from scouring to spinning yarns, both natural coloured and dyed. They also make quilts and wool batts.

Another important processor of raw fleece is Birkeland Brothers Wool Batts Ltd. of Vancouver, B.C. The year 1989 represents fifty years of service. No scouring is done, but both natural coloured and white wool are processed into batts.

Joanna Sleigh and Champion Coloured Ram Lamb at Pacific National Exhibition, 1988.

Ted Carson, formerly of Streetsville, Ontario, recently semi-retired and moved to Minden, Ontario. He has been known all over the continent of America for his expertise in the textile industry. He teaches and gives seminars and workshops on weaving and spinning, and has been associated with the Ontario Handweavers and Spinners Guild for more than twenty-five years. This guild conferred an Honourary membership of their Masters' program on Ted. He regrets having to part with his sheep when he moved to Minden. Previously he kept both coloured and white sheep, mainly Corriedales, as he had a preference for this breed's excellent handspinning wool. However, Ted always had a good use for every kind of wool produced, from the finest to the coarsest "braid" types. He has to admit he could not possibly count the number of seminars and workshops he has conducted during his lifetime.

Wetsa Products of Langley, B.C. is the name given the wool-processing business of the Langley Association for the Handicapped. This began about five years ago, with purchases of table and floor looms, a spinning wheel, picker, hand and electric carders. Wetsa offers complete custom service for both coloured and white wool. Scouring, picking, carding and spinning make up the daily routine. They offer for sale handcrafted woven saddle blankets, rugs, wall hangings, wearables, etc., mostly in beautiful natural earth tones. However, natural dyeing also produces colourful yarns. Kim Cripps is credited with supplying information about this very worthwhile endeavor.

The Canadian Co-operative Wool Growers Limited, with headquarters at Carleton Place, Ontario, was organized in 1918 to sell Canadian wool. Shipments of raw wool arrive there "in the grease"—that is, just as it comes off the sheep. Here it is weighed and graded, and baled for shipment to world markets. A branch at Stony Plains, Alberta, has a wool shop that supplies handcrafters. This wool co-operative also handles all necessary supplies for sheep breeders and shearers. erick Bjergso, general manager of the carleton Place operation, reports that two of the larger wool processors in Canada are Briggs and Little Woolen Mill at Harvey Station in New Bruswick, and Victor Woolen Products at Victor, Quebec.

Majorie MacDonald, editorial assistant of Canada Wide Magazines Ltd., and Carol Pope, editor of Westworld Magazine, kindly gave permission to quote from an article that appeared in the March–April 1976 issue of Westworld. Written by O. L. Johnston, the story is told of the history of the Cowichan Indian knitters at Duncan, on Vancouver Island, B.C. While the Native Indian women had woven blankets, baskets and ceremonial robes for centuries, using barks, grasses, dog hair and other materials, they were not intro-duced to sheep wool and the modern spinning wheel until 1885. A Scottish immigrant lady by the name of Jeremina Robertson from the Shetland Islands graciously made friends with the Natives and demonstrated her prowess with the spinning wheel she had brought with her. She also taught the Indian ladies how to make patterns on the knitted sweaters, such as she had fashioned in her homeland.

Very soon the Cowichan ladies began designing their own patterns, and thus the familiar killer whale, thunderbird, totem poles and others appeared on the world-famous sweaters. It has been traditional for the B.C. government officials to make visiting celebrities the gift of a Cowichan sweater.

Handspinning, weaving, nature-dyeing, lecturing and writing are only a part of what Paula Simmons has contributed to the natural coloured sheep and wool scene. Teamed up with her husband Patrick Green, they live in the Chilliwack area of B.C. and operate "Patrick Green Carders Ltd." Her workshops and expertise in the wool and fibre world are widely known on this continent.

The story of the Salish Weavers at Sardis, in the Chilliwack area of B.C. is a fascinating one. The Native Salish people of the western coastal region in Canada had been weaving for generations. Like other Indian bands they made use of cedar bark, grasses, hair from mountain sheep and goats, and other materials to fashion ceremonial garments, rugs, blankets and other useful articles. But when the Hudson's Bay Company established forts, and trading with the Indians took place regularly, the art of Native Salish weaving style gradually died out. "Hudson's Bay" blankets replaced the need for the Indian handcrafted articles. In 1961 Oliver Wells, affectionately known as "Uncle Ollie", of historic Edenbank Farms at Sardis, B.C. became interested in the culture of the Salish people. A search turned up two elderly Native ladies who were still adept at weaving in the Salish style of their ancestors. Oliver Wells had a kindly, soft-spoken manner which appealed to the shy Indians, and he was able to encourage them to revive their nearly lost art. These two ladies willingly taught many of the younger ones how to spin and weave with modern equipment, and soon a thriving business was born.

Mrs. Oliver Wells (Sara) and her friend, Miss Sadie Thompson, gathered together a mixed flock of both natural coloured and white sheep in order to supply wool for the Salish Weavers' project.

Besides sweaters, hats, socks and other wearables, colourful blankets, rugs, wall hangings and other items were offered for sale, and buyers came from all over the world to purchase these handcrafted goods. Many different motifs added artistic touches, one of the most outstanding being the "Flying Goose" pattern, which represented the return of the almost lost art of Salish weaving.

The Salish Weavers' operation was dealt a severe blow to its morale when their dear friend and much loved mentor was killed in a tragic automobile accident in Scotland in 1970, while researching for a book he was writing. Without his kindly presence there was a loss of heart, and gradual loss of interest in the business Oliver Wells had nurtured and developed into a highly successful endeavour. By the early 1980's it closed down; however, there is still a gift shop, the "Coqualeetza", which handles native arts and crafts, and

"Ladybird" and her quadruplets, Bo-Peep's Place in British Columbia.

there is a small library. It is hoped that one day there will be a museum to display native cultural articles. Several of the Salish Indian ladies still work with natural coloured wool, and keep the gift shop supplied with their handcrafts.

Oliver Wells wrote a number of books; the latest, "The Chilliwacks and Their Neighbors" was finally published in 1987, some eighteen years after his death. Through the efforts of his daughter, Marie Weeden, and her friends, the material Oliver had gathered was sifted through and presented in an attractive form, which includes many interesting photographs. Much of this story about the Salish Weavers was gleaned from Oliver Wells' books, with kind permission of Marie Weeden.

For those who may not be familiar with the story of the mythical "Salish" sheep, it is repeated here. The *Black Sheep Newsletter,* issue #10, Summer 1977, gives the details of how this fairy tale began. This story was also published in Sachiye Jones' book, *The Black Sheep Newsletter Companion, The First Five Years, Volume 1.* Mrs. Oliver Wells, who along with Miss Sadie Thompson owned the white and coloured sheep flock that supplied wool for the Salish weaving, corroborated the details as written, mainly that the Salish people never owned sheep. Mrs. Wells said she had never referred to her flock as "Salish" or "Sardis" sheep, and no attempt was made to develop a specific type of sheep—they were just a mixed commercial lot. In the last couple of years of the flock's existence some crossbred Karakuls were added. When the tragic accident took the life of Oliver Wells the flock was dispersed, some apparently going across the border into the United States. Just when, or where, the name "Salish" became attached to them is a moot question.

Further evidence of the myth is contained in the American Minor Breeds' Conservancy newsletter, March–April issue, 1987, Vol 4, No. 3. To Nancy Dickey of Salem, Oregon, goes the credit for sending along the article by Elizabeth Henson which explains her search regarding the so-called "Salish" breed of sheep, and of her eventual conclusion that, in her words, "The ghost of the Salish sheep has finally been laid to rest".

Miscellaneous Information

The Greater Vancouver Weavers' and Spinners' Guild celebrated its fiftieth anniversary in April of 1985. When it was formed in 1935, membership required one to possess a loom, and pay an annual fee of one dollar. While meetings were originally held in members' homes, the guild has met in several different locations, and for the past fifteen years its home has been Aberthau, in Vancouver.

Early in 1988 the Hooser Weaving Center opened in the historic Stewart Farm house, in White Rock, B.C. A group of weavers and spinners, many of them members of the Peace Arch Weavers' and Spinners' Guild, conduct workshops, and demonstrate their skills with looms and spinning wheels, for visiting tourists.

There are numerous small guilds in cities and small towns all across Canada, whose membership comprises people with varying skills. All are dedicated to helping one another in learning the basics of the hobby. Many members have gone on to become professional, designing and weaving their own creations, and marketing them.

The Canadian Weavers' Guild sponsors a program for those wishing to work toward earning a Master Weavers' certificate. Up to February of 1988, only eighteen have qualified for this honour, in Canada. Studies are conducted at Olds College, Olds, Alberta, in western Canada, and the Georgian College in eastern Canada, to name two that offer courses.

Conclusion

Twenty-five years of breeding coloured sheep, and supplying natural coloured wool for handcrafts, motivated preparation of this paper. However, it could not have been written without the kind co-operation of the many contributors, friends, sheep breeders and publishers who gave permission to quote from their sources. A special note of appreciation is directed to the staff of the Langley Library for their help.

My interest in sheep and lambs actually began in the 1920's, when my late father purchased a flock of about 125 commercial ewes, from a nearby neighbor in Alberta. Having had no previous experience with sheep, we soon learned that barbed wire fencing did not contain them. Hence much of my time that summer was spent herding the flock on horseback, with the help of a couple of dogs. A few months later, on Christmas morning, Father came into the house carrying a pail of milk in one hand, and what appeared to be a lifeless little lamb in the other. A ewe with Dorset blood had surprised us by lambing in a snowbank in zero weather.

Warmed in a cardboard box on the oven door of the coal and wood stove, the lamb revived and let out a feeble bleat, and very soon he rejoined his mother inside the cow barn. I thought he was the most beautiful little creature I had ever seen, and from that moment I was "hooked".

It has appeared to me over the years that the successful sheep farmers are not those who see only dollar signs, but rather they are people who care enough to go that extra mile to tend their flocks with loving care. These are the people who are richly rewarded for their efforts when they see healthy lambs frolicking at play in pastures, while their dams quietly graze, turning grass and weeds into meat and wool.

Who could say it better than William Blake, who was inspired to write his poem:

Little lamb, who made thee?
Dost thou know who made thee,
Gave thee life, and bid thee feed
By the stream and o'er the mead?

Jacob Sheep in the United States: To Preserve or to Improve?

Ingrid Painter
1607 232nd Avenue NE
Redmond, Washington 98052 USA

U.S. Jacob with British bloodlines.

This paper compares Jacob sheep in Britain (past and present) to Jacobs currently found in the United States. It discusses the history of the sheep in both countries and examines their appearance, traits, hardiness and behaviour. It questions the value of upgrading them to compete with modern sheep and whether they should be preserved as an old, unimproved breed.

The Jacob is a small sheep weighing under 200 lbs. (91 Kg.). It is a white sheep with varying amounts of black patches and spots distributed randomly over the entire animal. The breed is polycerate, meaning that it carries the trait for splitting the horn core into multiple horns. Both sexes can carry two, four or even six horns. The Jacob has been kept as an ornamentl park sheep for many centuries and in recent times is having difficulty competing with modern breeds.

History in Britain

The origins of Jacobs in Britain are obscure and controversial. The earliest known record of black and white spotted sheep was mention of them in a letter written in Warwickshire in 1756. There is evidence of their existence in Yorkshire. Both these flocks were two horned. Another flock dating to 1760 from Cheshire was known to be four horned. Despite the fact that various people imported spotted sheep from South Africa and from Spain there is no evidence to prove that these were not merely incorporated into existing spotted, polycerate flocks and are only part of the makeup of today's Jacob breed. There are records of spotted sheep imports to Britain from Portugal in 1755, South Africa in 1823, Spain in 1843, North Africa in 1880, as well as mention of them from Persia and Siberia. A flock was imported from Ireland in 1730. Legend has it that these spotted sheep swam ashore from the wreck of one of the fleet of the Spanish Armada. However, there is strong evidence to support that piebald sheep with multiple horns existed in Britain long before 1700.

There are many breeds of multi-coloured sheep throughout the world. Of the eighteen breeds mentioned in a paper by A. R. Werner, thirteen of them are known to have the polycerate characteristic. Many of the spotted, horned breeds of Europe had died out by the 1900's.

Jacob ram lamb flanked by two ewe lambs in the U.S. All are four months old.

History in America

Jacob sheep were first brought to Canada and the U.S. through zoos as early as the turn of the century. When it became necessary to reduce zoo populations, regulations restricting private sales were relaxed. Many of these sheep found their way into roadside zoos and exotic animal farms. They were regarded as freaks of nature only occurring occasionally. People lost track of what they were and where they had originated. They became known by different names. It is speculated that they were crossed to other breeds of sheep, including the Navajo-Churro which is a polycerate breed with origins dating to 1540 when the Spanish Conquistadors made their way Northward and Westward across America. They were generally called Four Horns or Jacob's. Breeding was not kept pure as evidenced by the appearance of many all black animals with white crowns and white tipped tails and one or more white socks. Black and white faces with white throats and bellies are often seen in crosses. Sheep like this with rams sporting huge horns often measuring two or three feet (90 cm) from tip to tip attracted the attention of owners of private hunting preserves. These people were interested in producing sheep which were either woolless or which would shed their fleece naturally in the spring. Four horned sheep were willfully crossed to Mouflon and Barbados Blackbelly to achieve this goal. The sheep commanding the highest price at the auction barns were the black and white spotted ones with four horns. "Jacobs" that are found today with excessively hairy (kempy) fleeces are more than likely to have a hair breed in their background.

In about 1976 several people became aware that there were Jacob sheep in the U.S. Gradually small pockets of surviving Jacobs cropped up in widely scattered areas. These sheep were snapped up by enthusiasts interested in preserving endangered domestic livestock. The Jacob Sheep Society of Britain formed a foreign branch of their association and recorded U.S. Jacob sheep, which passed their photographic inspection, in 1982. The American Minor Breeds Conservancy started recording Jacobs in 1985 and published the Foundation Flock Book in 1986. Jacob breeders have organized and become incorporated as the Jacob Sheep Breeders Association (JSBA) in 1989. Jacobs in the U.S. are made up of early imports dating to the turn of the century, imports in the 1950's and two in the late 1970's. These later imports provided the needed bloodlines though they were noticeably larger and heavier boned than the existing earlier population.

Jacobs in Britain prior to 1969, when the Jacob Sheep Society was formed, were also smaller and lighter boned than those one sees there today. British Jacobs can compete with commercial breeds for size and wool quality. They are no longer considered an endangered breed nor can they be classified as unimproved or primitive. Has this come about through selective breeding or the deliberate crossing with commercial breeds such as the Dorset Horn? There are many tell-tale signs that the latter is true. By "improving" on an existing breed it is easy to lose sight of the very reasons that first attracted one to them. They are small and hardy, economical to feed and they have very few, if any, veterinary problems. Pasture stocking rates for Jacob flocks are about six per acre (15 per hectare), as opposed to five per acre (12 per hectare) for larger breeds. The sheep are easily tamed and handled. The rams are generally good natured but this, like in any breed, should not be taken for granted. Their black and white facial markings and spotted or patched bodies attracts people into buying them.

Head

It is important when selecting breeding stock to look at horn direction—specifically in rams. If a two-horned ram is chosen his horns should be widespread. If they are not they will grow inward as they sweep round toward his cheeks. The horns get thicker as they grow, allowing less space, and can put pressure on the cheeks and jaw. This area becomes warm and moist and encourages infection and fly strike.

With four-horned animals the upper horns should sweep away from the face, preferably toward the back or sides. If the horns come out of the head directly above the eyes, they will grow forward and within two or three months will either be directed toward his eyes or will have grown longer than the length of his muzzle. Either way the horns should be cut and the lamb destined to be butchered.

The lower set of horns on a four-horned ram should sweep downward—the ideal would be with only a very slight curve to them. Tightly curled lower horns can also cause the same problems as in narrow two-horned rams. I have noticed that frequently the initial growth spurt of the lower horns directs them straight out at 90° to the lamb's head and within four to six weeks a second growth spurt swings them almost 45° toward the cheeks. I have found that by clipping off the initial 'baby' tips to where the angle turns, the horns steer clear of the face and jaw. Lambs breaking horns off in the first few months of life is commonplace. The horns grow back sturdy and straight with rarely a problem. Ewe horns are considerably smaller and finer than ram horns. The lower ewe horns are often knocked off and some never regrow. They are still considered four horned. I have rarely found ewe horns to be a problem.

Horns are generally black if the sheep's black eye and cheek patches extend toward the horn core. If the facial markings are white in that area, the horns are generally striped—the stripe being an extension of the spot. It is generally felt that all white horns are a sign of cross breeding. White horns coupled with them being heavily gnarled or convoluted is considered a fault. Smooth horns appear to be less 'domestic' or unimproved.

Black and white facial markings in this breed are another point where selective breeding has been overemphasized. Early pictures of Jacobs in Britain show that not all have the ideal black cheek patches which incorporate the eyes and extend to the ears. The sheep have a white blaze extending to the muzzle. This area is often black, looking as if it has been dipped in a paint pot, or it is white with only the nasal septum being black. There are many old pictures of Jacobs which have small, spectacle-like eye patches or lack one or more of the patches. These sheep give the impression of having broad white faces. The shape of the face of earlier Jacobs has also changed from being dished and triangular, extending to a narrow, deer-like muzzle, to the shape today of a much more convex head and broad muzzle. Rams today often have the large Roman nose seen in many modern breeds.

Another unusual feature I have observed in Jacobs both sides of the Atlantic is those which have blue eyes. I have seen Jacobs with two dark blue eyes as well as several with one eye the normal yellow/brown and one dark blue. The dark blue eyes do not seem to go with any particular facial colour. There is also a very light blue coloured eye which I have seen. This has more often been observed in Jacobs with less colouration on the face. These sheep have been observed to be light sensitive. This light blue colour may be linked to albinoism and is more than likely to be a different gene to the one for the dark blue eyes. The light blue eyed sheep have a normal black and white colour pattern on their body but lack one or both eye patches.

Body

Body weights of the British Jacob are conveniently left out of the Jacob Sheep Society breed standard. Pictures of the sheep prior to 1970 look considerably finer boned compared to those one sees today. The Jacobs found in the U.S. prior to the recent imports are also smaller and finer boned. The JSBA calls for body weights of mature rams to be 120 lbs. to 180 lbs. (54.5 to 81.8 kg.), and for ewes 80 lbs. to 120 lbs. (36.4 to 54.5 kg.). The U.S. Jacob gives a 30- to 40-lb. (13.6 to 18 kg.) carcass at seven months with little or no extra feeding. It is tasty and lean. The beautiful pelt commands a high price when tanned.

Less "improved" breeds will have a sloping dock. Through selective breeding in many breeds, the slope has been raised, allowing the leg muscle to increase in size. This is the most valuable cut of meat on the carcass. Altering the tilt of the pelvic bones means that the slope of the birth canal has also ben changed. This means that lambs entering the birth canal in "improved" breeds have to rise upward before they can be delivered outward instead of slithering downward and out unassisted as in sheep with sloping docks. For a few extra pounds of "leg o' lamb" modern breeds have had to sacrifice easy births. The biggest drawback in showing Jacob sheep is that all judges tend to place as winners the sheep with the larger leg muscle and carrying the tail well up on the chine. Judges are never present at lambing time! The change in the slope of the dock accounts for the most notable difference between the old and the newer Jacob.

Udders on Jacob ewes are small and fairly insignificant but they give plenty of milk and the ewes have no trouble raising twins or even triplets. Jacobs with huge, pendulous udders should be considered as not being purebred.

I have observed that the scrotal sack on Jacob rams is held high and tight to the body. The bottom of the scrotum rarely reaches below the hind knee, even in very hot weather, whereas in rams of more modern breeds it is long, pendulous and can almost reach the ground in highly bred animals. Jacob crosses show up with much longer scrotal sacks.

The colour pattern on Jacob sheep is something which can be selected for to a degree. Some people prefer more black and others mostly white with random small, black spots. I have noticed that pictures showing Jacobs in England prior to the early 1970's show more of them have large black or faded gray areas than those seen today. The ideal colour distribution is 40% black on a white body. I have never figured out how one can accurately judge the percentage of colour without laying out the fleece, separating the colours and weighing them. Hardly practical for a whole flock! If the distribution of colour is fairly even and the sheep has a white belly, my guess is that it would be considered 40% black.

Jacob ewes in the U.S.

One of the Jacob Sheep Society journals reported the birth in England of what was called a "lilac" coloured lamb. They asked if anyone else had ever seen one. These "lilac" or "dusty" coloured grey Jacobs are fairly common in the U.S. As far as I am aware the majority can trace their origins to one flock in the mid-west. These sheep are the "perfect" Jacob colour pattern except that where most Jacobs are black these lambs are born a lilac-grey, including their facial markings. Dr. Phil Sponenberg of Virginia-Maryland Regional College of Veterinary Medicine, has suggested that someone breed a group of these sheep to another breed—one which is not dominant for black or has the tendency for spotted offspring. If the lambs are born looking like the usual Jacob cross (i.e., mostly black with or without facial markings) it would suggest that the "lilac" lamb is a true Jacob and not just another form of a cross. The Jacob is dominant for black and recessive for spots, so crosses with breeds that do not carry a similar genetic colour makeup will be born black.

There are also Jacobs which have a very dark brown facial patch which fades to lighter brown around the eye. I have seen several Jacobs both in England as well as in the U.S. like this. The fleece colour on most has been the usual black and white. Only one have I seen which had a brownish tinge to the fleece. This was not the result of the sun bleaching the black on the surface.

Many Jacobs on both sides of the Atlantic grey with age. The black areas gradually become lighter with a mixture of black and white wool—the two areas seem to bleed into each other. The white areas also become less white but it is not as obvious until the sheep is shorn and it is seen that the animal has developed many freckles. These freckles increase with age. Some sheep develop this greying and freckling more quickly than others from as early as their first shearing to about five or six years of age. These mottled fleeces are popular with handspinners but very unpopular with the breeder.

In Merino sheep it has been found that freckles occur with age and may be due to exposure to the sun after shearing. These freckles produce the dreaded black fibre so for this reason Merinos are rarely kept beyond their sixth year. No studies have been done in the Jacob breed but it has been suggested that the fading gene is dominant. It is speculated that the fading colour or increased distribution of black fibres in Jacobs is a genetic trait and highly heritable.

In other breeds lambs are often born black. Depending on the breed the sheep will grey within varying lengths of time. Not all greys are the same. In some the fibres are grey but in others a mixture of black and white fibres are found throughout the fleece, making it appear grey. I believe that "faded" Jacobs are like the latter. I have observed in my own flock that the few ewes that have faded by their third to fifth year have actually become progressively darker with age, so that at age eight and ten they are darker than their daughters.

Ears

Jacobs have small ears, about three inches (8 cm) long in an adult ewe. They are erect and tuck neatly below and behind the horns. Long, pendulous ears would be indicative of cross breeding. At the other extreme I have seen quite a number of Jacobs in the U.S. which have no ears at all, as in the La Mancha goat. I believe that these sheep have stemmed from one flock and could possibly have obtained this characteristic from crossing to another breed, such as the Navajo-Churro where the condition has been reported.

An experiment was done at the New Hampshire Agricultural Experimental Station some years ago showing that by breeding earless sheep to normal eared sheep the resulting offspring have very small ears. If both parents carry the no ear or small ear condition then the lambs born will be earless. Since earless Jacobs have not been previously reported, it is not known if this would be considered a fault.

Legs

A noticeable difference between U.S. and British Jacobs today is the lack of spots on the legs of those in Britain. Early pictures of British Jacobs show the legs to be heavily spotted and indeed, even some with a leg that was almost entirely black. In the U.S., all-white legs on Jacobs are regarded with suspicion. Many people feel the legs must have spots to balance the spotted body. Jacobs that have been crossed to Dorsets rarely have spotted legs, but crosses with other breeds tend to throw legs that are predominantly black. The Jacob Sheep Society breed standard calls for "Legs of white colour, with or without black patches around the knee and pastern." The Jacob Sheep Breeders Association calls for "Legs white, with or without black patches."

Fleece

Fleece quality in this breed is important especially to handspinners who like the challenge of the variegated wool. It is soft, lustrous and has sufficient kemp in it to make it interesting, not harsh. When spun by hand or commercially, it makes up into beautiful knitting yarns. A fleece weighs 2 to 6 lbs. (1 to 3 Kg.), is 3 to 7 inches (8 to 17 cm.) long and has a large range in fibre diameter from 44's to 56's in the Bradford Count (34.4 to 27.84 microns). These figures have been taken from various British sources since no selection for wool quality and weights has been officially monitored in the U.S. The fleece is very high yielding due to low grease content. The Jacob Sheep Society breed standard calls for ". . . of fine quality, with little or no kemp." Pictures of Jacobs in the past show that the fleece was open, indicating that it was fairly coarse and would fall into locks like the breeds which have little or no crimp to the wool.

The wool on the ram is the most important fleece in the whole flock. Fleece quality and amount of kemp produced

Jacob Ram in the U.S.

is highly heritable so it is important never to buy a ram without first evaluating his wool. Jacob fleeces vary greatly from front to back. This is because they have undergone very little upgrading. The fleece quality is finer on the front half and gets coarser, more wiry and considerably more kempy toward the britch.

The ideal is to buy a Jacob after the first shearing when the fleece has grown out sufficiently to be able to evaluate it. Great changes take place in the type of wool on young sheep until they are twelve to eighteen months of age. The first fleece coming off a young Jacob is open, soft and very desirable to handspinners. Some fleeces change dramatically at this stage and become harsh, dull and spongy, resembling a poor quality grade mutton breed. When looking at fleece on a sheep, part it along a natural vertical line in at least three places—front, middle and britch. If a great deal of kemp is seen, or if the fleece appears wiry or has a harsh feel to it you may wish to eliminate if from your breeding program.

When looking at young Jacobs in England it became apparent that some have the open, long fleece with little or no crimp and others have a more compact and denser fleece. This latter type has to be parted manually in order to reveal the crimp along the staple length. If sheep were left unwashed and untrimmed for the showring it would be easier to tell the lambs which have had some upgrading from those which are more like the sheep of the past. It is speculation on my part that former Jacobs had coarser fleeces with little crimp.

U.S. ram showing good horn spread.

Split Upper Eyelids

Jacobs, being a polycerate breed, have a congenital defect which cannot be overlooked. The splitting of the horn core in some instances will continue right down to the upper eyelid and in rare occasions affect the skull formation also. The split eyelid deformity can show up as a mere break in the hairline along the edge of the lid or in severe cases the two halves of the upper lid are separate. The ragged edges of the lid turn under, or wool and hair can grow on the INSIDE of the lid. This leads to eye irritation, corneal ulcers or complete blindness. There are many intermediate forms of it which do not cause trouble.

I do not believe in perpetuating an injurious deformity by encouraging the retention of sheep which are born with a severe deformity but feel strongly that it is just as wrong to eliminate the polycerate condition by refusing to use multi-horned individuals for breeding. To reduce the incidence of split upper eyelids it is advisable to keep sheep which have two-horned ancestors and to occasionally use a two-horned ram on the flock.

The split eyelid has been given a grading system of "1" being a normal eyelid, to the worst split as being a "4" and skull involvement a "5." Each eyelid is evaluated separately so that the left eye is quoted first and then the right. All lambs should be evaluated at birth and the score noted on their record sheet. Someone commented that it seemed that all the Jacobs with the finest fleeces also had some degree of the split eyelid condition. I do not know if this could be true. There are many breeds of coarse wooled polycerate sheep where this same affliction shows up.

Conclusion

Jacobs lack the qualities that are strived for in modern breeds but make up for it in many other ways. Lambs from modern breeds have the ability to grow at a phenomenal rate when fed huge amounts of high priced protein concentrates. The Jacob lamb will only eat his fill then leave the feed trough. The carcass is lean and tasty with very little fat and bone waste compared to up to 30 lbs. of kidney fat in Suffolks in our area, plus the weight of their heavy bones. In England, where Jacob ewes abound, they are bred to large mutton rams to produce larger, early maturing, lean market lambs. The ewes are considered an economical flock to keep compared with the same number of a larger breed. Jacob rams have been recommended for use on yearling commercial ewes as a bonus, trouble-free lamb crop.

There is no doubt that Jacob sheep have many advantages as an unimproved old breed. It is important to remember that by striving for "bigger and better" we will have to compensate by sacrificing some or all of the instinctive traits already lost in modern breeds. Jacob sheep can think. They can walk through muck and mire without coming down with foot rot and, above all, they can birth their own lambs.

Bigger is not better. Specific spots or horn direction dictated by a breed standard do not a Jacob make! It is performance of the individual sheep that counts and not winning in the showring. I am not against showing because I feel that the showring can be used as an educational tool, but the judge has to be well qualified and know the breed well. The Jacob of the future is in our hands. We hope that they do not become too altered by modernization and greed.

I see a difference in British and American Jacobs. The difference is twenty years of upgrading. I know a breed does not stand still and will alter over the years due to selective breeding by each individual's preference, but it is up to every Jacob owner to chose what they feel a Jacob should be and not lose sight of what it was in the past. One ewe may not have the perfect horns or the perfect colouring but she has had trouble-free lambing all these years, has always had the hardest hooves around and never has overeaten or become overweight. We must preserve these attributes of the breed that are the very reasons that make them interesting and that have allowed them to survive for several centuries.

Bibliography

American Minor Breeds Conservancy, Box 477, Pittsboro, NC 27312. Breed information.

Alderson, L. The Chance to Survive. Cameron & Tayleur Ltd. London 1978.

British Wool Marketing Board, Oak Mills, Bradford. W. Yorkshire. U.K.

Clutton-Brock, J. & Hall, S. J. G., Two Hundred Years of British Farm Livestock, British Museum (Natural History) Cromwell Rd. London SW 7 5BD. 1989.

Elwes, H. J. Guide to Primitive Sheep and their Crosses, Redwood Burn Ltd. Trowbridge, Wilts, U.K. 1983.

Gosset, A. L. J. Shepherds of Britain, Constable and Co. Ltd. London. 1911.

Henson, E. British Sheep Breeds, Shire Publications Ltd., Rare Breeds in History Aztec Printers, Stow-on-the-Wold, Glos.

Jacob Sheep Society, 242 Ringwood Rd, St. Leonards, Ringwood Hants. Breed information.

Ryder, M. L. Sheep and Man, Gerald Duckworth & Co. London. 1983.

Vince, John, Old British Livestock, Shire Publications Ltd. 1974.

Werner, A. R. An Enquiry into the Origins of the Piebald or "Jacob" Sheep, Countryside Livestock Ltd.

References

E. G. Ritzman. Breeding Earless Sheep. Journal of Heredity, Vol XL, #6.

History and Development of the Sheep Industry, page 161, 162.

Art. LIII Notes on Coloured Sheep, by Taylor White, read before the Hawke's Bay Philosophical Institute, 11th Sept. 1888.

Letter from H. Dryden, Canons Ashby, Northamptonshire, 28th Oct. 1884.

Colours and Patterns of South Australian Sheep

Liz Lewis
161 Adelaide Rd.
Hahndorf, South Australia, Australia

It is only since 1976 that breeding of coloured sheep has been practised systematically in South Australia. Prior to this, either these animals were slaughtered or perhaps one was kept in the flock to indicate the degree of copper deficiency in the soil. In the mid 1970's there was a trend for people to become "self-sufficient" and spinning, weaving and knitting one's own garments became very popular among the more practical devotees. Consequently the need arose for naturally pigmented fleeces and fleeces with different patterns. Small acre farmers, looking for profitable means to use their land, quickly saw the benefit of growing coloured sheep and a whole new industry appeared virtually overnight. Many people were also raising coloured sheep for their own use, some to produce garments for their own family, others who sold their beautiful products to the handcraft shops.

Supplies of pigmented sheep could no longer meet the rapidly increasing demand for basic breeding stock and breeders found that they required lessons in sheep colour genetics. Fortunately the South Australian Department of Agriculture was able to call upon the knowledge of Scott Dolling and others, who were more than willing to give us the benefit of their years of experience.

Colour (Black, Brown)

The two forms of pigmentation which exist in South Australia are black, including all shades of grey, and brown, including chocolate, cinnamon and fawn. This refers to the colour at skin level, not near the tip where fading may occur. The tan or red fleeces seen in other countries are not seen here.

In both black and brown pigmented sheep we have four major patterns, each showing several minor variations.

Badger Face

Badger face pattern consists of a phaeomelanic body with black belly, black inside the front and back legs to under side of the tail and a black bar over each eye. Inside ears and lower jaw are darker than outside ears and upper jaw.

Badger-Face

This colour pattern is popular amongst breeders of skins as the black belly wool gives a black border on a paler skin.

Reverse Badger Face, Black and Tan

This pattern shows a sharp change from dark body to pale belly, inside of legs, underside of tail, throat, and a small brownish yellow bar over the eyes. The underside of the jaw is pale as are the inside of the ears.

As the bulk of the fleece is pigmented, breeders are willing to retain this pattern in a flock growing fleeces for hand spinning because they find an interesting range of silvers and greys in these fleeces.

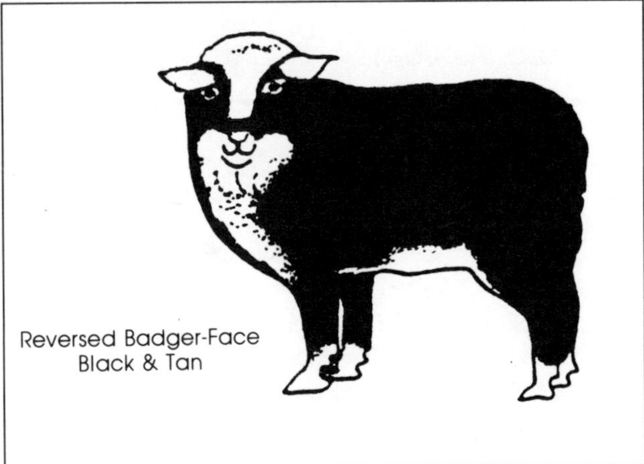

Reversed Badger-Face
Black & Tan

Self Colour

In South Australia we find that this allele displays three general variations in the self colour pattern:

Solid Self

A brown or black (including all shades of grey) fleeced sheep, with no white hair or wool. This pattern is very popular with breeders for both fleece and skins.

Baldy Self

These sheep have white wool growing in a symmetric pattern on the poll or face, which may also appear on the underside of the neck, tail and legs, forming the "head, socks and tail" distribution of white wool as described by J. J. Lauvergne (1979).

This pattern is popular with spinners who want a small quantity of white wool of the same count and quality as the coloured to use as contrast in a garment.

Spotted Self

These sheep have the white wool or hair on the face as in the "Baldy" pattern, while the body shows regular spotting in large rounded spots.

While popular with spinners, who make use of the different shades to blend colour or intersperse the different shades along the length of the thread, these sheep are popular for skins both for garments and car seat covers.

Spotted Self

Solid Self

Baldy Self

Lateral Stripe

These sheep are a rarity in South Australia and are basically dark with a pale band at the border of the belly and the body both of which are dark. The scrotum is pale. The ears are usually pale inside, black outside. A moustache extends to the tear ducts and then up over the nose. The chin varies from pale to dark.

Lateral Stripe

One flock of great interest to breeders was the "Richmond Park" flock. The owners of the flock, Joan and Mel Priest, planned their breeding schemes to produce sheep of either a specific colour or pattern. Their immaculately kept flock records have been a delight to sheep geneticists.

"Richmond Park" flock was a breeding group of some 43 ewes and 5 rams, situated on the sandy coastal plain of South Australia, with a Mediterranean climate.

The sheep were divided into five breeding groups:

Group 1 The aim with this group was to produce black, solid or baldy self patterned lambs.

Ram— Black solid self

Ewes—
2 black solid self
15 grey baldy self
1 grey spotted self
5 white heterozygous

Lambs born—
17 black solid self
9 black baldy self
3 black spotted self
3 grey spotted self
5 white

Group 2 The aim of this group was to produce grey sheep of the badger face pattern.

Ram— Grey badger face

Ewes—
1 grey solid self
1 grey baldy self
2 grey spotted self

Lambs born—
6 grey badger face
1 black baldy self
1 grey spotted self
1 grey badger face with spotted body
1 black badger face with a spotted body
3 white

Group 3 The aim of this group was to produce grey solid or baldy self patterned lambs.

Ram— Grey solid self

Ewes—
5 grey baldy self
1 grey spotted self
1 white heterozygous

Lambs born—
4 black solid self
5 black baldy self
1 black spotted self

Group 4 The aim of this group was to produce grey spotted self lambs.

Ram— Grey spotted self

Ewes—
1 grey solid self
1 grey baldy self
4 grey spotted self
1 lateral stripe
4 white heterozygous

Lambs born—
1 black badger face
1 grey reverse badger face
1 grey solid self
7 grey spotted self
3 black spotted self
2 grey lateral stripe
1 white with black eye patches and small grey spots all over body

Group 5 The aim of this group was to produce grey solid self lambs.

Ram— Grey solid self

Ewes—
1 grey solid self
1 grey baldy self
5 white heterozygous

Lambs born—
2 black solid self
1 black baldy self
1 black spotted self
3 grey spotted self
1 grey ? reverse badger face
1 white

South Australian breeders have, over the last 13 years, developed patterns and colours of naturally pigmented sheep to supply the demand in the handcraft market. Less favoured colours and patterns have diminished from breeding flocks, while the more favoured ones have increased. This always follows the market trend, one colour or pattern falling from favour as the fashion decrees. Fortunately breeders are now well aware of this trend and are able to breed toward the market's demand.

References

Carr, Peter. 1977. "Handbook for Breeding Coloured Wool."

COGNASAG Workshop, proceedings of the. 1988. "Standardized Genetic Nomenclature for Sheep and Goats."

Lauvergne, J. J. 1979. "Genetics of Breeding Coloured Sheep in France."

Lewis, Liz. 1984. "The Genetics of Breeding Brown Sheep."

Lundie, R. 1979. "The Genetics of Breeding Coloured Sheep in New Zealand."

Priest, Mel. (ed.) Flock Records of "Richmond Park."

Colour- and Island Sheep of the World

N. J. and G. J. Enzlin
Kleis 135
1911 Me Uitgeest, Netherlands

I ask your attention to the following sections, which will be completed in a book called "Colour- and Island Sheep of the World", published in due time, when all the information worldwide is gathered about the natural (original) coloured sheep races.

SHETLAND SHEEP
Country: United Kingdom
Area: Shetland Isles (Yell and Foula)
DESCRIPTION:
Colour: Shaela: grey; Emsket: blue grey; Moorit: reddish brown; Flecket: white, with large brown or black patches; Mioget: light "moorit" coloured; Katmollet: light coloured nose and jaws.
Bradford Count: 56/58

My most up to date information derives from Mr. Oliver Henry, wool-manager at the Jamieson & Smith Company in Lerwick on the Shetland Isles. This company buys approximately two-thirds of the Shetland annual clip, which covers in general three colours: grey, moorit and black (dark brown).

After World War II, the demand for the original Shetland wool dropped and the sheepmen had to change from wool to meat income. That's the reason the original coloured Shetland sheep were crossed with larger and "white" breeds like Romney, Dorset, Suffolk and most of all Cheviots. This cross-breeding had a severe effect on the quality of the wool and changed the typical original Shetland fleeces.

Still, there are some areas left where you can witness the original flocks on the islands of Yell and Foula.

Shetland Sheep.

JACOB SHEEP

Country: United Kingdom
Area: All parts of the United Kingdom
DESCRIPTION:
Head: Clear of wool forward of horns—dark nose preferred. Dark bold eyes with no tendency to split-eyed deformity
Horns: Any number (2-4-6)) may be carried, providing they are well balanced
Neck: Strong medium length; well set on shoulder
Body: Broad shouldered; straight back with well sprung ribs
Legs: Set at each corner so that the animal stands firm and well balanced
Fleece: White, of a fine quality with little or no "kemp"; with well-defined dark patches in proportion 60/40. Skin beneath white area to be good pink, whilst black beneath the dark wool
Bradford Count: 48/54

There are different opinions about where these sheep originally came from. To quote from the 1599 edition of the Bible: Laban's shepherd said, "I will passe through all thy flock this day and separate from them all the Sheeps with little spots and great spots . . ." Another one is said, that the Spanish Armada was destroyed at the coasts of Britain and some of their spotted sheep managed to reach the British shores.

Flock of Jacob sheep.

The British Jacob Sheep Society was founded in 1969, in order to foster the breeding and to promote the Jacob sheep in the U.K. and abroad. New members are provided with a handbook of the society.

Jacob sheep are very attractive, due to their multi-coloured fleeces; the lambs are born with distinctive black and white patches, which change as they get older into brown and white. The average fleece weight is about 2–3 kg.

MANX LOAGHTAN

Country: United Kingdom
Area: Isle of Man and parts of the United Kingdom
DESCRIPTION:
Head: Light boned; straight nose
Eyes: Bright and alert
Ears: Small, short and strong
Face: Clean, with minimum of wool
Neck: Medium length
Shoulders: Ram showing strength, Ewe light
Back: Straight
Feet: Very small and neat
Horns: Very dominant in rams; Ewe has four finer curled horns
Fleece: Very soft, close textured and lustrous; heavily oiled and excellent for handspinning
Colour: At birth black or "bitter chocolate", changing to lighter brown or coffee colour
Name: Loaghtan is Manx Gaelic for brown

The Manx Loaghtan sheep are an ancient breed, native to the Isle of Man and there is no evidence that they are imported originally from Scandinavia by the Vikings.

The Manx Loaghtan Sheep Breed Society was founded in 1976 by its president member, Mr. H. C. Kerruish, O.B.E., S.H.K., C.P., in order to gather people who were concerned about the future of this breed and to lay down its flock book standards.

A close look shows that there were three distinct types of sheep, with a variation of colours and there is no doubt that several changes resulted from cross-breeding in order to produce more meat and better fleeces for spinning.

The fleece of the Manx Loaghtan sheep is of fine texture and rich brown colour. It makes extremely attractive and hard-wearing garments.

HERDWICK

Country: United Kingdom
Area: West Cumberland
DESCRIPTION:
Head: Carried high; broad across the forehead; deep in the jaw
Ears: Well set up
Horns: Rams almost always horned; Ewes are hornless
Body: Deep and round
Fleece: Start at birth from black through shades of brown and fawn to almost white in old age

The origin of this sheep breed is unknown. Some people say that they are descendants of a flock of forty small sheep who went ashore from a wrecked Spanish vessel, others say they came from Scandinavia, Sweden or Denmark.

These semi-wild sheep, now domesticated, have a unique ability to survive in the most rigorous climate areas of the United Kingdom. They are surviving on the poor scant grazing typical of mountain terrain at 3,000 feet above sea level. The reason they survive is because their waterproof coat, which contains a reasonable amount of "kemp".

By selecting the shorn fleeces, it is possible to weave tweed in the natural dark and light shades.

BLACK WELSH MOUNTAIN

Country: United Kingdom
Area: Wales
DESCRIPTION:
Head: Rams, masculine, wedge shaped; Ewes, feminine, wedge shaped
Eyes: Black around the eyes
Ears: Small to medium
Face: Broad forehead
Horns: Rams, fairly strong; Ewes, hornless
Neck: Short
Skin: Blue; well-mellowed to touch
Fleece: Black, short and firm; No kemp, no white

The mutton of black-fleeced Welsh Mountain sheep was highly appraised for its richness and excellence during the Middle Ages. Only a century ago Welsh shepherds started to select black lambs in order to produce a purebred Black Welsh Mountain sheep.

In 1920 the Black Welsh Mountain Sheep Breeders' Association was founded at the same time the Royal Smithfield Show was held.

The wool from these sheep is suitable for making into cloth, which is durable, light and warm.

CASTLEMILK MOORIT

Country: United Kingdom
Area: Twelve districts in England
DESCRIPTION:
Fleece: Light tan or moorit, with white underparts
Horns: Both rams and ewes are horned
Tail: Short
Legs: Long

These sheep are considered by some to not be an original breed because of crossing with Shetland, Soay and Manx Loaghtan sheep. The home of this breed is the Buchanan-Jardine family, which started this race in the early 1900s.

The present population is bred from a surviving group of 10 ewes and 2 rams back in 1970. Furthermore, there is a breeding society called the Castlemilk Moorit Sheep Society that will see to the future of this coloured sheep race.

OUESSANT

Country: France
Area: Island Ouessant
DESCRIPTION: Dwarf sheep
Head: Fine and regular
Eyes: Very alert and bright
Ears: Fine and small
Horns: Rams, curly shaped; Ewes are hornless
Height: Rams, 49 cm; Ewes, 46 cm
Fleece: Black (indeed black), chocolate-brown, white (only a few), and moorit

Ouessant ram.

These sheep are dated to the time Christ was born and were bred from the race called "Mouton noirs du pays de Galles", but smaller. In those ancient days there were about 3–4000 of these sheep kept on the island Ouessant, most of them black and some white. Most of the inhabitants of this island were fishermen and they only kept these sheep for an extra income, and because of the little care that was spent, the number of these animals declined rapidly.

During the 18th century a member of the French royal family kept a pure flock on one of the islands in the River Seine.

In 1920 a group of people took interest in this neglected sheep race and started to create a flock standard, and now you can find these sheep outside of France in Belgium and the Netherlands. Even though they are dwarf sheep, almost every mature animal produces more than 1 kg. of coloured fleece.

SOLOGNOTE

Country: France
Area: Cher and Val de Loire (Sologne)
DESCRIPTION:
Head: Fine chestnut coloured
Ears: Medium, horizontally placed
Fleece: Moorit coloured, starting just behind the ears to tail
Weight: Rams, 80 kg; Ewes, 55 kg

The Solognote is an ancient French sheep race which was bred from the valleys of the Loire up to Normandie. In 1949 they created the Flock Book Solognot, and today there are about 2000 ewes, of which 500 are cross-bred with other French sheep races.

It is a very docile and hardy sheep race and most of them are kept in difficult climatological areas the year round.

After 1968 members of the Flock Book created a special "log book" for circulating the breeding animals among their members (often done in France). Part of the beautiful coloured wool is used by handspinners (about 2.5 kg. per ram and 1.5 kg. per ewe).

NOIRE DU VELAY

Country: France
Area: Mountains of Massif Central
DESCRIPTION:
Head: Fine, long shaped
Ears: Fine, medium and horizontally placed
Fleece: Black, with a brown reflection

This sheep race dates back from the time when the Celts set up their camps in the mountains of the Massif Central in France and bred them on the volcanic slopes of the Velay.

Formerly called "Noir de Bains" (by a local shepherd from the village Puy-en-Velay), people founded a society of fellow shepherds in 1931. In 1970 they changed the name to "Noire du Velay" and expanded their flock up to about 3500 ewes in 1980.

As some of these areas are highly inaccessible, some of these flocks are kept with cows, and it is very common for the rams to stay with the ewes year round.

DRENTS HEIDESCHAAP

Country: Netherlands
Area: Province of Drente
DESCRIPTION:
Head: Small, short and lightly bent, not Roman shaped
Legs: Fragile
Horns: Rams, spiral-shaped; Ewes, short, curved
Weight: Ewes, 35 kg
Fleece: Spotted, light brown, dark brown and black
Weight of Fleece: 1, 5–3.0 kg
Wool: 3 parts: a: short fine fibre; b: long hairy fibre; c: "kemp"

This kind of sheep has been kept for many centuries on the poor "moors" and in that respect they contributed to the amelioration of the agricultural areas in the Netherlands. During the day a shepherd, with his dog, wanders through the moors. After this daily routine, the sheep are driven into the "sheepcotes" until the following morning. The collected sheep manure in these sheepcotes was mixed with "sods", to be spread over the pastures.

There are still specimens of the ancient type of Drents Heideschaap, but by developing certain parts of the "moors", people changed to other sheep races and created by cross breeding the so-called new-type Drents Heideschaap.

BLAUWE TEXELAAR

Country: Netherlands
Area: Four northern provinces of the Netherlands
DESCRIPTION:
Body: Medium built
Horns: Both rams and ewes are hornless
Fleece: Almost black to grey

People discovered this variant of the original "white" Texelaar back in 1977. They informed Mr. P. Hoogschagen, working at the Department of Agriculture, and he stated the presence of the so-called "blue" colour. This definition derives from opening the fleece, which then shows near the skin a steel blue coloured fibre.

During the summer of 1983 the Dutch "Fokclub Blauwe Texelaars" was founded and nowadays consists of 95 breeders with 1000 mature animals.

SCHOONEBEKER

Country: Netherlands
Area: Southeast part of the province of Drente
DESCRIPTION:
Body: Medium built, larger than Drents Heideschaap
Ears: Long, horizontally placed
Horns: Both rams and ewes are hornless
Tail: Long and wooly
Fleece: Piebald, or black and brown

It is quite an ancient sheep race, kept on the "moors" in the province of Drente in flocks, the same way they kept the Drents Heideschaap.

In 1900, in the eastern part of the province of Overijssel, breeders imported sheep from Germany (Münster and Hannover) and it is suggested that a cross breeding between these races and the Drents Heideschaap created the Schoonebeker.

Nowadays, one of the largest flocks is kept in Westerbork (Drente).

DUTCH ZWART-BLES

Country: Netherlands
Area: All provinces of the Netherlands, the larger numbers in the provinces of Groningen and Gelderland
DESCRIPTION:
Exterior Marks: White spot between the eyes and minimal both hindlegs white
Head: Long, nose slightly bent
Ears: Horizontally placed
Horns: Both rams and ewes are hornless
Fleece: Black-brown coloured
Total Number in 1988: 1420 registered animals

The breed started in 1979 under the protection of the Dutch Rare Breed Society, and in February 1980, the first meetings were held with about 82 members, who kept 1250 animals.

In 1985 the breeders decided to found the Dutch Zwart-Bles Flock Book.

COBURGER FUCHSSCHAFE

Country: West Germany
Area: Bayern, Baden, Rheinland-Pfalz
DESCRIPTION:
Body: Medium built
Head: Small and Roman shaped
Horns: Neither rams nor ewes are horned
Fleece: Light moorit coloured, the colour of newborn lambs is reddish brown, turning lighter when exposed to the weather
Weight: Ewes, 55–60 kg, Rams, 50–60 kg
Fleece Weight: Ewes, 3.5–4.5 kg; Rams, 5 kg

There are only a few sheep left of this race, and they are kept mainly on the mountain slopes of Southern Germany. Their beautiful coloured wool is especially sought after by handspinners.

Graue Gehornte Heidschnucken sheep.

GRAUE GEHORNTE HEIDSCHNUCKEN

Country: West Germany
Area: Lüneburger "moors"
DESCRIPTION:
Head: Fine, prolonged
Horns: Ewes, lightly curved; rams, heavily curved
Fleece: Greyish, top of the fleece (wool) is rather "kempy," beneath is the fine quality.
Legs: Fine

Most of this sheep race is kept on the Lüneburger "moors". In the year 1867, because of the vast areas where these sheep could be kept, there were about 2 million animals in the Northern part of Germany.

Their ancestors were probably the Mouflons, who migrated from Sardinia upward north to Scotland, North Germany and even to Scandinavia and Siberia.

Besides the Graue Gehörnte Heidschnucke, there is also a smaller amount of the white coloured species. There was a decline in breeding after World War II, but today the number of these sheep increased up to 15,000 mostly raised in Germany, but also exported to the Swiss Alps.

In 1949 the "Verbandes Lüneburger Heidschnucken-züchter E.V." was founded, and by 1950 it consisted of 62 members with 82 flocks, containing 31,000 sheep.

Today people are more attracted to the tanned skins than to the wool.

SKUDDEN

Country: West Germany
Area: East Preusia
DESCRIPTION: Dwarf sheep
Head: Fine and small
Horns: Rams, heavily curved; Ewes, hornless
Weight: Rams, 35 kg; Ewes, 25–30 kg
Height: Approx. 50 cm—about the same size as the other European dwarf sheep, the Ouessant.
Fleece: Colour: white, brown, black (rare), and moorit
Ears: Small

This sheep race dates back many centuries and was discovered near Preusia and in some Baltic states. Today there are about 600 in West Germany, of.which only 280 are ewes, so these sheep are becoming a rare breed.

POMMERSCHES LANDSCHAF

Country: West Germany
Area: Pommern, Baden
DESCRIPTION:
Head: Medium long
Ears: Horizontally placed
Weight: Rams, 70–75 kg; Ewes, 50–55 kg
Fleece Weight: Rams, 6 kg; Ewes, 4 kg
Fleece Colour: Greyish to blue grey; Lambs are born with a black fleece

This sheep is also from an ancient breed and has been cross-bred with several other races kept on the poor moors, like the Skudden. There are about 100 of these sheep, so like the Skudden they are also becoming a rare breed.

BRAUNES BERGSCHAF (Brown Mountain Sheep)

Country: West Germany
Area: Bayern
DESCRIPTION:
Body: Medium built
Head: Roman shaped
Ears: Long and down placed
Fleece: Cognac coloured
Weight: Rams, 75–100 kg, Ewes, 55–70 kg
Fleece Weight: Rams, 4.5–6 kg; Ewes, 4.0–5.5 kg

These sheep are also part of an ancient race that wandered along the borders of West Germany, Switzerland, Italy and Austria. Their wool is very long and they are sheared twice a year. Most of the wool is used for the special folklore clothing of the people, who live in these mountain areas.

The "brown" coloured individuals are a quite recent variety, deriving from the older "white" breed.

KARAKUL

Country: Austria
Area: Breitenfurt, Laab im Walde, Sulz
DESCRIPTION:
Head: elongated face
Body: Medium built
Ears: Placed downward
Fleece: Colour: Black (Arabi), Schiraz (dark blue), Sur (gold, platin), and Guligas (rosa)

The breeding of Karakul sheep in Austria started around 1900 and had a dual purpose: a. skin; and b. meat.

The skin of one-day-old Karakul lambs is prized in the fur industry and carries different names, like Persianer, Karakul, Astrachan, or Bucharalamb. The original areas where these sheep were bred, are the deserts of Usbekistan and Turkmenistan (Buchara) and the northern parts of Persia and Afghanistan.

By 2000 BC people in these areas already tanned these skins and used them for clothing. Nowadays you can find these sheep in the following countries: U.S.S.R., Afghanistan, Namibia, Republic of South Africa, Rumania, Persia, Argentina, and the U.S.A.

In Austria at the moment there are about 1000 Karakul sheep, of which 95% are kept in three flock books. Most of these animals are sold abroad for stud breeding because of their fine quality and colours.

NAVAJO-CHURRO

Country: United States of America
Area: Southwestern U.S.A.
DESCRIPTION:
Body: Narrow built
Horns: Multi-horned (2, 4–6) rams; Ewes mostly polled
Legs: Long
Weight: Rams, 160 lbs; Ewes, 100 lbs
Fleece Weight: 4–5 lbs
Fleece: Colour: Spotted, piebald, brown, black and white

The name Navajo-Churro is composed of the names of the inhabitants of this area in the U.S.A., namely the Navajo Indians and the sheep imported from Spain, Churro. This name was originally given to the Spanish flock of neglected Merinos, after the Fall of the Roman Empire.

Navajo Indians possessed quite a flock of these sheep and used their "coarse" wool for their traditional weaving patterns. During the 1800's the flock of Navajo-Churros decreased, both by introducing foreign improved sheep races, and by the destruction of the flocks by Colonel Kit Carson on behalf of the U.S. government, due to "overgrazing" problems.

In 1987 the Navajo-Churro Sheep Association was founded and in 1988 the first flock book was issued.

Much has been done for this breed by Dr. Lyle McNeal, Dept. Animal Dairy and Vet. Sciences, USU, Logan, Utah, who is leading the Navajo-Churro Sheep Project.

TEXAS BARBADO

Country: United States of America
Area: State of Texas
DESCRIPTION: Hair Sheep
Body: Medium built
Head: Small elongated face with black (two) stripes
Ears: Horizontally placed
Horns: Hornless
Skin: Black belly, leg markings and tail markings
Fleece: Hair, some crossed with wool descendants

The Barbados Black Belly, which lead to the Texas Barbado, was imported about 100 years ago and reached a number of 250,000 animals in 1980. They were frequently crossbred with Merino's and the European Mouflons, which eliminated sometimes the black markings on their bellies.

FLORIDA NATIVES

Country: United States of America
Area: Florida
DESCRIPTION:
Body: Small and refined bone structure
Face: Wool free
Fleece: White and dark brown

The creation of the Florida Native sheep dates back from the import of Spanish sheep to the U.S.A. and started about 1565, when Admiral Pedro Menendez de Aviles arrived with a fleet of vessels, which also carried sheep from Spain.

These sheep are very adapted to their environment and ewes raise at least one lamb each year.

Nowadays you can see the cross-breeding with American Rambouillets.

Black Cotswolds of North America
Evolution of a Sub-Breed

Patricia Frisella
282 Meaderboro Road
Rochester, NH 03867 USA

Several years ago we had a ewe we called Bonita, who on her first lambing threw twin ewe lambs. We called the lambs jointly "Bonita's Bananas," and separately "Chiquita" and "Spike." Spike did not have a typical Cotswold fleece and did not live a typical long Cotswold life. The following year we bred Bonita to our quadruplet ram, Peter, and she produced triplets. I had been keeping an eye on her as she lambed and was initially pleased as she bore two lovely ewe lambs. It soon became apparent she had another to come, so I watched her more closely. Out came a black nose, then a black head—A Black Head—I was so excited I ran to the house to inform my husband we were expecting a black lamb. On the way I had a flash that the lamb might be strange, part black, part white. Well, it was a beautiful coal black ram lamb Bonita presented to us. In time he faded to a lovely, lustrous silver grey, was named "Silver Cloud," and went to California to live with Evelyn Elledge.

The arrival of Silver Cloud forced me to do some soul-searching. As Secretary of the Cotswold breed association I had reeceived a few reports, and even a strong complaint, about black Cotswold lambs. In each instance I had dismissed the lambs as being crossbred . . . this breeder also had black-faced rams, that breeder was totally disorganized, etc. . . . Now I was confronted with a black lamb out of purebred parents of my own breeding?!?

A few years later, I bought a few aged Cotswold ewes from another breeder, primarily because they were related to my favorite old ewe, Constance, mother of Pete, who has raised not only that set of quadruplets, but also numerous sets of triplets. One day while looking at one of these aged ewes, I thought, "God, she is so ugly, I'll cull her after this lambing." Well, darned if she didn't have twin lambs, one white, one black, hence reminding me that you can't judge a book by its cover. The black was a beautiful coal black ewe lamb who faded to a shimmering grey. She is named Alexandra and lives with Robert and Lydia Hale in Massachusetts.

Both of these black lambs had excellent conformation and breed character, aside from color. Both were coal black at birth, with black pigmented nipples, tongues, lips and blue-black skin. Both lightened to silver-grey with the fore and hind quarters darker than the center, but with black leg and face hair remaining. Whatever their ancestry, these lovely sheep merit the attention and enthusiasm lavished on them by their owners.

Although the exact origin of the Cotswold breed is lost in the "mists of history" it is generally felt that they were developed from sheep brought to England by conquering Romans over 2000 years ago. The breed was well established by the 15th century, and became the cornerstone of England's wealth throughout the Middle Ages. During the 18th century, hardy "Cotswold lions" were used in a system known as "Golden Hoof" husbandry, an early version of intensive rotational grazing where sheep were moved frequently with the aid of wooden hurdles, to build up the fertility of the wolds or hills of Gloucestershire. The Cotswold remained a popular English breed throughout the 19th century, when it provided the rams that were sent for crossing on Suffolk ewes.

The modern Cotswold was developed between 1780 and

1820 through the introduction of Leicester blood, and has been bred purely since that time.

The first major importation of Cotswolds to the United States was made in 1832 by Christopher Dunn of near Albany, New York. Some forty plus years later, Cotswold breeders became the first Association of sheepmen to register purebred sheep in this country, and they published their first flock book in 1878. The Cotswold became immensely popular in the United States because of its value as a sire breed on range Merino ewes.

Currently, the Cotswold is a rare breed and is classified as such by both the Rare Breeds Survival Trust in England and the American Minor Breeds Conservancy in the United States. What precipitated this change is not clearly documented, but Norman Dunn in the March 1982 *Furrow* cites "growth in the Australian Merino wool trade, and a shift in demand to smaller, earlier maturing lambs" as two possible factors.

In contrast to the long established history of the Cotswold breed, black Cotswolds are a recent and a rare phenomenon. They are unheard of in England, and only recently documented in the United States. A recent survey of the Association's membership resulted in pedigree information on only six black lambs, with several others being mentioned.

Since Cotswolds are a rare breed, both because the population is small and because there are few, if any, unrelated blood lines, it would seem initially reasonable to presume that these black lambs are the result of inbreeding causing various recessive genes to be expressed. However, there are several arguments which refute this view.

First, Cotswolds have experienced in their native England, genetic pressure similar to what they have experienced here, with no black lambs resulting.

Second, black lambs resemble white lambs in terms of the degree of inbreeding which produced them. Using a formula involving four generations of ancestors, one can establish a ratio of inbreeding:

$$\frac{\text{number of ancestors}}{\text{30 possible ancestors}} \times 100 = x\% \text{ inbreeding}$$

Using this formula, it was found that the black lambs were 0%–60% inbred. A random sampling of the flock book indicates that these figures hold true for the general Cotswold population as well. Of course, as breeders try to reproduce black lambs by deliberately breeding back and line breeding, inbreeding ratios on the black lambs could be expected to increase.

Third, as the population and number of rams in service has grown, thus relieving some of the inbreeding pressure, the number of black lambs has grown. Bob Gillis who has raised Cotswolds for 22 years, longer than any other current Cotswold breeder, has produced only one black Cotswold. In the 10 years I have raised Cotswolds, I have produced 2 and heard of perhaps a dozen, including the above mentioned six. It would seem that the black gene is of relatively recent origin.

Speaking about colored sheep in Australia, Lawrence Alderson says in his book, *The Chance to Survive,* "the majority of the rams were whole colored, mainly black, and were derived from a variety of white breeds, in the same way that the Black Welsh Mountain breed was derived from a white population. Black is generally a recessive factor . . . " (p. 143). Because black is recessive to white, both parents of black lambs must be presumed to have passed along the gene for color. Thus it should be possible to trace the sire or dam lines on the six black lambs and come up with a common ancestor. This has not been possible. Going back 10 generations did not reveal an ancestor common to all six of the lambs. Unfortunately, it is not possible to fully evaluate this outcome as it was not possible to establish complete pedigrees on all the lambs because some of the pedigrees disappear into the Canadian Registry. There has always been lively trade between United States and Canadian Cotswold breeders so it is entirely possible that the disappearing pedigrees would reappear in the United States flock book if they could be traced through the Canadian Registry. At any rate, where it is not possible to establish a proven source of the black gene, it becomes necessary to speculate. It is possible that the black gene was introduced through miscegenation with a white non-Cotswold, probably a ram, carrying the black gene. The implications of this are manifold, the most important being that any white Cotswold may be carrying outside genetic material, white and/or black. Thus, whatever the origin of this introduced material, it is now an inherent part of the American Cotswold, and black lambs must be considered to be as pure as white lambs. This does not mean, however, that the best interests of Cotswolds in general, and black Cotswolds in particular would be best met by registering black lambs in the standard flock book.

The Black Welsh Mountain Sheep Breeders' Association was formed in 1921 to register the black lambs produced from the white population. Flockmasters began about 100 years ago to select and breed the black lambs, producing a pure black strain or sub-breed. Taking inspiration from the Welsh, it would seem beneficial to establish a separate flock book for recording black lambs. Indeed, this idea has been received favorably by the Cotswold breeders. The major dilemma has been, what to do with the parents of the black offspring, whom we earlier presumed to be carrying the recessive black gene. If they remain in the white flock book, they will continually contribute black genes to the general population. If they are placed in the black flock book, their recessive black genes would have a greater opportunity to be expressed. However, their white lambs would be caught in limbo, and not be registerable as white lambs. Sensing a reluctance on the part of the membership to removing the white parents of black lambs from the standard flock book, it may be necessary to start a black flock book with the proviso that any lambs recorded there be descended from registered Cotswolds.

New Sheep Breeds of Australia

B. C. Jefferies
44 Ferguson Ave.
Myrtle Bank, 5064 S. A., Australia

The purpose of this paper is to define a breed, describe how it is formed, its characteristics and its relative importance in the Australian sheep industry. For the sake of completeness and because the Merino is such a dominant breed in Australia, the strains of Merino and some other breeds in Australasia to which reference is made in the text, will also be described in summary form in the table beginning on page 76.

> Prof. Coop stated that:
> "Formation of new breeds involves combining the virtues of two or more existing breeds by crossing and then interbreeding those animals agreeing with the new breed-type definition or breeding objective. The parent ewes usually come from a breed numerically very strong and the parent rams from a breed numerically weak. The new breed therefore expands from the base ewe breed." (Coop 1974)

There have been exceptions to this rule as in the case of the Corriedale in Australia and New Zealand. Merino rams were used on Lincoln ewes in Australia and vice versa in New Zealand.

Definition of a New Breed

We must first of all define what we mean by a new breed of sheep. I shall quote from a letter written in 1973 by Dr. Helen Newton Turner to Dr. Maximo Gamarra Rojas of S.A.I.S. "Tupac Amaru", Pachacayo in the Peruvian Andes. Dr. Gamarra is managing director of this huge Cooperative employing 800 Indians, and carrying 150,000 sheep of the Junin breed, derived from several (at least 7) breed types (viz. Spanish Merino, Corriedale, Columbia, Panama, Targhee and Warhill) with selection for over 30 years toward a defined breed-type suitable for grazing in high mountains at 3000 to 5000 metres altitude, 10 degrees south of the equator. Dr. Turner wrote in 1973:

> "If there has been no further introduction of outside animals for 19 years, and, if during this time, there has been continued selection towards a defined goal, then in my opinion the sheep can certainly be considered as belonging to a definite breed."

Breed associations in Australia such as the Polwarth require three generations (each averaging four years) from the first interbreeding of ¾ Merino–¼ Lincoln x ¾ Merino–¼ Lincoln before the progeny can be regarded as of the new breed, i.e., 14 years if we allow two years after the flock were closed before progeny are interbred.

However, new breed-types have been developed by simpler means than this. We shall refer to several where one pair of genes has been replaced by another pair producing different characteristics. There must still be the establishment of a clear breeding objective and selecting vigorously toward the new type and rejecting or culling all animals which do not agree with this definition.

Genetic progress will depend on the intelligent and efficient use of genetic differences between the new breed-type and the average of the breeds or flocks from which it was derived, the heritability of the main characters under selection and the genetic correlations between them. Then it is largely a numbers game. With large numbers of animals under selection, wide genetic variation between the extremes and accurate selection of economically important, measurable characters quite rapid genetic progress is possible (i.e., about two percent per year). Finally, the breeder must have a clear appreciation of the type of animal in mind.

Relative merit of new breeds can only be assessed by comparing their progeny objectively with the progeny of established breeds from randomised groups of ewes bred and run together under the same environmental conditions.

Prof. Coop states that interbreeding experiments should be interpreted with care as firstly, selection during interbreeding is usually at random as it should be in an experiment, whereas in breed formation it is not, due to rigorous culling of off-type, and unproductive sheep. Secondly, the two parent breeds, the F_1 with hybrid vigour, and the interbred generations, with or without hybrid vigour, are unlikely to be equally adapted to the environment where the experiment is conducted. The loss of hybrid vigour during interbreeding can be offset by selection, if numbers allow, usually by F_4, so that interbred generations need not necessarily have a lower performance than the F_1, and may even have a superior performance if selection is efficiently applied (Coop 1974).

Dr. Raul Ponzoni, a geneticist in the South Australian Department of Agriculture, defines five logical steps in the design of a breeding programme, in order to maximise monetary returns from a flock. These steps could be used as an aid in breed formation:

Definition of a Breeding Objective

This step can be regarded as the development of a model describing the relationship between various sheep production characteristics (traits) which the breeder wishes to improve, and the economic returns to the producer. It involves specifying the traits to be improved and assigning relative economic values (REV's), to each of them.

Even though prices for wool and meat may fluctuate considerably and so affect REV's, they are unlikely to change the rate of genetic gain. In general, the ranking of animals is relatively unchanged by moderate changes in the REV's. Flock composition may change, such as a higher proportion of breeding ewes and selling wethers at two years of age or less to the live sheep trade, compared with mixed ages of wethers, but it does not alter the breeding objective.

The Australian national performance recording scheme (WOOLPLAN) for woolgrowers has a breeding objective to improve:

- clean fleece weight (CFW)
- fibre diameter (FD)
- reproduction rate (RR)
- body weight (BWt)
- weight of cast-for-age sheep (MW)

Options are available to hold fibre diameter or reproduction rate at their present genetic levels in a flock while other traits are improved.

These traits indicate the sources of profit or loss (income - costs) in wool growing flocks.

Some traits such as feed intake might have a major influence on costs, but we cannot yet obtain selection criteria on which to select which could be measured economically. This procedure highlights the gaps in our knowledge due to lack of adequate genetic parameters.

Choice of Selection Criteria

Once the problem of definition of a breeding objective has been clarified it is necessary to choose selection criteria which we actually measure on the sheep or its relatives to give information about the traits in the breeding objective.

Examples of selection criteria (some of which resemble the traits above) CFW (derived from GFW x Yield % of clean wool), Staple length (SL) if it affects wool prices, Fibre Diameter (FD), Dam's number of lambs born, one record (NLB), Dam's number of lambs weaned, one record (Dam's NLW); 16 month old bodyweight (16mthBW); Skin Wrinkle Score (Wr) and Face Cover Score (FC). The use of any of these selection criteria will depend upon how much genetic information they provide and the cost of measurement.

Alternatively, a system of independent culling levels can be used. Two or more selection criteria may be combined into an Index that is an estimate of overall breeding value for a chosen objective.

WOOLPLAN offers a choice of sets of characters that can be included in the Index.

The alternative combinations of characters are:

CFW and FD	GFW and FD
CFW, FD and BWt	GFW, FD and BWt
CFW, FD, BWt, and dRR	GRW, FD, BWt, and dRR

(dRR = a single record of the dam's reproductive performance)

The addition of dam's reproduction rate dRR will add about 2% to the accuracy of selection so must be weighed against the cost of measurement.

Performance Recording

Some form of performance recording is now required to organise collection and analysis of data on sheep performance in the most efficient way. (Computer programmes such as WOOLPLAN in Australia and SHEEPLAN in New Zealand have been designed to do this.)

Use of Recorded Information

This recorded information must then be used as an aid in decision making in relation to selection of flock ewe and ram replacements. Departmental Extension Officers, geneticists and consultants in sheep breeding can help breeders choose the most appropriate combination of selection criteria and index for their situation.

A Mating Plan

The final step is made by the breeder in deciding how to mate the selected animals to achieve the breeding objective and marketing goal most efficiently.

Now, let us look at the new sheep breeds in Australia. For the sake of completeness, some earlier breed-types developed in Australasia have been included.

The most common breed of sheep in Australia is undoubtedly the Merino (see Table 1). Various crosses between Merino and fat tail breeds from South Africa and India were made in the period 1788–1820, but the Merino was developed from the 1820's as the predominant breed.

A number of strains of Merino were developed after the initial Spanish and Saxon superfine Merino. The strong SA Merino was developed since 1840 and the Peppin Merino was developed in N.S.W. since 1860. In 1986 the Merino comprised 63% of the sheep population of 58 million in New South Wales, 57% of 27 million in Victoria, 97% of 14 million in Queensland, 84% of 18 million in South Australia, 94% of 33 million in Western Australia, 13% of 5 million in Tasmania and averaging 72.8% of the Australian sheep population of 156 million. Some 60% of Merinos are of Peppin origin and over 30% of South Australian strong wools.

The rate of genetic progress will depend on the accuracy of selection and the age structure of the flock which determines the generation interval.

Using measurement of economically important characters as an aid to selection, quite rapid genetic progress (about 2% per year) is possible for the first 2 or 3 generations of a selection programme. Then progress begins to decline or plateau, as shown by research at Trangie by Dr. R. B. Dun. This emphasizes the need to select high producing animals at the beginning of a breeding programme and to select the most productive and profitable replacement stock.

It was in an attempt to break through this genetic plateau that some nucleus and group breeding schemes were developed in Australasia.

Booroola Merino (B Mo)

The Seears Bros. at Cooma, N.S.W., selected among medium non-Peppin ewes of the "Egelabra" strain for multiple births. They have produced many twins, numerous triplets and even some quadruplets, quintuplets and a sextuplet. It is thought that the recently identified F gene which is a simple recessive may have come from the fecund Bengalee fat-tailed sheep imported by Governor Phillip from India in 1791. These small, harsh, hairy sheep with a fine undercoat were able to produce four lambs in 13 months and undoubtedly contributed to the rapid increase in sheep numbers from 526 to 6,000 in six years.

The current lambing percentage of the Booroola flock at CSIRO, Armidale, N.S.W., is in excess of 220%, which ranks this flock among the top four high fecundity breeds of sheep in the world. The sheep are relatively small with mature ewes weighing 40–45 kg. The mature ewe fleece weight ranges between 3 and 4 kg of 19–21 micron wool.

The Booroola is being used to transfer the fecund F gene into other less fecund breeds but particularly into the prime lamb mother breeds and crosses. In New Zealand large numbers of B x Romney and B x Border Leicester rams has meant an immediate lift of 25–30% in lambing percentage of those flocks. This increase would take more than 20 years by selection for fertility within a flock. Mr. Jim Pocock has increased the lambing percentage of his "Panlatinga" Merinos by 13% in 15 years.

The AMS Merino

In 1967, Mr. Jim Shepherd at Kwolyin, in the eastern wheatbelt of Western Australia, started the huge group breeding scheme, based originally on his "Bungaree" blood stud of 4500 ewes, but ultimately drawing on a base of some three million Merino ewes of all of the Merino strains in

TABLE 1 Numbers and Percentages of Sheep Breeds in Australia (31/3/86)

Name of breed	*Number*	*Percentage of total*
Merino	113,191,000	72.76
Corriedale	4,587,000	2.95
Polwarth	2,805,000	1.80
Bond	268,000	0.17
Cormo	231,000	0.15
Zenith	32,000	0.002
Border Leicester	1,035,000	0.67
Romney Marsh	579,000	0.37
Perendale	72,000	0.005
Drysdale (Carpet Wool)	44,000	0.003
Tukidale (Carpet Wool)	60,000	0.04
Poll Dorset	983,000	0.63
Dorset Horn	560,000	0.36
Suffolk (incl. South Suffolk)	210,000	0.13
Southdown	161,000	0.10
Merino Comebacks	7,005,000	4.50
Crossbreds	14,409,000	9.26
Unspecified*	9,052,000	5.82
Other breeds	277,000	0.18
	155,561,000	100.00%

*Incomplete reporting of the breeds of sheep items in the agricultural census.

Australia. When he became expelled from the Australian Stud Merino Sheep Breeders' Association, he formed the Australian Merino Society (AMS), which was growing rapidly, and currently has some 1500 members in most mainland States.

There is a pyramid structure with a Central Nucleus of 4200 ewes, presently situated at Kojonup on the edge of the rainfall zone (500 mm rainfall) in W.A. Then there are about 80 Ram Breeding Co-operatives (RBC) for multiplying high performance rams from the Central Nucleus, for distribution to the 1500 commercial breeders. There is a movement downward of genetic material as in traditional stud breeding, either by natural mating or Artificial Insemination (AI). But unlike the traditional stud system, there is also a movement upward of the elite ewe hoggets from the commercial flocks to the RBC's (about half to one percent) and from the RBC's to the Central Nucleus (CN) (about 1 to 2 percent). Half of the CN ewe replacements come from the above-average ewe progeny born within the CN and the other half comes from the flocks in the pyramid below. All selection is made on visually acceptable ewe hoggets, selected with the aid of objective measurement and the use of a selection index: GFW x 10 + BW = I.

Final selection is now being made using WOOLPLAN indices. The gains in clean fleece weight and bodyweight over the past 18 years have been small (1.1% per annum) with a small reduction in fibre diameter (Winter 1987 regression analysis).

In a control flock at the AMS RBC at "Ramsyn", Jamestown in S.A., a measure of genetic progress in the AMS was made. (Jefferies unpublished). After some 10 years of selection in the CN in W.A. and use of top CN rams at "Ramsyn", either by natural mating with the ram allocation from CN or by AI with some of the top 18 rams at CN, there was a gain of 0.4 kg in GFW and 4.6 kg in BW while at the same time fibre diameter was decreased by one micron.

In addition to this huge group breeding scheme, Jim Shepherd has also established a pyramid of superfine Merinos to capitalise on the higher prices being paid for finer wool.

Then in a further attempt to break through the apparent plateau in selection, he has established a series of eight intercross flocks using every breed and type of sheep in Australia from black faced Suffolks to medullated carpet wool breeds. He aims to obtain significant recombination of genes that will respond further to selection in the development of the Meridale breed of sheep.

Comeback Breed Types

Comeback sheep are basically Merino-type with an admitted outcross to a British Longwool breed or breed derived from a British Longwool Merino cross.

The wool is *predominantly* that described in the wool trade as *comeback,* though it may also be acceptable to the trade as Merino.

A comeback stud ram is a sire of the *required standard of type* and performance, either bred in the ram breeding nucleus of an established comeback stud, or obtained from a registered stud of another breed.

In 1874, the New Zealand Corriedale was developed by interbreeding male and female F_1, or first cross progeny of Lincoln rams x Merino ewes to produce a dual purpose sheep for the higher rainfall areas (500 to 800 mm).

In 1882, the Australian Corriedale was developed reciprocally by mating Merino rams with Lincoln ewes and interbreeding the acceptable progeny of both sexes to produce a farmer's type dual purpose sheep for higher rainfall country (500 to 800 mm).

When reciprocal crosses are made between a long wool or dual purpose breed and the Merino, it is surprising that the progeny of the Merino ewes perform better in survival, vigour, growth rate and production than the progeny of the reciprocal cross. It is postulated that this is due to a sire effect which overrides the superior mothering ability of breeds other than the Merino.

The Corriedale type was fixed as a dual purpose meat and wool sheep in each case by line breeding within the progeny which agreed with the defined breed type. There are now about 4.6 million Corriedales in Australia.

Bond Corriedale (or Bond)

In 1909 Thomas Bond of "Yarren" Lockhart, N.S.W., fixed the type of the "Commercial Corriedale" which he called his breed-type, developed by mating Saxon and Peppin Merino stud ewes to stud Lincoln rams. He selected and interbred progeny suited to the Riverina environment. The breed has more open faces and often pink noses and white feet not tolerated by the traditional Corriedale breeders. In 1932 the fame of Thomas Bond and his successful Commercial Corriedales came to the notice of the Australian Corriedale Association. His flock was officially inspected and around 3000 ewes were registered as stud Corriedales.

In 1979 the name was changed to Bond Corriedale and more recently to just BOND because of the pressure from traditional Corriedale breeders who feel threatened by the new breed-type.

Recent research by Mr. Roger Foulds of CSIRO and Mr. Malcolm Fleet of the S.A. Department of Agriculture has shown a strong correlation between isolated pigmented fibres in the fleece and pigmented fibres at the horn bud sites, on the legs, around the mouth and nostrils and to a lesser extent, around the eyes and on the ears. This has tended to favour the Bond sheep over traditional Corriedales which must have black nostrils.

In a weaner trial, conducted by the N.S.W. Department of Agriculture in 1980 at Lockhart, lambs sired by Bond rams filled the first 12 places among 27 entries, and all major placings in a 62-team wether trial in 1981. Of Corriedale rams sold in N.S.W. in recent years, 65% came from seven Bond studs and 35% from 65 traditional Corriedale studs.

Bond wool sells at a premium over crossbred wool of similar micron, and the sheep average 6–8 kgs of 24–30 micron wool. The fleece tends to have a denser, more blocky tip than traditional Corriedales.

Comeback types (coming back to the Merino) with long stapled, open fleece and higher reproduction rates than the Merino, have been used in particular environments of higher rainfall (over 600 mm) and cold, wet winters. Apart from Tasmania where the Polwarth breed predominates with about 40% of the sheep population of 5 million, they have not replaced the Merinos in the major sheep areas of Australia.

The Polwarth breed was founded in the Polwarth county near Colac in Victoria, Australia in 1887 by mating Lincoln rams with Merino ewes and mating the acceptable ewe progeny back to Merino rams. The ¾ Merino–¼ Lincoln male and female acceptable progeny, agreeing with the new breed-type, were then interbred and line bred to fix the type as a dual purpose meat and wool sheep.

Cormo

While working in the Tasmania Department of Agriculture in the 1960's it became evident to me that it was very difficult to obtain suitable comeback sheep of the quality, production and price required by commercial breeders. Therefore, an attempt was made to develop a comeback type of sheep suitable for the cold, wet mountainous areas of central Tasmania. This comprises altitudes of 700 to 1300 metres and 500 to 800 mm rainfall, with snow every winter and often at lambing time in September–October.

Polwarths tended to be too open in the backs for this bush country. Corriedales were open in the back with their more tippy stapled, fine crossbred fleece, while Saxon Merinos had lower lambing percentages, low fleece weights and lower value for surplus sheep.

So, big framed, heavy cutting Corriedale rams, were mated with 1200 Saxon superfine Merino ewes to produce a very acceptable comeback type of sheep with 21–23 micron wool. These sheep had excellent conformation and could rear over 100 percent of lambs (up to 130% from mature ewes) in this cold climate. Unfortunately, it was not possible to obtain rams selected on measured performance at the time.

The F_1 male and female, visually acceptable progeny were classed rigorously to the breeding objective and interbred with up to 30 percent of the off-types being rejected or culled.

A nucleus flock of some 2000 selected ewes were established within a much larger commercial flock of some 10,000 ewes by selecting on greasy fleece weight, bin class and bodyweight among visually acceptable progeny at 15 to 18 months of age. The rams were selected with the aid of clean fleece weight, bin class, fibre diameter and bodyweight with culling on high wrinkle score offshears.

This was a pioneering effort at nucleus ram breeding in a commercial flock in Australia. There was tremendous social pressure and vested interests against changing from traditional breeding practices. (As Mr. Lance Lines said in South Australia, "If you wish to break with tradition, you must be prepared to walk a very lonely road.")

We shall use the Cormo breed, established here in Tasmania at "Dungrove" in co-operation with Mr. I.K. Downie and Sons, to illustrate some principles. Techniques were further developed and improved in Patagonia in the development of the Corino ((1/2 Mo – 1/2 COR) and the Argentine Cormo (3/16 Mo – 13/16 COR, including seven Cormo rams from Tasmania), and later in Victoria with the development of the Bundoran Comeback.

When genetic progress in the Cormo was less than expected an experiment was initiated to test whether clean fleece weight could be increased by introducing high performance rams of the SA Merino, Poll Peppin Merino, or redundant worker rams from Bundoran Comebacks. The aim was to compare the performance of the progeny of these introductions with pure Cormos in the same environment using four randomised groups of Cormo ewes.

In the first two dry years the progeny of both the SA Merinos and Bundoran Comebacks produced more clean wool than the Cormos and of the same fibre diameter. The Peppin progeny cut less wool and it was a micron finer in diameter than the other three groups and they were slower maturing than the Cormos or SA Merinos.

However in a wet year there was much more fleece rot and flystrike in the progeny of SA Merino or Bundoran Comeback rams than either of the other two groups and the Cormos had least wool faults of all groups. During a fly wave some 30% of the SA Merino progeny became flystruck compared with 20% in the progeny of Bundoran Comebacks, 6% for Peppin progeny and only 5% for the pure Cormos.

If the project was repeated with present knowledge and experience, I am sure that we could screen a much larger population of ewes. I was able to do this in 1967–70 in Argentinian Patagonia in flocks of 10,000 to 115,000 Merinos and Corriedales, to select a high producing nucleus flock of the top 5 to 10 percent of ewes. Then we chose the highest producing rams from wherever they could be obtained. Gains of as much as 1 kg of greasy wool were made in the present generation which with 40 percent heritability would give a genetic gain of the order of 300 to 400 grams in the next generation, allowing some swing back toward the average of the unselected flock from which they came.

At a major field day in March, 1986 results of these comparisons were presented. It was concluded that under the cold, wet conditions of Tasmanian central highlands, the Cormo performed best in terms of production, freedom from faults and returns per head. (See Tables 2 and 3 for results, and Reid, 1987).

TABLE 2 October 1985 Hogget (One Year Old) Shearing (F2 Ewes and Wethers Combined)

Source of Rams	Breed or Strain	GFW kg	Yield %	CFW kg	Fibre Diam. u	Price cents/clean kg	Fleece Value $
Dungrove	Cormo	3.14	77.9	2.45	19.9	652	15.95
Bundoran	B.Cbk x Cormo	3.17	77.8	2.47	19.9	652	16.08
Gumhill	SAMo x Cormo	2.92	76.5	2.23	20.3	632	14.12
Pylara	Peppin Mo x Cormo	2.67	74.1	1.98	19.6	667	13.19

I believe that we could have gone further by selecting all of those very well grown two- and three-year-old ewes of these three crosses, which had demonstrated their adaptability and performance in that environment and culling all unacceptable ewes. They could then be added to the nucleus to provide some more genetic variation from which to select. However, the decision was made to sell all of these three experimental groups and to continue with the pure Cormo, the wool of which is still attracting a premium of 70 cents per kg over similar wool sold on the open auction system. The wool is bought by Itoh for the Toyoba mills in Japan for specialty Donicormo cloth.

Some introductions are still being studied in Cormo flocks on Kangaroo Island and at Casterton in Victoria where the owners wish to increase fleece weight, decrease fibre diameter, without losing the excellent characteristics of the Cormo for thrift in cold wet environments. Some notable gains have been achieved, particularly on Kangaroo Island.

Two breed-types with which the author has been associated since their formation in 1971 are the "Fernleigh" and the "Bundoran" Comebacks (coming back to the Merino). They were both developed from Polwarth x Fine Merino with various infusions of Polwarth (3/4 Mo - 1/4 L) Zenith (7/8 Mo x 1/8 L) and numerous other strains of Merino, to produce a type of sheep suited to the cold, wet, Western District of Victoria near the Grampians (700–800 mm rainfall and many wet days). The wool of adult sheep averages around 22–24 microns.

There are over 230,000 Cormos. Relatively low fleece weights and low sale values for surplus sheep have limited the expansion of the Cormo breed.

Beddale

One of the successful steps in the large AMS intercross programme was a cross between fecund AMS Merinos, Border Leicesters (BL), English Leicesters (EL) and Poll Dorsets (PD), which, when interbred have about 7/8 Merino and 1/8 (BL x EL x PD). It was called the Beddale and is a comeback type of sheep with 23-25 micron wool, good fertility (over 100% lambing) and excellent thrift. Under dry summer conditions with pasture and crop residues, Beddales will be able to maintain one fat score higher than Merinos running with them.

Unfortunately, all of the breed-types resulting from the crosses were retained so there is not a distinct breed-type as yet. It is usual to screen large numbers of sheep in the formation of a new breed, but then to cull rigidly to the new breed-type with line breeding to fix the new type. There are only several thousand Beddales.

Dual Purpose Breeds

The live sheep trade has helped to increase the demand for sheep meats and New Zealand work has shown the advantages of high reproduction rates as a major contributor to returns per head from sheep, particularly in the higher rainfall areas.

Two new breeds were developed in New Zealand to fill an economic gap which the predominant Romney could not fill.

Sir Geoffrey Peren at Massey College, Palmerston North, N.Z., in the 1960's crossed fecund Cheviot rams with Romney ewes with interbreeding of the F_1 progeny to form the Perendale breed which could walk and graze more efficiently than the Romney on the steep hill country of the North Island of N.Z. The Romney at the time comprised about 90% of the N.Z. sheep population. Perendales average about 150% lambs marked to ewes mated with autumn mating so they need a long growing season such as in N.Z. The wool is an excellent, basic carpet wool with harsh handle, helical crimp, poor lustre and of about 30 microns.

The Perendale Genetic Development Group (PGD), group breeding scheme was established in 1969 by eight breeders in New Zealand, after screening about 100,000 Perendale ewes to form a nucleus of 750 elite ewes of high performance.

Simultaneously, in the South Island, Prof. Ian Coop crossed BL rams with Romney ewes and selected strongly for twinning ability. He interbred highy productive twin-bearing F_1 ewes and twin or triplet-born F_1 rams to form the new Coopworth breed in 1968. Now it is one of the few breeds in which ewes have to rear twins to be registered in the breed. The average lamb marking percentage for the breed is 180%.

Both Perendales and Coopworths have higher performance and returns than the Romney (R) and so are increasing rapidly in popularity both in New Zealand and Australia to which they have been imported. There are now over 11 million of each breed in New Zealand, where they are better suited than in many areas in Australia because they are very seasonal breeders and need to be mated in late autumn for best results. There are some reservations about the Coopworth at Lincoln College New Zealand and some breeders are reverting to the first cross BL x R.

Australian Perendale

Mr. L. T. Ross-Anderson on Flinders Island, Tasmania, had a very productive Romney stud with high wool production and reproduction rate with full pedigrees. He had maiden ewes producing over 160% of lambs.

He selected the finest Romneys with about 30 micron wool and above-average fleece weight and mated them with triplet-born Cheviot rams to form a fecund, highly productive Perendale in Australia. The entire stud was purchased by Mrs. Flora Richardson of "Boorana," Woorndoo, RSD, Mortlake, Victoria, 3272.

There are only about 70,000 Perendales in Australia and it is confined to high rainfall areas (over 700mm).

The prime lamb industry of Australia depends on two major breed-types: the Merino (Mo) for early, out-of-season mating and the BL x Mo in the higher rainfall areas, because of its considerable hybrid vigour (over 25% more lambs than the average of the two parent breeds). These F_1 ewes produce 75% of the 16 million prime lambs slaughtered each year in Australia. BLs are so successful in this role that they now represent 87% of the longwool sheep in Australia.

Despite vigorous efforts by Mr. A. J. Vasey of the Animal Production Committee in Australia, no research work was done for many years to improve the often low reproduction rate and poor survival of pure BL. Many flocks in New South Wales only average 70–80% lambing which resembles the average for Merinos in Australia (75%), yet the first cross (F_1) ewes frequently produce 120–150% of lambs marked to ewes mated (averaging 90–120% depending on time of mating). There are individual studs with 120% lambs marked.

Attempts at Condobol Research Station N.S.W. to fix the BL x Mo as a breed failed because most of the hybrid vigour was lost when the F_1 progeny were interbred and all by F_4.

However, the N.S.W. Department of Agriculture did eventually conduct a research project to improve the reproductive rate of BL sheep genetically.

Border Leicester Improved (BLI)

The improved Border Leicester (BLI) was generated by repeated back-crossing of selected, highly fecund BL x Mo ewes and their progeny to BL rams until it stabilised at approximately 82% BL and 18% Mo genes. The major emphasis is on selection for increase reproduction which has resulted in a high reproduction rate of 128% lambs weaned to ewes joined in the BLI flock.

There are only a few thousand of this new breed and it will only increase slowly because of its similarity to the established Border Leicester of which there are one million.

Gromark

Former Principal Livestock Officer (Wool) in the N.S.W. Department of Agriculture, Mr. Arthur Godlee, aimed to produce a highly productive, self-replacing, dual purpose, prime lamb mother breed. In 1965 he commenced interbreeding the progency of 12 twin and triplet BL rams and 430 selected Corriedale ewes and applying heavy selection pressure for measured production within three family groups using syndicate, or group matings. The new breed was called the Gromark and their Breed Society was formed in 1979 at Hawkesbury College.

Growth rate was given the heaviest selection pressure, with twinning and fleece weight, carcase and wool quality receiving less selection pressure. They produce a lean carcase with only 4mm fat coverage over the eye muscle in an 18kg carcass. Stud rams had to be in the top 20% for body weight, above average in fleece weight and, preferably twins. Production records are only taken on rams to keep costs down.

Like the Perendale and Coopworth, the Gromark can only express its genetic potential for reproduction rate if mated in mid to late autumn so it will be confined to the high rainfall areas (over 700mm). There are only a few thousand Gromarks and there does not seem to be a good niche in the industry for it as the wool is too coarse.

Daldale

There is another breed of prime lamb mother which was displayed at a field day at the University of N.S.W. in 1978. It was formed at Wellington N.S.W. by crossing BL rams on Merino ewes, including highly fecund Booroola Merino ewes and the progeny mated to a Poll Dorset before interbreeding. The wool type is fine crossbred of 25–30 microns. It is a prime lamb mother with a high lambing capacity, long breeding season and high wool production. There are only a few Daldale.

TABLE 3 Description of Australasian Breeds of Sheep—Strains of Merino

Breed	*Colour of head and points*	*Body Weight (mature) kg*	*Wool Quality number (microns)*	*Greasy fleece weight (kg)*	*Staple length (mm)*	*Purpose*	*Adaptations*	*General*
Superfine Merino (Saxon Merino strain)	Soft white face, pink skin, legs and hooves white. Horned rams.	Rams: 40 to 60 Ewes: 32 to 45	80s and fine (< 18) 74s (18–20)	3 to 4 3 to 4.5	70 to 80 75 to 90	The Saxon Merino superfine wool is used for luxury garments of very finely woven cloth.	The superfine Merino was developed for the cooler high rainfall areas, particularly in the hilly areas of Eastern Australia and Tasmania on unimproved & semi-improved pastures free of dust & vegetable fault.	This strain is derived from importations mainly from Saxony in Germany & the Spanish flock of King George III. Tasmanian rams were sent to every state of Australia until the 1860's when more productive strains replaced them in the hotter dry dusty areas of Pastoral Australia.
Fine Merino strain	Soft white face, pink skin, white legs and hooves. Horned rams.	Rams: 40 to 65 Ewes: 35 to 45	70s to 64s 20 (19–21)	3 to 5	75 to 90	Fine Merino wool is used for high quality suiting materials and billiard felt cloths.	The fine Merino has been adapted to the cooler high rainfall areas of the Western District of Victoria and along the Great Dividing Range of Eastern Australia.	This strain has received infusions from other strains of Merino from France, the Spanish Merino Flock of King George III in England & other Australian Merino strains.
Medium non-Peppin Merino strain	Soft white face, pink skin, white legs and hooves. Horned and polled rams.	Rams: 45 to 80 Ewes: 35 to 50	66s to 60s 21 (20–23)	4 to 5	90 to 100	Medium Merino wool (very white) is ideal for paste) shaded fabrics of high quality.	The medium non-Peppin strain has been adapted to the drier mountainous areas and western slopes of the Great Dividing Range in N.S.W. mainly.	This strain has been developed from the Saxon and fine wool strains of Merino.
Medium Peppin Merino strain	Soft white face but mottles tolerated on the the non wool growing areas. White legs and feet. Horned & polled flocks.	Rams: 50 to 90 Ewes: 38 to 55	66s to 60s 22 (20–24)	4.5 to 7.0	90 to 105	AS ABOVE	The Peppin Merino strain was adapted to the cereal and pastoral zones of N.S.W., Victoria and Queensland with some spread to W.A. Wrinkles were tolerated in this strain to increase the wool growing area.	This strain was developed in the 1860's by the Peppin Brothers by introductions of Rambouillet rams from U.S.A.

TABLE 3 Description of Australasian Breeds of Sheep—Strains of Merino (*continued*)

Breed	*Colour of head and points*	*Body Weight (mature) kg*	*Wool Quality number (microns)*	*Greasy fleece weight (kg)*	*Staple length (mm)*	*Purpose*	*Adaptations*	*General*
S. A. strong wool Merino strain	Soft white face, some mottles tolerated on non-wool growing areas of face. White legs & feet. Horned & polled flocks.	Rams: 55 to 100 Ewes: 45 to 70	64s to 56s 24 (21–26)	5 to 7.5	100 to 120	Developed in S.A. for the semi arid pastoral zone as a dual purpose Merino of high fleece weights & high body weight for surplus sheep.	The S.A. strain was developed for the dry pastoral zone but has been adapted to the cereal zone and more recently to the high rainfall zone. Most prominent in S.A. and W.A. but increasing in the Eastern states.	This strain was derived from N.S.W. Merinos into which infusions of English long wools were carefully made to make this the most productive dual purpose strain of Merino in Australia.
Fonthill Merino strain	Soft white face, pink skin, white legs and feet. Horned rams.	Rams: 55 to 85 Ewes: 40 to 60	70s to 60s 22 (19–23)	4 to 5	85 to 100	Founded in 1954 by Mr. Jim Maple-Brown to produce a prolific fine Merino with good body weight to improve sales of surplus sheep.	The Fonthill Merino was developed near Goulburn in the Southern Tablelands of N.S.W. as a dual purpose Merino.	This strain was developed from the original "Camden Park" Spanish Merino with infusing of Saxon Merino from Tasmania and Rambouillet Merino rams from U.S.A. in 1950.
AMS Merino	Soft white face, with no pigmentation preferably on non-wool growing areas of face, legs or feet. Horned & polled, but polled sheep preferred.	Rams: 60 to 95 Ewes: 45 to 65	70s to 58s 22 (20–24)	4.5 to 7	90 to 120	Developed in W.A. by Jim Shepherd in a huge group breeding scheme to capitalise on genes of every Merino strain to produce a highly productive easy-care sheep.	Rapid genetic progress was obtained by elevation of high performance ewes to Ram Breeding Co-operatives and thence to Central Nucleus and distribution of rams downward to flocks or by use of A.I. from top rams.	Genes from every strain and environment have been recombined to break through the genetic plateau which develops within a flock after 8–10 years' selection. A highly profitable Merino type has evolved.
Booroola Merino	Clean white silken face, pink nostrils. Legs & feet white. Rams horned.	Rams: 45 to 80 Ewes: 35 to 50	66s to 60s 22 (20–23)	3 to 4	60 to 95	This high fecundity breed is being used to transfer the F Fertility gene into other breeds particularly crossbreds and breeds used for prime lamb production. High mortality and poor mothering ability limits its use in the Merino.	This breed was developed by the SEARS in the mountains at Cooma, so is useful for higher rainfall areas.	The F gene possibly came from the fecund Bengalee fat tail breed imported from India to Australia in 1797. Ewes have litters of 2–6 lambs but only 2 teats.

TABLE 3 Description of Australasian Breeds of Sheep

Breed	*Colour of head and points*	*Body Weight (mature) kg*	*Wool Quality number (microns)*	*Greasy fleece weight (kg)*	*Staple length (mm)*	*Purpose*	*Adaptations*	*General*
Corriedale	White face, dark skin on nostrils, legs white & hooves preferably dark. Polled.	Rams: 90 to 110 Ewes: 50 to 70	56s to 50s 28 (25–32)	4.5 to 6.5	150 to 180	A good dual-purpose breed for wool and mutton. The ewes make good prime lamb dams. This breed is also crossed with Merinos and long-wool breeds.	Hardy and adapted to a wide range of conditions. Best suited to improved pasture conditions in Australia. Some selection against fleece rot and body strike required in high rainfall areas.	A inbred Lincoln x Merino half-bred. Lamb marking percentages of more than 100% are common. Corriedale crossed with B.L. in NSW have been interbred and fixed as the new Gromark Breed.
Bond (Corriedale)	White face, with dark or pink skin on nostrils. Prefer pink nose, white legs & feet.	Rams: 90 to 110 Ewes: 50 to 70	58s to 54s 27 (25–30)	5.0 to 6.5	150 to 180	A good dual purpose breed for wool and mutton. The ewes make excellent prime lamb mothers. This breed is also crossed with Merino and long-wool breeds.	Hardy and adapted to a wide range of conditions, particularly of the cereal and higher rainfall areas. The aim is to reduce number of individual black fibres in the fleece.	An inbred Lincoln x Merino half-bred with some later infusions of high performance, open faced, plain bodied, fecund Merinos to increase wool cut, and reproduction rate.
Polwarth	Soft white face but may be black mottles on nose. White legs & feet. No kemp on face or legs. Horned or polled.	Rams: 60 to 90 Ewes: 45 to 60	60s to 56s 24 (22–26)	4.0 to 5.5	100 to 140	Mainly for wool. Better mutton conformation than the Merino. Developed as a wool producing sheep for the high rainfall zone because of greater resistance to fleece rot and fly strike.	In cold, wet conditions of 500mm rainfall and over, pure Polworth progeny thrive better than pure Merinos. Lambing percentages of over 100% are common.	This is a fixed 3/4 Merino – 1/4 Lincoln "comeback," based upon the white-woolled fine Victorian Merino which is high in wax content. Should show pronounced Merino characteristics.
Zenith	White, open face & legs. Large ears. Polled.	Rams: 60 to 80 Ewes: 45 to 60	60s to 58s 24 (22–25)	4.5 to 5.5	100 to 120	Wool. Better mutton conformation than fine Merino.	Bred in the cereal zone of Victoria to produce a more dual purpose type of sheep than the fine Victorian Merino.	Developed by Mr. Barrett at Donald, Victoria. Origins not clear but thought to be a 7/8 Merino – 1/8 Lincoln.
Cormo	Clean white open face, soft pink skin but some mottles on nose. White legs. Polled.	Rams: 70 to 90 Ewes: 45 to 60	64s to 60s 22 (21–23)	4.0 to 5.5	100 to 130	A dual purpose sheep suitable for higher rainfall and mountainous areas. Established in Tasmania by crossing Corriedales on Saxon superfine Merinos by Mr. I. K. Downie.	Withstands wet and cold conditions.	Good fertility (100%) Excellent comeback wool sought by Japanese mills.
Fernleigh & Bundoran Comebacks.	White face, legs & feet white. Polled. preferred; some scurs.	Rams: 70 to 95 Ewes: 45 to 60	64s to 58s 23 (20–24)	4.0 to 5.5	100 to 130	A dual purpose sheep suitable for the higher rainfall areas (over 500mm). Developed in Victoria using Polwarth, Zenith and various Merino strains.	Selection has been directed toward sheep which thrive and perform well under cold, wet conditions.	The Japanese seek this excellent springy, comeback (coming back to Merino) wool. These comebacks usually mark over 100% of lambs.

TABLE 3 Description of Australasian Breed English Longwool—Dual Purpose Breeds

Breed	*Colour of head and points*	*Body Weight (mature) kg*	*Wool Quality number (microns)*	*Greasy fleece weight (kg)*	*Staple length (mm)*	*Purpose*	*Adaptations*	*General*
Coopworth	White face, dark nostrils and feet. Polled.	Rams: 90 to 120 Ewes: 50 to 80	46s to 40s 34 (32–36)	5.5 to 6.5	200 to 240	A self-replacing, fecund breed for the high rainfall areas with a 9 month growing season.	Developed in New Zealand by Professor Coop by interbreeding BL x RM and selecting for fecundity and growth rate.	This fecund breed is the only breed in which ewes cannot be registered unless they rear twin lambs. The average lambing percentage is 180%.
Perendale	White open face, dark skin on nostrils & feet. Legs white. Polled.	Rams: 80 to 100 Ewes: 50 to 65	54s to 48s 30 (28–32)	4.5 to 6.0	120 to 180	A dual purpose sheep bred for hilly country in New Zealand by Sir Geoffrey Peren who crossed Cheviot rams with Romney ewes.	Hardy sheep, a good walker. Easy care sheep. Excellent for prime lamb mothers.	Good fertility (more than 100% of lambs marked to ewes mated.
Improved Border Leicester (BLI)	Clean white head, face and legs, all free of wool. Polled.	Rams: 90 to 115 Ewes: 55 to 75	46s to 40s 33 (30–36)	5.5 to 6.5	200 to 250	Wool and mutton; used extensively in the first cross with the Merino; pure-bred ewes and first-cross ewes make good prime lamb dams.	Good; withstands warmer and drier conditions better than any other longwool breeds.	First crosses prolific, (highly fertile with lambing percentages commonly of 120 or more) good mothers, early maturing. The BLI was formed by crossing BL rams with fixed fecund BL x MO ewes. The BLI has 5% higher fertility and 20% more lambs weaned/ewes formed than the BL.
Gromark	White, open face & legs, dark skin on nostrils. Polled.	Rams: 90 to 115 Ewes: 65 to 80	54s to 46s 30 (28–34)	5.0 to 6.0	170 to 210	A good dual purpose coarse wool breed for wool & mutton. This breed is also crossed with Merino ewes to produce first cross prime lamb mothers.	Scientific breeding principles and measurement of body weight, growth rate, fleece weight, fibre diameter and fecundity have aided the development of a highly productive breed for the later high rainfall districts.	Developed by Mr. Arthur Godlee at Tamworth in NSW. Emphasis has been placed on rapid growth rate and high fecundity. The breed was developed from triplet Border Leicester rams crossed with twin Corriedale ewes. The purebred ewes need to be mated after the first of February for best lambing percentages.

TABLE 3 Description of Australasian Breeds—Dual Purpose Comeback Prime Lamb Mothers

Breed	*Colour of head and points*	*Body Weight (mature) kg*	*Wool Quality number (microns)*	*Greasy fleece weight (kg)*	*Staple length (mm)*	*Purpose*	*Adaptations*	*General*
Beddale	White face, preferably pink nostrils, white legs & feet.	Rams: 60 to 110 Ewes: 50 to 70	60s to 56s 23 (22–25)	4.5 to 6.5	90 to 120	An efficient comeback type of sheep which thrives in the cereal zone (one fat score higher than Merinos).	Selected for production on dry summer pasture residues in W.A. cereal zone.	An inbred 7/8 Merino—1/8 (English Leicester [EL], Border Leicester [BL] and Dorset [D]). Good mothering ability (marks over 100% of lambs).
Daldale	White face, may have dark nostrils and feet. Prefer white legs & feet. Polled.	Rams: 80 to 110 Ewes: 50 to 70	58s to 50s 27 (25–30)	4.0 to 6.0	90 to 130	A dual purpose fecund, prime lamb mother.	Developed at Wellington NSW as a self-replacing prime lamb mother.	Interbreeding BL x MO (including Booroola fecund Merino) x D. Lambing percentage of 150.
Glenara Improver	White face, pink nose, white legs & feet. Polled.	Rams: 80 to 110 Ewes: 50 to 70	60s to 56s 26 (24–28)	4.0 to 5.5	80 to 110	A dual purpose, fecund, fast growth rate, prime lamb mother or terminal sire.	Developed at Langkoop, Victoria as a self-replacing prime lamb mother or terminal sire to produce heavy lean lamb.	A very long bodied, fast growing lamb produced by crossing BL x MO with DH and interbred to produce 1/2 DH – 1/4 BL – 1/4 MO.
Hyfer	White face, pink nose, white legs & feet.	Rams: 80 to 100 Ewes: 50 to 70	60s to 56s 25 (23–27)	4.0 to 5.0	80 to 100	A fecund, dual purpose, self-replacing prime lamb mother or terminal sire. Aiming for 2 lambs per ewe every 8 months.	Developed in NSW as a high producing, fecund breed which can be joined at any season. Reciprocal matings with Poll Dorsets (D), Booroola (B) and high fertility Trangie (T) Merino ewes.	First cross prime lamb mothers have had low gross margins. This fecund breed with a finer fleece and greater flexibility should rectify that. The breed is 1/2 D – 1/4 B – 1/4 T.
Dormer	White face, pink nose, white legs and feet.	Rams: 70 to 95 Ewes: 45 to 65	64s to 58s 23 (22–26)	4.0 to 5.0	85 to 100	A self replacing, fecund, flexible breed as a prime lamb mother or wool producer.	Multiple birth Merino rams joined with polled or horned Dorset ewes and the progeny joined with fine Merino ewes to produce 3/4 Mo – 1/4 D.	This flexible breed can be joined in almost any mother to produce good lean prime lamb and a profitable comeback type fleece.

TABLE 3 Description of Australasian Breeds of Sheep—Carpet Wool Breeds

Breed	*Colour of head and points*	*Body Weight (mature) kg*	*Wool Quality number (microns)*	*Greasy fleece weight (kg)*	*Staple length (mm)*	*Purpose*	*Adaptations*	*General*
Drysdale	White face, dark skin on nostrils, legs white & hooves preferably dark; tan fibres may occur on poll of lambs. Rams usually horned. Some N/N ewes horned.	Rams: 80 to 100 Ewes: 55 to 70	>40s (35–45)	5.0 to 7.0	300 to 450	Carpet wool & mutton, run as a pure breed to produce specialty carpet wool. Cull & cast-for-age ewes for mutton. Wether lambs for carpet wool or meat. Ewes make good prime lamb mothers.	Good, well adapted to colder and wetter areas of the hills & lower S.E. of S.A. Adult fleece variable in carpet wool quality.	Derived from the Romney in New Zealand due to a genetic mutation of one pair of genes (N/N) governing carpet wool. Resistant to internal parasites and footrot. High fecundity (120% lambs marked).
Tukidale	White face, dark skin on nostrils, legs white & hooves preferably dark. Rams usually horned. Some T/T ewes horned.	Rams: 80 to 100 Ewes: 55 to 70	>40s (35–45)	5.0 to 7.0	300 to 450	AS ABOVE	Good, well adapted to high rainfall areas. Adult fleece more uniform in carpet wool quality than Drysdales (3 different wool types).	Derived from the Romney in New Zealand due to a genetic mutation of one pair of genes (T/T) governing carpet wool. Resistant to internal parasites and footrot. High fecundity. (120% lambs marked).
Carpet-master	White face, dark skin on nostrils, legs white & hooves preferably dark. Rams usually horned. Some N/N ewes horned.	Rams: 80 to 100 Ewes: 50 to 70	>40s (35–45)	4.5 to 6.5	300 to 450	AS ABOVE	Good, well adapted to high rainfall areas. (3 different wool types.)	Derived from the Perendale in New Zealand due to a genetic mutation of one pair of genes (N/N) governing carpet wool. Resistant to internal parasites & footrot. High fecundity (150% lambs marked).
Elliotdale	White face, dark skin on nostrils, legs white & hooves preferably dark. Rams horned or polled. Ewes usually polled.	Rams: 80 to 100 Ewes: 55 to 70	>40s (35–45)	5.0 to 6.0	300 to 450	AS ABOVE	Good, well adapted to high rainfall areas. Fleece more variable than other carpet wool breeds.	Derived from the Romney in Tasmania due to genetic mutation of one pair of genes (N/N) governing carpet wool. Resistant to internal parasites and footrot. High fecundity (120% lambs marked).

TABLE 3 Description of Australasian Breeds of Sheep—Terminal Sire Breeds

Breed	*Colour of head and points*	*Body Weight (mature) kg*	*Wool Quality number (microns)*	*Greasy fleece weight (kg)*	*Staple length (mm)*	*Purpose*	*Adaptations*	*General*
Polled Dorset	White face no horns. Legs and hooves white.	Rams: 90 to 110 Ewes: 55 to 75	56s to 50s 28 (25–30)	2.3 to 2.7	80 to 100	Mutton; prime lamb sire or may be mated to the Merino for the production of prime lamb dams for mating in late spring. Lamb carcase weight 15 to 18kg.	Good; withstands warm and dry conditions comparatively well. Wool used for hosiery and felts.	Very early maturing, good mothers, long breeding season. First crosses prolific, but cut about 1 kg less wool than Border Leicester x Merino.
South Suffolk	Smoky to black open face, some muffled. Clean brown to black legs. Polled.	Rams: 80 to 100 Ewes: 50 to 65	58s to 56s 25 (24–27)	2.0 to 2.6	50 to 80	Excellent sire for prime lambs of 14 to 18kg. carcase weight.	Fair; withstands warm and dry conditions but thrives best on good green pastures.	Intermediate between Southdown and Suffolk (used to produce the breed in New Zealand)
White Suffolk	Clean, white face, head & legs. Polled.	Rams: 90 to 120 Ewes: 60 to 80	56s 28 (26–32)	2.3 to 2.7	70 to 90	To produce a heavy prime lamb with minimal fat cover and no pigmented fibres in the fleece.	Good: crossbred lambs require good feed for best results but this breed type can recover better after a check than most. Useful in cereal zone over Merino ewes.	Largest down breed, rapid growth rate. It is a prolific breed, ideal for the present markets favouring heavier, leaner lamb.
Wiltshire Horn	White open face and points. Rams & Ewes horned.	Rams: 90 to 120 Ewes: 55 to 80	50s to 40s (40–60)	0.8 to 1.5	20 to 30	An ideal prime lamb sire for heavyweight lean carcases of 16 to 20 kg.	The sheep are very hardy. Could be used to produce easy-care sheep as crosses shed wool from head, legs, belly and crutch in varying degrees.	High fecundity (150%) An ancient breed thought to be introduced to Britain by the Romans. The pure-breed sheds all its wool in late spring and summer. The wool is very short, coarse and kempy.

Dual Purpose Prime Lamb Mothers

Glenara Improver

Mr. George Stewart of Langkoop, Victoria aimed to produce a prime lamb sire breed which could also be used as a self-replacing breed for prime lamb production. The Glenara Improver breed was developed from a nucleus of 500 ewes in co-operation with the Victorian Department of Agriculture.

He used the findings of Dr. T. S. Ch'ang (1983) of CSIRO, that hybrid vigour is generated if F_1 rams are mated to F_1 ewes of the same two breeds rather than just crossing the two parent breeds. This policy gave a 66% improvement in lamb production over the purebred parent rams and ewes of the same breed.

Selection pressure in these Glenara Improvers is applied to growth rate and twinning ability. These rams produce some of the longest bodied lambs that I have ever seen with a raised and rounded loin and they finish at heavy weights without excess fat and carry a good skin. The new breed consists of 50% Dorset horn, 25% BL and 25% Mo.

His aim to produce a self-replacing breed was to reduce costs in the purchase of prime lamb mothers at $40–$50 per head and to reduce the risk of introducing footrot and worm resistance to his property. Glenara Improvers will mate earlier than breeds derived only from British breeds.

Ewes are divided randomly into four sire groups. Each group is mated with a syndicate of rams at 4% (5 rams with 125 ewes) to test each sire line and restrict inbreeding. Lambs are all mothered and pedigreed. Ewes are side branded for quick identification. Reproduction rate is similar to F_1 BL x Mo ewes on the place. The development of high withers has reduced lambing difficulties in this breed, as in Perendales, and Lines blood Merinos in S.A.

There are only about 2,000 Glenara Improvers and it is not sufficiently distinct to create a great demand for it.

Hyfer

The N.S.W. Department of Agriculture is also developing a prolific prime lamb dam breed called Hyfer to allow higher production and greater flexibility under good management than the traditional BL x Mo. (Hall and Fogarty, 1982.)

This fecund breed can be joined in almost any month of the year with the ultimate aim of rearing two lambs per ewe every eight months.

The Hyfer is being developed from Dorset (D), Booroola Merino (B) and Trangie high fertility Merinos (T). In 1978–81 there were 1780 D ewe single-sire joinings to 24 B and 26 T rams, and in 1981, 270 T ewes to 9 D rams produced 630 surviving F_1 ewes. F_1 sheep were bred reciprocally at Cowra in autumn 1980–84 to produce F_2 progeny (½ D–¼ B – ¼T). Fertility of F_1 ewes was 99% and 98% with litter size 1.59 and 1.79 in 1980 and 1981 respectively.

Litter size of 2- and 3-year-old BD and TD ewes in 1981 was 2.11 and 1.48 respectively. Olvulation rate in February 1981 ranged from 1.36 for eight month old to 2.31 for 31 month old F_1 ewes which was 0.42 to 1.23 higher for BD that TD ewes. There were 71% of eight month old and allolder F_1 ewes ovulating.

Following the 1981 spring lambing, 61% of 2- and 3-year-old F_1 ewes exhibited oestrus with in eight months of the previous joining, whilst still rearing lambs. (Few ewes ever mate within a month of lambing due to involution of the uterus.) From 1982, F_2 ewes have been joined 8-monthly in February, October and June over two years prior to selection into a nucleus flock to breed replacements.

Selection will be based on total weight of lamb weaned. Dam performance is the primary trait for ram selection, within sire lines to minimize inbreeding. Growth rate, appearance and absence of dystocia and horns are also considered in ram selection as they are important in industry acceptance and production.

This breed has certain unique characteristics which could enable it to multiply in certain areas of the cereal zone, but numbers are only small at present.

Dormer

Another flexible breed is the Dormer, which is a dual-purpose ewe that can be mated to a prime lamb sire or a wool-producing Merino sire, depending on economic conditions at the time. It is a fixed ¾ Merino-¼Dorset. Prof. E. M. Roberts developed the breed by crossing inbred high multiple birth Merino rams with polled or horned Dorset ewes. In 1972, second and third generation Dormer rams of 22.3 to 28.7 microns, 73–74% yield were joined with Ledgeworth blood 20 microns, 73% yield Merino ewes.

Mr. J. Garnock of "South Bukalong", Bombala, N.S.W., used Poll Dorset rams and 1200 Merino ewes of 22–23 microns wool. The best ewe progeny were mated with three Merryville Poll Merino rams of 20–21 micron wool. The resultant progeny were then inbred to produce Dormers.

Dormer ewes mated to Dorset rams can produce 115% lambs marked to ewes mated, and 14–18 week-old lambs dressing 17.8 kg. Some 46% of the mob could be sold in the first draft as suckers in early August and the second draft of 34% of the mob in mid-August of pastures. Dormer ewes cut 5 kg of 22.7 micron wool yielding 69.8%.

This breed may create interest in parts of the cereal zone but so far numbers are small.

Terminal Sires

Poll Dorset

The Poll Dorset is an important new Australian breed of prime lamb sire. It causes much less problems in management than the Dorset Horn, the lambs of which may be born with horns up to a cm long, causing both lambing and shearing difficulties.

Mr. Wilson of Howlong, N.S.W., crossed polled, white-faced Ryeland rams with Dorset Horn ewes and bred back to the Dorset horn for seven generations while selecting for polled progeny of good conformation and growth rates. The new breed thus became 98.75% Dorset Horn with all of its excellent characteristics as a terminal sire, except that the dominant PP genes for polling were introduced to replace the Dorset Horn P'P allelic genes, which produced horns in both rams and ewes.

Concurrently, Mr. W. J. Dawkins, a famous Dorset Horn breeder at Gawler River, South Australia, observed the result of a chance mating of a Corriedale ram with his Dorset Horn ewes, and the lambs were polled. So he also introduced the true PP poll gene into his Dorset Horn ewes with which he had won prizes in Dorset County in England.

The Poll Dorset breed spread rapidly and is now the most numerous terminal sire breed in Australia with its own flock book equal in size to that of all other shortwool breeds put together with almost one million.

South Suffolk

In New Zealand the Southdown was for many years the predominant breed of prime lamb sire (over 90% of all terminal sires). However, changing markets forced breeders to produce a heavier, leaner lamb.

A cross between the larger, leaner Suffolk and the excellent conformation Southdown with interbreeding of the acceptable male and female F_1 progeny, led to the formation of the South Suffolk breed in N.Z. in the 1920's. This breed was developed in Australia since 1958 as an alternative terminal sire breed but is not sufficiently distinct and productive to justify rapid expansion. Numbers are small.

White Suffolk

A problem has arisen in marketing prime lambs sired by black-faced Suffolk rams due to the dark points and individual pigmented fibres through the fleece. Lamb skins are penalised by around $2 compared with white-faced breeds such as the Dorset.

At the sucker stage there is very little difference between the progeny of Suffolk and Dorset rams, but at the carryover stage the Suffolk sired lambs tend to grow heavier, leaner carcases which are in greater demand today. In S.A. prime lamb producers in marginal areas claim that the Suffolk is able to withstand checks during dry periods better than the lambs sired by other terminal sires.

Consequently, the University of N.S.W., at their Hay Field Station, are aiming to produce a white-faced Suffolk. Suffolk ewes were mated with Dorset and Border Leicester sires and the F_1 progeny interbred.

More than one set of genes is involved in the black points. It is debatable whether the effort expended on this difficult project is justified when there are excellent long bodied Poll Dorset rams available in reasonable numbers.

Of 30 Dorset x Suffolk F_2 lambs, 10 had full black faces and points, 15 had broken colouring, 2 full white and 3 slightly grey. This breed may attract interest due to more valuable skins than the black-faced Suffolk. So far numbers are small.

Carpet Wool Breeds

In the 1960's and again more recently, the returns to prime lamb producers using first cross ewes has been low compared with other sheep enterprises due to low prices for both crossbred wool and lamb. So an attempt was made in Tasmania to produce an alternative sheep enterprise for the high rainfall areas of N.W. Tasmania (800 to 1800 mm rainfall).

Unlike the three genetic mutations of the Romney and Perendale (Cheviot x Romney) from N.Z., i.e., the Drysdale, Tukidale and Carpetmaster, with the allelic gene pair for carpet wool production (Nd, Nt, and Nj respectively) Elliottdale lambs had small horns, scurs or no horns. The three N.Z. mutations all had full horns on the rams and horns on many ewes and a complete cover of harsh, chalky white, medullated halo hair birthcoat on the lambs at birth. Some also had tan fibres on the back of the neck. There was a small patch lacking harsh halo hair near the shoulder of heterozygote lambs of the Drysdale and N.Z. Carpetmaster but not the Tukidale or Australian Carpetmaster (derived from the Drysdale).

Elliottdale

The new breed developed in Tasminia was called the Elliottdale. It was regarded as a separate breed from the other three genetic mutations, because of the different expression of the N gene for carpet wool and small or no horns.

All four breeds are being run in the high rainfall areas (over 700 mm rainfall) of all States as a profitable alternative to pure longwool breeds or first cross prime lamb mothers. The carpet wool sheep thrive in cold, wet areas and are easier to manage than the pure Merino. They have to be shorn twice a year because this coarse, straight, harsh, hairy wool grows at about 2 to 3 cms per month (more in spring and summer, so it is usual to shear them at intervals of 5 and 7 months).

Some 30% of the fleece is medullated and the average fibre diameter is 35 to 45 microns with no lustre or crimp. The average fleece weights are in the order of 6 to 8 kg per head. The Gross Margins at December 1987 for Carpet wool production was of the order of $28 per Dry Sheep Equivalent (DSE) compared with $30 to $40 for pure Merinos, and $23 for first cross BL x Mo ewes joined with terminal sires. The current price for carpet wool of AAACW grade was 650 cents per kg in December 1987. There may be 20,000 Elliottdales and numbers are increasing.

Conclusion

Despite the diversity of breed types developed in Australia during the last 200 years, none of them have been able to displace the Merino from the wide range of environments (100 to 800 mm rainfall) in which it is run, with the possible exception of some of the high rainfall areas. Even there, special types of Merino are being developed which can thrive under wet conditions and produce a profitable fleece. See Table 1 for comparative numbers of different breeds.

Generally it is the high fleece weight and fine apparel wool in the fibre diameter range of 18 to 25 microns which makes the Merino so profitable.

More work is still required to improve the feet and fleece of Merinos to handle wet conditions and to increase their resistance to fleece rot and flystrike and internal parasites.

References

Australian Society of Breeders of British Sheep (1975). British Breeds of Sheep in Australia.

Barwick, S. A. (1985). The BLI Project in N.S.W. Dept. of Agric. Agdex 430/37, October 1985.

Ch'ang, T. S. (1983). Crossing strains boosts performance of Merinos in Farm, April 1983, p. 57.

Coop, I. E. (1974). Crossbreeding, interbreeding and establishing a new breed of sheep in Proc. N.Z. Soc. An. Prod. pp. 11–13.

Coop, I. E. (1978). The Coopworth Breed of Sheep. World Review of Animal Production Vol. XIV, No. 3, pp. 9–15.

Downie, I. K. (1986). A History of the Cormo Breed. Presented at a Field Day 12/3/86.

Dreaver, R. W. (1964). Tasmanian Sheep Breeds. Tas. J. of Agric. 17/64, pp. 114–124.

Elliott, K. H., Rae, A. L., and Wickham, G. A. (1979). Analysis of Records of a Perendale Flock. N.Z. J. of Agric. Res. 22: Part 1, 259–265, Part 2, 267–272.

Elliott, K. H. (1981). The Development of the Drysdale Breed. Paper presented to Carpet Wool Workshop, Burnie, Tasmania.

Godlee, Arthur. Why the Gromark is a "Good Sheep" in Landscene No. 3, June 1981.

Hall, D. G., and Fogarty, N. M. (1982). Development of a prolific breed for intensive lamb production in Proc. Aust. Soc. Anim. Prod. 14:651.

Howe, R. R. Breed Development—Elliottdale. Carpet Wool Workshop, Burnie, Tasminia, June 1981.

Kajons, A. (1981). Breed Development—Tukidale. Paper presented to Carpet Wool Workshop, Burnie, Tasmania, 2–5 June, 1981.

Kelly, R. W., Wilson, K. R., and Trotter, R. W. Evaluation of Sheep Breeds at Invermay, N.Z. (Report Experiment.)

Maple-Brown, J. (1970). Fonthill faces the fibre challenge. Brochure for Field Day 8/2/71 with report on wool by H. B. Carter, formerly of CSIRO.

Mason, I. L. (1969). A Dictionary of Livestock Breeds. Commonwealth Agricultural Bureau, Farnham Royal, Bucks, England, p. 262.

Ponzoni, R. W. (Pers. Com) Objectives in Merino Breeding.

Reid, R. D. N. (1987). Merino Improvement Programs in Australia, Lewra.

Ryder, M. L. and Stephenson, S. K. (1968). Wool Growth, Academic Press (London), p. 44.

Stewart, G. G. (1981). Design of Selection Programs for Sheep Meat Production. Proc. Second Conf. AAABC, Melbourne, Victoria, pp. 200–203.

Thatcher, L. P., and Pascoe, H. J. (1973). The Wiltshire Horn in Journal of Agriculture, Victoria, Jan. 1973.

Wigan, R. L. (1980). Alternative Sheep Breeds. Proc. Seminar at Glenormiston Agricultural College, Glenormiston South, Victoria, 20/6/80.

Winter, J. D. (1988). Consultative Committee Report to Australian Merino Society (A.M.S.) in A.M.S. News, WAD 1371, Vol. 5, No. 3, Feb. 1988.

Wool Technology and Sheep Breeding Sept./Oct. 1978, pp. 5–18. New Australasian Breeds of Dual or Special Purpose Sheep.

British Coloured Sheep Breeding and Wool Production

Michael J. Woodhead
Coloured Sheep Breeders' Association
White Alice Farm
Carnmenellis, Redruth, Cornwall TR16 6PH, UK

This paper gives information on developments in British coloured sheep breeding since 1984; discusses some factors which affect British coloured sheep breeding and wool production; and describes what the (British) Coloured Sheep Breeders' Association has done since it formed in November 1985.

British Coloured Sheep Breeding Since 1984

British coloured sheep breeds were described in detail at the last two Congresses (Alderson 1979[1], Alderson 1984[2], Ryder 1984[3], Macdonald 1984[4], and I will not repeat these descriptions. It may be more useful to review developments over the last five years, and particularly the growing interest in coloured sheep other than from the traditional native coloured breeds. Some of these developments were stimulated by the New Zealand Congress in 1984 and the formation and activities of the Coloured Sheep Breeders' Association.

In 1984, British coloured sheep could be described as being mainly of the primitive or improved primitive and Northern short-tail breeds. While these may still be the most numerous, this description would not now do justice to the present interest in a much wider range of coloured sheep. Specialist coloured sheep breeders and wool producers have emerged, and are more interested in the wider range of fleece types, colours and patterns; greater wool weights and generally higher productivity that can be found in coloured sheep from white breeds. Such breeders may not be interested in the rare breeds conservation concerns associated with British primitive breeds, and be more motivated by higher earnings from wool and skins or products made from them.

Coloured sheep in Britain may be classified as follows:

Traditional native coloured breeds. These range from rare and primitive types: Hebridean, Manx Loghtan, Castlemilk Moorit, North Ronaldsay, Soay; to "improved" breeds increasingly used commercially: Jacob and Shetland; and include hill breeds: Herdwick and Black Welsh Mountain.

Coloured sheep from white breeds. Originally derived from recessive colour pattern genes in white breeds, these include the Badger Faced/Torddu, Torwen and Balwen Welsh Mountains, and more recent recognitions of coloured longwool sheep: Black Leicester, Lincoln and Wensleydale. Coloured Down type ram breeds in use include the Ryeland and Oxford Down. Many other instances of coloured individuals occurring in purebred white flocks have been reported, and it is assumed that coloured sheep could occur in any modern British breed.

Coloured crossbred sheep. Sheep specially crossbred for wool colour and quality include those using dominant black genes such as Jacob X Lincoln and Black Welsh X Wensleydale; and recessive colour genes as in Friesland X Black Leicester.

Imported coloured breeds. Imported recognised coloured breeds include Gotland and Icelandic flocks, and coloured sheep from imported white breeds include Friesland, Texel and Zwartbles.

While all the above are rare or minority breeds, and some are very scarce, the range of breeds and crosses available to British coloured sheep breeders is very extensive. Between them they offer a wide range of fleece types, colours and patterns; and enable choices of different production systems adapted to the range of climates and altitudes found in Britain.

Fleece types available from coloured sheep include the coarse weaving fleeces from Herdwick (28s), Dartmoor, and Devon and Cornwall Longwools (36–40s); through the lustre and longwool English Leicester, Lincoln and Wensleydale, with finer longwool available from the Gotland, Romney and Bluefaced Leicester. Down type coloured fleece is available from the Ryeland, and the primitive and improved breeds offer a range of medium wools of differing length, fineness and handle. The finest fleece available from a British coloured breed is said to be from the Shetland (up to 60s)[5]. For reasons which are discussed below, some breeds' fleeces lack uniformity, and crossbreeding produces intermediate types which can be difficult to classify.

The range of *colours and patterns* available from British coloured sheep is much wider than many users realise—many handspinners and others still seem only to have heard of Jacob sheep and get their colours from mixtures of its black and white. Besides the black, brown (moorit) and grey colours of the primitive sheep described by Ryder[3]; silver, blue and dark greys are available from the blue pattern in coloured longwools and other white breed sheep, which appear to resemble the Netherlands blue Texels[6]. An all-moorit breed—the Manx Loghtan—exists, but this colour has yet to be reported in British modern breeds. Badgerface (Torddu), reverse badgerface (Torwen), and black with white head/socks/tail (Balwen) patterns exist as variants of the Welsh Mountain breed. The range of colours and patterns discovered in Shetland sheep is extensive[7], and research is needed to establish to what extent these are genetically identical with those found in modern white breeds. Work is also needed to establish what *A* locus genes are present in British modern white breeds, and whether these are the same as those identified in Scandinavian and New Zealand sheep.

All the above colours and patterns are recessive, except for the dominant black of the Jacob and Black Welsh Mountain. Large numbers of black offspring from Jacob crosses are being produced, some from breeders who wish to introduce colour into a specific fleece type, while many others are simply using a white Down ram to produce heavier carcases from a Jacob ewe flock, without any special interest in the resulting black wool. Some breeders are interbreeding dominant black crossed sheep with the intention of creating new coloured breeds.

Production systems using coloured sheep vary according to flock size, climate and altitude, and not least to the extent to which the financial performance of the coloured flock is important to the owner. Breeders' aims and policies are stated or may be inferred from information provided by members in the Association's Flock List which records ewe breeds and rams used. Many breeders with small flocks of primitive or traditional breeds aim to sell their lambs primarily as pedigree breeding sheep. As flock sizes increase, the problem of getting acceptable returns from some types of coloured wethers leads to breeders using a white Down type ram, thus enabling the finished lambs to qualify for subsidy payments, while the flock ewes provide the coloured wool. More commercial systems involve using prolific coloured longwool cross ewes, put to a coloured Down type ram. This produces one or two fleeces and/or a pelt from the lambs as well as an acceptable carcase. Some breeders are using coloured Friesland or Friesland crosses in dairy flocks, and thereby producing coloured fleeces as well as milk. Many breeders are reviewing their management practices to improve wool quality and profitability, and pre-lamb shearing is carried out in some flocks.

Some coloured sheep entrepreneurs base their businesses on marketing the appeal of their chosen breed or breeds, aided by the selective rare breeds exemption from the British Wool Marketing Scheme (see below), or by buying the wool used for their products from the Board. The obstacles which coloured sheep face in competing with other fibre producing "alternative enterprises" to traditional farming are discussed below.

Factors Affecting British Coloured Sheep Breeding

Choices of coloured breeds and production systems are also affected by a number of historical, legal and economic factors, some of which are unique to Britain, which discourages specialised wool production.

The most far-reaching of these is the *legal requirement of the British Wool Marketing Scheme* (1950) which requires wool producers with more than four sheep to register and sell all their wool to the British Wool Marketing Board. The Board has powers to enter the premises of producers and impose fines. The only exemptions to this requirement are for some rare breeds wool, and since May 1987, following the Association's representations, sales of raw fleece "direct to a user for the purposes of handspinning by that user only". Both these exemptions are made for a 12 month period only and there is no certainty that they will be continued to be renewed. The breeds included in the rare breeds exemption vary from one year to the next, and breeds have been taken off the list as well as being added to it. While some coloured breeds have negotiated a special identity and prices from the

Board, in general prices paid by the Board for coloured wool reflect those paid by commercial buyers and do not offer any incentive to specialised wool producers.

The effect of these very limited exemptions is to confine wool producers (except for some rare breeds) to selling raw fleece only to individual handspinners, and to prevent them from selling through cooperatives or to dealers. They are prevented from adding value by selling wool in carded form or by having the wool spun on commission and made into finished products. The fact that the exemptions are made only for 12 months at a time must deter breeding programmes for wool improvement, and investment in a specialised coloured sheep enterprise. The discriminatory exemption in favour of some rare breeds distorts competition between breeds and leads to a narrowing of the range of coloured wools that can be offered to buyers. Other competing textile fibres are not affected by any such restrictions and therefore for example goat and llama fibre producing enterprises can be established without marketing restrictions. This situation makes it difficult to establish the credibility of coloured sheep as a commercially competitive activity, and limits the Association's potential funding and activities.

The longstanding existence of the British Wool Marketing Board as a centralised monopoly buyer and seller of virtually all wool has had other effects which hinder the development of small-scale vertical coloured sheep businesses. Most of the small local mills which once served rural communities have disappeared, and—even if it was lawful—it is difficult and expensive to find mills to carry out commission spinning or other services except for large quantities of fleece. Almost all wool expertise and resources have been concentrated in a single system devoted to handling the entire British wool clip. This has provided a generally efficient service to British farmers disposing of a by-product, but since the Board is legally obliged to buy all wool offered to it by registered producers, there is little incentive for nonspecialist producers to raise standards despite the Board's exhortations.

Economic factors also affect the viability of British specialist wool producers. Average lamb sales per lowland ewe (including variable premium) are worth approximately eighteen times as much as wool sales[8]. With the need to conform to a tight carcase sepcification to obtain the vital variable premium, it is hardly surprising that wool characteristics are low on most flock owners' lists of selection objectives. By the same token a specialist wool breeder risks losing the subsidy if selection for wool characteristics leads to loss of conformation—or if the chosen breed is the wrong shape in the first place. Coloured sheep breeders may therefore be faced with the dilemma of increasing wool earnings at the expense of what could be earned from lamb sales. Impending changes to both wool and lamb sale subsidies are unlikely to change this basic situation.

Until recently, *users' knowledge of coloured wool* available has been restricted. The general lack of interest in wool quality, and prohibition of private wool sales, has limited the quality and variety of wool available to handspinners. While private wool sales took place despite the ban, producers of specialist handspinners fleeces could not advertise, and many handspinners bought their fleeces from the nearest farmer without knowing that anything better existed, and without knowledge of the variety of fleece types and colours that might be chosen for specific projects. With direct contact between producer and user discouraged by the wool sales ban, little could be done to create awareness and a market for specialised fleeces. Furthermore, lacking critical and demanding handspinning customers, producers were not encouraged to remedy their own inexperience and possible defects in their wool. This Association's activities are gradually improving this situation (see below).

There are signs however that the recent exemption may bring about an *oversupply of fleeces,* since the producers of large numbers of black fleeces previously sent to the Board are now trying to sell them privately to handspinners. Although the Board does not publish separate figures of British coloured wool produced (unless this is what they mean by "Oddments"!) they handled about 500 tonnes of coloured wool in 1985[9] and the amount is unlikely to have lessened since. The most recent published analysis showed 156 tonnes of Herdwick and 86 tonnes of Jacob fleece to have been produced in 1987[10]. It is likely that much of the remaining quantity of coloured fleece would be from commercial crossbred ewes.

Attitudes to coloured sheep are sometimes an obstacle to their availability and development. Some white breed societies have reacted positively to coloured fleece versions appearing in the breed—influenced perhaps because they make higher prices at pedigree sales—and the English Leicester, Lincoln and Wensleydale breed societies have all provided for the registration of coloured sheep. Other white breed societies have done the opposite and have not welcomed coloured sheep or their breeders. It is often alleged that any coloured sheep appearing in a pedigree flock must have resulted from crossbreeding. This tendency is not helped by the practices of breeders who cross white longwools or other breeds with a dominant black breed, and then identify the progeny as pure coloured versions of the breed in question. Even where no deception is involved, there is concern that the use of dominant black breeds to introduce colour into flocks could lead to loss of breed identity in a large proportion of the national coloured sheep population, and thus threaten the identity and reputation of all pedigree coloured sheep and their breeders.

Support for coloured sheep breeders is fragmented between organisations with differing objectives. The Rare Breeds Survival Trust is concerned with the conservation of native coloured primitive breeds and provides registration facilities for them. Other recognised coloured breeds have their own breed societies which register sheep and publicise the breed. The Coloured Sheep Breeders' Association's aims and activities (described below) are intended to serve needs not met by other organisations, but the scale of its activities and potential membership is limited by the existence of well supported organisations already providing services for some traditional coloured breeds.

Formation and Activities of the Coloured Sheep Breeders' Association

The Association was formed in late 1985 to promote the interests of coloured sheep breeders and to provide a means of communication between them. The initial impetus was a widely shared urge to remove the legal obstacle to the private sale of fleeces, previously only attempted by individuals. No other organisation is primarily concerned with the breeding, management, and promotion of all coloured sheep and their products, and the Association is a focus of support and mutual help to breeders interested in wool or colour in sheep, regardless of breed. This has been particularly needed by breeders who are located in remoter areas, or who have felt isolated or lacking support because their interests were not being catered for in other organisations.

Membership is around 250 flocks and includes sheep from all the categories and breeds listed above. These are located in all parts of the country, but with many in central Southern and South-West England. This inevitably leads to most events taking place in those areas, although we are encouraging local groups to form and run their own events. Flock sizes of members who completed returns for the Flock Book in 1987 showed a predominance of small flocks below 20 coloured ewes, with another grouping around 40–60. Only one flock exceeded 100. Many members have said that they are increasing their flocks, and it is likely that the 1989 figures now being collected will show increases.

The Association's *activities and policies* may be summarised as addressing the problems which the factors set out above have caused for coloured sheep breeders and specialist wool producers; and helping members to sell coloured sheep and their products. A key event in the Association's life was obtaining the exemption for legal sales of handspinners' fleeces described above. Although the exemption is very limited, and to an extent only made lawful what was already happening, it was an important morale booster to have our activities legitimised and out in the open. The Association will continue to seek legal freedom to sell our products in whatever form we choose.

Promotional and educational activities have included displays at major shows and at the Association's own events. These have showed what kinds of coloured sheep are available, their products, and the opportunities for increased earnings from coloured sheep. Through our own technical events, shows and fleece competitions, we have tried to educate ourselves to seek the highest possible standards in both breeding and wool production, and have used technical assessors and judges from the commercial world to help with this as well as experts with a craft background. From the outset the Association has sought to make contacts with handspinners and their groups, and any interested non-breeder can become an Associate member and is welcome at meetings and activities. These efforts have created wider interest in coloured sheep and wool, and generated helpful Press coverage. We have been given much support and cooperation by other organisations including the British Wool Marketing Board. All these things have improved the culture and climate for British coloured sheep breeding.

We help members to sell sheep and wool by arranging sales events which are then publicised to the appropriate public. *Coloured sheep sales* have been held in conjunction with other organisations. This gives breeders and buyers opportunities to meet each others needs which might not otherwise exist and creates a market for coloured sheep. Prices have been generally higher for the small numbers of white breed sheep that are available for sale, with primitives and most crossbreds cheaper except for show quality sheep. The Association's Flock List, published bi-annually and given to intending buyers, gives members a shop window for sheep sales, and an identity to some flocks outside the scope of a breed society, besides being a way of collecting and sharing information among members.

Because of concern over buyers possibly being misled over claims about the heritability of crossbred ram's fleece and other characteristics, the Association has decided not to allow them to be entered at our sales or be advertised in the newsletter—a decision strongly objected to by some members.

Fleece sales are held at various centres, sometimes together with sheep sales, and the Association also publishes a list of members having fleece for sale which is sent to handspinners' guilds and to other likely buyers. Concern over the quality of fleece offered to buyers is reflected in our Code of Practice, which makes recommendations about fleece condition, and the preparation and offering of fleeces for sale; and by previous inspection of fleeces offered for sale on behalf of members at Association events. Because of the former legal obstacle there has been no conventionally agreed way of preparing and presenting fleeces for sale to handspinners, and we have been experimenting with various forms of packaging and preparation. Some producers skirt fleeces before sale, but handspinners have been accustomed to buying unskirted fleeces and some prefer them that way. Our only attempt to auction fleeces was unsuccessful, perhaps because British buyers are not used to this way of selling fleeces.

Our evolving *marketing strategy* needs to take account of the large volume of coloured fleeces produced by non-specialists which overhangs the handcraft market. We are recognising that specialist producers should therefore aim at a high quality product, attractively packaged, informatively labelled for specific uses. Higher prices which justify the costs of specialised wool production can only be obtained by wool which is seen to be different from that offered from by-product flocks.

We communicate with each other through our quarterly newsletter *Coloured Sheep News* which besides giving information and reports on future and past events, aims to collect and circulate the views and needs of members. It contains material on genetics and management practices relating to coloured sheep and wool, sometimes drawing on information from coloured sheep industries in other countries. The newsletter is taken by many handspinners and their guilds, and is therefore also a means of communication between producers and users.

The Association and our coloured sheep industry is still in its early days compared with those of other countries, and we are only too well aware that we still have a long way to go and many obstacles to overcome. We are grateful for and heartened by all the help, encouragement and example that has come from other countries.

References

[1]Alderson, G. L. H. (1979). Coloured sheep in Great Britain. In: "Breeding coloured sheep and using coloured wool". Peacock Publications, Frewville, South Australia.

[2]Alderson, G. L. H. (1984). Breeds of coloured sheep found in Great Britain and their potential value. In: "Coloured sheep and their products". Ed. H. T. Blair. New Zealand Black and Coloured Sheep Breeders Association, RD 12, Masterton, N.Z.

[3]Ryder, M. L. (1984). The coloured sheep of Europe. In "Coloured sheep and their products".

[4]Macdonald, W. F. (1984). Coloured sheep—their commercial use in the United Kingdom. In "Coloured sheep and their products".

[5]National Sheep Association. (1988). British Sheep (Seventh edition). National Sheep Association, Malvern, Worcs, UK.

[6]Hoogschagen, P. (1984). Blue sheep in the Netherlands. In "Coloured sheep and their products".

[7]Bowie, S. U. (1987). Colours and patterns in Shetland sheep. The Ark. *XIV*: 194.

[8]Meat and Livestock Commission. (1988). Sheep Yearbook 1988. Meat and Livestock Commission, Bletchley MK2 2EF.

[9]British Wool Marketing Board. Personal communication.

[10]British Wool Marketing Board. (1988). Report and Accounts, 30 April 1988. British Wool Marketing Board, Bradford, West Yorkshire BD14 6JD.

The Karakul Sheep: Its Wool and Its Uses

Therese Huneger Ohlson
35698 Phillips Lane
Philomath, Oregon 97370 USA

A type of broad-tailed sheep is the Karakul, also known as fat-tailed sheep, originated in Central Asia and the Middle East. The name Karakul apparently originated from the town of the same name in Turkistan. Karakul are best known for their production of Persian lamb pelts, which were popular in the fashion industry in the early 1900's.[3,8,9]

The majority of Karakul lambskins are produced in Bokhara (USSR), Afganistan (Herat), Southwest Africa, Bessarabia, Shiraz (Iran), Baghdad and Salzfelle (Iraq), and India. The highest grade lambskins come from Bokhara. When demand for lamb fur first came from Europe, the Karakul industry was centered in Bokhara and the surrounding Central Asiatic territories where it has remained more or less ever since.[4,3]

There are about 30 different varieties of broad-tail (*Ovis platyura*) or fat-tail sheep (*Ovis steotopyga*). The Karakul is thought to be tied the closest in origin to the broad-tailed species called the Arabi. C. C. Young believes that the Arabi owes its great hardiness, wonderful fat accumulating properties and black pigment to the Karakul.[9]

The small Arabi, which weighs on average 130 lbs (59 kg) is believed to produce the highest quality lamb fur. They are found only in Bokhara between the city of Bokhara and the Amu-Daria river.[9]

The most valuable Persian pelts are produced from Karakul lambs 1 to 5 days of age. After the 5th day the desired pipe curl begins to open decreasing the value of the pelt. Pelts can be taken up to 10 days of age as the value of the pelt increases with size, but it is essential that the curl remain tight to hold the value. Some breeders believe that by crossing a Karakul with a long wool luster breed, such as a Lincoln, the desired tight curl and larger pelt can be obtained.[2,9]

The Karakul sheep industry in North America began in 1908 when the first 15 individuals were brought in from Russia. Five years later, in 1913, another 17 Karakuls were brought in again, from Russia. A year later 21 Bokharan Karakuls were imported. In 1914, 130 Karakuls from European Russia arrived in Canada and in 1929, 10 individuals arrived from Germany. The Karakul industry in the United States was forced to make its start from 87 head: 53 from their native lands and 34 from other countries. The largest herds of Karakul in the US today are located in the southwest, primarily in and around New Mexico.[3]

Fat-tailed sheep are probably the most important in the world in terms of overall contribution to mankind. As much as 33% of the world's sheep are fat-tailed and are an important source of milk, meat and wool in the developing countries. The tail fat is considered the "butter of Central Asia" and is highly valued for cooking.[8]

The two most prevalent breeds of fat-tailed sheep are the Karakul, known for their Persian pelt production and the Awassi, known for its superior milk production. All fat-tailed breeds of sheep share a number of common characteristics. All are adapted to arid regions, produce carpet wool, have well developed flocking instincts, can travel long distances, have lower twinning rates than most mutton breeds, and in many areas people show a distinct preference for the meat of these fat-tailed sheep.[8]

Karakul rams are generally horned. The horns are outwardly curved spirals with well developed rippling. They are similar to those of their cousin, the big horn sheep. About 30% of the rams are hornless and almost all the ewes. Karakuls generally have long ears that always point downward and slightly forward. Their heads are long and narrow. They stand tall with a long narrow body that is lower over the shoulders with the back sloping upward, the highest point at the loin. They are humpbacked in a stately sort of way. The rump is long, sloping and free from excessive fat. The tail will vary in size but is generally flat, U-shaped with the tip often recurved giving it an S shape.[3,5]

An adult, apricot-colored Karakul ewe, 2 months after shearing. (Photo by author)

There is a long boney appendage sometimes found at the base of an undocked tail. Many breeders do not dock this "tail." For sanitation reasons, I believe that this appendage is best docked at the point the floppy skin meets the boney appendage.

When compared to Rambouillets, Karakuls are more hardy, especially under adverse range conditions. They have a lower lambing percentage (117% compared to 143%), but are easy lambers with the lambs having long narrow heads and slender shoulders. The ewes are very protective and attentive mothers resulting in higher lamb survival rates. This makes the Karakul an "easy care" breed of sheep ideal for the first time sheep owner or gentleman farmer.[8]

There are 2 theories believed to be the reason Karakul meat is preferred over other commercial breeds of sheep. These are: 1) the lack of mutton odor and flavor. This is thought to be associated with the types and amounts of fatty acids present in Karakul fat, which differs from commercial breeds; and 2) fat deposits differently in the Karakul, leaving the muscle tissue with less intramuscular fat.[8]

Karakul ewes are heavy milkers and the lambs gain quickly and up to 3 months of age will average higher in body weight than mutton sheep. This makes the Karakul lamb ideal for a "hothouse" lamb operation. Karakuls also breed out of season making it possible to manage for two lamb crops every 3 years.[7]

The hearty nature of the Karakul makes it long lived and a ewe will produce well past the age of 10. Some have been known to live into the 20's. The harsh desert conditions under which the Karakul evolved has given the Karakul equally strong and long lasting teeth. Their feet are soft and trim easily. The black Karakul is resistant to foot rot, more so than the commercial sheep in my flock. The recessive brown and red Karakuls seem to be slightly more prone to foot rot. Chronic foot rot, however, is not a problem in these hearty sheep.

Characteristics of Karakul Wool

Karakuls are considered a long wool, luster breed of sheep. Typically, the wool is considered carpet quality and indeed it produces a superior carpet yarn, but is often similar to the long wool luster type. Their wool is typically 2 coated with a fine down under coat and a coarser guard hair. Karakul typically has no crimp. The staple will generally grow 1″ (2.5 cm) per month and will shear 6 lbs (2.7 kg) of skirted wool annually on average. The Karakul does not produce a "greasy" fleece which accounts in part for the lighter shearing weights. Because there is less lanolin in Karakul wool there is also less shrinkage with washing. I have found that shrinkage averages 15% with the Karakul fleeces.

Since lanolin coats the wool fiber protecting it from the weather, Karakul wool is more vulnerable to cotting in wet climates. This does not mean that the Karakul needs to be limited geographically. I live in the coastal foothills in Oregon's Willamette valley, which receives an average of 60″ (150 cm) of precipitation annually. The key to success is to make sure the sheep are sheared as close to the onset of the wet season as possible. If the sheep have more than 3″ (7.5 cm) of fleece going into the rainy season, cotting may result from the sheep shaking excess water from their fleeces. During the rainy months it is best if sheep have access to dry shelter.

The open straight locks of the Karakul hold little or no vegetable matter. If reasonable care is taken, beautifully clean fleeces are generally the rule rather than the exception. The dirt washes away easily with water and a summer shower will leave a clean, shining sheep behind.

Within any breed of sheep there are variations in fleece type. The same is true with the Karakul. The Karakul has a wide variability in its fleece type. Fleece characteristics may range from horse tail coarse to silky soft. Some of the silkier

Karakuls may produce a very fine, silky guard hair with the typical 2-coated, crimpless characteristics. Others may produce heavier amounts of down with a guard hair of about the same length which is quite soft. There may be evidence of crimping at the withers which disappears gradually down the back toward the legs. I have not found this variation in crimp to affect the yarn that is produced from these fleeces when worsted spun. The contrary is true. The yarn is springy with lots of loft and excellent for knitting.

The diameter of Karakul wool averages 29.2 microns and is similar to that of mohair, which ranges 22–40 microns. This is considerably more than Rambouillet, which is 20–22 microns in diameter. Because of the similarities to mohair, Karakul is an ideal fiber for blending with mohair for either rug weaving or garments.[8]

Karakuls are a true colored breed with the white color gene being recessive to black, brown and red. Black Karakul lambs are unique because their wool will not bleach in the sun and they remain jet black. Because of the tendency for most black sheep to bleach in the sun, black Karakul is highly prized by spinners and brings a premium price. Animals don't begin to grey until generally their 3rd shearing, around 18 months.

The red and brown colored lamb fleeces are the darkest at birth and lighten with age. In some cases a lighter shade of red or brown will begin to darken at about 6 months with the development of the guard hair. Most brown lambs are born black with a pink nose and tongue, turning a chocolate color as they age. At the first shearing at about 6 months these lambs typically have developed a creamy colored down with chestnut brown guard hairs. The red lambs follow a similar pattern in that the down will lighten and the guard hair will darken to a deep red color. Lamb fleece in most cases is quite suitable for next to the skin garments as the guard hair doesn't begin to develop until about 5 months.

Karakul Wool for Spinning, Weaving and Felting

The distinctive two coated qualities are similar to Icelandic wool in that the down and the guard hairs may be separated and spun separately or spun as one. The staple length and great tensile strength make Karakul wool indispensable in the production of strong, long wearing fabrics and rugs. Karakul wool remains to this day one of the important ingredients of Oriental rugs.[3]

Since Karakul wool has less lanolin than most other breeds of sheep, cleaning and washing the fleece takes less energy. Dirt can easily be removed by soaking in cold water. Mild pH balanced soap, such as shampoo, and warm water (less than 95 °F/35 °C) will sufficiently clean a fleece leaving some lanolin behind. If the wool needs to be fully scoured for processing use very warm (95–120 °F/35–49 °C), but not hot, water with a grease cutting soap, such as Dawn dish soap. Take note that the luster may be damaged if temperatures higher than 95 °F (35 °C) are used to wash a fleece.[6]

Karakul wool can be spun with very little preparation. It can be spun into a fine worsted yarn by combing or gently flicking the locks open. Worsted is very well suited to the Karakul fiber because it is ideal for displaying the elegance of the long-stapled lustrous fibers.[6]

The best method of picking a Karakul fleece, if other than by hand, is with a triple picker designed for kid mohair. This is because the fleece is so similar to mohair and locks are often missed when fed through a standard wool picker. When planning to pick a fleece right out of the wash it is helpful to add some detangling conditioner to the final rinse, especially with the longer stapled fleeces. This final rinse will help speed up the picking process and help prevent the long fibers from tangling.

If sending fleece off to be commercially prepared for spinning or felting, be sure to inquire as to the type of carding equipment being used. Some carding machines that are designed to pick and card fibers all in one step may separate the 2 fiber coats, discarding the fine down coat. This will result in a much coarser spinning fiber or felted garment.

Karakul wool blends beautifully with other wool and of course mohair. A yarn spun from Karakul or predominantly a Karakul blend does not seem to pill in a kintted garment. Instead, Karakul will improve its qualities with wearing. As the garment is worn the down fibers surface and the garment takes on a "brushed" appearance. The fine down fibers will then begin to fuse together forming a more wind and water resistant garment. This will increase the warmth and durability of the garment. Because of these attributes, Karakul wool is ideal for socks that are knit a bit larger then fulled to the desired size and for outer wear garments. The greater the percentage of wool that is blended in with the Karakul, the less of these characteristics will surface. When blending with mohair the "brushed" appearance will be enhanced as mohair also takes on a "brushed" look with wearing.

Karakul wool is an ideal wool for felting. It felts quickly, forming a sturdy, dense fabric which is especially well suited for 3 dimensional pieces needing to stand on their own. It is an ideal wool for teaching children to felt with. The quickly felting wool will hold the children's attention throughout the felting process.

The wool may also be felted from woven or knitted pieces. The different processes produce felt with different properties, so one needs to know what qualities are needed for the finished piece. Felt from wool batting produces a felt with no give, but it will have a tendency to be the most wind and water resistant, making it ideal for felt boots. Because the fabric is created by individual fibers fusing, felt from wool batting will wear more quickly than a felt made from a spun, woven or knit fabric. Thus felt rugs made from batting would not withstand heavy foot traffic as well as a woven felt rug. However, the Karakul wool felt rug would withstand heavy foot traffic much better than the wool from most other breeds because of its durable nature.

Felt made from the woven or knit fabric is generally thinner, has some give and stretch. It may not be as dense or as sturdy a finished product. The traditional French berets are a perfect example of this. They are knitted first, then felted. The hats are thin, pliable and have some stretch.[4]

Karakul wool when felted can produce some interesting textural effects. The coarser Karakul when felted will produce a hairy texture. As the fine undercoat begins to felt first, the coarser fibers surface, giving it the unique texture. This "hairy" type of felt is especially interesting when the guard hair and down coat are different colors. If this type of texture is undesirable simply take a razor and shave it off or alter it to fit the design of the project.

The History Behind Karakul Wool

Along with the contribution Karakul wool has made in the Oriental rug industry, it is the wool from which the art of felt making evolved.[3]

It is from the Central Asian Steppes that the oldest examples of felt have been found so far. The oldest known felt came from the ancient site of Catal Hüyük in Turkey. Here felt was discovered dating back to the Neolithic Age (6500/6300 B.C.). Another important discovery was in the Altai mountains of Siberia and these felt pieces date back to the early Bronze Age (1600 B.C.).[1]

The stone burial chambers in the Altai mountains of Siberia uncovered by far the most important surviving example of this ancient art of felt making. The great Pazyryk felt was nearly intact because of the icy conditions of permafrost under which the material was well preserved.[1]

Although the origin of felt is unknown, in all probability it is the earliest form of textile known to man. This is because felt is not a woven material. However, like most textiles, it too has a low survival rate in archaeological conditions. There are many natural causes for its invention but it is likely we will never know for sure the true origin.[1]

Felt is still used today by the Asian nomads for their shelter, floor coverings and clothing. Felt wall coverings are an important part of the nomadic yurt. Yurts are circular dwellings with domed roofs. The thick felt is lashed onto a diagonal wood lattice supporting a willow frame which is mobile. The felt provides warmth, is resistant to strong wind, repels water and yet is light weight for easy mobility.[1,4]

The traditional felt garment of these nomads is the "kepenek." They provide their owners with a cape as well as a portable tent. The "kepenek" is made from a long, narrow blanket with the front half narrower than the back. It is an extremely simple garment and variety consists of making the shoulders broader or narrower. The very thick, solid nature of these "kepeneks" provide their owner with a free standing tent and are big enough for their owner to sit well sheltered within. These "kepeneks" are still an important part of the traveller's wardrobe in the more rural populations where foot travel is still the primary means of transportation. And of course we can't forget the other traditional garment that is known world wide. This is the felt hat. The "kepenek's" equivalent of the hat is a hood called the "baslik" which is worn in inclement weather.[1]

The felt rugs are considered the "poor man's art" in Asia. This is because only the richer upper class own looms. Just because the nomadic people do not own a loom does not mean that their skill in designing beautiful felt does not rival the best tapestries. Great care is taken by the felt maker in construction techniques which involve applique, mosaic, inlay and embroidery. Since the days of the Neolithic Age these techniques have been improved upon as with the great Pazyryk felt, of the early Bronze Age, and are still widely used by both the nomads of Asia and the fiber artist.[1]

With a growing interest in the fiber arts, there has been an increased interest in the Karakul sheep. The Karakul sheep is considered a rare breed in the U.S. and most likely will remain so. As the interest in colored genetics continues to grow, I see a growing interest in the Karakul because they are one of the few colored breeds. Their unmatched hardiness is also a desirable attribute that has earned them recognition.

Literature Cited

[1]Burkett, M. E.; The art of the felt maker, Titus Wilson & Son Ltd, 1979.

[2]Ensminger, M. E.; Sheep and wool science. Interstate Printers & Publishers; Danville, Illinois; Fourth edition, 1970.

[3]Hagerman, L.; The karakul handbook. Published by the Karakul Fur Sheep Registry, under agreement with the author, 1951.

[4]Hyde, N.; "The fabric of history: wool," National Geographic, Vol. 173, No. 5, May 88, pgs 552–591.

[5]O'Neill, J.; The Karakul. Information sheet for the American Karakul Sheep Registry; Rt. 1, Box 179, Rice, WA 99167; Date unknown.

[6]Ross, M.; the essentials of yarn design. Published by the author; Spinningdale, Crook of Devon, Kinross, KY13 7UU, Scotland, 1983.

[7]Simmons, V.; Comparison of growth in Karakul and mutton breeds of sheep. Journal of Animal Science, Vol. 3, No. 1, Feb. 1944.

[8]Shelton, M., et al. Investigations relating to fat-tail sheep. Texas Agricultural Experiment Station; San Angelo, TX; PR 91, Date unknown.

[9]Young, C. C.; Concerning the fat-tail and the broad-tail sheep. American Breeders' Magazine, Vol. no. and date unknown, pgs 181–200. Article references information about broad-tail Karakuls: American Breeders' Magazine, Vol. ii, No. 1, 1911.

The C. V. M. Breed

Karen Eidman
President, Natural Colored Wool Growers Association
Rt. 1, Box 53
Strong City, Kansas 66869 USA

During the early 1960's a badger patterned ewe lamb was born in the large commercial Romeldale flock of J. K. Sexton. Mr. Sexton placed heavy emphasis on producing high yielding white fleeced wool within his ewes. This funny colored ewe lamb with wool as gray as soot was sorted into the commercial feeder lamb group without question. She was born from a top producing Romeldale ewe, but that made no difference. Down the one-way road to market she was to go.

Mr. Sexton's partner, Glen Eidman, spotted the odd colored ewe lamb and pulled her off the truck and into his own feed yard. The two ranches were several miles apart, so he decided he would keep her hidden, should Mr. Sexton come to visit.

The first year Mr. Eidman bred her to a Suffolk ram, and she produced crossbred speckle-faced offspring. It wasn't until her third lambing that he was able to sneak a badger-patterned ram lamb home to breed to his secret. Surprise! The colored ewe had twin ewe lambs looking just like their mother. The C. V. M. breed was born.

The C. V. M.'s were developed from the Romeldale breed. Romeldales are a breed of sheep begun by A. T. Spencer, whose dream was to raise a white faced breed with wool quality and carcass desirability. In 1915 he purchased the entire New Zealand consignment of purebred Romneys exhibited at the Pan American Exposition in San Francisco. He was sure that this muscular, longwooled breed, when crossed with his Rambouillet flock of ewes, would improve carcass traits, and increase staple length. After many years of rigorous selection and crossing, the Romeldale breed was developed.

During the 1940's and 1950's, the Stone Valley Ranch, owned and operated by the J. K. Sexton family, developed the largest and highest quality Romeldale flock in the world. The sheep were selected for twinning, mothering ability, wool softness, fleece yield and fleece grade of 60's to 64's. All of the wool from this impressive flock has been marketed directly to the Pendelton Woolen Mill for many years.

The new colored Romeldales were christened the C. V. M.'s, following the birth of the first colored twins. The secret was out, so any of the rare colored lambs, born from the white flock of Romeldales, were made at home on the Eidman's ranch. As their numbers increased, Mr. Eidman placed heavy replacement emphasis on the regularity of the color pattern, fleece color, softness and spinnability of the wool, twinning, mothering ability and lambing ease. Throughout the fifteen years he spent developing the breed, not a single breeding animal was sold, so only the highest quality genetics were used to replace the nucleus.

In 1982, Mr. Eidman hosted the "First Ever" C. V. M. Production Sale. He sold the entire herd to over a dozen buyers from throughout the state of California. Within the last seven years, the purebred numbers of C. V. M.'s have remained stable, with approximately seventy-five to one hundred head of mature breeding ewes. Because of the relatively few numbers, breeders became concerned and wanted to protect the numbers currently remaining. The American Minor Breeds Conservancy was asked to help in establishing a registry. Previously, all the C. V. M.'s had been registered with the Natural Colored Wool Growers Association, but a central nucleus of records was deemed necessary. Presently, and still in development, the American Romeldale–C. V. M. Registry is at home in the state of Washington.

The color patterns of the C. V. M.'s vary widely, especially the darkness of the body wool. The typical and original pattern of the C. V. M.'s is the Badger Face. Within the Badger Face pattern, the body wool variegations will range from creme to dark gray or silver. Since the dispersal in 1982, the newly organized flocks have developed and kept new color patterns within the breed. These patterns include: solid self, spotted self, baldy self, piebald black spots, and a limited number of moorits. The color pattern may vary, but when a C. V. M. is bred to a C. V. M., color always results. Current emphasis is still placed on producing a soft handling, high yielding, long stapled fleece, grading 58's to 62's. Fleece colors will darken with age, particularly from the first to second shearing. Twinning and lambing ease are still of utmost priority. In fact, if ewes are left with a ram year-round, they will usually breed while suckling lambs. Rams are aggressive, virile breeders, able to cover more than the standard number of ewes. C. V. M.'s combine the two most talked-about traits: a soft handling, high yielding fleece and a lean, muscular carcass.

Major Coloured Sheep Breeds in Australia

Marree Vinnicombe
'Rochford Park'
RMB 1040
Mitiamo 3573
Victoria, Australia

The beginning of the sheep and wool industry in Australia coincided with the landing of the First Fleet in 1788. After 200 years this industry is well established as a stable and significant national industry.

The Merino more than ever remains the dominant breed and has evolved into four basic breed types. Superfine, Saxon, Peppin and South Australian, each different in many salient features from others and from the original Spanish and other European Merinos that contributed to the early breeding developments in Australia. As Australia has no native sheep, it can be assumed that the early imported sheep into Australia carried the recessive colored gene. Because of this an industry some 180 years later has developed. Prior to this, the coloured sheep or lamb was of little interest. Interest in spinning and weaving had not developed.

It is possible that the first Merino landed from the Cape of Good Hope in 1797. The first flock consisted of 5 rams and 8 ewes and as these sheep bred up and spread across the Australian country, the need for different breeds to suit the different climates and terrain was evident. As these breeds developed and multiplied, the genetic material for the coloured sheep was also spread across the country. This genetic material has been gathered to breed quality coloured sheep in Australia over the last 15 years. Many breeders are now running purebred flocks, with the Corriedale being the most popular breed. Other breeders are using the crossbred sheep and achieving the results of good spinning wools they require.

It is interesting that whilst Merino sheep breeders have been breeding only for white sheep for many decades, coloured sheep do sometimes appear. These coloured sheep, if bred together, produce coloured offspring. The tendency to coloured wool persists regardless of the effort to eliminate it, and the fact that the black and coloured sheep breeds true, leads to the conclusion that the original type of domestic sheep was pigmented.

The breeders of coloured sheep are essentially wool producers, who aim to produce specialty fibres. The type and shade of wool produced will vary widely from fine wool suitable for the finest and softest yarn to coarse wool suitable for rug making. The wool produced will range from silver through the shades of grey, black and brown.

The Australian Merino Types

The history of the Australian Merino goes hand in hand with the growth of Australia as a colony. In the evolution of the Australian Merino sheep, four distinct types have sprung into being. These range from the South Australian drier climate where a strong wool, large framed Merino is found, to a higher rainfall area with better pastures like Western District of Victoria where a fine wool Merino thrives without showing colour deterioration of its fleece caused by rain.

Merino wool provides a compactness and softness in yarn which other breeds cannot produce. These qualities are due to the Merino wool's small fibre diameter and its inherent softness. The Merino has been bred to produce a much heavier cutting fleece, and the frame of the animal has become larger with an increase in staple length of fibre.

Fine wool Merino ewe. Smallest of the Merinos. Notice the white top knot—something that is typical of the black Merinos.

The Saxon Merino

The Saxon Merino (superfine and fine wool type) is a special type of Merino sheep descended directly from Merino flocks from Saxony and Silesia in South-East Germany. The wool produced by these sheep is prized for its softness, color and fineness. The sheep are found in high rainfall areas. These fine-woolled types are distinguished by a comparatively smaller frame. They carry a very dense fleece with length of wool ranging from 51 to 90 mm. The fibre is fine, ranging with a fibre diameter from 17 to 19 microns. The crimps are numerous yet well-defined. The wool has only a small amount of tip, which is generally black, due to the richer condition of the wool, together with the cooler, more moist climatic conditions. The Saxon Merino has the lowest cutting weight for fleece, usually 3.5 to 5 kg. but the quality of this fleece is incomparable: softness, silkiness and brightness.

The Peppin Merino

'The Wanganella' sheep stud was established by the Peppin brothers near Deniliquin in the Riverina in 1861. Merinos of both Spanish and French origin were introduced. The influence of a single French 'Rambouillet' ram, called Emperor, had a great bearing on this type of Merino. About 70% of todays Australian Merinos are said to be directly descended from the Peppin developed sheep. The breed prefers a much drier country, where its large frame and long legs make it an efficient forager. Its face and horns will also have a more robust growth. A heavy producing fleece which falls in the mid range of Merino wool qualities: 20–21 microns with wool length varying from 90–102mm. This wool rarely carries as much condition as the fine wool, nor is it quite as soft in texture.

South Australian Merino

The South Australia Merino (strong woolled type) was specifically bred to thrive in the arid pastoral condition. It is the biggest of the strains of Merino in Australia, generally being longer, taller and heavier than the other types. The wool is the strongest. It carries a higher proportion of natural grease, has an open face, plain body and produces a heavy cutting fleece around 5.5 to 18 kg. Length of fibre is around 100 mm and fibre diameter of 23–25 micron.

If there are so many Merinos in this country, then what is its wool used for? From the finest of baby wear and shawls using the fine Merino wool to the fine wear jackets and fashion garments that show the softness that is associated with wool of the fine micron Merino. Fine cloth for men's suits and evening wear for women are all the end products of this wool.

Within the coloured sheep industry, although the coloured Merino appears in higher numbers, its popularity is not all that great. The sale of this fine fibre does not fetch the highest prices from the Spinners and Weavers. Many handspinners say the wool is too short and fine and is too sticky. Yet the commercial buyers are prepared to pay higher prices for the coloured Merino than any other type of wools. This has been seen in the last three auction sales in Melbourne Victoria conducted by Websters for The Black and Coloured Sheep Breeders Association.

The development of The Poll Merino is relatively new. Recessive poll genes are believed to have existed in the breed for many years and infusions of hornless sheep during the development of the Merino bred in Australia also left some poll genes within normal Merino flocks.

Whilst precedence has been given to the Merino it does not by any means follow that this breed of sheep should be the only one to be considered. The more our country is developed, the more inducement there will be to keep sheep of the British Breeds and their crosses.

British Long-wool Breeds

The British Breeds of sheep may be divided into 3 main divisions, viz the Long-Woollen group, the Short-Woollen group and the Mills Breeds. I will limit my discussion to the Long-Woollen group.

As the name implies the distinctive feature of the different component breeds is their comparatively long-stapled fleeces, ranging from 180–300 mm for a 12 month growth. In the main, they are large framed, hardy and carry a heavy fleece of relatively strong quality wool. The colour is also distinctive: it is more or less lustrous, whilst the character is generally wavy or curly. The markings are also peculiar to the group: they have, in most instances, flat or coloured faces and legs (in the coloured sheep). The hooves are in all cases black.

Lincoln

The Lincoln breed is claimed to be the parent of all modern long wool breeds. The first Lincolns were imported to Australia in 1880. The old Lincoln was crossed with the English Leicester, then known as Dishley Hall breed. Lincoln rams were mated with English Leicester ewes and English Leicester rams were mated with Lincoln ewes. The progeny was crossbred and by careful selection, the present day Lincoln was evolved and was first recognized as a purebred in 1862.

The Lincoln is a big sheep with great weight and substance. Liveweight mature sheep-rams, 120 kgs, ewes 90 kgs. Lincoln ewes cut an average of 8–9 kgs of fleece per head with rams shearing 14–15 kgs per head. No breed of sheep can match the Lincoln for length of staple, whilst as regards weight, it is only excelled by the Merino.

Lincoln wool is highly valued for its length of staple, lustre and resilience which imparts special properties, a thick durable fibre mainly used by weavers for rugs and wall hangings. It is the strongest woolled of the British Long Wools used in Australia, with long, deep, wide locks of wool, and a forelock of evenly-waved wool hanging over the face. The head is rather massive, entirely free from horns and is broad between the ears. The wool must be free from kemp and not less than 180 mm long for 1 years growth, measuring 35–38 microns. The Lincoln is found in high rainfall area with lush pastures.

The English Leicester

A native sheep of the English Midlands, Leicesters have been in Australia since 1838. It actually is the only breed that has not been outcrossed to produce an improvement in size and conformation. The English Leicester has a top knot instead of a forelock, face rather wedge shaped, locks of wool rather thin and spiral. There are no signs of horns. The face is dense, free, even and lustrous, locks medium width, showing small, well defined wave or crimp from skin to tip. Fleece weight should be 5 to 6 kg, length 200–250 mm with a suggested wool count 32–38 micron.

The Border Leicester

The Border Leicester is an English long wool breed, and ewes were first introduced into Australia in 1871. They are the lineal descendants of the Dishley Leicesters. The Border Leicester differs from the English Leicester in many respects: the head differs both in shape and character, its head is more bony and cleaner—the face is decidedly Roman with a full muzzle and wide nostrils. The ears are carried erectly and the head is carried high and actively, giving the sheep a most alert appearance. The head is free from wool and horns.

The wool is a lustre type, most attractive and stylish. It is soft handling. The fiber measures 32–34 micron with a fleece weight lighter than the English Leicester, owing chiefly to the lack of wool on the legs and head. No kemp or coarse fibres in the fleece are allowed.

As the Border Leicester has soft handle wool it is used for knitting yarns and apparel.

The Romney Marsh

The Romney Marsh or Kent sheep has been bred from time immemorial. The first record of importation of this breed to Australia took place in 1872. Romneys have been found to thrive under conditions that would be regarded as being quite unsuitable for most kinds of sheep. The breed is a typical English long wool that is well adapted to the temperate higher rainfall areas. The Romney Marsh is a polled breed, large in frame. Wool is long, at about 200 mm, with a fibre diameter of approximately 33 microns. It is lustrous and soft and handles well, with a weight of 4 to 5 kg per fleece.

The main use at present is to provide the base ewe flock for a newly emerging carpet wool industry. Apart from its hardiness it produces a bulky, heavy fleece of good lengthy wool, bright colour and free from kemp. The crimp should be fairly well defined.

This wool is again used for knitting yarns, rugmaking and wall hangings.

The Perendale

The Perendale is the progeny of second and subsequent crosses of Cheviot/Romney Cross sheep developed in New Zealand in the late 1950's. This sheep was introduced by Massey University in response to the demands for a sheep that would give higher lambing percentages on the drier hill country of the North Island of New Zealand. The Perendale is particularly hardy, very active and does well on poorer type feed. The Perendale is a medium sized sheep, has an open face, free of wool below the eyes. A medium top knot is desirable but not essential. The clean face shows unmistakably the Cheviot ancestory. The Romney is represented in its type of fleece and to a lesser extent in its size. It produces a fleece of high yield and bulk which is keenly sought by many different users, particularly carpet users.

The Perendale fleece has a length of 100–125 mm and a quality from 28–33 micron. The fleece is well defined, has limited lustre, handle should be full, springy and fairly soft, free from hair, kemp and rust. The Perendale wool is sometimes used where a thicker knitting yarn is required. It is used also in woollen upholstery fabrics.

Australian Breeds of Sheep

We now come to these sheep which are of Australasian origin. They are the Corriedale, the Polwarth and the Bond. These breeds were evolved by Australian breeders to suit conditions in particular localities; the Corriedale was established as a purebred in New Zealand, whilst the Polwarth was established in Victoria and the Bond originated at Lockart in New South Wales.

The Corriedale

The Corriedale was simultaneously evolved in both Australia and New Zealand about 1874 by selectively breeding from crossbred progeny of pure Merino and Lincoln sheep; the breeding programme took seven generations to establish the Corriedale breed. The original Australian Corriedale sheep were referred to as Moorale Sheep and the official history of the breed in Australia started in 1882.

The Corriedale has a very heavy frame, symmetrical conformation, good mutton qualities and is a prolific breeder. Its wool production is heavy, with a fine to medium crossbred quality wool. The spinning quality is good, and when manufactured, the wool possess good, hard wearing qualities coupled with fineness and soft handling. Fabrics from it compare closely with those made from fine wools.

The modern Corriedale is a large framed sheep, well woolled yet a plain bodied polled sheep with a large top knot.

Corriedale ram showing a good fleece and leg cover with the traditional top knot.

The fleece is bright with good style, length and handle, with a fibre diamenter range from 25–30 microns. This may be as low as 20 micron in lambs and hoggets.

The Corriedale is bred in a wide range of conditions, its popularity ranking second only to the Merino.

The Polwarth

This breed is purely Australian and mainly found in Victoria, Tasmania and Southern parts of New South Wales. It was developed in 1880 in the Western District of Victoria at a place called Tarndwancoort near Colac by the Dennis family. It was the (extreme) climate and wet underfoot for much of the year that persuaded Richard V. Dennis to cross a Lincoln ram with Merino ewes.

The Polwarth is a large sheep similar to the Merino in appearance, but without the neck development. The sheep is really an inbred quarter-bred sheep of Lincoln-Merino foundation stock.

Polwarth ram with very dense fleece carried well down on the belly and points.

They are a large framed robust sheep, producing a high yielding, soft handling fleece of 22.25 micron fibre diameter. The rams can either be horned or hornless and this is a breeders preference.

The wool shows the character of the strong Merino style but long (12 mm) dense of 22.25 micron. It also shows a good colour throughout the fleece.

The Polwarth wool has been used extensively by spinners, but a mill in Victoria near Bendigo known as the 'Mohair Farm' uses 85% Polwarth wool. This is blended with mohair to produce yarns and fabrics, shawls and pullovers to sports coats. The wool is used as it has a good handle and character, more lustre than fine wools but not the harsh touch of coarser wools.

The Bond

This fixed type, originally famous as the 'Commercial Corriedale' originated at Lockart in New South Wales in 1909 when Thomas Bond mated his Saxon/Peppin Merino stud ewes to stud Lincoln rams and subsequently selected progeny to better suit the harsh Riverina environment. Bond sheep produce a high fielding, very long stapled, bright, 23–27 micron wool of excellent style with a soft end tip. The outstanding feature of the wool is the handle. The average fleece weighs 6–8kg. In recent years Bond sheep have been selected to handle all types of climatic conditions with rainfall variations of 14–45 inches.

Conclusion

Whilst many other breeds of sheep are bred successfully in Australia, and their wool used for the craft industry, I must stop at what I feel are major wool breeds found amongst members of The Black and Coloured Sheep Breeders Association.

Over the last 5 years an increasing interest in carpet wool sheep has sprung into light. The berber carpet trade and the normal carpet trade (which were once dominated by the Lincoln and Perendale breeds) have used most of this wool.

As years go by the fashions change and so does the interest in coloured sheep breeds. Some 10–15 years ago the Lincoln was a popular breed. This changed to the Border Leicester and the Romney. With the interest in spinning and weaving coming to a pinnacle in the last couple of years, and these craft persons becoming more experienced in their trade, the demand for finer wools became popular. The Corriedale, Polwarth and strong wool Merino are now popular breeds producing fleeces with soft handle, good colour and style.

References

D'Archy, J. B.. Sheep Management & Wool Technology.
Harmsworth, T. B. & Day, G. Wool & Mohair.
Harmsworth, T. B., & Sharp, J. P. Sheep & Woolclassing.

Pigmentation in the Australian Merino

C. H. S. Dolling
Sheep Breeding Consultant
P. O. Box 74,
McLaren Vale, South Australia 5171, Australia

1. Genetic Nomenclature

The names and symbols of the colour genes in sheep did cause some confusion to coloured sheep breeders at the 1979 Congress in Adelaide and the 1984 Congress in New Zealand. So a Committee on Genetic Nomenclature of Sheep and Goats, COGNOSAG, was formed at Palmerston North, New Zealand, in 1984.

Three workshops of COGNOSAG have been held in 1986, 1987 and 1988 at Les Deux Moulins, Gontard, Provence, France. Significant inputs have been made by COGNOSAG members from Iceland, Scotland, England, the U.S.A., France, Italy, Bulgaria, the U.S.S.R, New Zealand and Australia.

Names and symbols for the genes affecting coat colour in sheep have been agreed upon by COGNOSAG at the July 1988 workshop. Here is a list of the gene or allele names agreed upon, together with the names and symbols which are currently used in Australia to describe those alleles for which genetic evidence has been reported from Australian flocks.

TABLE 1 Loci and Alleles for Coat Colour in Sheep. As Agreed at COGNOSAG Workshop at Gontard, France, in July 1988

COGNOSAG Names and Symbols				Names and Symbols in use in Australia	
Locus		*Allele*		*Allele*	
Name	*Symbol*	*Name*	*Symbol*	*Name*	*Symbol*
Agouti	A	white or tan	A^{Wt}	white	W
		badger face	A^{b}	badger face	w_1
		black and tan	A^{t}	reversed badger face	w_2
		grey	A^{g}	No genetic evidence has been reported in Australia for the presence of these alleles.	
		light badger face	A^{lbf}		
		blue	A^{bl}		
		light blue	A^{lbl}		
		grey and tan	A^{gt}		
		light grey	A^{lg}		
		gotland grey	A^{gg}		
		swiss	A^{s}		

Continued . . .

TABLE 1 Loci and Alleles for Coat Colour in Sheep *(Continued)*

COGNOSAG Names and Symbols				Names and Symbols in use in Australia	
Locus		*Allele*		*Allele*	
Name	*Symbol*	*Name*	*Symbol*	*Name*	*Symbol*
Agouti (*con'd.*)		lateral stripes eye patch wild non agouti	A^{ls} A^{ep} A^{+} A^{a}	No genetic evidence has been reported in Australia for the presence of these alleles.	
				solid self baldy self spotted self	w_3 w_3 w_3
Albino	*C*	wild albino	C^{+} C^{c}	No genetic evidence has been reported *in Australia.*	
Black spots	*Sp*	wild black spots	Sp^{+} Sp^{sp}	white piebald	*Sp* *sp*
Blue		No recommendation; insufficient evidence.		No genetic evidence has been reported in Australia.	
Brown	*B*	wild brown	B^{+} B^{b}	black or grey *moorit* (Includes the colours of chocolate, cinnamon, fawn.)	*B* *b*
Extension	*E*	dominant black blackish brownish yellowish wild	E^{D} E^{Bl} E^{Br} E^{Y} E^{+}	No genetic evidence has been reported in Australia.	
Pigmented Head	*PH*	afghan lethal turkish persian wild	PH^{Al} PH^{T} PH^{P} PH^{+}	No genetic evidence has been reported in Australia.	
Red	*Rd*	No recommendation; insufficient evidence.		No genetic evidence has been reported in Australia.	
Roan	*Rn*	lethal roan wild	Rn^{Rn} Rn^{+}	No genetic evidence has been reported in Australia.	
Spotting	*S*	wild spotted	S^{+} S^{s}	No genetic evidence has been reported in Australia.	
Sur Bukhara	*SuB*	wild bukhara	SuB^{+} SuB^{s}	No genetic evidence has been reported in Australia.	
Sur Surkhandarya	*SuS*	wild sur surkhandarya	SuS^{+} SuS^{+}	No genetic evidence has been reported in Australia	
White Collar	*WC*	No recommendation; insufficient evidence.		No genetic evidence has been reported in Australia.	
White Saddle		No recommendation; insufficient evidence.		No genetic evidence has been reported in Australia.	
White Spot Modifier	*J*	No recommendation; insufficient evidence.		No genetic evidence has been reported in Australia.	

2. Patterns

In considering the types of pigmentation in Australian Merino sheep, we shall see how the genotype of a solid self black sheep (A^a/A^a; or w_3w_3) can be changed or modified to produce lighter fleeces; and how the genotype of a white sheep (A^{wt}/A^{wt}; or *WW*) can be modified to produce several different types of pigmentation.

a. *Black sheep, fleece lightened by alleles on the* Agouti *locus* (Brooker and Dolling 1965, 1969b; Dolling 1979).

$A^a/A^a(w_3w_3)$. A solid self; black or grey wool all over the body.

$A^t/\text{-}(w_2\text{-})$. Reversed badger face; non-pigmented wool on the belly, with a sharp change from non-pigmented wool on moving from the belly to the flank and side.

$A^b/\text{-}(w_1\text{-})$. Badger face; lightly or non-pigmented wool on the body, darkly pigmented wool on the belly, with a sharp change from one to the other.

A small percentage of lambs born show characteristics of more than one of the three patterns, most of them fitting the phenotype one would expect by super-imposing any pattern on any other and assuming that the darker portion of each patterns predominated over the lighter portion of the other.

b. *Black sheep, fleece lightened by unidentified alleles* (Brooker and Dolling, 1969b; Dolling 1979).

A^a/A^a. Baldy self; a self colour of black or grey over the body, but with a white poll and sometimes any or all of a white dewlap, white on the lower legs and white tail tip.

A^a/A^a. Spotted self; a baldy self with areas of the body, also, not pigmented. Pigmented areas of the body are, in the main, circular or eliptical; the extreme is a sheep with an apparently white fleece all over but with circles of pigmented skin and hair around the eyes and the muzzle in a symmetrical fashion.

c. *Black sheep of baldy self or spotted self patterns.* A^a/A^a. A pigmented multiple spotting over the white area of either the baldy self or spotted self patterns (please refer to photographs and amplification below).

d. *White sheep, the fleece changed by a gene not on the* Agouti *locus.* (Brooker and Dolling, 1969b; Dolling 1984).

$A^{Wt}/\text{-}\ Sp^+/Sp^+$ and $A^{Wt}/\text{-}\ Sp^+/Sp^{sp}$. No black spots of skin and wool in the wool-growing area.

$A^{Wt}/\text{-}\ Sp^{sp}/Sp^{sp}$. One or more rounded areas of pigmented skin and wool in an otherwise white coat. The location of spots, on what has been termed the piebald sheep in Australia, does not conform to any recognisable pattern. Spots may be in the hair-growing areas.

The *Agouti* locus is not involved in the transmission of this type of black spots; the recessive allele Sp^{sp} hypothesised to explain their inheritance penetrance of approximately 25%, and varies between sheep.

e. *White sheep with pigmented skin spots bearing pigmented fibres—appearing with age.* (Kelley and Shaw, 1942; Fleet and Forrest, 1984).

$A^{Wt}/\text{-}$. Skin spots are black or grey and often give rise to a sparse population of black or grey wool fibres which are not usually identified at shearing.

The incidence of these spots increases with age and with the number of shearings and consequent successive doses of U/V radiation on the skin of the back.

From the test mating of 10 black or piebald rams to 8.5 year old ewes with pigmented spots appearing with age, it has been concluded that there is no association between the inheritance of recessive black or of piebald with the occurrence of spots of pigmented wool fibres on the ewes (Fleet *et al.*, 1985).

f. *White sheep with fleeces with isolated pigmented fibres.* (Fleet, Stafford and Dolling, 1984; Fleet, 1985).

$A^{Wt}/\text{-}$. Isolated pigmented fibres in white fleece wool have been described in young Corriedale and Merino sheep. The mean diameter of the pigmented fibres (28 microns) is not much greater than that of the wool samples of Corriedales from which they are extracted (25.6 microns).

It has been suggested (Adalsteinsson, Dolling, and Lauvergne, 1980) that selection for a pigmented nose in the Corriedale may be contradictory to selection for a white fleece. A degree of support for this hypothesis was recorded by Fleet, Stafford and Dolling (1984).

The ability to produce isolated pigmented fibres in the Merino is transmitted to some degree from parent to offspring, but a satisfactory explanation of the mode of inheritance is lacking (Fleet, 1985).

On Corriedale sheep, the types of pigmentation on the non-wool-growing area which may be useful indicator traits of isolated pigmented wool fibres are pigmented fibres on the horn sites and spots of pigmented fibres on the legs (Fleet, Stafford and Dawson, 1985).

g. *White sheep with tan spots in the hair of the face and/or legs.* (Brooker, 1968).

$A^{Wt}/\text{-}$. A flock of 1.5 year old medium Peppin white Merino ewes was scored for presence of tan hairs on the legs (Brooker, 1968).

The flock consisted of three large and sixteen small closed selection groups, all breeding replacements coming from within each group. The selection groups and the number of ewes with tan hairs on the legs are shown in Table 2.

Individual sheep in the +Body weight group were characterized by a high incidence of tan spots in the hair of the face and legs.

h. *White sheep with tan hairs in the birthcoat.*

In a number of $A^{Wt}/\text{-}(W\text{-})$ white Merino lambs, long tan hairs grow in an area on the back of the neck, or elsewhere in the birthcoat.

TABLE 2 Numbers of 1.5 Year Old White Merino Ewes with Tan Hairs on the Legs (Brooker, 1968)

Selection Group	*Number of Ewes*	
	Total	*With Tan Hairs on Legs*
Control	50	8
S*#	51	3
MS*#	48	0
+ Bodyweight	10	5
− Bodyweight	7	1
+ Fibre Number	10	0
− Fibre Number	9	0
+ Staple Length	7	0
− Staple Length	7	1
+ Clean Wool Weight*	6	0
− Clean Wool Weight	10	0
+ Wrinkle Score	8	0
− Wrinkle Score*	10	2
+ Fibre Diameter	9	0
− Fibre Diameter*	9	1
+ Clean Wool/Unit Area*	8	0
− Clean Wool/Unit Area	8	0
+ Clean Scoured Yield*	9	1
− Clean Scoured Yield*	8	0
Total	284	22

*At least one Black (ww) lamb born in this group.
#Selected on high clean wool weight (Further details, see Turner, 1958).

The growth of these hairs is such that they are eventually shed. At first adult shearing, however, one can still find pigmented fibres, even though the lamb birthcoat spot is not evident (Fleet, personal communication).

3. Colours

a. *Brown Sheep.*
The presence of a recessive allele producing brown as opposed to black or grey wool was demonstrated in a series of matings in the Australian Merino by Gregor, Bennett and Bennett (1984). The B^b/B^b *(bb)* sheep—which they termed moorit—grew chocolate brown, cinnamon or fawn wool, and completely lacked black or grey wool or hair fibres.

In the white Merino flocks which produced the black or grey ewes joined in the matings, the frequency of *b* would have approximated 0.1. One would therefore expect that 99% of the A^a/A^a (ww) sheep born in these white flocks would be black or grey and that 1% would be moorit.

4. Melanocytes in Merinos

The three pre-requisites for the production of pigmented wool fibres—melanocytes being near the papilla, producing melanin, and transferring melanin granules to the fibre—were summarised at the 1984 Congress (Dolling,1984).

In Merino sheep which are A^a/A^a (w_3w_3), melanocytes are localised in three regions of the skin—the epidermal-dermal border, the outer root sheath and the follicle bulb (Forrest, Fleet and Rogers, 1985).

Melanocytes in these regions were found to be actively producing melanin, had numerous dendritic extensions and were able to transfer melanin to adjacent keratinocytes.

In white Merino sheep, in both A^{Wt}/A^a and A^{Wt}/A^{Wt} animals, melanocytes were observed only at the epidermal-dermal border. They appeared also to contain less melanin, to be less dendritic and to have a reduced ability to transfer melanin granules to adjacent keratinocytes.

A^{Wt} (W) in the Merino appears to regulate pigmentation, at least in part, by controlling the location and activity of melonocytes in the wool-bearing skin (Forrest, Fleet and Rogers, 1985).

5. Pigmented Multiple Spotting

Pigmented multiple spotting (pims) is a phenotype which has been observed in coloured Australian Merino sheep. To date, it has been observed only on the white-woolled parts of the fleece of A^a/A^a (w_3w_3) sheep with either the baldy self or the spotted self pattern.

The pigmented spots are of the skin and the wool fibres growing from the skin spot. The spots, up to 1 cm or so in diameter, extend over the white wool of these patterns. Those seen to date have been black, not chocolate or cinnamon or fawn.

Pims on the wool-growing area has not been recorded on sheep in white Australian Merino flocks. Pims differs in appearance from black spots (piebald) in that the latter has many fewer spots over the body and the spots vary considerably between themselves in size. It differs from pigmented spots appearing with age in that the latter become evident at 5 years of age or so.

The photographs 1 to 5 are of pims on a spotted self castrated male at 7 months of age, with 2 months' wool growth. The large black spots and the black neck and belly are part of the spotted self pattern—they are not manifestations of black spots (piebald).

Photograph 6 is of pims on the restricted area of white wool on the head of a 7 month old ram with the baldy self pattern. The sheep photographed are half-brothers whose sire is of a baldy self pattern with pims, like his ram offspring.

It is hypothesised here that pims result from the action of a gene which can express itself in the white-woolled areas of $A^aA^a(w_3w_3)$ sheep, but not in the white fleeces of $A^{wt/-}$ (W) sheep.

Although the gene is hypothetical at present, it has been given a name and symbol in accordance with the guidelines of COGNOSG. The phenotype symbol for pigmented multiple spotting is *pims*; the allele name is *pigmented multiple spotting*; the allele symbol is *pims*; and should the allele be on a locus other than *Agouti*, that locus would be *PiMS*.

If *pims* is on a locus other than *Agouti*, it would be segregating independently of the *Agouti* alleles; and it would remain to be seen if it could express itself in the white fleece wool of badger face sheep or the white belly wool of reversed badger face sheep. The gene would always be 'imprinting' its effect on one or another of the *Agouti* alleles other than A^{Wt}; it would be epistatic to these alleles, hypostatic to A^{Wt}.

If the gene were on the *Agouti* locus it would be dominant to $A^a(w_2)$. It would remain to be seen if it should be dominant to $A^b(w_1)$ or to $A^t(w_2)$. Furthermore, a A^{pims}/A^{pims} sheep would be expected at some time, with pims over the whole of the otherwise white wool-growing area.

That the *pims* observed to date have been black simply reflects that it has been seen in B^+/B^+ or B^+/B^b (BB or Bb) sheep only. One would expect pims to appear one day in brown B^b/B^b (moorit bb) sheep, producing chocolate or cinnamon or fawn spotting.

I encourage breeders to consider if pims exists in their flocks; against which genetic background it exists and how it could be inherited; and if a choice could be made as to whether the hypothetical gene *pims* is on the *Agouti* locus or on another locus.

References

Adalsteinsson, S., Dolling, C. H. S., and Lauvergne, J. J. (1980). Breeding for white colour sheep. *Agricultural Record* 7:40.

Brooker, M. G. (1968). Pigmentation in the Australian Merino. M. Sc. Agr. thesis, University of Sydney, Australia.

Brooker, M. G. and Dolling, C. H. S. (1965). Pigmentation of sheep I Inheritance of pigmented wool in the Merino. *Australian Journal of Agricultural Research* 16:219.

Brooker, M. G. and Dolling, C. H. S. (1969a). Pigmentation of sheep II The inheritance of colour patterns in black Merinos. *Australian Journal of Agricultural Research* 20:387.

Brooker, M. G. and Dolling C. H. S. (1969b). Pigmentation of sheep III Piebald pattern in Merinos. *Australian Journal of Agricultural Research* 20:523.

Dolling, C. H. S. (1979). The genetics of coloured sheep in Australia. *Proc. Congress on Breeding Coloured Sheep and Using Coloured Wool*, Adelaide, South Australia. Peacock Publications pp. 42–50.

Dolling, C. H. S. (1984). Spotting in Merinos. *Proc. Congress on Coloured Sheep and their Products*, New Zealand pp. 232–236.

Fleet, M. R. (1985). Pigmented fibres in white wool. *Wool Technology and Sheep Breeding* 33:5.

Fleet, M. R., Flavel, P. F., Stafford, J. E., Smith, D. H., and Dolling, C. H. S., (1985). The incidence of pigmented lambs among the progeny of pigmented rams and Merino ewes which have a high frequency of individuals with spots of pigmented fibres. *Agricultural Record* 12:23.

Fleet, M. R. and Forrest, J. W. (1984). The occurrence of spots of pigmented skin and pigmented wool fibres in adult Merino sheep. *Wool Technology and Sheep Breeding* 32:83.

Fleet, M. R., Stafford, J. E., and Dawson, K. A. (1985). Association between non-wool pigmentation and the occurrence of isolated pigmented wool fibres in fleeces from Corriedale sheep. *Proc. 5th Conference Australian Association of Animal Breeding and Genetics*. Sydney, pp. 220–222.

Fleet, M. R., Stafford, J. E., and Dolling, C. H. S. (1984). A note on the occurrence of isolated melanin pigmented fibres in the white fleece wool of Corriedale sheep. *Animal Production* 39:311.

Forrest, J. W., Fleet, M. R., and Rogers, G. E. (1985). Characterisation of melanocytes in wool-bearing skin of Merino sheep. *Australian Journal of Biological Sciences* 38:245.

Gregor, N. R., Bennett, A. R., and Bennett, N. P. (1984). The moorit gene in Australian sheep. *Agricultural Record* 11:19.

Kelley, R. B., and Shaw, H. E. B. (1942). A note on the occurrence and inheritance of pigmented wool. *Journal of Council for Scientific and Industrial Research* 15:1.

Turner, H. N. (1958). Relationships among clean wool weight and its components I Change in clean wool weight related to changes in the components. *Australian Journal of Agricultural Research* 9:521.

Pims on spotted self wether.

Pims on spotted self wether.

Pims on right side of backline of male.

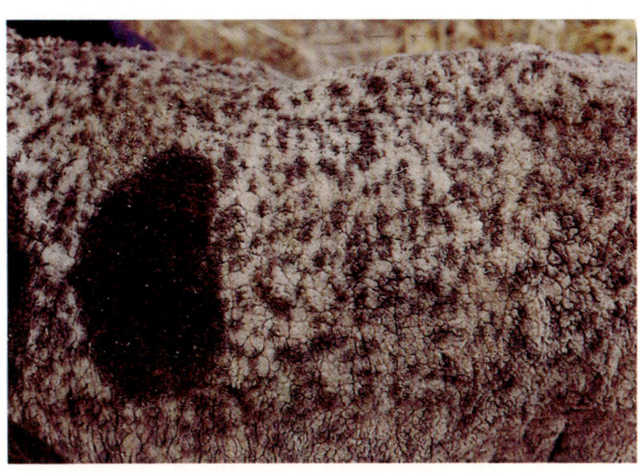

Pims on left side of backline of male.

Pigmented Multiple Spotting.

Pims on poll of baldy self ram.

Top—Rugged ewe and lamb.

Middle—Samples from unrugged fleeces.

Bottom—Samples from rugged fleeces.

From the article "The Rugging of Sheep in Australia" on page 191.

Status of Shahabadi Sheep in West Bengal

D. N. Maitra
Professor of Animal Production & Management
Faculty of Veterinary & Animal Sciences
Bidhan Chandra Krishi Viswavidyalaya
West Bengal, India

West Bengal, a state of north-eastern India being hot and humid, is not considered to be the ideal location for sheep husbandry. However, the south-west portion of the State has many alluvial depressions and mountain ranges where a particular type of sheep have adapted through generations. A majority of the species in the country do not belong to any specific breed, since most of them lack in specified common characters. There are no breeding societies or agencies to register animals of a particular breed or to maintain flock books or ensure purity of breed. A population of sheep in a given locality, with characters distinct from other populations in the vicinity and with a distinct local name, has usually been considered as a breed. Such a type is the Shahabadi, named after the place of origin in the adjoining State of Bihar. Very little information is available about the breed or the breed-type.

The habitat of the breed is very harsh. Climate, soil and other features are very unfavourable. It occupies 20.2 percent of the state's area having human population density of 365 per Km^2. Organic matter in the soil is low and pH ranging between 5.5–6.5. Due to undulating terrain, soil condition, and lack of irrigation, the area is agriculturally ill developed. The community adheres to the traditional nomadic husbandry practices. In the present investigation an attempt was made to objectively assess the breed for production potential under free range conditions.

Physical Feature and Biometry of the Breed

It is a medium sized breed. Usual coat colour is white or grey, sometimes with black spots which may be scattered over the various parts of the body including face, forehead and tail. They have long legs, distinct roman nose, long and thin tail of about 35.25 ± 0.32 cm in length. Ears are medium sized and drooping. Both sexes are polled. Fleece is extremely coarse and hairy. Legs and belly are devoid of wool.

It was found that the animals could not maintain progressive increase in weight throughout the year irrespective of physiological status. However, the overall performances could not be designated unsatisfactory. It may be presumed that the animals could have recorded steady increases in body weight provided the nutrition from the field would have been optimum.

Birth Weight and Growth Rate

The birth weight and the rate of growth at different stages are presented sexwise in Table 2. Average weight at birth of the breed has been found to be 2.70 kg. At birth, the weight

TABLE 1 Biometry of the Breed

	Adult Male [15]		Adult Female [136]	
	Average	Range	Average	Range
Body weight	30	21–35 ± 2.51	27.6	21–30 ± 2.29
Body Length (cm)	41.5	38–45 ± 0.46	46.0	45–55 ± 0.46
Height at wether (cm)	51.5	46–57 ± 1.62	48.0	45–55 ± 1.20
Chest girth (cm)	80.0	75–85 ± 1.92	76.8	75–80 ± 1.15

Figures within brackets indicate number of observations.

TABLE 2 Body weight of Lambs at Different Stages of Growth (kg)

	Male	Female	Average
Birth weight (kg)	2.80	2.60	2.70
Weaning (90 days)	11.20	11.40	11.30
Gain in weight (kg) during the period	8.40	8.80	8.60
Gain in weight per day (gm)	93.3	97.77	95.53
Weight at 180 days (kg)	15.60	16.80	16.2
Gain in weight (kg)	4.4	5.40	4.9
Gain in weight per day (g)	48.88	60.00	54.22
Weight at 12 months (kg)	27.80	23.88	25.84

of the male was higher (2.80 kg) as compared to female (2.60 kg). Average daily gain in weight during the period (0–90 days of age) was 95.53 g and that between 90–180 days was 54.22 g. The female had higher growth rate during the period but the differences were found to be statistically insignificant.

Reproductive Performance

Reproductive performances as indicated by age at first freshening, freshening weight and inter-freshening intervals are given in Table 3.

Greasy Fleece

The greasy fleece production of the type is negligible ranging from 250 gm to 300 gm per year. The coat is a mixture of wool, hair and kemp. The medullation percentage was 84–85 and that of average fibre diameter was 46μ. The yield was far less as compared to the findings of other workers.[5, 14]

Livestock agriculture in this dry land is the primary occupation for the economically depressed, but the existing nutritional status appears to be the major constraint. The present experiment revealed that even on the natural pasture the animals gained body weights in most of the months of the year. There were some periods when the animals lost body weights. This was due to inadequate availability of nutrients from the pasture. It is true that the natural pasture under the present experimental conditions failed to meet the nutrient requirement.[11] The situation can be improved by adequate improvement of grassland or by reducing the stocking rate of which the former is desirable at least in definite seasons. The condition is expected to improve further with additional supplementary feeding. The breed, so far the productive and reproductive capabilities are concerned, can not be underestimated. There is a lot of variability among the animals in terms of physical features. Culling of animals below a certain standard may not also be considered effective as the farmers still value increased flock size as most viable, irrespective of productive ability.

TABLE 3 Some Features of the Breed

1. Age at 1st freshening (days)	622 days ± 26
2. Body wt. at 1st freshening (kg)	26.2 kg ± 0.82
3. Freshening interval (days)	405 days ± 15
4. Foot steps (cm)	38.4 ± 1.21 [5]
5. Average distance travelled (Km)	11.500 [5]
6. Annual greasy fleece (g)	50–296 ± 16.28
7. Medullation (%)	84–85
8. Average fibre diameter (μ)	46 ± 9.06

Figures within brackets indicate numbers of observations.

References

[1]Acharya, R. M., Swain, N., Jain, P. M., Mishra, R. K., Mathur, P. B., Sharma, M. M. and Mudgal, R. D. (1980). Final report—Relative productivity of sheep and goat on free range grazing/browning management on semi-arid range lands of Rajasthan. CSWRI.

[2]Acharya, R. M. (1982). Sheep and Goat breeds of India. FAO Animal Production & Health Paper, 30, Rome.

[3]Anderson, D. M., and Kothman, M. M. (1977). Monitoring animal travel with digital Pedometer. *J. Range. Manage. 30,* 316–317.

[4]Bhattacharya, A. N., and Hussain, F. (1974). Intake and utilisation of nutrients in sheep fed different levels of roughages under heat stress. *J. An. Sc. 38,* 877–886.

[5]Bhatt, P. P., and Bhat, P. N. (1981). Sheep. In animal genetic resources in India, pp. 45–46. Ed: Sachdeva, K., NDRI, Karnal.

[6]Canfield, R. M. (1941). Application of line interception method in sampling range vegetation. *J. For 19,* 388–394.

[7]Hafez, E. S. E. (1953). Ovarian activity in Egyptial (Fat tailed) sheep. *Caira Fac. Agric. Bull.* No. 34.

[8]Hill, F. W., and Anderson, O. L. (1958). Comparison of metabolisable energy and productive energy determination with growing chicks. *J. Nutr. 64,* 587–603.

[9]McDowell, R. E. (1972). Problems of forage production in warm climates—Nutritive value of tropical forages. In *Improvement of Livestock Production in Warm Climates,* p. 127. W. H. Freeman and Co., San Francisco.

[10]Mishra, H. R., Prasad, R. J., Biswas, S. C. and Jaruhar, H. K. (1970). A study of breed characteristics of Shahabadi strain of sheep in Bihar. *Bihar J. of Vet. Sci. and Animal Husb. 1,* 20–27.

[11]Patnayak, B. C. (1981). Research in India on nutrient requirements for sheep. Seminar on Indian Society for Sheep and Goat Production & Utilisation, pp. 1–18. CSWRI, 10–14th April, 1981.

[12]Sahni, K. L., and Roy, A. (1967). A study on the sexual activity of Bikaneri sheep. (*Ovis Aries* L.) and conception rate through artificial insemination. *Indian J. Vet. Sci. 37,* 327–334.

[13]Sharma, V. V., and Murdya, P. C. (1974). Utilisation of Berseem hay by ruminants. *J. Agri. Sci. Camb. 83,* 289–293.

[14]Swain, N., Maitra, D. N., Monohar Singh and Acharya, R. M. (1986). Relative performance and economics of sheep and goat under production remuneration managemental system. *Indian J. Anim. Sci. 56,* 597–605.

[15]Taneja, G. C. (1966). Fertility in sheep I. Conception rate in relation to lamb survival. *Indian Vet. J. 43,* 727–734.

Sheep Breeding and Local Sheep Breeds in Bulgaria

Tzeno Hinkovski, Prof. Dr.
Agricultural Academy, Sofia, Bulgaria

Snejana Alexieva, Ph.D
Institute of Animal Husbandry, Kostinbrod, Bulgaria

At the end of World War II Bulgaria had a large sheep population (12 million) mainly consisting of local aboriginal breeds with coarse, untrue wool. All sheep were used for milk production—a practice of long lasting tradition in East and South Europe.

A nationwide breeding program changed drastically the breed composition and sheep productivity for a period of 30 years. This process was accomplished by the introduction of small foreign populations of different breeds.

In recent years a new structure of the sheep population was formed to satisfy the requirements of the economy and of the separate ecological regions of the country. Generally, it can be characterized as follows: Merino sheep breeding—38%, crossbred sheep breeding—17.7%, Tsigai sheep breeding—27.9%, dairy sheep breeding—12.18% and local aboriginal coarse wool sheep—4.18%.

The Merino group shows the highest wool yield. Flocks of the Askanian and Caucasian Merino breeds were created via introduction. They are about 10.5% of the finewool population.

Four more Merino sheep breeds of similar characteristics were also created in Bulgaria—Northeast Bulgarian, Dounavska, Thracian and Karnobatska. The method of the creation of the finewool Bulgarian sheep breeds was complex reproductive crossing of the different local sheep strains with the participation of Merinofleisch and the Soviet finewool breeds such as Caucasian and Askanian as well as Stavropol.[11]

The Merino sheep breeds are grown mainly in the plain regions of the country where the production of grain and green feed is concentrated.

The wool of the sheep is white, 8.5 to 9.4 cm long with a fibre diameter of 20.6–25.0 microns. The Merino wool is basically 60's–64's, the proportion of animals with 70's being relatively small. In the conditions of our country the wool is of alkaline nature with a pH of the water extract 7.0–8.5, the grease is subject to yellowing.[8] This has a negative effect on the strength and level of whiteness and does not allow dyeing the wool white. The investigations made during the last few years on the reasons of wool yellowiong in different breeds showed that the genetic factors form about 25% of them and for this reason it is possible to select successfully on the level of whiteness.

The clean scoured yield of Merino wools is 38.8% to 45.0%. All Merino wool sheep breeds in Bulgaria have a high prolificacy—up to 135–140%. It is economical to fatten lambs up to 33 kg. Above 37 kg. fattening is not economical. Milk yield for 120 days for the different Merino wool sheep breeds varies in wide limits—from 60 to 101.

Recently we started creating coloured Merino wool sheep flocks. Two of the new breeds were created on the basis of aboriginal breeds with coloured wool—the copper-red Shumen and Karnobatska.

Crossbred sheep breeding in Bulgaria was created on the basis of the New Zealand and Australian Corriedale, the

North Caucasian breed, Romney Marsh and Lincoln. Upgrading was the main method of crossing. There are purebred flocks of all breeds used as improvers, around the country.

Crossbred sheep breeding is settled in the hilly regions of the mountains and the high plains. The system of management is similar to that for the Merinos. It is a characteristic feature of the animals of the crossbred type that their feed to wool conversion efficiency is very high. The crossbred wool is white and is characterized by good length—about 9.0 cm, fibre diameter of 25–31 microns, typical curliness and shine. The wool of the elite flocks has a high clean scored yield—54.0 to 59.0% and very good topographic uniformity, making it look distinct among the other wool types.

Tsigai sheep breeding is concentrated only in the mountainous regions of Bulgaria—the Rodopa mountain, Rila, Pirin, Stara Planina, etc. These are local sheep breeds crossed with the Tsigai breed, imported from the USSR[5]. The wool is uniform, with lower levels of whiteness compared to crossbred wool, higher shrinkage rate and clean scored yield—59.8% to 66.2%. The productivity of these sheep is not high, but they are extremely well adapted to the severe mountainous environment.

For ages on end sheep in Bulgaria have been used for three purposes—wool, milk and meat production. Selection has also been carried out for these three traits.

The reason for the high milk yield of the local Bulgarian sheep is the great importance placed on sheep milk and sheep milk products. Two of the local breeds from the central parts of the country—the Blackhead Pleven and Whitehead Stara Zagora sheep—are selected for high milk productivity.

The changes in the breed composition of sheep in recent years and the creation of the Merino, crossbred and Tsigai sheep breeding lead to the decrease of milk production. This was the basic reason for the creation of specialized dairy sheep breeding in Bulgaria[9,10].

Two methods were used:

1. Restoration of the two local dairy sheep breeds—the Blackhead Pleven and the Starozagorska.
2. Creation of high yielding synthetic sheep populations.

The creation of the synthetic populations was based on two models for the utilization of components of the above mentioned local breeds and of the high yielding dairy breeds such as the East Friesian, Awassi and Chios. The East Friesian is recommended as a final form for the creation of a synthetic dairy sheep type[13]. On this basis a synthetic population and suitability for machine milking was formed, with high milk yield, prolificacy, (of over 250,000 sheep). Machine milking is suitable for the conditions of Bulgaria and is economically efficient above an average level of milk yield 100–120 l.

The existing tendency for a narrower specialization in productive directions in sheep breeding and the maximum utilization of breeding rams leads to the decrease of genetic variability, to loss of genetic plasma and to limitations of the possibilities to obtain animals with specific features for the needs of the different regions and the separate stages of breed formation.

The activities for the preservation of genetic resources in sheep breeding lead to the conservation of a number of valuable local sheep breeds and strains in Bulgaria. Now we maintain flocks of 16 such breed groups amounting to 21614 ewes in the frames of the state farms. More than 150000 ewes of these breeds are kept not only as gene pool but also used for the production of coarse and semi-coarse wool, high quality leather and for the production of tasty meat.

The animals of the gene pool are kept in the respective ecological high mountainous regions under the traditional conditions of nutrition and utilization[1,12,15]. The values of the basic productive traits of these breeds are low. The body weight is 38.0 kg for the Rodopa sheep and 55–57 kg for the Dabene and Sredenogorska sheep breeds. The greasy wool yield is between 2.2 and 3.48 kg, prolificacy is 100–115%, milk yield per lactation—26.1–53.0 l.

The local sheep strains, preserved as a gene pool, are characterized by a great diversity in their wool characteristics. Thus they are distributed in two groups; strains with clearly defined two- or three- level fleece and such with not so great differences of fibre morphological characteristics to form independent levels in the fleece. Our invesigations show[4] that the coarse and semi-coarse mixed wools, white or coloured (obtained by the breeds Strandjanska, Karakachanska, Rodopa and Weststaroplaninska) have wool fibre length in the upper level of 17.2–26.2 cm, their fineness rates 43.2–76.8 microns for the upper level and 24.2–30.5 microns for the lower level of the fleece. Awned and thin downy fibre is predominant in the staple, dry fibres are rare.

The wool of the local strains has a high clean scored yield—49.57–67.13% and the quantity of clean wool is between 1.410–1.833 kg. The wool of some breeds is semi-fine and pigmented, such as the Copper-red Shumen and the Karnobatska and in the other strains the animals with white fleece comprise about 60–70% and those with light brown, brown to dark brown and black colour of the fleece are between 30–40% (Karakachanska, Kotlenska, Rodopa and Srednogorska). The availability of natural wool colours in a wide spectrum allows the manufacturing of fancy dresses and souvenirs. Annually, 1.5 thousand tons of coloured wool are purchased in the country, which is 5% of the total wool purchases.

In collaboration with geneticists from France, New Zealand and Iceland we are creating a system for the complete description of the pigment colours of Bulgarian sheep breeds and identification of the genes, providing for the colour of the pigment, according to the international gene nomenclature of sheep and goats[3].

New scientific information was obtained on the heritability of the colour of the pigments on the head of the Karakachanska sheep[14].

TABLE 1 Number and Productivity of the Local Sheep Gene Pool Herds

		Average Values			
Breeds	*n*	*Body Weight (kg)*	*Wool Yield (kg)*	*Prolificacy (%)*	*Milk Yield per lactation (1)*
1. Copper-red Shumen	790	50.8	3.48	105.1	51.9
2. Karnobatska	788	51.5	2.86	115.0	47.3
3. Svishtovska	260	46.0	2.80	104.8	51.9
4. Replianska	2112	43.5	2.22	102.7	46.9
5. Sakarska	1560	44.7	2.84	106.9	45.4
6. Srednostaroplaninska	4875	48.9	2.15	100.0	37.3
7. Dabene	302	55.4	2.50	105.3	30.6
8. Kotlenska	1541	43.8	3.05	101.9	38.7
9. Srednogorska	2735	57.2	2.83	102.0	73.5
10. Strandjanska	804	38.4	2.95	104.0	28.4
11. Tetevenska	1200	46.5	2.22	100.0	46.6
12. Weststaroplaninska	985	40.3	2.73	108.3	53.0
13. Rodopa	258	38.0	2.86	102.1	35.7
14. Breznik	858	53.0	2.94	102.0	69.9
15. Local mountainous	535	44.2	2.74	111.1	26.1
16. Karakachanska	1959	39.6	2.44	102.8	39.2
Total:	21614	47.1	2.58	103.2	47.3

The research on the genetic control of pigmentation and the interrelations of the genes from different colour loci is necessary for the successful management of these interrelations and their application to obtain animals with desired pigmentation.

The system of management of sheep in Bulgaria is intensive. This is true not only for the plain regions, where pasture is limited, but also of the mountains, where there are enough meadows and pastures.

Recently, a successful two-fold shearing of lambs is applied—at 7 and at 18 months of age. As a result of this the intensity of wool growth increases as well as the clean wool yield. In the case of two-fold shearing of lambs the wool is long enough for manufacturing[6].

References

[1]Alexieva, S., 1979, Comparative characteristics of some local sheep populations in connection with conservation of sheep genetic resources, Thesis, Agricultural Academy, Sofia.

[2]Alexieva, S., 1987, Possibilities for the conservation of the local sheep breeds as a gene pool in Bulgaria, J. Selskostopanska nauka, 3 (in Bulg.).

[3]Alexieva, S., Djorbineva, M., Lauvergne, J. J., Hinkovski, T., 1986, Visible genetic profile of Karakacan sheep in Bulgaria, Colloque de l'INTRA, Goutard.

[4]Alexieva, S., Nedelchev, D., et. al, 1988, Production and processing of the wool of local sheep strains, Harakteristika, J., i ratsionalno izpolzvane na produktsiata ot mestnite otrodia i porodi ovtse za mliako (in Bulg.).

[5]Dontchev, P., Tsotchev, I., Iliev, M. 1979, Problems of the mountainous and semi-mountainous sheep breeding, Institute of Mountainous Animal Breeding and Agriculture, Troyan (in Bulg.).

[6]Nedelchev, D., Alexieva, S. et. al, 1988, Investigations on the properties and the suitability for processing of lamb wool from the different purpose directions, I-Dairy, Jivotnovadni, J., nauki,4 (Bulg.).

[7]Ryder, M. L., 1984, Sheep representations, written records and wool measurements, in "Animals and Archaeology":3, Early Herders and their Flocks, B.A.R., International Series, 202.

[8]Stoyanov, A., Nedelchev, D., et. al, 1986, Efficient production and processing of fine wool, Efektivno, J., proizvodstvo i prerabotka na produktite ot ovtsevadstvoto, Sofia (in Bulg.).

[9]Hinkovski, T. et. al, 1979, Dairy sheep breeding and sheep management technologies, Zemizdat, Sofia.

[10]Hinkovski, T., 1980, Specialization in sheep breeding—a prerequisite for high efficiency, Jivotnovadstvo, J., 8 (in Bulg.).

[11]Hinkovski, T., Stoyanov, A., Nakev, S., 1980, Northeast Bulgarian Finewool Sheep, Zemizdat, Sofia (in Bulg.).

[12]Hinkovski, T., Alexiev, A., 1981, Conservation of Animal Genetic Resources in Bulgaria, in FAO/UNEP Technical Consultation on Animal Genetic Resources Conservationa and Management, Rome, FAO, Anim. Prod. Health Paper, 24, 77–83.

[13]Hinkovski, T., 1982, Creation of a synthetic dairy sheep population in Bulgaria, Symposium on Dairy Sheep Breeding, Thessaloniki, Greece.

[14]Hinkovski, T., Alexieva, S., 1989, Investigations upon the heritability of pigmentation in Karakachan sheep, Selskostopanska, J., nauka (in press in Bulg.).

[15]Hlebarov, G. S., 1940, Researches on the Bulgarian Land Sheep and possibilities for its improvement, Coll. lap. Bulg. Acad. Sci., B, 1–188.

BREEDING, CONSERVATION AND GENETICS OF NATURAL COLORED SHEEP

The Erosion of the Future

H. P. Jorgensen
C. S. Fund
469 Bohemian Hwy.
Freestone, CA 95472 USA

My job is to direct a livestock reserve focusing on two examples of endangered U.S. livestock. The reserve and its educational programs are a part of the C. S. Fund, a private foundation with its headquarters in Northern California.

The foundation makes grants to non-profit organizations working to promote sustainable environmental practices through projects of public education, research, or public interest litigation. Grants are made in four areas:

Peace—steps toward the elimination of mass warfare.

Poisons—steps toward the reduction of present environmental contamination to assure a healthy future world.

Dissent and Diversity—toward the preservation of people's rights to live, act and speak in ways that may run contrary to popular or governmental view.

Genetic Diversity—steps toward the preservation of the earth's dwindling pool of genetic material.

The livestock reserve, spotlighting Navajo-Churro sheep and American Bashkir Curly horses, complements the granting with a hands-on, boots in the muck, daily chore-time reality.

Like all ranchers and farmers, we try to keep the alfalfa out of the floppy disk ports . . .

What is Genetic Diversity?

Being genetically diverse means that a group of organisms or an ecosystem has the ability to respond to changing conditions by embracing a large number of options.

The more options the more likely one of those options will be a match to the changed conditions and thus survive to reproduce.

The more differing individuals within the system, the more likely that some individuals will be well-matched to the environment and produce the next generation.

This applies to differing individuals within a species and to differing species within an ecosystem.

Sometimes a species does not contain enough diversity and it dies out when the environment changes. Mother nature encourages diversity.

Wise agriculturalists would do the same.

How Do We Know that Livestock Breeds are Declining in North America?

Who Keeps Track of the Numbers?

Every country has its own method of keeping track or not keeping track of its germplasm. Great Britain is often espoused as the example others should follow in terms of livestock. They have a network of registered farms that see to the conservation of livestock strains. Those of you from other nations; I would be most interested to hear about germplasm conservation efforts in your country.

In the U.S. & Canada the American Minor Breeds Conservancy (AMBC) is the only organization that is trying to keep track of all the various North American breeds of livestock and fowl. It is a member-based organization with no government support which is located in Pittsboro, North Carolina.

In 1985 they published the first-ever North American livestock census. There's an update planned in 1990.

Breeds are compared by numbers of yearly registrations for 1970 and for 1985 and fall into four categories. Starting with the:

Watch category: from 5000 down to 1000 registrations a year, or are showing a 25 year decline, through

Minor: from 200 or 500 (depending on type) up to 1000 per year and

Rare: less than 200 registrations per year.

They also have a category for feral populations where stocks have been known to be running wild for at least a hundred years.

The census lists over 170 breeds of livestock.

Eighty of those breeds are low enough in registrations per year to be in the Watch, Minor, or Rare categories. Forty-five percent of our livestock breeds are moving toward extinction.

People say, "So what? If they're not needed any more, if they're not competing in the markets, who needs 'em? Let 'em die out. Extinction's natural, isn't it?"

Let's Put It in Perspective

All 170+ livestock breeds in North America constitute but a few small shelves in the huge genetic library of species that encompasses the earth. Tens upon tens of thousands of shelves. Millions upon millions of volumes.

When we selectively breed animals for high production you might say we "edit" those books. We cut out the parts that nobody's paying us for and make copies of the parts they are paying us for.

As long as we're only cutting up a copy, we can always go back to the original and get parts if we decide things have changed and we need those parts.

But if we don't conserve the original material . . . we are in big trouble.

The fact is that extinction rates of wild and domestic animals are rising alarmingly. A certain amount of extinction is natural, but the geologic record shows nothing comparable to the rate occurring in our time.

Extinctions for the remaining eleven years of this century may be greater than in the age of the dinosaur, the last time there was a great species die-out, with at least 10,000 species going extinct and maybe as many as a million. Only this time it may only take decades instead of centuries.

Even such projected losses in wild systems (as much as 15%) barely compare with threatened losses of up to 45% in agricultural crops and animals.

It is against this background that we consider the potential of losing eighty breeds of livestock and thousands of strains of all the crops.

The diverse genetic populations of these older breeds and seeds represent the genetic capital of civilization—They are the fortune on which our agriculture is based. Those diverse older populations, unlike the shallow genetic pools of modern hybrid stocks, contain all the genetic treasure necessary to select and develop the foods and fibers of the next century and beyond if we will only take the short and obvious steps to conserve them.

We are beginning to realize that our current system of agriculture, developed over the last 50 years, operates at a huge environmental deficit. Business agriculture has come to rely on a few landraces of high production livestock and hybrids of a handfull of high production plants, all propped up by huge quantities of temporarily cheap oil and water. Get big or get out is what we're still being told.

The diversity of crops and animals on farms has disappeared as the small farmer has disappeared.

Giant pharmaceutical companies have bought up little seed companies, picked out the material they want to use for their products and discarded the rest.

Additionally, the "bank" from which to get new species or strains, the world's forests, have been cut nearly in half by the last 50 years' logging, largely to make way for short-term agriculture. Last century it was slash and burn. This century it is slash and sell.

The combined assault amounts to nothing less than a wholesale loss of the genetic heritage of civilization, as well as the makings of an ecological disaster.

We are at the threshold of losing the precious common heritage of worldwide agriculture. Society's future ability to produce food in a changing environment lies in the diversity of its agricultural gene pool. The pool has a leak.

The Future is Eroding

Yet, it is the losses in our agricultural genetic bank that we are in the best position to do something about.

I believe we can save our agricultural future. It is certainly not something we can defer. No culture has ever survived without an ecologically sustainable system of agriculture.

Conservation genetics is an essential part of that picture.

Governments and universities are not currently in the forefront of this vital effort. Great bureaucracies that they are, they scurry in the wings while on stage crops are being planted and birthing pens are cleaned out. There will undoubtedly come a day when they will play an important part.

But for now the story of genetic conservation in America is being played out by ordinary gardeners and the dwindling number of small farmers who, at their own pleasure and expense, take care that these genes are replicated year after year, decade after decade.

They have organized grass-roots groups to track and communicate one another's efforts. I have mentioned the American Minor Breeds Conservancy in regard to livestock. It counts roughly 400 members who breed animals and about 1600 supporting members.

The most significant among groups keeping track of vegetable and grain stocks in the U.S. is the Seed Savers Exchange, of Decorah, Iowa, directed by Kent Whealy. Kent's network of seven or eight hundred gardeners across the United States are keeping alive well over 5000 varieties of vegetables no longer produced in mainstream agriculture.

Other seed-saving groups are contributing significantly to the effort.

The literal future of American agriculture lies in those few thousand hands.

The cast is still small. The stage is not crowded. There is a part for you.

Programmes for the Conservation and Improvement of Numerically Small Coloured Breeds

Lawrence Alderson
Colonsay, Hampton Lovett
Droitwich, Worcs WR9 OLZ England

Flock size is a factor of great importance in determining the ability of a flockmaster to follow a constructive policy of improvement for his sheep. The majority of flocks are single-sire units, and this precludes any opportunity for the comparative testing of rams. Thus, minority breeds, which because of their limited numbers are maintained in small flocks, often have no effective improvement policy. On the other hand, they do face the problem of inbreeding, both within the flock and throughout the breed, and a proportionately greater effort is likely to be devoted to survival than to improvement.

Paradoxically, if an improvement policy is implemented, it may accentuate the problem of inbreeding. The performance and qualities of a breed are enhanced by the selection of superior individuals as the parents of the next generation. Particularly outstanding animals are likely to increase their influence within the breed to an even greater degree, and this inevitably leads to a higher level of inbreeding. Thus, in minority breeds, the formulation of any improvement programme must take into account the potential dangers of a reduction in the genetic base of the breed leading to unacceptably high levels of inbreeding.

Inbreeding need not always be a problem. I have practised it with the Colerne flock of Poll Dorset sheep (see pedigree of K513), and the result was the highest performance flock in the breed. However, with the British Milksheep, where the average performance of the whole breed is at a much higher level (3.07 average litter size; 450 1. of milk per lactation), I have only linebred to outstanding individuals, and at a more moderate intensity (see pedigree of AAA 88011). Similarly, with White Park cattle, inbreeding closely to the bull Whipsnade 281 was very successful, but with other lines problems of inbreeding depression occurred. Unfortunately, it is not possible to predict which lines will tolerate inbreeding and benefit from it. The basic principles which should be observed when inbreeding are:

1. Only inbreed to outstanding individuals.
2. Only linebreed through high-quality descendants.
3. Stop as soon as inbreeding depression occurs.

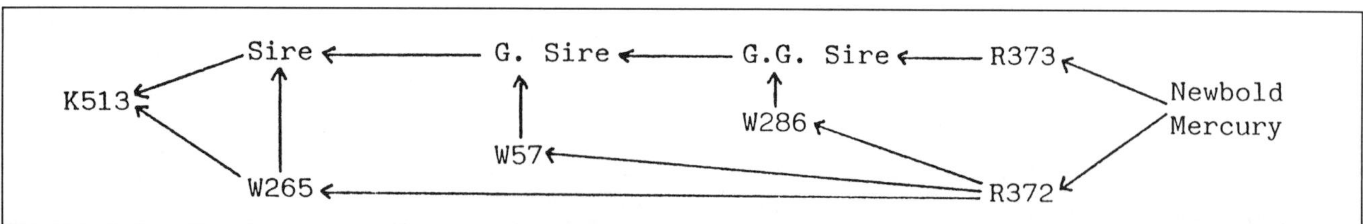

Figure 1 Linebreeding in the Colerne flock of Poll Dorset sheep.

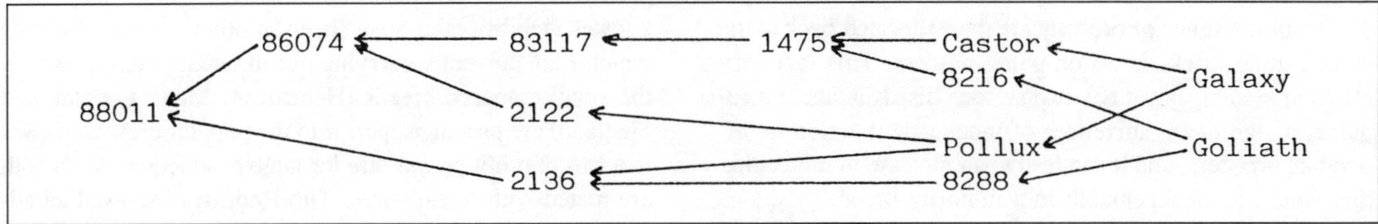

Figure 2 Linebreeding in the Gallowshieldrigg flock of British Milksheep.

One of the advantages of linebreeding is that it can give the breeder a greater ability to predict more accurately the result of matings in his flock. Inbreeding reduces variation by increasing homozygosity, thus making each succeeding generation more uniform. Purebreeding, even without inbreeding, gives the same advantage over crossbreeding, and programmes which are not based on pure breeds, or at least on carefully recorded breeding groups, are liable to produce unexpected and undesirable results. In polite terms it is 'a mixed bag', but it is probably more accurately described as mongrelisation. For example, in Great Britain, as part of the established system of stratification within the sheep industry, Swaledale ewes from the hills are mated with Bluefaced Leicester crossing rams to produce Mule ewes as the next link in the chain of lamb production. The performance and type of the Mule, being a cross between two well-established breeds, is reasonably predictable, but if a Mule ewe is mated with a Mule ram the result is entirely unpredictable.

but fashion then required light to medium-grey sheepskin coats with a small curl, and the selection programme was changed accordingly.

2. A knowledge of the principles of genetics and animal breeding are essential to enable a breeder to follow the optimum selection programme for each characteristic. Wool colour is a relatively easy characteristic to control through a simple breeding programme, but wool quality and the major production factors of prolificacy, milk yield and carcase quality, require more sophisticated programmes and testing procedures.

3. A breeder must have the ability to select the correct animals to use in whatever programme is chosen. This will involve a combination of an ability to keep and interpret records, and an 'eye' for a sheep, as many factors rely on subjective visual judgement. Assessment of wool quality, conformation, constitutional soundness and breed type all rely in most circumstances on the 'eye' of the breeder.

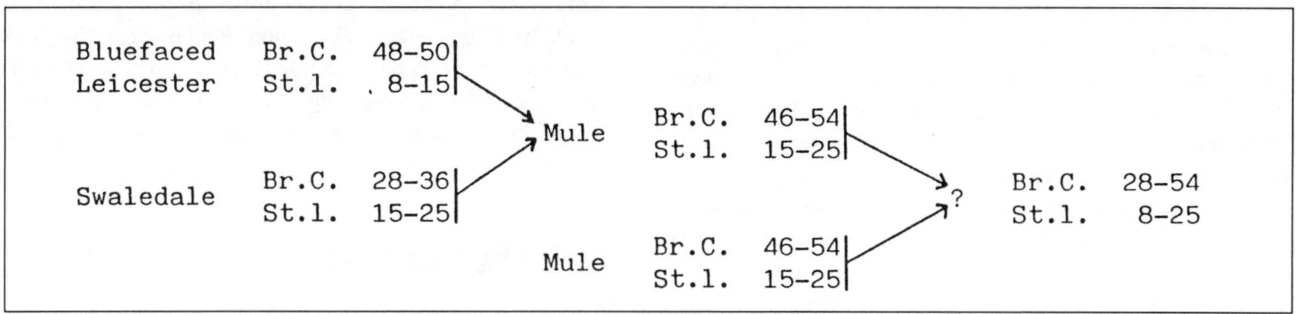

Figure 3 Wool Characteristics of Bluefaced Leicester, Swaledale and Mule sheep.

There are opportunities for breeders of minority breeds and owners of small flocks to improve the quality of their stock with purebreeding and without running the risks of excessive inbreeding, but first there are three basic requirements that need to be fulfilled.

1. The objectives of the breeding programme must be clearly defined. Otherwise there is no end-point on which to focus the selection criteria. There must also be a willingness to change the objectives if circumstances so dictate. For example, a Gotland flock that I advised was selected to breed true for medium to dark-grey wool with a medium curl,

These factors give the breeder the basic tools to create the desired end-product, but he still lacks the necessary structure to put the programme into effect. He can obtain this in two ways, namely by increasing the effective size of the breeding unit, or by using extra facilities outside the normal range of his breeding programme.

The size of the breeding unit can be increased by co-operating in a group breeding scheme. There are two systems which have been proven. One, which was first devised in New Zealand, uses an elite central flock as the focus of improvement. The best ewes in each co-operating flock are transferred to the central flock where they are mated to superior rams. The resulting superior progeny are then allocated back to the co-operating flocks in an on-going process. This is a very efficient system, but it has two serious disadvantages. It requires a significant surrender of independent action by individual breeders, and it can lead to an increase in inbreeding that would be unacceptable in a minority breed.

The second system is one which I devised specifically for use in minority breeds, and which has been used successfully with breeds such as Caspian horses, Norfolk Horn sheep and Portland sheep. All the animals within the breed are divided, according to type and breeding, into at least five groups. The best females in each group are mated with males in the same group (linebreeding), while the other females in the group are mated with males in the next group (cyclic crossing). The groups are arranged in a sequence which ensures that the qualities in each group compensate for any weaknesses in the previous group. This system prevents any rapid increase in the level of inbreeding within the overall breed.

A breeder can also increase the efficiency of his system by using facilities outside his immediate control. For example, the genotype of wool colour in any sheep can be tested relatively easily by a non-comparative progeny test. If a white Shetland ram sires all white progeny when mated to twenty moorit ewes, he is almost certain to be homozygous white. However, if a ram is bought from another flock he is an unknown quantity unless records from that flock can be examined, and these may enable the buyer to obtain a good idea of his genotype. If he is a white ram out of a moorit ewe, he will carry the moorit recessive.

For other factors it is not so easy, but a rapid feedback of information from consumers of any product—handspinners or gourmets—will help to give a progeny test of a ram before he is dead. Progeny testing is essential to properly assess the breeding worth of an animal for many important qualities, and for characteristics such as prolificacy or milk yield a ram will be at least five years old before his true worth is known. Thus, the other valuable facility which can assist a breeder is the long-term storage of semen and/or embryos. These can be brought out of storage when the donor animal is proven, which may be long after it is dead.

The principles of animal breeding detailed above apply not only to wool production, but also to all other products of sheep. Wool is simply one product within any sheep enterprise. The other products make a significant contribution to the total output, and they should not be ignored. In some cases superior performance for some other characteristics seems to be associated with coloured wool. Black Wensleydale and black Leicester Longwool sheep seem to have a leaner carcase than that of their white flock-mates, but this has not yet been properly tested. Coloured Icelandic sheep have superior prolificacy. On the other hand, fawn self-colour Soays have a lower viability than Soay sheep of other colours. In trials which I am currently carrying out in Great Britain, ewes of the small coloured breeds (Hebridean, Manx Loghtan and Shetland) are proving superior to the popular crossbred ewes in terms of profit per hectare for lamb production, when both are mated with 'meat' sires. The Hebridean showed an advantage of 6.7 per cent over the Mule.

TABLE 1 Performance of Hebridean and Mule Ewes Mated to Suffolk Rams for Lamb Production

Breed	*Lambing % reared*	*Daily liveweight gain of twins to 100 d.(kg)*	*Sales of lamb per flock (£)*
Hebridean	156	0.19	7239.50
Mule	174	0.25	6782.00

There is a continuing need to identify the advantages of coloured sheep, and to integrate a policy of improvement for characteristics other than wool into the overall breeding programme by means of a selection index. A selection index is the most effective method of selecting for several characteristics simultaneously within the same programme. In a programme that I used for Hebridean sheep, the lambs were evaluated at 50 days of age for weight (corrected for age of dam, litter size and sex), fleece quality, breed type and conformation. From the records the breeding worth of the ewes was compared by an index which gave a weighting of 40 per cent to litter weight, 40 per cent to fleece quality and 20 per cent to breed type and conformation. Ram progeny tests were calculated on the same basis. Different circumstances might require a different weighting, but in each case the system should be based on clear objectives and sound genetic principles.

References

Alderson, G. L. H. Work done on the conservation of Animal Genetic Resources in the U.K. Paper to the UNEP/FAO Technical Consultation on Animal Genetic Resources Conservation and Management, Rome (1980).

Alderson, G. L. H. A Breeding Policy for Minority Breeds (International Caspian Stud Book, Volume III). Countrywide Livestock Ltd., Droitwich (1983).

Alderson, G. L. H. Selection Methods and Breeding Programmes used in the Conservation of Rare Breeds of Coloured Sheep. Paper to the World Congress of Coloured Sheep, New Zealand (1984).

Alderson, G. L. H. The Chance to Survive (in press).

Conserving Rare Sheep for the Future: The American Minor Breeds Conservancy

D. P. Sponenberg, DVM, PhD
Virginia-Maryland Regional College of Veterinary Medicine
VPI and SU
Blacksburg, VA 24061 USA

The goal of the American Minor Breeds Conservancy (AMBC) is to conserve rare genetic resources for the future. The AMBC is a nonprofit organization that is composed of interested breeders, scientists, and many other individuals with no direct livestock connection but interest in the work.

Part of the work of AMBC is the identification and monitoring of breeds of livestock that are rare or are declining in numbers. Many sheep breeds fit this category, and sheep breeders can easily be very helpful in saving some breeds from extinction. Some of the rare breeds are occasionally (or even usually!) colored, and many of the breed registries now allow registration of colored individuals.

The breeds fall into various broad groups. One of these is the breeds that were imported, experienced popularity but are now in decline. This includes the Cheviot, Dorset Horn, Oxford, Shropshire, Southdown, Lincoln (none of these is rare yet, but a close eye is kept on numbers). A rarer example of this group is the Delaine Merino. Other breeds were never numerous here, but have a long history of having been in the USA: Border Leicester, Cotswold, Karakul, Leicester Longwool. Information on all these breeds is fairly readily available since they tended to be in the main stream of sheep production in the USA.

Some breeds are rather recent imports that are rare in the country of origin as well as rare here: Wiltshire Horn (a white hair sheep with horns), Shetland (a small fine wooled short tail sheep with a wide range of colors), and American Jacob (a black and white spotted sheep that tends to have multiple horns).

A very interesting group is those breeds that have largely been developed in the USA. The Barbados Blackbelly (usually known as the Barbado) is an hair sheep that originated in Barbados from West African sheep brought in during the slave trade. In Barbados these sheep are polled and have multiple lambs at each lambing. In the USA this base has been used, but to it have been added varying amounts of mouflon or wooled sheep breeding. The result has added horns to most lines of the Barbados. They are still used, though, for fecundity and the vigorous lambs they have. The sheep are usually tan or red with a black belly.

The Gulf Coast native (consisting of the Louisiana native and Florida native) has a Spanish churro background. These are rare today, but are very well adapted to the Gulf Coast climate and as such represent an important component of sheep adaptations. They are somewhat resistant to parasites. They also vary somewhat in color and in the presence of horns. The wool tends to be of secondary importance, and can be short and hairy or kempy (a consequence of the churro heritage).

The Navajo-Churro is a desert-adapted breed with a Spanish churro background. These vary in color, and come polled, horned, or with multiple horns. The fleece is double and very useful in the manufacture of the traditional Navajo rugs. They are a very exciting breed since the adaptations, colors, and wool type combine to make them very unique.

Another breed that has been useful in the American South is the American Tunis. As the name suggests, this breed began with Tunisian sheep, but other wooled types have entered into it. Once fairly common in the South, its numbers there were greatly reduced by the War Between the States. It is now primarily found in the North as a productive sheep that really needs to be rediscovered. Tunis sheep are polled, have fairly fine wool of medium length, and traces of the fat rump ancestry from Tunisia. They are born red but the wool quickly fades to white leaving the head and legs a pretty red color. They have semi lop ears and are really a pretty sheep.

A few feral strains have also been identified and are being conserved. The Hog Island sheep come from a coastal island off of Virginia. These are few in numbers, but well adapted to coastal mid-Atlantic conditions. Santa Cruz Island sheep come from the island of that name off of the California coast, and are more adapted to arid conditions. These represent a resource that is very rare worldwide, since feral sheep are less common than feral strains of other species.

The Activities of COGNOSAG Since Its Foundation in 1984

A. L. Rae
Massey University, Palmerston North, New Zealand

J. J. Lauvergne
Laboratoire de Génétique INRA, 78350 Jouy-en-Josas, France

C. H. S. Dolling
Box 74, McLaren Vale, South Australia 5171, Australia

P. Millar
12A Riselaw Crescent Edinburgh, EH10 6HL, Scotland

B. Denis
Ecole Nationale Vétérinaire BP 527, 44026 Nantes, France

The field of genetics is concerned with investigating and understanding the variation which occurs within a species and how that variation is transmitted from parent to offspring. Consequently, a consistent and widely accepted system of naming the genes or units which are transmitted from one generation to the next is an indispensable aid in understanding the genetics of a species.

This paper gives an account of the progress which has been made in the establishment of a system of naming genes in sheep and goats by the Committee on Genetic Nomenclature of Sheep and Goats (COGNOSAG) in the last four years. Because the origin of COGNOSAG resulted from the activities of the first two Congresses on Coloured Sheep (in Adelaide and New Zealand), this contribution may be considered as a progress report of a committee to the parent body which established it.

History of COGNOSAG

The origins of COGNOSAG go back to the National Congress on Breeding Coloured Sheep and Using Coloured Wool which took place in Adelaide, South Australia from January 30-February 3, 1979. At this meeting, two aspects were clear: (i) the naming of the colour genes in sheep was confused, contradictory and in an unsatisfactory state as far as the geneticists were concerned, and (ii) the upsurge of interest in the colour and pattern variants in sheep which had resulted from the commercial demand for naturally-coloured wools in the craft industries required the establishment of a systematic and generally accepted nomenclature for the colour genes in sheep.

Following further discussion it seemed that the second in the series of Congresses on Coloured Sheep to be held in New Zealand in January 1984 would present an opportunity to further the project. Consequently, A. L. Rae who was a member of the Scientific Programme of the Congress asked J. J. Lauvergne to prepare an introductory review to focus attention on the main issues involved. This review was subsequently published as Technical Bulletin No. 38 of the Department de Génétique Animale of the Institut National de la Recherche Agronomique (Lauvergne, 1984) and discussed at a meeting held on January 25, 1984 at Massey University, Palmerston North, New Zealand during the World Congress on Coloured Sheep and Their Products. The meeting concluded by founding COGNOSAG.

In late 1986, it became clear that a more formal organisation of the Committee was desirable. Hence, COGNOSAG was registered as an association under the Law of 1901 at the Sous Préfecture at Antony (Hauts-de-Seine) on April 27, 1987 under the name of COGOVICA/COGNOSAG with A. L. Rae being named as President and J. J. Lauvergne as Secretary/Treasurer. Since that time, the Committee has conformed with the requirements of this registration.

In the intervening years, the Committee has been concerned with the organisation of three Workshops on the genetic nomenclature for sheep and goats. The first workshop was held on July 3–5, 1986 at the Deux Moulins de Gontard, near Manosque in the Department of Alpes-de-Haute Provence in France. Two further workshops have been held at the same place on July 21–26, 1987 and July 11–16, 1988.

The Objectives of COGNOSAG

The most important objective of the Committee is to establish rules for the naming of identified genes in sheep and goats, and to review these rules whenever this may be required by advances in knowledge of the genetics of the two species. The system of nomenclature needs to meet the requirements that it be: (i) sufficiently flexible to accommodate loci with widely different functions and effects on the organism, e.g. from genes with visible effects through to those which are identified only by their DNA sequences, (ii) able to take account of the established homologies of genes in different species, and (iii) simple from the viewpoint of typing, printing and computerisation.

Further objectives include publishing and circulating lists of loci, alleles and chromosomal variants; listing breeds and lines which carry identified genes and chromosomal variants and giving information about the location of these stocks; identifying breeds, types and lines which warrant genetic evaluation and preservation and to foster the training of research workers to undertake this work; collaborating with other organisations which have related interests.

Participation

The foundation committee of COGNOSAG established at the meeting at Massey University in 1984 comprised 10 members from 6 different countries. They were: S. Adalsteinsson (Iceland), T. E. Broad (N.Z.), Melinda J. Burrill (U.S.A.), C. H. S. Dolling (Australia), P. Hoogschagen (Netherlands), J. J. Lauvergne (France), R. S. Lundie (N.Z.), A. R. Quartermain (N.Z.), A. L. Rae (N.Z.), G. A. Wickham (N.Z.).

At the 1986 Workshop, reviews prepared by 12 different authors were presented. In all, 17 people from 9 countries (Bulgaria, Italy, Portugal and U.K.) were represented.

The 1987 Workshop was attended by 11 members, and the 1988 Workshop brought together 18 participants, including two members from the U.S.S.R. and a representative from the International Livestock Center for Africa (I.L.C.A.).

Overall, the total membership of COGNOSAG, including full and corresponding members, is more than 30 scientists from 10 different countries or international organisations.

Sponsorship

Since its inception, COGNOSAG has received financial support for its work from the following organisations:

Academy of Agricultural Sciences of Bulgaria
Academy of Sciences of the U.S.S.R.
Academy of Sciences of Tajikstan (U.S.S.R.)
B.R.G. (Bureau des Ressources Génétiques, France)
European Economic Community, Brussels (Programme AGRIMED)
Fondation de France (France)
GRD PAGE PACA (Groupe de Recherche et de Développement sur le Patrimoine Végétal et Animal de la région Provence Alpes Côte d'Azur)
Fondation Ophthalmologique Adolphe de Rotschild, Paris
ICAMAS (International Center for Advanced Mediterranean Agronomic Studies)
ILCA (International Livestock Center for Africa)
INRA (Institut national de la Recherche Agronomique, France)
INSERM (Institut national de la Santé et des Recherches Médicales, France)
ITOVIC (Institut Technique Ovin et Caprin, France)
New Zealand Wool Board

The support of these organisations is gratefully acknowledged by COGNOSAG.

Activities

1986 Workshop

At the outset of its deliberations, the Committee's objective was to establish nomenclature for the colour genes in sheep because, as already noted, this was the initial stimulus for the formation of the Committee and because the system was likely to be of immediate use in breeding of coloured wool. However, in the planning of the 1986 Workshop, it was quickly appreciated that it was not possible to consider a standardised nomenclature for a single category of genes. It was necessary to encompass the whole genetic complement or genome of the sheep, just as it had been found in the mouse and in man.

Also in the planning stage for the 1986 Workshop, the question of including goats in the programme arose. Because goats are closely related genetically to sheep and also produce fibres which have similarities and complementary uses to wool, it was decided to include them in the operations (and in the name) of the Committee.

The preparation for the 1986 Workshop also showed that, in order to systematise the work involved, it was convenient to consider three categories of genes—those concerned with coat colour and pattern, those controlling other visible traits and abnormalities and those controlling biochemical polymorphisms. A further category was devoted to chromosomal anomalies. Preparatory reviews were presented according to this structure and as a result of discussion at the Workshop, definitive reports and recommendations by COGNOSAG as a whole were agreed.

In particular, the 1986 Workshop accepted a set of proposed rules for karyotypic nomenclature and a list of karyotypic variants in sheep and a similar proposal for goats. Also, proposed rules for genetic nomenclature in sheep and goats were considered and agreed for further discussion with workers in the field. It was not possible, however, to complete for publication, summaries and lists of the colour and pattern genes, the visible traits and biochemical polymorphisms.

1987 Workshop

At this Workshop, the rules for nomenclature of genes for sheep and goats were revised for the first time and modifications were made to take into account constructive comment from members and others, especially on the need to achieve better conformity with the already established nomenclature for biochemical polymorphisms. The rules which were finally accepted are presented in Appendix 1.

In addition, there was discussion on general procedures for listing loci and alleles and on the assessment of evidence in the literature for the existence and mode of inheritance of the alleles described. These procedures were formally adopted by COGNOSAG.

Substantial progress was made in this Workshop by the sub-committees established to summarise and list the information on each locus and its alleles. The sub-committees were responsible for: (i) coat colour and pattern genes of sheep; (ii) other visible traits and abnormalities of sheep; (iii) biochemical polymorphisms of sheep and goats, and (iv) coat colour and other visible traits of goats. These sub-committees were chaired respectively by D. P. Sponenberg (U.S.A.), C. H. S. Dolling (Australia), Elizabeth M. Tucker (U.K.) and P. Millar (U.K.). The work of the second, third and fourth sub-committee was completed to the stage where it could be submitted in 1987/88 to other interested people for comment.

1988 Workshop

At the 1988 Workshop, the Committee was pleased to welcome Academicians G. A. Aliev and M. L. Rachkovsky from the Republic of Tajikstan, U.S.S.R. Their research with and knowledge of the genetics of the many fur breeds of Southern U.S.S.R. was of great importance, especially in the discussion on the colour and pattern genes which was a major feature of this Workshop and will be discussed in a paper at this Congress.

The 1988 Workshop also finalised the lists and summary of the loci and alleles for visible traits (other than colour) in sheep and in goats and biochemical polymorphisms of blood and milk in sheep and in goats. These will be published in early 1989 and will be designated as COGNOSAG 1987.

It is hoped that a COGNOSAG report on the colour and pattern genes in sheep and goats will be published later in 1989.

Discussion

The crucial decisions in the work of COGNOSAG have been those concerned with establishing the rules for naming genes for sheep and goats. At present, two main models for gene nomenclature exist. One is from the genetics of the mouse which had its origin in 1940 (Dunn et al., 1940) with revisions and extensions in 1963, 1973, 1979, and 1984. The other applies to the genetic complement of name which was standardised in 1979. In the end, after much discussion, the Committee opted for an intermediate solution, with a tendency to adhere more closely to the rules for the mouse. This came about mainly because of the presence in the mouse of many variants with visible effects (especially the colour and pattern genes which have homologies with the sheep and goat).

Another element in the work of the Committee has been the fact that, wherever possible, it was important to achieve unanimity, not just a majority vote for the major decisions which had to be made. This, along with the need to give people outside the Committee opportunity to comment has slowed the work but has helped to ensure that the conclusions reached were soundly based.

Progress toward completion of the first overall list of genes in sheep and goats is such that publication should be completed toward the end of 1989. It is noted that the colour genes, which were the category which COGNOSAG was originally set up to resolve, have taken the most time. This is because in this area there is a large number of papers and observations going back more than 80 years. Moreover, the question of the homology of genes in sheep and other species is more important for the colour genes than other categories. Hence more possibilities had to be considered.

Future operations of COGNOSAG will include an ongoing review and updating of the lists of loci and alleles for sheep and goats as new research information becomes available. Completion of information on sources and location of different genetic variants and its publication has still to be completed. Current planning is for a meeting in Tajikstan (U.S.S.R.) for a programme proposed by the Academy of Science of U.S.S.R. and the Academy of Sciences of Tajikstan in the region where the main fur-producing breeds of sheep of the world were developed.

In terms of organisation, it will be necessary in the future to consider the question of the relationships which COGNOSAG should have with the International Society for Animal Genetics (previously the International Society for Animal Blood Group Research), which has as an objective the coverage of the genetics of all domestic animals.

Whatever may be in the future, however, the work done already provides a sound basis for the further development of genetic investigation by scientists and for improvement by breeders of sheep and goats, two species which are so useful to man.

References

COGNOSAG 1988. Standardized genetic nomenclature for sheep and goats. 1986. Proc. COGNOSAG Workshop, July 1986. Gontard Manosque. Bureau des Ressources Génétiques et Lavoisier, Paris, 112 pp.

Dunn, L. C., Gruneberg, H., Snell, G. D. 1940. Report on the committee on mouse genetics nomenclature. J. Hered. *31*:505–506.

Lauvergne, J. J. 1984. A project for standardizing genetic nomenclature in sheep. Bull. Techn. Dept. Génét. Anim. INRA, No. 38, 59 pp.

Appendix 1

Proposed Guidelines for Gene Nomenclature in Sheep and Goats 1987

(By a COGNOSAG Committee comprising in alphabetical order: Snejana Alexieva, B. Denis, C. H. S. Dolling, J. J. Lauvergne, R. S. Lundie, P. Millar, A. L. Rae, C. Renieri, D. P. Sponenberg and Elizabeth M. Tucker.)

(Reproduced in extenso from COGNOSAG, 1989: Standardized nomenclature for sheep and goats, 1987. Proc. COGNOSAG WORKSHOP, Gontard/Manosque, 1987. Bureau des Ressources Génétique et Lavoisier, Paris, 17–21. With the kind permission of the Publishers.)

I. INTRODUCTION

Proposed guidelines were drawn up at the meeting of COGNOSAG in July 1986 (COGNOSAG WORKSHOP, 1986, 1988), and were reviewed at the COGNOSAG WORKSHOP of July, 1987.

The revised guidelines are set out here.

II. LOCUS

A. The Locus Name

1. Choice of name

The name should be brief and either convey as accurately as possible the character affected by the alleles segregating at the locus or reflect the effect of the function by which the locus is recognized. The name may indicate a morphological or disease character (*Ear Length,* for example) or, alternatively, a name should be developed to indicate the exact biochemical property or nucleotide sequence.

As far as possible, the principle of interspecific homology will be applied to biochemically and serologically detected loci as well as to loci with visible effects.

If a newly described allele has the same or similar effect to that of an allele already named, but no evidence for identity has been obtained, then a new allele or locus may be named according to its breed, location, or flock of origin.

2. Printing the name

All the letters will be in italics.

In order to distinguish the locus name from an allele which may be identically written, the initial letter of the locus name will be capitalized.

If the locus name is of two or three words, each word will begin with a capital Latin character. Examples: *Agouti; Ear Length; Fecundity Java.*

B. The Locus Symbol

1. Choice of symbol

The locus symbol will consist of one to four letters or a combination of letters and numbers. The initial character will always be a capital Latin character which, if possible, will be the initial letter of the name of the locus, to facilitate alphabetical listing; the following letters may each be either lower case or upper case.

If the locus name is of two or three words, and the initial letters are used in the locus symbol, then these letters will be in Latin capitals.

All characters in a locus symbol should be written on the same line; no superscripts or subscripts, no Greek letters may be used as initial characters and no Roman numerals may be used at all. . .

Where appropriate, the symbol should indicate the exact biochemical property or nucleotide sequence.

The rules of mammalian interspecific homology already used in the choice of the name of the locus will be applied to the choice of the symbol.

2. Printing the symbol

The locus symbol will be in italics if possible or underlined if not. Example: the symbol of the *Agouti* locus: *A or* $\underline{\text{A}}$

III. ALLELES

A. Allele Name

1. Choice of the name

A brief name should be used which should be chosen to convey as accurately as possible the variation caused by the allele.

Alleles at biochemical and serological loci need not be given names but must be given symbols as described in section IIIB.

2. Printing the name

In order to distinguish the allele name from the locus name which, in some cases, may be the same, the initial letter of the allele name will always be lower case, all the letters being in italics if possible and underlined if not. In case the name is reduced to a group of graphic symbols the use of capital Latin characters is allowed provided they are not used as numerals. Example: the allele for polled at the *Horns* locus: *polled*; the allele *D* a the *Albumin* locus.

B. Allele Symbol

1. Choice of symbol

The allele symbol will be a combination of the invariant locus symbol as defined in II.B.1., followed by the allele superscript consisting of a maximum of 4 characters (letters or numbers).

As far as possible the allele superscript will be an abbreviation of the allele name and start with the same letter. In the polymorphic loci it may even be the allele's name itself.

Greek letters must be avoided because they are not taken into consideration by the computer when a classification according to the alphabetical order is asked for. Roman numerals must also be avoided because they are considered by the computer as letters. The symbol + can be used alone for identification of the wild type or standard allele for alleles having a visible action. Both + and − symbols may be used in the polymorphic series.

Alleles which are known to be dominant or co-dominant (i.e. those having consistent effects in the heterozygote) should have a capital initial letter in the superscript, with the rest being lower case. All other alleles should have a lower case initial letter.

2. Printing the symbol

Whenever possible the allele superscript should be printed as such; otherwise it should be separated from the locus symbol by an asterisk (*).

The allele symbol will be in italics if possible and underlined if not. Examples: The recessive allele *hornless* at the *Horns* locus will be symbolized:

in italics, Ho^{hl} or *Ho*hl*

underlined, $\underline{\mathrm{Ho}^{\mathrm{hl}}}$ or $\underline{\mathrm{Ho{*}hl}}$

the dominant allele for *polled* will have the symbol:

in italics, Ho^{P} or *Ho*P*

underlined, $\underline{\mathrm{Ho}^{\mathrm{P}}}$ or $\underline{\mathrm{Ho{*}}^{\mathrm{P}}}$

IV. GENOTYPIC TERMINOLOGY

The genotype of an individual should be shown by the relevant allele symbols for the two homologous chromosomes concerned, separated by a slash, e.g. Ho^{P}/Ho^{P}.

V. PHENOTYPIC TERMINOLOGY

The phenotypic symbol should have the same characters as genotypic and allelic symbols. The difference is that they will not be underlined or in italics and will be written on one line but without any asterisk. Square brackets [ ] may also be used in phenotypic terminology.

VI. RECOMMENDATION

COGNOSAG recommends that these guidelines be used in the naming of loci and alleles discovered in the future.

VII. REFERENCES

COGNOSAG, 1988: Proposed Rules for Genic Nomenclature in Sheep and Goats, Proc. COGNOSAG WORKSHOP, Gontard/Manosque, July 1986. Bureau des Ressources Génétiques/Lavoisier, Paris, 107–110.

The Present State of Standardization of the Nomenclature of Sheep Color Loci

D. P. Sponenberg, S. Adalsteinsson, C. H. S. Dolling, P. Millar, S. Alexieva, J. J. Lauvergne, M. Rachkovsky, R. Lundie, C. Renieri

VA-MD Regional College of Veterinary Medicine
VPI & SU Blacksburg, VA 24061 USA

For the past three summers the Committee on Genetic Nomenclature of Sheep and Goats (COGNOSAG) has met in the beauty of the Southeast of France to review the scientific literature pertaining to the genes described in sheep and goats, to standardize the names applied to these genes, and to assess the scientific evidence for the genes. One of the working groups of COGNOSAG is concerned with sheep color, and this group has made great progress over the past three years.

The method chosen for the work of this group has been to review all scientific literature pertaining to the color of sheep. This was done chronologically so that the history of the nomenclature could be easily assessed. This proved to be useful in the process of standardizing the nomenclature, since any group would be ill advised to run counter to history and established usage unless a very compelling reason for such a change were involved.

The evidence for the existence of the various genetic variants was also assessed. This is necessary because the scientific reports vary considerably. Some of these involve description and hypothesis only, which is very shaky ground for the proposition of a new gene. Other involve data which clearly show the mode of inheritance as well as the loci involved. This is very good evidence but is sometimes difficult to attain in all situations.

The process of evaluating the past yielded a host of tables in which names of loci, alleles, names used, symbols used, and literature citations all appeared. The committee then tried to formulate a standard name and symbol for each locus and allele. This process has not yet been completed with a uniformly accepted list. The reason for this is that different philosophies of standardization are present, and all are equally defendable. These include 1) historic usage should take precedence, 2) gene homology should take precedence (using the mouse nomenclature as the most accepted model), or 3) gene action should be described in the name. Any system has pitfalls. The first series of names is likely to confuse comparative geneticists who work with coat color across species. The second can easily end up confusing sheep raisers. The third can likewise confuse sheep raisers, and indeed can also confuse some comparative geneticists.

This rather lengthy preamble is to illustrate the controversies that can surround something like sheep color. As a result, the end product of deliberation can change. In order to reflect this, I have chosen to make a table of the attempts to standardize the nomenclature. This has to be somewhat abbreviated, but includes some historic references and then the COGNOSAG recommendations. Most of these are from 1986 and 1988, and are not yet final. Still, it is obvious that in 1986 the attempt was to define gene action, while in 1988 the historic usage tended to win out. Opinions are as numerous as experts, but the final form will probably reflect a lot of the names in the 1988 list.

Locus	Symbol	Allele	Symbol	Reference
Agouti	*A*			Adalsteinsson 1974
Agouti	*A*			COGNOSAG 1986, 1987,1988
		badgerface		Heller 1915
		dark belly (badgerface)	A^{db}	COGNOSAG 1986
		(badgerface)	A^{b}	COGNOSAG 1988
		reversed badgerface		Roberts 1930
		light belly (black and tan) (reverse badgerface)	A^{t}	COGNOSAG 1986
		black and tan	A^{t}	COGNOSAG 1988
		grey		Dry 1927
		grey	A^{g}	COGNOSAG 1986, 1988
		black		Davenport 1985
		no agouti pattern	A^{a}	COGNOSAG 1986
		nonagouti	A^{a}	COGNOSAG 1988
		white		Davenport 1905
		white or tan	A^{Wt}	COGNOSAG 1986
		white/tan	A^{Wt}	COGNOSAG 1988
		wild mouflon	A^{+}	Lauvergne *et al.* 1977
		wild	A^{+}	COGNOSAG 1986, 1988
		grey badgerface		Sumner 1980
		grey belly (light badgerface) (grey badgerface)	A^{gb}	COGNOSAG 1986
		light badgerface	A^{lbf}	COGNOSAG 1988
		blue	A^{bl}	Hoogschagen *et al.* 1978
		blue	A^{bl}	COGNOSAG 1986, 1988
		light blue		Lundie 1984
		light blue	A^{lbl}	COGNOSAG 1986, 1988
		grey mouflon	A^{6}	Adalsteinsson 1970
		light bellied grey (agouti mouflon)	A^{gt}	COGNOSAG 1986
		grey and tan	A^{gt}	COGNOSAG 1988
		light grey	A^{lg}	Adalsteinsson *et al.* 1978
		light grey	A^{lg}	COGNOSAG 1986, 1988
		gotland grey	A^{gg}	Adalsteinsson *et al.* 1978
		dark grey (gotland grey)	A^{dg}	COGNOSAG 1986
		gotland grey	A^{gg}	COGNOSAG 1988
		swiss marking		Lundie 1984
		light marks (swiss marked)	A^{lm}	COGNOSAG 1986
		swiss	A^{s}	COGNOSAG 1988

Continued . . .

Locus	Symbol	Allele	Symbol	Reference
		lateral stripes		Lundie 1984
		lateral stripes	A^{ls}	COGNOSAG 1986, 1988
		red eye	A^{re}	Lauvergne, Adalsteinsson, 1976
		pale cheek/eyering	A^{pc}	COGNOSAG 1986
		eye patch	A^{ep}	COGNOSAG 1988
B	*B*	*black*	B^{1}	Adalsteinsson 1960
Brown	*B*	*standard (black)*	B^{+}	COGNOSAG 1986, 1987
		wild	B^{+}	COGNOSAG 1988
		moorit		Elwes 1913
		brown/chocolate	B^{b}	COGNOSAG 1986
		brown	B^{b}	COGNOSAG 1987, 1988
Albino	*C*			Adalsteinsson 1977
		nonalbino	C	Berge 1976
Albino	*C*	*standard*	C^{+}	COGNOSAG 1986, 1987
		wild	C^{+}	COGNOSAG 1988
		albino	C^{a}	Adalsteinsson 1977
		albino	C^{a}	COGNOSAG 1986, 1987, 1988
Extension				Lawrence 1975
Extension	*E*	*dominant black*	A	Adametz 1917
		dominant black	E^{D}	*COGNOSAG 1986, 1988*
		blackish	E^{bl}	Aliev, Rachkovsky 1986
		blackish	E^{Bl}	COGNOSAG 1988
		brown	E^{br}	Aliev, Rachkovsky 1986
		brownish	E^{Br}	COGNOSAG 1988
		yellow	E^{y}	Aliev, Rachkovsky 1986
		yellowish	E^{Y}	COGNOSAG 1988
		nonblack	b	Duck 1921
		wild	E^{7}	*COGNOSAG 1986, 1988*
Persian and Afghan Pied	*Wh*			Adalsteinsson 1979
		black headed persian		Darcy 1927
Pigmented head	*Ph*	*persian*	Ph^{P}	COGNOSAG 1987, 1988
		white	c	Vasin 1928
		afghan lethal	Ph^{Afl}	COGNOSAG 1987
		afghan lethal	Ph^{Al}	COGNOSAG 1988
		piebald akaraaman	L^{ak}	Lauvergne 1987
Akarman Spotting	*L*	*akarman spotting*	L^{ak}	COGNOSAG 1986
Pigmented head	*Ph*	*turkish*	Ph^{T}	COGNOSAG 1987, 1988
		self	P^{m}	Diomidova, Murvey 1935
		wild	Ph^{+}	COGNOSAG 1987, 1988

Locus	Symbol	Allele	Symbol	Reference
Roan	*R*			Adalsteinsson 1979
		grey	*We*	Vasin 1928
Roan	*Rn*	*lethal roan*	Rn^{Rn}	COGNOSAG 1986, 1988
		nongrey	*c*	Langler 1949
		wild	Rn^{+}	COGNOSAG 1986, 1988
S	*S*			Adalsteinsson 1974
Spotting	*S*	*self*		Roberts 1924
		standard	S^{+}	COGNOSAG 1986
		wild	S^{+}	COGNOSAG 1988
		spotted		Roberts 1924
		spotted	S^{s}	COGNOSAG 1986, 1988
Black Spot				Brooker, Dolling 1965
Black Spot	*Sp*	*normal*	*Sp*	Brooker, Dolling 1969b
		standard	*Spt*	COGNOSAG 1986
		wild	Sp^{+}	COGNOSAG 1988
Sur Bukhara	*g*			Vasin 1968
		surbukhara	*a*	Vasin 1968
Sur Bukhara	*SuB*	*surbukhara*	SuB^{s}	COGNOSAG 1988
		solid color	Sb^{Sc}	Aliev, Rachkovsky 1986
		wild	SuB^{+}	COGNOSAG 1988
Sur Surkhandarya	*Ss*			Aliev, Rachkovsky 1986
		solid color	Ss^{Sc}	Aliev, Rachkovsky 1988
Sur Surkhandarya	*SuS*	*Wild*	SuS^{+}	COGNOSAG 1988
		intensive expression	Ss^{i}	Aliev, Rachkovsky 1986
		middle expression	Ss^{m}	Aliev, Rachkovsky 1986
		sur surkhandarya	SuS^{s}	COGNOSAG 1988

Moorit Sheep Production—Field Trial Report

Allan R. Bennett
"Parana Park" Coloured Sheep & Wool Producers
24 Seventh St.
Gawler, South Australia 5118, Australia

The aim of this paper is to report on field trials of a breeding programme for moorit sheep, which I outlined to the World Congress on Coloured Sheep in New Zealand in 1984. However, because of the breed and genetic differences in sheep flocks of the United States, I have restated some of the genetics involved in breeding coloured sheep in Australia.

Background to Australian Coloured Sheep Breeding

Lorry Dunning in his paper "Coloured Sheep in North America," reported to the 1984 Congress on a number of breeds of sheep in the United States which are coloured. These included Karakul, Navajo, Barbados Blackbelly, and the newer California Red, California Variegated Mutant and Morlam. Karakul sheep are mentioned as having "dominant black breeding characteristics being used to produce colour in finer wool breeds."

Such is not the case in Australia, where all breeds of sheep are white. In defining the breed standard, the white colour is often included. Hence coloured sheep as we know them are not recognised as being of that breed type by white breed standards. But, of course, we do have coloured sheep from most breeds for as we have learnt from Icelandic geneticist, Stefan Adalsteinsson, "white sheep are just black sheep in white pyjamas." Black lambs continue to turn up in white flocks despite long term efforts to select against them.

We have no breed like the Karakul with its dominant black gene. Colour genes in Australian sheep are recessive to white. Hence any breeding programme using white sheep is two generations away from producing coloured lambs. The possible matings to produce coloured sheep in Australian flocks can be illustrated in the following manner:

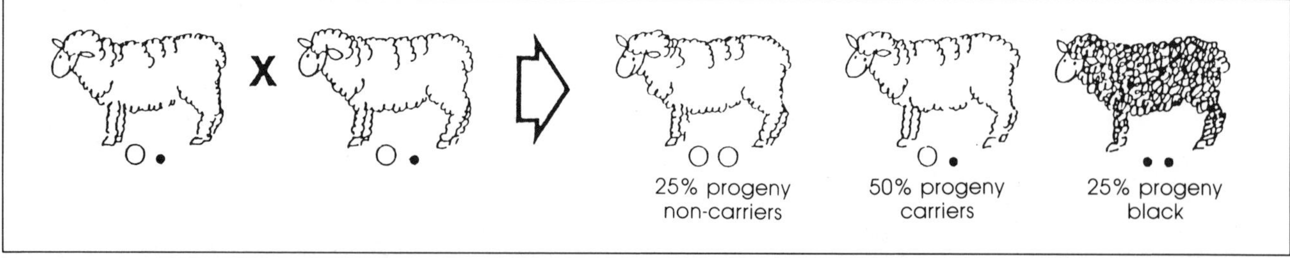

Figure 1 Carrier x Carrier.

This is the mating which produces the occasional black lamb in white flocks. Both ram and ewe carry the gene for the black, and the lamb has a one in four chance of being coloured.

Coloured sheep breeders often use a black ram over white ewes to produce a drop of carrier lambs. Unlike the dominant black Karakuls, such lambs are white in Australia. These carrier ewes, mated to a black ram will produce 50% black lambs (see Figure 2). Thirdly, a mating of black ram to black ewes produces 100% black lambs (see Figure 3).

Identification of Moorit Brown Sheep

In 1976, Scott Dolling, then Principal Livestock Research Officer with the South Australian Dept. of Agriculture, identified a Merino ewe which phenotypically appeared very different from the familiar black sheep. She carried a caramel coloured fleece, with brown hair on the face and legs, and brown skin and mucuous membranes of the mouth. Unfortunely, both she and her black ram lamb were lost in a dog attack before her genetic potential could be explored.

In 1978, Bennett & Gregor purchased a young ewe of similar appearance and later that year discovered a ram in Victoria. These two sheep were on display at the first World Congress on Coloured Sheep held in Adelaide in 1979, where they became known as moorit genotypes, from the red/brown colouration of their fleece and points. The term moorit comes from the Icelandic "moorut" meaning as red as the moors.

We have since learned to manipulate this moorit gene in much the same way as the black gene was used to produce coloured sheep from white stock. In this way we can produce a range of brown fleece colours from a deep chocolate through cinnamon to fawn and beige.

By mating a moorit ram to moorit ewes, moorit lambs are produced (see Figure 4).

This mating parallels the black by black mating (see Figure 3).

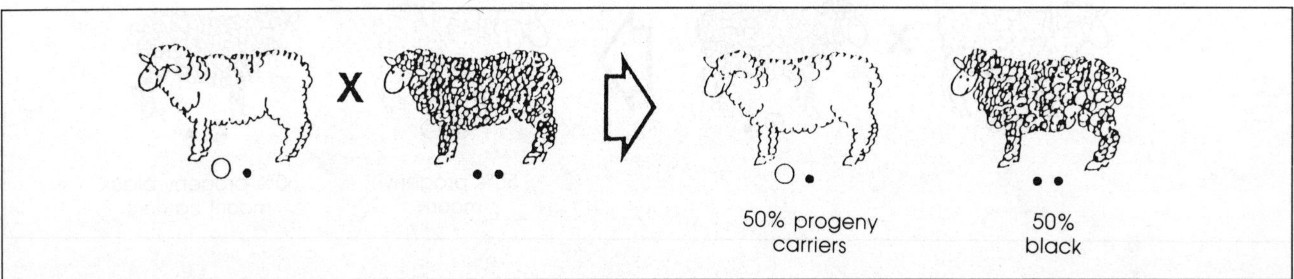

Figure 2 Carrier x Black.

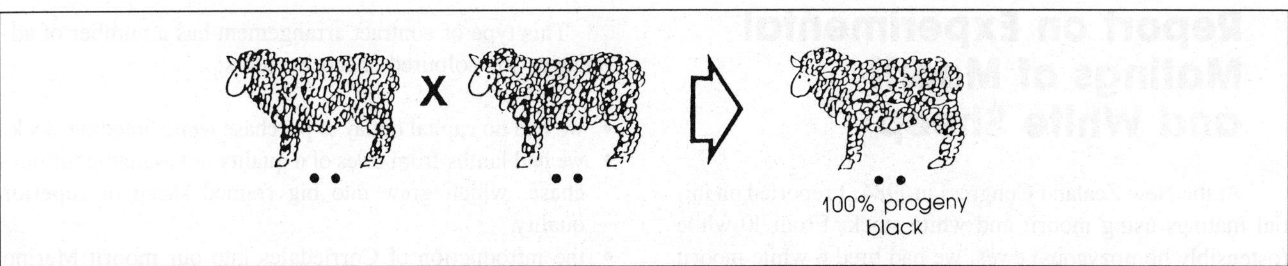

Figure 3 Black x Black.

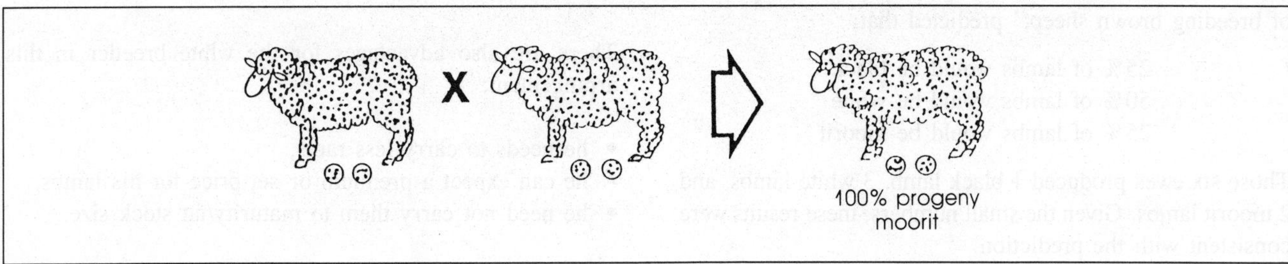

Figure 4 Moorit x Moorit.

The moorit ram was mated to black ewes also, to produce carriers of the brown gene. As expected, these lambs were black, the moorit gene being recessive.

By mating a moorit ram to black moorit carrier ewes produced from mating 5, it was expected to produce black and moorit lambs.

Rather than buying a large mob of white ewes, we approached an empathetic breeder of white Corriedales. He agreed to use four of our moorit rams over some 160 selected white ewes. We contracted to buy all of the ewe lamb progeny at a premium over market prices, while he sold the wether progeny to the livesheep export trade to the Middle East.

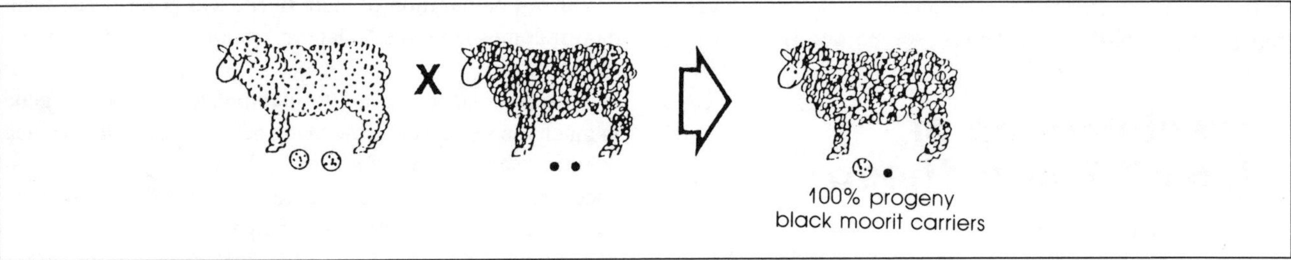

Figure 5 Moorit x Black.

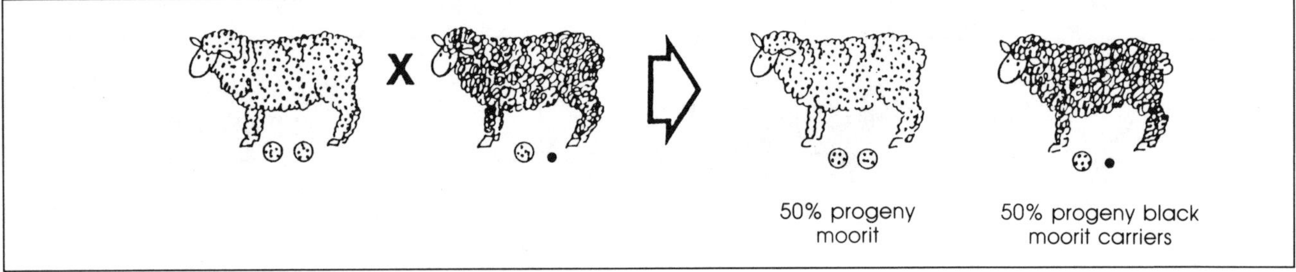

Figure 6 Moorit x Black Moorit Carrier.

Report on Experimental Matings of Moorit and White Sheep

At the New Zealand Congress in 1984, I reported on initial matings using moorit and white stock. From 10 white (ostensibly homozygous) ewes, we had bred 6 white moorit carrier ewes using a moorit ram. These six ewes were mated back to a moorit ram to see if we could produce moorit lambs.

Liz Lewis, in her 1984 Congress paper "The genetics of breeding brown sheep," predicted that;

25% of lambs would be black
50% of lambs would be white
25% of lambs would be moorit

Those six ewes produced 1 black lamb, 3 white lambs, and 2 moorit lambs. Given the small numbers, these results were consistent with the prediction.

We embarked on a programme to rapidly increase our moorit sheep numbers by using white foundation stock.

This type of contract arrangement has a number of advantages for coloured sheep breeders:

- we had no capital outlay to purchase white breeding stock,
- we had lambs from ewes of a quality not available for purchase, which grew into big framed sheep of superior quality,
- the introduction of Corriedales into our moorit Merino flock has meant a strengthening in wool count to the more desirable 56's quality sought by handcrafters in South Australia.

There are also advantages for the white breeder in this arrangement:

- he needs to carry less rams,
- he can expect a premium or set price for his lambs,
- he need not carry them to maturity/fat stock size.

This white flock is run separately from our coloured flock in order to limit the amount of cross contamination of black fibres in the white fleece which occurs as black and white sheep rub together during mustering and yarding. It has proved to be most valuable as a source of:

- excellent quality Corriedale type moorit, black and white breeders,
- large framed, heavy cutting coloured wethers for hand-spinning fleeces,
- white wether lambs for slaughter,
- excellent white handspinning fleeces,
- white fleece for production of our own carded spinning sliver,
- white lambs fleece for the S.A.C.S.O.S. Sable Wool yarn project

Field Trial Results—Moorit Lambs from White Ewes

The following figure gives details of lambing results from our white ewes bred by moorit rams. White ewe lambs bred within the flock are added to it as they mature.

TABLE 1

	Number of ewes lambed	Number of black lambs	Number of white lambs	Number of moorit lambs
1986	24	9	11	6
1987	32	14	21	10
1988	37	16	22	18
Totals	93	39	54	34
Total No. of lambs	127	30.7%	42.5%	26.8%

Explanation

From Table 1, given that the number of matings is still small, the percentages of black, white and moorit lambs are of the order of 1:2:1 as predicted by Lewis (1984). Of particular interest is the increasing percentage of moorit lambs recorded in 1988.

A moorit ram used over white ewes produces a drop of white lambs which are heterozygous for self colour and moorit, WwBb, (Lewis, p. 164). Using a moorit ram over these first generation ewes produces white lambs of two genotypes:

(a) 25% heterozygous for self colour and moorit WwBb
(b) 25% heterozygous for self colour and homozygous for moorit Wwbb

The remainder of the drop will be 25% black and 25% moorit.

In 1988, the first of these second generation ewes were included in the white flock for mating. Half of them are of the same genotype as their dams, and would produce similar percentages of black, white and moorit lambs. The others, being homozygous for moorit, would produce:

(c) 50% heterozygous for self colour and homozygous for moorit Wwbb. These lambs are white.
(d) 50% homozygous for self colour and moorit wwbb. These lambs are moorit.

No black lambs would be produced from a mating of these ewes.

Two points emerge. Firstly we can expect an increasing proportion of moorit lambs as more second generation ewes are introduced to the breeding flock. Secondly, any white ewe which drops a black lamb is not homozygous for moorit, and could be culled from the flock to further increase the percentages of moorit lambs dropped. In the Bennett & Gregor flock however, we have made no attempt to identify the two genotypes, as we value the black progeny as much as the moorits.

Conclusion

The aim of this breeding programme was to see if we could breed moorit (brown) sheep using white sheep as a breeding base.

There were several reasons for taking this course of action. Many of the moorit sheep within coloured sheep flocks were related, and in the Bennett & Gregor flock we were looking to improve the quality of our sheep by outcrossing. By using white ewes, we could take advantage of the many years of breed improvement achieved by white sheep studmasters. At the same time, we could develop a line of stronger woolled sheep to suit the demands of handcrafters in South Australia.

These aims have been satisfied. Selection in future will be made in an endeavor to increase greasy wool production per head, and to fix a strain of moorit sheep which will maintain a darker brown fleece colour.

Footnotes

In this paper I have used the terms white, black and moorit to describe sheep of a particular genotype. These terms are familiar to Australian breeders and are also descriptive of the phenotype of the sheep.

At its meeting in July 1988 in France, the Committee on Genetic Nomenclature of Sheep and Goats, COGNOSAG, agreed on a set of names and symbols for the genes affecting coat colour. The terms used in this paper should therefore be listed along with their COGNOSAG equivalents:

moorit, b, is equivalent to brown, B^b
black, B, is equivalent to wild, B^+
white, W, is equivalent to white or tan A^{Wh}

Realising that moorit genotypes produce a range of colours of brown fleece, the South Australian Coloured Sheep Owners' Society has adopted the recommendation of Gregor, Bennett & Bennett (Agricultural Record 1983), that the colour names of chocolate, cinnamon, and fawn be used to more accurately indicate the phenotype of moorit genotype sheep, i.e. moorit sheep are identified in South Australia as moorit chocolate, moorit cinnamon, or moorit fawn.

Bibliography

Dolling, C. H. S. (1970). Breeding Merinos. Rigby. Adelaide, South Australia.

Dolling, C. H. S. (1979). Breeding Coloured Sheep and Using Coloured Wool. Peacock Publications. Adelaide.

Gregor, N. R., Bennett, A. R. and Bennett, N. P. (1983). Agricultural Record, The Moorit Gene in Australian Sheep. Department of Agriculture, Adelaide. South Australia.

Lewis, E. A. (1984). Coloured Sheep and Their Products, The Genetics of Breeding Moorit Sheep, p. 164. New Zealand Black and Coloured Sheep Breeders' Association.

The Genetics of Coat Colour in New Zealand Sheep

R. S. Lundie
"Eidnul Park", Sutherlands, R.D. 13,
Pleasant Point, South Canterbury, New Zealand

Since 1979, research into the genetics of coat colour in New Zealand sheep has been carried out by the author on his farm. Progress of this research was reviewed in 1984.[8] At that stage most of the work was involved with colour patterns and while no results were given it was indicated that the evidence available pointed to some of these being *Agouti* locus alleles. A recent paper[9] presents the evidence used in establishing a number of patterns as *Agouti* locus alleles—some of these alleles have not been established before.

While there is still a lot of work to be done with the *Agouti* locus alleles and possible alleles, the completion of many of these experiments has allowed the start of more detailed studies into other colour traits.

The aim of this paper is to update the 1984 review by detailing the results so far obtained, giving an idea of the work in progress and if possible their likely outcome and pointing out further work that it is hoped to undertake.

The *Agouti* Locus in New Zealand

In 1984[8] eleven patterns were detailed and three others were briefly mentioned. Now enough evidence is available to establish many of these as *Agouti* locus alleles.[9] Many of these alleles have already been established in other breeds elsewhere in the world, but some are established here for the first time. COGNOSAG, an international committee set up to standardise the genetic nomenclature for the sheep and goat, has recommended new names for some of the established alleles[5] and these are used. The following alleles have been established.

White/tan — A^{wt}

The sheep used were of the Romney, Coopworth (interbred Romney-Border Leicester cross) and Drysdale (basically a Romney which is homozygous for the N^d gene which results in the production of coarse medullated primary fibres-giving a hairy outer coat and fine inner coat—referred to as N-type) breeds. These sheep are all white with dark hooves and flesh of the nose and mouth. The Drysdale at birth usually has tan fibres present—normally in a patch on the back of the neck and occasionally elsewhere on the body.

The *white/tan* allele is the most dominant allele at the *Agouti* locus. This agrees with the findings in many other breeds including the Icelandic and Australian Merino.[1, 4]

The cause of the difference in colour between the breeds used and the Down breeds (born dark but later becoming white on the body with the points remaining dark) and the Merino and Dorset Horn breeds (white wool with white hooves and pink nose) has not been investigated.

Black and tan — A^t

This allele was previously referred to as both mouflon and reverse badgerface. *Black and tan* lambs heterozygous for *nonagouti* have white belly and off black/black upper body. As well, there is a white bar in front of each eye, white inside the ears, white under the jaw and up the centre of the chin. The white under the jaw extends back to the bib area. In homozygotes there is a broad band, without clearcut edges up the underside of the neck. By weaning age the body of homozygotes appears grey rather than black and often there is a progressive greying at the end of the nose.

Blue — A^{bl}

In heterozygotes (with *nonagouti*) the body and belly are black with some greying on the side of the body and across the rear of the back. In homozygotes the area of grey is more extensive so that only the neck and front of the shoulders is dark and the grey is a much lighter shade. Varying amounts for grey can occur in the moustache and there is usually white in the centre of the inside of the ear and along the back edge and tip outside. The tear ducts are always white.

Lateral stripes — A^{ls}

Lambs are born with a white band running along the edge of the belly. The white runs around the front of the brisket and is scattered across the nipple area. The belly is jet black or there may be some white fibres scattered on it. The purse is white, especially the tip. The body and neck are usually black—within this heterozygotes (with *nonagouti*) have a little grey on the upper mid side and another area on the rear of the flanks. In homozygotes these grey areas are more extensive so that there is an indistinct dark area left in the centre of the hip. The ears are mainly white inside and black outside, although some have little white on the outer back edge. There is a very obvious white moustache. The tear ducts are white. At birth the chin varies from black to almost white and by the hogget stage varies from grey to white.

Badgerface — A^b

The *badgerface* that has been studied in New Zealand originated in the South Suffolk breed. They have a near white body, clear black bars on the head and black jaw, chin, inside the ears and belly.

Adalsteinsson[1] when describing the *badgerface* in the Icelandic breed states that the upper body is grey (dirty white or tan outercoat and black undercoat). The head markings are not so clearcut. The same phenotype has been seen in Merinos in New Zealand. So the *badgerface* already established has a grey body whereas that from the South Suffolk and studied here has a white body. There is a chance that there are two different alleles involved—a set of experimental matings using the badgerface from the Merino will help to answer this question. If they are found to be different than the allele in the Merino (if it is the same as the Icelandic one) will be *badgerface* (A^b) and the other could probably best be called "whitish badgerface".

Hip spot badgerface — A^{hbf}

A type of badgerface has been found in the Corriedale breed in New Zealand[8] which is similar to a second type of badgerface that Adalsteinsson[1] described in the Icelandic breed. The belly is black. Behind the shoulders the body is almost white, except that in the centre of the hip there is a dark spot of vague outline. The neck and shoulders are the same dark colour. All the evidence indicates that it is a separate allele.

Light blue — A^{lbl}

Same as blue with the addition of a grey brisket, purse and around the penis and navel at birth. The chin is grey at birth or soon after. When homozygous, all the belly quickly becomes grey at the base. This effect that is "added" to the blue is termed the "lightening" effect.[9]

Light badgerface — A^{lbf}

A badgerface with the same "lightening" effect affecting the belly and chin as occurs in *light blue.* The body is virtually white.

Fawn eye — A^{fe}

Lauvergne and Adalsteinsson[6] described a pattern red eye (red eye patch[2]), seen in Corsica, which they postulated as being an *Agouti* locus allele. It is now known as *eye patch* — A^{ep}.[5] It is a completely eumelanic animal with a phaeomelanic (red/tan) patch around the eye and on the cheek. Similar looking animals have been seen in Merinos in both Australia and New Zealand.

The pattern which Lundie[8] described as being red eye was in fact different and has since[9] been named *fawn eye. Fawn eye* has all the characteristics of the *blue* allele with the *eye patch* as well. Instead of being red the patch is more fawn at birth and later becomes white.

Tan eye — A^{te}

Previously Lundie[8] only briefly mentioned this pattern and noted that it came from the Hampshire breed. It is made up of all the characteristics of the *lateral stripes* allele combined with those of the *eye patch* allele. Once again the patch of the eye and cheek is not red but a tan/fawn colour fading to white with age. (Photos 1, 2)

1. Tan Eye. Both heterozygous for non-agouti. N-type fleece on left.

2. Tan Eye. Same lambs as in 1.

3. Non-agouti locus grey. Twin lambs at 3 weeks.

4. Non agouti locus grey. Lightest and darkest half siblings at 6 months.

5. Transverse Stripes. None visible from a distance at 6 months.

6. Transverse Stripes. Same lamb as in 5, at 8 months—shorn.

7. Brown. Homozygous (left), dark brown heterozygote (center), black lamb (right)—at 5 months.

8. Same three genotypes as 7, at 8 months—shorn.

9. Dark brown. Heterozygote (right) and suspected homozygote (left). Note slightly browner face—5 months.

10. Dark brown. Same lambs as 9, shorn next day. Homozygote is silvery (right), darker along the back.

11. Types of brown: Left top—black; middle—lighter brown (moorit); bottom—moorit brown has retained its color. Right—wool from lambs in 9 and 10. Top and middle—wool from back and lower side of homozygote; bottom—from the dark brown heterozygote.

12. Swiss. N-type lamb with characteristic features at 19 days.

Swiss — A^a

This pattern has also been briefly mentioned previously.[8] It is derived from the Southdown breed. It has some similarities to the *lateral stripes* allele in that heterozygote (with *nonagouti*) has some grey on the upper mid-side and also on the rear of the flank. When homozygous these two areas are a whiter shade and more extensive, so that a dark spot of vague outline remains on the centre of the hip. It has white tear ducts, white chin and moustache, and usually a broad white band going back from the moustache, above the eyes to the horn buds. (Photo 12)

Nonagouti — A^a

In the Icelandic breed Adalsteinsson[1] described this pattern as having no white whatsoever with the possible exception that some lambs heterozygous for *spotted* may have a few white fibres on the crown. Here anything that has absolutely no sign of any of the characteristics of any of the patterns has been classified as *nonagouti*. Most of the individuals are completely black or brown but a few could be classed as dark grey or similar to the "Gotland grey" described in the Gotland breed of Sweden.

The *Agouti* Locus — A Complex Locus

Lundie[9] has postulated that the *Agouti* locus, instead of being a single locus, actually consists of two loci which are very closely linked (close to each other on the same chromosome). This explains why some alleles, such as *tan eye* and *fawn eye,* appear to be the combination of two other alleles. The component alleles of these two loci, Pattern factor and Greying factor, combine together to form the *Agouti* locus alleles. The Table 1 is taken from that proposed by Lundie.[9]

Grey and *grey and tan* (grey mouflon) are two alleles that have been established in the Icelandic breed.[1] *Wild* is the allele present in the wild Mouflon sheep.[7] Some Navajo have a pattern which appears a combination of *eye patch* and *light belly* which appears to be passed on as a whole to some of their offspring, but no experimental matings have been carried out to substantiate this.[10]

The component alleles:

light belly — seen in the *black and tan allele*
lateral stripes — seen in the *lateral stripes* allele
whitish rear — seen in the *blue* allele
darker badgerface — seen in the darker type of badgerface of the Icelandic breed

TABLE 1 *Agouti* Locus Alleles and Their Component Alleles

	Pattern factor component locus					Greying factor component locus				
Agouti locus allele	*light belly*	*lateral stripes*	*whitish rear*	*darker b.f.*	*non agouti*	*grey*	*lightening*	*eye patch*	*hip spot*	*non agouti*
grey					x	x				
grey and tan	x					x				
black and tan	x									x
(darker) *b. f.*				x						x
hip spot b. f.				?x?					?x?	
"whitish b. f."?				?x?		-?	-?	-?	-?	-?
light b. f.				?x?			x			
blue			x							x
light blue			x				x			
eye patch					x			x		
fawn eye			x					x		
tan eye		x						x		
lateral stripes		x								x
non agouti					x					x
"some Navajo"?	x							x		
wild	x					-?	-?	-?	-?	-?

nonagouti — completely eumelanic (black or brown)
grey — seen in the *grey* of the Icelandic breed (the outercoat fibres mainly black and the undercoat fibres white but with a great variation between animals).
lightening — the effect seen in the *light blue* and *light badgerface* when compared with *blue* and the other badgerfaces.
eye patch — seen in the *eye patch* allele of Corsica
hip spot — the effect seen in the *hip spot badgerface* that has removed the colour from the upper body of the *darker badgerface* so that it only remains on the shoulders, neck and vague spot on the hip.

There are probably further component alleles which would be present in the *white/tan*, *light grey* and *swiss* alleles.

The *Agouti* Locus — Related Topics

Grey and Light Grey

Previously[8] a grey was described (derived from the Romney) that appeared similar to that seen in Iceland (*grey* A^g) and a light grey (derived from the Coopworth and also seen in the Corriedale) which appeared similar to the *light grey* seen in the Gotland breed[3] and postulated as being an *Agouti* locus allele.

Both types were studied and appear to be alleles at the *Agouti* locus. Whereas, at first, they both seemed to have completely different phenotypes, it was later found that the grey type when mated to Coopworth sheep produced animals resembling the light grey type, and that the light grey type when mated to fine woolled (Merino/Corriedale) sheep produced animals similar to the grey phenotype. While both types appear to be *Agouti* locus alleles, experiments have not finished yet in deciding if they are the same or two different alleles.

The Badgerfaces

Three badgerface alleles have been studied: *light badgerface, hip spot badgerface* and "whitish badgerface". The darker *badgerface* as described in Iceland has been seen in the Merino in New Zealand. It is the intention to start experimental matings with the darker *badgerface* and at the same time cross it to the coarser woolled (Romney, Coopworth) type to determine if it is in fact a different allele.

Eye patch

This phenotype has been seen in the Merino in New Zealand and it is hoped in the future to obtain one to begin experimental matings to study this allele.

The "Self Colour" Group of Animals

While the other *Agouti* locus alleles were being studied everything that didn't contain any of the other patterns was classed as "self colour". Most of these were completely eumelanic but it was noticed that a few did differ slightly and these are not being investigated. The various types seen within this group could be due to additional *Agouti* locus alleles varying slightly from the *nonagouti*, or the *nonagouti* being altered by a single gene at another locus, or the *nonagouti* being modified by a number of genes.

In a number of cases an experiment has been set up to look at a particular variation. Firstly, matings are carried out to produce animals that are homozygous for the same allele at the *Agouti* locus—normally it is assumed this will be *nonagouti* (unless a new *Agouti* locus allele is involved). If there is any variation within these "homozygotes" then it can be assumed this is caused by a gene or several genes at another locus/loci.

The phenotypic traits being investigated are: Gotland grey types, animals which grey around the muzzle with increasing age, animals which are grey immediately around the ears and animals which by weaning age are quickly greying at the base of the staple. Some preliminary results are available for one of the types being studied:

A "Grey" — Not Situated at the *Agouti* Locus

In one of the experiments enough data has been collected to show that the type of grey being studied is definitely not caused by an *Agouti* locus allele but probably a single gene at another locus. In 1987, 10 lambs were produced from the same source, so they were all homozygous for the same *Agouti* locus allele—assumably *nonagouti*. Two were virtually jet black, five were described as dark grey and three as light grey although they could not be put in obvious classes.

There was a range in the amount of grey in the dark grey group. The darkest were off-black, with only a little grey on the shoulder and grey or dark grey on the brisket. Others were grey to dark grey, with the belly being darker around the navel and purse and across the nipples between the hind legs and the body lighter on the lower shoulders and lower neck.

The lightest animal showed an obvious difference at the body/belly border. The belly (in front of the nipples) and brisket were a grey/light grey colour with an off black area around the navel. The wool of the four legs below the belly height was almost black (with the hair jet black) and this colour continued up to meet between the hind legs behind the nipples: this dark area also extends back to the anus to form a triangle. At the edge of the belly there is a clear-cut change and on the body the wool was mainly off white. There was a slight variation at the base of the staple (4 months) with most being off white/light grey, but along the backline it was grey at the base. The head from behind the jaw and crown was dark grey with the bare hair areas jet black. (Photos 3, 4).

The other two animals that had been grouped in the light grey group were not as light as this animal, and at birth they appeared much darker. But they displayed many of the same characteristics: the darker wool of the head and legs and this extending on to the belly behind the nipples where it forms a triangle back to the anus, and a grey to light grey belly and brisket with a darker area around the penis and navel. On the body, the sides of the body and neck are light grey/off white but along the broad backline it begins dark grey at the back of the neck and is light grey from behind the shoulders back.

It is to be expected that the blackest group lack the grey gene completely. There is a group which is heterozygous and finally some belong to a homozygous grey group. If only the very lightest animal belongs to the homozygous grey group then it is an obvious group, but the heterozygous group shows a very wide range in its expression. If, however, the three lightest lambs were all homozygotes then both these and the heterozygotes show a range of expression.

All that has really been established so far is that the grey in question is not caused by a gene at the *Agouti* locus. Once it has been established that only one gene is involved, it will then be necessary to establish definite grey homozygotes, and then the range of phenotypes within each genotype.

Types of Brown in New Zealand Sheep

Moorits

The brown of the moorit (brown) coloured sheep that occur in New Zealand is inherited as a recessive to black. Unpublished experimental data long with the observations of breeders of moorit sheep in New Zealand agree with this. The conclusion is reached as the result of research carried out in other countries.[1] Sheep of this colour are homozygous for B^b—the *brown* allele that occurs at the *Brown* locus. The other allele at this locus is B+—the *wild* allele and its results in black animals.

Lambs that are homozygous *brown* are a very obvious brown colour at birth, on both the wool and the hair on the points. As they age, the wool of many, especially those in breeds of British origin, seems to fade in colour so that after a year or two many are almost white. Most moorits being bred in New Zealand are Romney or Romney cross sheep and one of the big challenges facing these breeders is going to be to breed sheep which will retain their brown colour with advancing age. The Merino moorits that are bred by some breeders usually seem to retain their brown colour better.

Dark Brown (Photo 7–11)

For the last three years experimental matings have been carried out involving sheep that are a dark brown colour. The original dark brown ram was obtained from a mainly Corriedale (?) flock (Peter O'Connor's at Kurow). When mated to a number of black ewes, he left both black and dark brown offspring. In 1987 a dark brown son of the first ram was used over a large number of unrelated ewes. The *nonagouti* offspring were inspected—there were 14 present: 7 were classed as dark brown and 7 as black.

There seems to be a range of dark brown shade—from some very obviously dark brown to others that are getting close to black. At birth they all appeared black and none could at that stage be classed as dark brown although it was noted that some of them appeared a dark pastel colour. Even at three months of age two were classed as "black" and "doubtful" who later, at five months, were classed as dark brown. When the dark brown lambs were shorn they appeared a different colour from those with the black phenotype (Photos 7–11). The brown of these dark brown lambs is darker than that of the moorit sheep where it is caused by the *brown* allele. The points of a dark brown are a dark, dark brown—getting close to black.

These results indicate that this dark brown colour is dominant to black—the ram used would have been heterozygous (his dam was black) and when mated to black ewes, half his offspring were black (homozygous) and half dark brown (heterozygous).

The question that now arises is—if the heterozygous appears dark brown, then what is the phenotype of an animal that is homozygous for this gene? In 1988 two dark brown ewes, daughters of the original ram, were mated to their dark brown half brother. Three *nonagouti* lambs were produced. One developed the dark brown phenotype and the second was black. The third lamb appeared different to all the other lambs so far produced.

At birth it appeared a uniform pastel grey/dark grey all over the body and belly. The hair on the legs was black but that of the head was a definite dark brown shade. This lamb at five months of age had wool that was best described as being a slightly pastel or silvery shade of dark brown. The darkest colour (along the backline) was much lighter than that of the heterozygous dark browns. The wool gradually became lighter over the sides and appeared its lightest on the belly. The dark brown appeared to get lighter and lighter, rather than being brown (as in homozygous *brown*) and getting lighter. When newly shorn, the body was a pastel/silvery shade—again darker along the backline (almost grey) and gradually becoming lighter down the side so that the belly was almost a light grey shade. The hair on the bare areas of the head was an obvious dark brown about midway in colour between black and a homozygous *brown*.

The mating that produced this lamb would be expected to produce homozygotes (for the gene that causes dark brown as a heterozygote) in a quarter of the offspring. The fact that this lamb was so different from any of the other lambs is a good reason to strongly suspect that it is in fact homozygous. Further matings of this type will need to be made in the future. Experimental matings will also have to be carried out to see if this gene is an allele at the *Brown* locus or located at another locus.

Transverse Stripes

Mention was made of these previously.[8] Since then, another unrelated ram with transverse stripes has been obtained and mated to the daughters of the first transverse stripes ram. There haven't been enough numbers produced to decide much about its genetics, although results do point to it being caused by a recessive gene. One of the problems is deciding, when the weaning descriptions are recorded, which lambs actually have the stripes and which don't. No sign is visible on the woolly lambs from a distance. The wool of all the offspring is opened up along the length of the body to see if any stripes are visible, but in some cases the only way to be sure is to shear the lambs as the stripes are then quite obvious—this is now being done. The stripes are grey on a black body, and mainly occur on the side of the body and side and under the neck. A lot more work is required to be done in studying this trait. (Photos 5, 6).

References

[1]Adalsteinsson, S. (1970). Colour inheritance in Icelandic sheep and relation between colour, fertility and fertilization. J. Agric. Res. Icel. *2*(1):3–135.

[2]Adalsteinsson, S. (1979). Genetics of breeding coloured sheep in Iceland and Scandinavia. In: "Breeding coloured sheep and using coloured wool." Peacock Publications, Adelaide. 17–25.

[3]Adalsteinsson, S., Lauvergne, J. J., Boyazoglu, J. G. and Ryder, M. L. (1978). A possible genetic interpretation of the colour vairants of the Gotland and the Goth sheep. Ann. Genet. Sel. anim. *10*(3):329–342.

[4]Brooker, M. G. and Dolling, C. H. S. (1969). Pigmentation of sheep. II. The inheritance of colour patterns in black Merinos. Aust. J. agric. Res. *20*:387–394.

[5]Committee on Genetic Nomenclature of Sheep and Goats. (1988). Proceedings of the workshop of July 1988 at Gontard, Manosque in France (in press).

[6]Lauvergene, J. J. and Adalsteinsson, S. (1976). Genes pour las couleur de la toison de le brebis Corse. Ann. Genet. Sel. anim. *8*(2):153–172.

[7]Lauvergne, J. J., Denis, B. and Theret, M. (1977). Hybridation entre un Mouflon de Corse (Ovis ammon musimon, Schreber [1872]) et des brebis de divers genotypes: genes pour las colouration pigmentaire. Ann. Genet. Sel. anim. *9*:(2):151–161.

[8]Lundie, R. S. (1984). Colour patterns in New Zealand sheep. In: "Coloured sheep and their products." World Congress on Coloured Sheep, New Zealand, 1984. 128–138.

[9]Lundie, R. S. (1989). Colour inheritance in New Zealand sheep—the *Agouti* locus (in preparation).

[10]Painter, Ingrid (1984). Personal communication.

Genetics of Wool and Hair Color in Some Asiatic Breeds

G. A. Aliev.
M. L. Rachkovsky
Academy of Sciences of Tajik SSR
Lenin Av. 33
Dushanbe 734742 USSR

Asia is a Motherland of some ancient breeds which have practically remained the same during a number of centuries (e.g. Karakul and Hissar breeds). In this region there are some newly developed breeds which are important for scientists because they descend from aboriginal ones. To study their melanogenesis and genetic control is very interesting, for the factors of intensive solar irradiation and very high summer temperature would have dramatically affected the character of melanogenesis in Asiatic sheep. Investigation of pigmentation in aboriginal breeds is also important because such breeds as Hissar have a very coarse hair-covering with a great amount of dead hairs approximate to the hair-covering of wild rams: it is obvious that there arises a rare opportunity to study genetic and physiologic aspects of melanogenesis in domestic sheep as they represent a model of this process in their wild relatives. We should underline an extremely good opportunity to study pigmentation in the Karakul breed because it was bred from ancient times especially for lamb pelts of definite colors. The genetics of some of these colors is not fully investigated and the character of genetic determination of sur and shades of brown and red colors was unknown until recently. In this paper we'll discuss genetic problems which have not yet been investigated in western countries.

Genetics of Brown and Red Color

The majority of sheep colors are determined by interaction of loci A and E alleles. It is generally accepted that there are two alleles in locus E: gene E^D determines intensive melanogensis—"dominant" black color epistatic over white color of fine-wooled sheep; allele E^+ doesn't have its own effect and allows full expression of alleles of locus A[1]. But genetic interpretation of brown and red colors, which are wide spread among domestic and wild sheep of Asiatic and African origin, was not given until 1984. In our paper[3] the presence of 3 more alleles E^{Bl}, E^{Br} and E^Y in locus E were postulated. These alleles like E^Dsupress the expression of alleles of locus A; the effectiveness of their action is reduced from E^{Bl} to E^Y. The action of these alleles is additive; the final result is that the function of the melanocytes and their interaction with keratinocytes is determined by interaction of alleles of gene E and A: alleles of locus E induce melanogenesis and alleles of locus A inhibit it with different effectiveness.

According to our interpretation colors of coarse-wooled sheep of Asiatic origin are determined by alleles of locus E in the following way (in interaction with gene A^{Wt}): E^{Bl}/E^{Bl}, E^{Bl}/E^{Br}—black (third genetic variant of black color) or blackish brown color, E^{Bl}/E^Y, E^{Br}/E^{Br}—dark brown, brown or dark reddish, E^{Br}/E^Y, E^Y/E^Y—reddish, light reddish, or tan color. In Asiatic sheep the frequency of allele E+ is very low. Obviously its presence in the heterozygous state substantially increases the variety and accounts for the possibility of extremely different and unexpected segregations: e.g. tan (or light reddish) parents of genotype E^{Bl}/E^+ would give 25% black lambs (E^{Bl}/E^{Bl}) in the F_1 generation.

This hypothesis explains well how the wide spectrum of brown and reddish color shades in different domestic sheep is genetically controlled. The results of many authors can be interpreted with its help.[2,9,19] Besides according to the analysis of crosses between wild rams and domestic sheep[6, 10, 12] brown and red color polymorphism in different races of wild rams is also determined by the action of locus E alleles. The most widely spread, in wild sheep populations are combinations of loci A and E alleles which determine a comparatively low intensity of melanogenesis. This type of melanogenesis was formed in the process of evolution and is optimal for wild sheep inhabiting the mountains for it provides a protective coloration.

There is a great variety in the expression of wild color both within different races of wild sheep and between them: from light tan shades to dark brown. The presence of 5 alleles at locus E is a good adaptation which gives an opportunity for population migration into a new region in which lighter or darker protective color can increase in frequency through natural selection in subsequent generations.

More intensive melanogenesis in most Asiatic domestic coarse-wooled sheep in comparison with wild rams may be accounted for by continuous artificial selection, first of all for E^D, E^{Bl} and E^{Br} alleles, as from ancient times shepherds preferred black and dark brown sheep, considering them more adapted for the hot climate of valleys and foothills. Natural selection, on the contrary, prevents the increase of E^D and E^{Bl} allele's frequency in the populations of wild sheep, because their presence deprives animals of protective color.

Genetics of Sur Color

Sur color is expressed only in new-born Karakul lambs. Sur hairs have a light distal part and a dark basal part. The color of the distal part fluctuates from almost white to brown and the color of basal part from tan to black. The color and length of the transitional part of the hair can change considerably. In the Soviet Union we differentiate 3 types of sur within the Karakul breed: bukhara, surkhandarya and karakalpak. In each of them there are several color shades.

A genetic interpretation of sur color formation was given in our book.[5] It is formulated as a result of our analyses of Soviet Karakul-breed experiments, genetic experiments on mice and other laboratory mammals and our morphological investigations of melanogenesis in hair follicles of sur lambs. In this paper we'll touch upon it briefly.

Agouti color (which resembles sur) is a wild type in many species of mammals and is controlled by homologic alleles of locus A.[17, 20] It is also shown that in the presence of gene E^D, agouti (or sur) color can't be expressed. Animals of this genotype are "dominant" black. This color can't be expressed either in the recessive black genotype (A^a/A^a). So, according to our theory for sur color development, two things are necessary:

1. The intensity of melanogenesis must be decreased as a result of interaction of alleles A and E loci.

2. In embryogenesis, at the beginning of hair development, the factors which change the conditions of hair formation must act. These factors (hair growth acceleration, diameter decrease, change in biochemical environment, etc) acting together inhibit melanogenesis and determine weak pigmentation of the distal part of the hair.

The action of some of these factors has been investigated in mice: e.g. hair growth acceleration as a result of mitotic activity increase of follicle cells induces the formation of yellow zone in hairs.[7,8] Local administration of colchicine (mitosis inhibitor) on the contrary decreases the number of hairs with yellow band or fully reduces the length of this band. Subcutaneous injections of DL-5-methyltryptophan to yellow A^Y/a mice cause measurable depression in growth rate and give rise to a great increase of hairs containing black pigment.[15] Glutathione or other sulphydryl compounds inhibit the formation of eumelanin as they react with tyrosinase, stopping melanogenesis at an earlier stage, which results in the formation of phaeomelanin instead.

Thus we should underline that the formation of agouti color in wild mammals occurs in hairs with cyclic growth, when such phases as anagen, catagen, telogen are differentiated distinctly. It is known that conditions for the development of the yellow band arise in anagen. Centuries of selection for wool production in domestic sheep dramatically transformed the function of hair follicles: e.g. in mice, the phase of hair growth (anagen) continues during 3 weeks, but in Merino sheep wool can grow during 8 years.[21] Distinct cycles of growth in sheep of wool type are practically absent.

So for the development of sur color in addition to the combination of alleles of the A and E loci determining the lower intensity of melanogenesis (in comparison with E^D/E^D or/and A^a/A^a genotypes) a short-term change of hair follicle growth function is necessary.

Thus analysis of numerous Soviet Karakul-breeders experiments[9,11,13,14,16,18,22] (we refer only to some papers, detailed information is given in our monograph[5]) gives us an opportunity to develop the following genetic interpretation of color determination:

1. The change of hair follicle function at the beginning of hair formation is determined in lambs of sur bujkara by allele SuB^{s*} of the SuB locus and in sur surkhandarya by allele SuS^s of the SuS locus. Both alleles are recessive and expressed only in the homozygous state. Heterozygotes $SuB/SuB^sSuS/SuS^s$ don't express sur color. We've got only one paper[22] where the results of crossing sur karakalpak rams with sur bukhara ewes have been presented. There was no sur color among F_1 lambs. This test and personal communications with Karakul-breeders support the hypothesis that surkhandarya and karakalpak sur are determined by the same gene SuS^s. It is understandable that there are two genetic types of sur color, as many different factors inhibiting melanogenesis at the beginning of hair follicle function could act. These two

different genes Sub^s and SuS^s were fixed by selection in different geographical regions which are separated from each other by thousands of kilometers.

2. Color and length of basal, transitional and distal parts of hairs are determined by interaction of loci A, E, SuB and SuS alleles. For example, the genotype of sur bukhara goldish color is presented by $A^{Wh}/A^{Wh}E^{Br}/E^{Br}SuB^s/SuB^s$, and surkhandarya bronze shade by $A^{Wh}/A^{Wh}E^{Br}/E^{Br}SuS^s/SuS^s$. (We use nomenclature according to symposium of COGNOSAG 1988.)

In crosses of sur rams with black ewes Karakul-breeders usually get 4% sur lambs in the F_1. According to our theoretical analyses (considering usual segregation of black "recessive" lambs—A^a/A^a) in the F_2 31, 8% sur lambs would be expected and in the F_3, 41, 9%. Using the above described hypothesis, there seems to be a good agreement between theoretically expected data and experimental results.

Obviously, along with the interaction of genes mentioned, sur color formation may be affected by other genes: modifiers, which determine wool structure, the way of curling, lustre, etc. It is necessary to take into consideration the effect of environmental factors which can develop some variations in color even in lambs of identical genotype.

Inhibition of Melanogenesis by Genes-Modificators

Brown and red protective colors of wild sheep are comprised of a mixture of extremely coarse black, dark brown, red and white hairs; very thin (7–20 mkm) and short underhair is usually brown or light brownish-grey. Only representatives of aboriginal breeds have a similar structure of hair-covering (Hissar breed). Wooled type breeds are usually white and sometimes black ("dominant" and "recessive"). Until recently a very interesting question wasn't explained: why do wild rams and aboriginal coarse-wooled sheep have a wide spectrum of wool-covering color, from intensive black through numerous shades of brown and red to light tan, while at the same time melanogenesis in sheep of wool type breeds is either very intensive (black, badger-face, reversed badger-face) or completely inhibited (white color)? In other words, why has the brown-reddish color usual for wild rams almost completely disappeared in domestic sheep of wooled type?

Being guided by the results of hybridologic analyses in numerous and various breeds, and by morphologic and physiologic investigations of melanogenesis in the ontogenesis of Tajik and Karakul lambs[4] we can develop the following interpretation:

1. Interaction of the A^{Wh} gene with E^{Bl}, E^{Br} and E^{Y} alleles determines shades of brown and reddish color only in wild and aboriginal coarse-wooled sheep. The interaction of controlling melogenesis genes through gene-modification of the whole complex of wool-type factors (high density, speed and evenness of wool and hair growth, absence of thick hairs and clear cycles of growth etc.) determines the development of depigmented wool covering.

2. The inhibition of melanogenesis is more effective when gene-modifiers determine wool type. Genotype $A^{Wt}/A^{Wt}E^{Br}/E^{Br}$ controls the formation of dark brown in wild and aboriginal sheep,brown in improved coarse-wooled breed (like Karakul), reddish in breeds with carpet wool (Tajik), and white (complete depigmentation) in fine wooled and half-fine wooled breeds.

Lambs of the Tajik breed are a very suitable model for investigation of the mechanism of gene-modifier action. At birth they are pigmented in different shades of brown and reddish, and during the first weeks or maybe months, melanogenesis in hair follicles is inhibited as a result of gene-modifier action.

Light microscope study of the depigmentation process in the ontogenesis of Tajik lambs has shown that the supression of melanogenesis is expressed in inhibition of melanocyte function, inclusion of melanocytes into the growing hair (fig. 1–3) and gradual exhaustion of follicles. At the moment of full depigmentation we couldn't find melanocytes in the region of follicle papilla even using the DOPA-hystochemical reaction. The more active the function of melanocytes (their activity is controlled by interaction of loci A and E), the later they leave the region of papilla and the later the period of depigmentation comes. Thus in hair follicles of tan and light reddish lambs, melanocytes are inhibited by the age of 4–6 days of their birth. They practically don't provide hair with melanosomes and are incorporated into the hair during the first month. Meanwhile melanocytes in dark brown lambs are usually active during 2–4 months.

It is important to underline that in the largest follicles melanogenesis will continue in lambs of all shades (except tan and light reddish); in this case there are some thick, pigmented hairs in the wool-covering of adult sheep and this is the key: all events of melanogenesis act on morphologic, physiologic and biochemical backgrounds of the individual follicle and specific conditions in this follicle determine the fate of melanogenesis. The next example is a good argument in favour of our interpretation. In adult carpet wooled sheep (Tajik breed) and English short wooled blackface breeds (Suffolk or Hampshire) the wool on the body is white but protective hairs on head and legs are pigmented. They are short and thick, and their growth is cyclic. Such hairs are similar to those of wild sheep in many of their characteristics.

Obviously supression of melanogenesis in sheep of wooled type can act in different ways. Probably one of the main factors is the high concentration of cysteine—an amino acid which is necessary for keratin synthesis (SH-groups of cysteine are inhibitors of tyrosinase). Another factor is a sharp

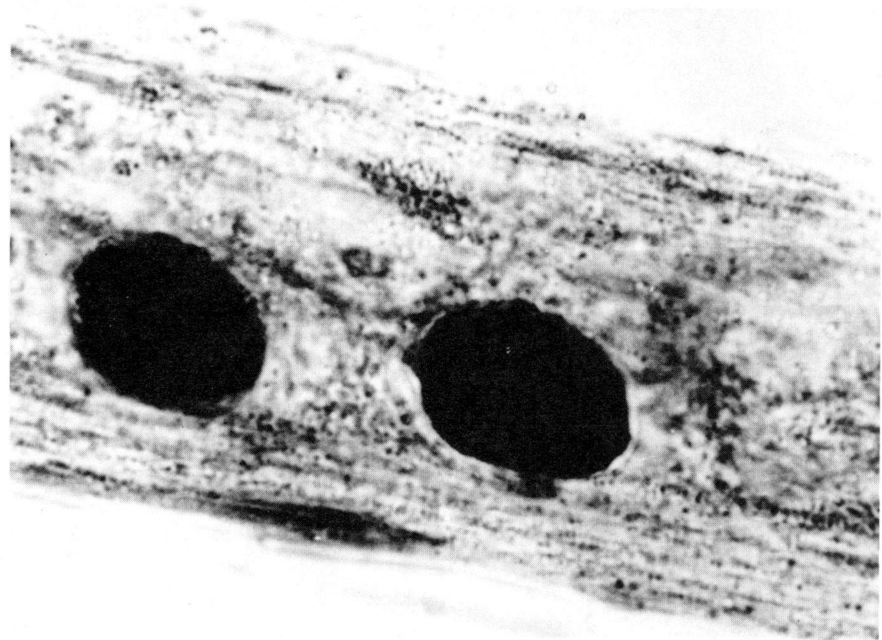

Figure 1 Two melanocytes without dendrites in the hair of dark red Tajik lamb (x 330).

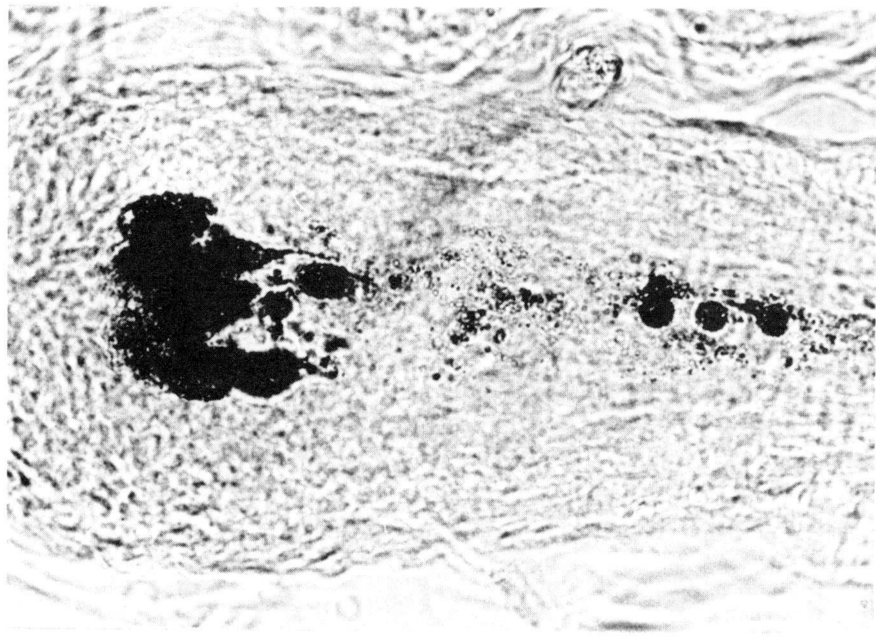

Figure 2 Melanocytes migration in the hair follicle of brown Tajik lamb (x 100).

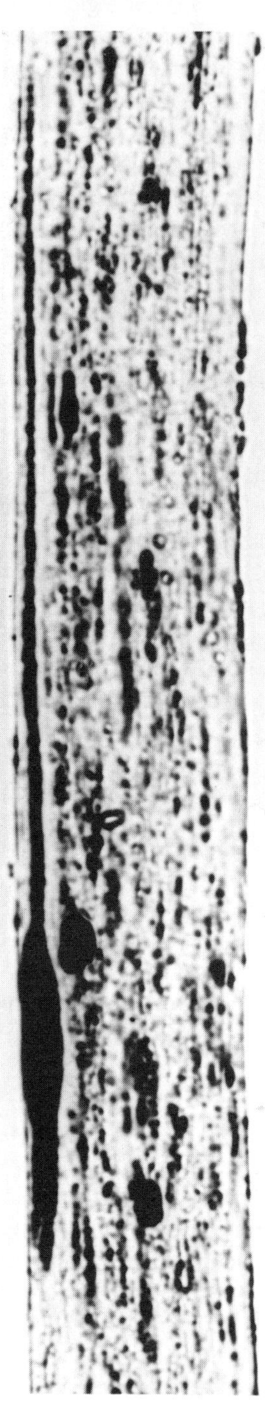

Figure 3 Long dendritic melanocyte in the hair of dark red Tajik lamb (x 100).

competition between melancyctes and keratinocytes multiplying intensively in wooled sheep. In follicles of wooled sheep, hair papillae are comparatively small and amino acids and other biochemical material come through the outer sheath of the follicle: in such conditions the metabolism of melanocytes becomes more difficult.

We see that investigations in the field of physiologic genetics can give wide prospects for better understanding of the action and interaction of color genes.

Literature

[1]Adalsteinsson, S. Inheritance of colors, fur characteristics and skin quality traits in North European sheep breeds: A review.—Livestock Production Science, 1983, v. 10, pp.555-567.

[2]Aliev, G. A. Tajikskaya myaso-salno-sherstyanaya poroda ovets. Dushanbe: Irfon, 1967, 346 s. (Russian).

[3]Aliev, G. A., Rachkovsky M. L. Genetica okraski sherstnogo pokrova ovets. Soobtsh. III. Varianty okraski sherstnogo pokrova ovets, opredelyaemye allelyami genov A i E i ih vzaimodeistviem.—Genetika, 1984, t.XX, 6, s.895-906 (Russian).

[4]Aliev, G. A., Rachkovsky, M. L. Vzaimodeistvie regulyatornyh genov A i E i genov-modifikatorov v protsesse ingibirovaniya melanogeneza u ovtes sherstnyh porod.—Doklady AN SSSR, 1986, t.287, 4, s.990-994 (Russian).

[5]Aliev, G. A., Rachkovsky, M. L. Geneticheskie osnovy pegmentatsii sherstnogo pokrova ovets, Dushanbe: Donish, 1987, 200s. (Russian).

[6]Butarin, N. S. Otdalyennaya gibridizatsiya v dzivotnovodstve (Arharomerinos i gibridnaya svinya) Alma-Ata: Nauka, 1964, 210s. (Russian).

[7]Cleffman, G. Agouti pigment cells in situ and vitro.—Ann. N.Y. Ac.Sc., 1963, v.100, p.749-761.

[8]Galbraith, Donald B. The Agouti Pigment Pattern of the Mouse: A Quantative and Experimental Study.—J. Exp. Zool., 1964, v.155, p.71-89.

[9]Gigineishvili, N. S. Plemennaya rabota v tsvetnom karakule-vodstve. M: Kolos, 1976, 191 s.(Russian).

[10]Gigineishvili, N. S. Sozdanie novogo tipa seryh karakulskih ovets s povyshenoi dzizneustoytsivostyu putyem gibridizatsii s muflonom.—V kn.: Selektsia, gibridizatsia i akklimatizat-siya s-h dzivotnyh. M.: Kolos, 1983, s.224-235 (Russian).

[11]Gigineishvili, N. S., Ukbaev, H. I. Vnytriporodnoe scretshivanie dvuh tipov sura. Ovtsevodstvo, 1983, 4, s.31-32 (Russian).

[12]Ivanov, M. F. Unasledovanie kachestv shersti gibridami ot muflona i merinosovyh ovets. Polnoe sobr.soch., t.2, M.: Kolos, 1963, s.579-592 (Russian).

[13]Jilyakova, V. S. Vyvedenie karakulskyh ovets sur na osnove raznorodnogo po okraske sparivaniya. Tr. Uzb.NII dzivotnovodstva, vyp.2. Tashkent: Gosizdat Uzb.SSR, 1957, 114-122 s. (Russian).

[14]Mukhamedov, K. Harakter nasledovaniya odraski sur u karakulskig ovets. Genetika, 1976, t.XII, 2, s.88-96 (Russian).

[15]Nachmias, V. T. 5-Methyltryptophan and Darkening of the Hair in Yellow A^Y/a Mice.—Nature, 1951, v..191, 4791, p.934-935.

[16]Pogodin, V. N., Polomoshnov, P. M., Suharkov, S. I., Isataev, J. Metody plemennoi raboty s ovtsami okraski sur i tsvetnymi. Tr. Kaz. NII karakulevodfstva, t.2, 1974, s.33-39 (Russian).

[17]Searle, A. G. Comparative Genetics of Coat Color in Mammals. London—N.Y.: Logos Press, Academic Press, 1968, 291 p.

[18]Taganov, B. Opyt sozdaniya stada bronzovogo sura. Ovtsevodstvo, 1968, 12, s.16-17 (Russian).

[19]Vasin, B. N. Evolyutsia sherstnogo pokrova ovets. Novosibirsk: Nauka, 1969, 283 s.(Russian).

[20]Vatti, K. V., Alekseevich, L. A. Sravnitelnaya genetika i ontogenetika okrasok u dzivotnych.—V kn.: Fiziologicheskaya genetika, L.: Meditsina, 1976, s.326-350 (Russian).

[21]Vsevolodov, E. B. Volosyanye follikuly. Alma-Ata, Nauka, 1979, 192 s.(Russian).

[22]Zakirov, M. D. Nasledovanie rastsvetok sur pri odnorodnom i raznorodnom sparivanii. Ovtsevodstvo, 1965, 9, s.8-9 (Russian).

Inheritance of "Face and Leg Colour" in the "Mouton Vendeen" Breed

X. Malher and B. Denis
Service de Zootechnie
Ecole Nationale vétérinaire de Nantes,
Case Postale 3013, 44087 Nantes Cedex 03, France

No recent experimental evidence has been presented to explain the genetic determination of the colour of the Down Breeds: there is only WOOD's data (1905, 1909), dealing with the Suffolk breed, which was given a genetic interpretation by ROBERTS (1928).

This paper presents data and the interpretation of crosses between the "Mouton Vendeen" breed (a recent French breed with Southdown origin, phenotypically looking like the Southdown or Dorsetdown in colour) and the "Solognot" breed (a red breed, with black variants, whose colour genetics have been known since LAUVERGNE's (1969, 1975 publications).

Experimental Crosses

All matings took place in the experimental flock of the Zootechnie service of Nantes National Veterinary School. Close inspections ensured there was no doubt of the parental relationships. A description and photograph of each lamb were made at birth.

The "Mouton Vendeen" Pattern

As in other Down breeds, the "normal" Mouton Vendeen sheep has pigmented head and legs with grey to dark brown color.

Pigmentation is not always uniform. The hair may be white, fawn, tan, agouti or dark color in various proportions, mixed together or forming dark and light patches, temporarily called "brindles." The pigmented areas vary in extent on the neck, the shoulders, the back line, the belly and the thighs, producing a type of lateral depigmentation at birth. Only the hair of the head and legs remains pigmented after the growth of the wool. Sometimes lambs have a uniform black phenotype. Their fleeces remained of a solid black color with age.

"Normal" Vendeen ewes (n = 11) crossed with a black Vendeen ram gave us 100% "normal" Vendeen lambs (n = 18). From the F_1 generation, back-crossed with the black ram, we obtained 50% black Vendeen and 50% "normal" Vendeen (5 Vs 3).

Testing with the Solognot Breed

To explain the genetic origin of the more or less dark pigmentation on face and legs, we crossed Vendeen ewes with two rams of the Solognot breed. This Solognot breed is a uniform red breed (with recessive black variants) whose color genetic has been described by LAUVERGNE (1969, 1975) and is considered as a standardized coloured breed segregating for dominant red (A^{WT}) and recessive black (A^a). At the other loci, the genetic formula for color genes is B^+/B^+, E^+/E^+, S^+/S^+ (COGNOSAG, 1986).

At birth, the Solognot lamb is uniformly red (fleece grows with a reddish white color) and the black variant is uniformly black at every stage.

When crossing the red Solognot to the Vendeen ewes ("normal" or black, n = 8), all lambs (n = 17) at birth were red with a lateral depigmentation and "brindles" mainly on the face and the legs.

The back cross F_1 (Vendeen x Red Solognot) ewes x Red Solognot ram gave a similar pattern but 50% (n = 6) have no brindles vs 50% (n = 5) with brindles.

When crossing Black Solognot ram x Normal Vendeen ewes (n = 6), we obtained exactly the same pattern: red with lateral depigmentation and brindles (n = 7). No distinction can be made between the F_1 with Black or Red Solognot origin.

Interpretation

At the *a* locus, A^{WT}(White or Tan) is dominant to A^a("black"-non agouti) (COGNOSAG, 1986). So the genotype of the Red Solognot is A^{WT}/A^{WT}or A^{WT}/A^a and the Black Solognot is A^a/A^a (LAUVERGNE, 1969, 1975).

From Normal Vendeen/Black Vendeen cross, we can assume that black, in Mouton Vendeen, is recessive to the normal color and the genotype of the blacks is A^a/A^a.

For the interpretation of the Vendeen/Solognot crosses, we must propose the action of two loci. Black Vendeen ewes must transmit "brindles" to the F_1-lambs ("brindles" is not present in the Solognot breed) and this cannot be explained by an allele at the *A* locus because they are A^a/A^a.

It is proposed that an allele at the *E* locus produces "brindles"' in the F_1 and a darker head and legs or brindles in the pure breed ("brindles" disappears with age to give a more extensive, more uniform and often darker pigmentation). The segregation in the back cross confirms this interpretation. This gene is dominant to the wild allele present in Solognot breed.

At the *A* locus, the results lead to the conclusion that the allele present in "Normal Vendeen" is the same as in the Red Solognot, that is to say A^{WT}, because the same result is obtained in Normal Vendeen x Red Solognot and in Normal Vendeen x Black Solognot crosses. In this last cross, it is proposed the red color is due to the action of modifiers, inherited from the Black Solognot, which intensify the expression of the phaeomelanin but can only be expressed when A^{WT} is at the *A* locus.

Discussion/Conclusion

BERGE (1974) and ADALSTEINSSON et al. (1980) both proposed that the colour of the Downs occurred in the presence of A^{WT}but no experimental data was given. Neither was an explanation given for the presence of dark pigmentation on this theoretically phaeomelanic genotype.

The varying amounts of dark pigmentation on the face and the legs appears to be determined by a gene at another locus. This gene, with A^{WT} at the *A* locus, induces "brindles" in F_1 (Solognot x Vendeen) lambs and in some pure Vendeen lambs. With age, the contrast usually disappears and a more regular, and often darker, pigmentation takes place. It is proposed that this gene is at the *E* locus because of numerous homologies in other species ("Mask" and "Brindles" in dog, "Brindles" in cattle, "Japanese brindles" in rabbit and pig.).

A name for this gene could be "Japanese Brindles": E^J(to be homologous) or "Smoky" E^S (to describe the expression of the gene in an adult of a Down Breed).

It is still not known if the variation of pigmentation which is observed between Down breeds and also within Down breeds (for example, see the description of NICHOLS, 1927, for the Suffolk Breed) is due to a quantitative variation or to the action of different alleles at the *E* locus (for example, "Blackish," "Brownish" and "Greyish"). So further investigations are needed.

The Black Solognot appears to be a good line for testing different white breeds for colour genes because it induces a better expression of phaeomelanin in the F_1 and can give indications to determine whether a white fleece is induced by a dilution of phaeomelanin (F_1 appears reddish at birth) or by the action of a piebald gene (F_1 appears red and white) or the union of the two mechanisms.

We can now assume that A^{WT} at the *A* locus is the principal gene which determines the white fleece in the Down Breeds. There is sufficient evidence to propose that a gene at another locus, likely at the *E* locus, induces a varying pigmentation, more or less homogenous and dark, of the hair on the head and the legs. At birth, this pigmentation may be more extensive on the body but disappears with the growth of the wool.

References

Adalsteinsson, S., Dolling, C. H. S. & Lauvergne, J. J., 1980. Breeding for white colour in sheep; *Agric. Rec.*, 7, 40–43.

Berge, S., 1974. Sheep colour genetics; *Z.Tierz. Züchtungsbiol*, 90(4), 297–321.

COGNOSAG, 1988. Standardized genetic nomenclature for sheep and goats 1986, *Proceedings of the COGNOSAG Workshop 1986*, J. J. Lauvergne et Bureau des Ressources Génétiques éd., Paris 1988, 112 pp.

Lauvergne, J. J., 1969. Hérédité de la couleur blanche du moutonj Berrichon croisé à des Solognots; *Ann. Génét. Sél. anim.*, 1(3), 219–226.

Lauvergne, J. J., 1975. Génétique de la couleur de la toison des races ovines francaises Solognote, Bizet et Berrichonne; *Ann. Génét. Sél. anim.*, 7(3), 263–276.

Nichols, J. E., 1927. On the occurrence of dark fibers in the Suffolk fleece with particular reference to the birth coat of the lamb; *J. Text. Inst.*, 18, T395–413.

Roberts, J. A. F., Colour inheritance in sheep. III-Face and leg colour; *J. Genet.*, 19, 261–268.

Wood, T. B., 1905. Note of the inheritance of horns and face colour in sheep; *J. Agric. Sci.*, 1, 364–365.

Wood, T. B., 1909. The inheritance of horns and face colour in sheep; *J. Agric. Sci.*, 3, 145–154.

Comparative Pigmentation of Sheep, Goats, and Llamas—What Colors are Possible through Selection

D. P. Sponenberg
Virginia-Maryland Regional College of Veterinary Medicine, VPI and SU
Blacksburg, VA 24061 USA

Shosuke Ito
Fujita-Gakuen Health University School of Hygiene
Toyoake, Aichi 470-11
Japan

It is possible by looking at the biochemistry of the natural pigments in fibers and the modifications of fleece type to discover what range of colors are possible in the various species we use for fiber.

Natural mammalian pigments are generally considered to be of two types biochemically: eumelanin and pheomelanin. While overlap of these is now documented, the concept of two basically different pigments is helpful in understanding mammalian pigmentation. Eumelanin is generally black, but can be chocolate brown. Pheomelanin varies from pale yellowish tan through to rich, dark, reddish brown. As a result, eumelanin is usually referred to as black/brown, while pheomelanin is red/yellow. The brown eumelanin and dark pheomelanins can be very confusing to separate visually, although they are distinct genetically and biochemically.

Most wild mammals have combinations of eumelanin (black, except for very rare exceptions in wild animals) and pheomelanin (tan) areas. Wild sheep and goats, for example, are generally tan on the body, pale to white on the belly, and have patterns of black stripes along the spine, legs, and face. This is an indication that a single animal can express both pheomelanin and eumelanin color. In addition, the pheomelanin usually varies in shade among various areas of the body. Eumelanin usually does not vary, and the choice is generally black *or* brown. Those animals genetically programmed to be black do not form brown, those brown do not form black. The browns are the moorit shades of color.

If we take each sort of pigment separately it is possible to discuss the ranges of colors available, and potentially available, from that pigment.

Black eumelanin is perhaps the most common pigment of colored fiber producing species in North America. If black fibers are mixed with white fibers or if the black pigment granules are clumped then the result is the range of greys and blues and silvers. So black eumelanin is responsible for the entire range from coal black, grey, blue to silver. This range is already fully expressed in most species available to us: sheep, Angora goats, Angora rabbits, llamas and alpacas.

One sidetrack to take now is to discuss the architecture of mammalian coats. This account is simplified, but in general hairs are of two types: primary and secondary. Primary fibers are usually the thick guard hairs, secondary fibers are the soft, fine undercoat. In fiber producing species these have been modified to varying extents. In the relatively unmodified fleece the primaries are the most likely to be heavily pigmented (dark) and the secondaries are usually poorly pigmented (light). Domestication and selection have modified the fibers, and so it is possible to have uniform fine fleece of mostly secondary fibers, such as merino sheep, and indeed these can be coal black. In the luster longwools the primaries, have been made thinner and the secondaries made thicker so that all fibers are roughly the same diameter. The same change has occurred in Angora goats. Again, the entire range of black to silver is possible. In other fiber types the fleece is less changed. Cashmere goats are an example.

They have a coarse over coat and the fine, valuable, cashmere undercoat. This cashmere is usually very pale, even on a black goat. Llamas also usually have a double coat (coarse primaries, fine secondaries) but these can be uniformly dark (through the whole range black to silver).

The animals that have black eumelanin commonly fade throughout the life of the animal. As a result animals born coal black can go through several shades to end up silver or near white. This is much more pronounced in some breeds than in others.

The brown eumelanins are responsible for the chocolate browns and beiges. This variant is rarer than black, and is commonly termed moorit in sheep. The color varies from dark chocolate brown through milk chocolate to paler beige. This type of eumelanin seems to be more "fragile" than black. Animals that are brown generally fade more than do blacks, and so the darker end of the range is somewhat more rare. This could also be due to the fact that browns are rare to begin with. The colors caused by brown eumelanin are available in sheep, Angora goats (but very, very rarely), Angora rabbits, and possibly in llamas and alpacas although this is yet to be completely documented in North America.

The black range of colors has been used effectively by breeders of animals and a wide range of choices is possible. The brown range is much more limited currently, but the entire range will no doubt be available in the near future.

The pheomelanin group of red to tan colors is one of the more fascinating biologically. This is for a variety of reasons. White sheep are usually genetically programmed to be all red, but selection has removed all the red pigments, leaving a white (or ivory) fleece. This is the reason that tan fibers or tan spots on the face are so common in sheep. It is difficult to find an intensely red sheep that retains its color because so much selection has been accomplished to get rid of all the red. Another reason for the difficulty of attaining a red fleece is that the red enters the secondary fibers poorly at best, so the fleece ends up pale.

Some intensely red breeds of sheep do exist: the Barbados Blackbelly is one, but of course its "fleece" is useless since it has a hair coat like a common goat. Many breeds worldwide are both red (can even look dark brown, hence the difficulty in classifying all animals correctly) but the fleeces fade quickly. Karakul (some colors) and Tunis sheep are the best examples. These retain the red in the haired areas of face and legs. The Karakul is a double coated type which can also retain the red in the primary fibers. Still, since the secondaries are pale, the overall fleece is generally pale.

In Angora goats we have found a pheomelanic color that has not faded. This is a head scratcher, since we were thinking that the Angora goat would be similar to sheep. While it is too early to be sure of the exact mechanism it is possible to make some educated guesses. The mohair fleece has come about in a manner similar to the luster longwools of sheep. That is, the secondaries and primaries are all selected toward a middle degree of diameter that is very uniform. So, secondaries are thicker than the wild types, primaries are finer. It could be that this allows pheomelanin to fully pigment both types and result in a uniformly dark red fleece. If this is the case, then through selection (and probably *much* selection) it might be possible to achieve the same colors in the luster longwools. The range of pheomelanins is dark red-brown through reddish tans to yellowish tans and ivory. The similarity between these shades and the eumelanic browns can be remarkable, but the pheomelanins generally have a reddish tinge to them.

One point that is especially important for some more primitive breeds is that the reds can occur with black in the coat. The Barbados Blackbelly is the best example of this. So, if a part red "brown" and black lamb is born then the brown areas are indeed pheomelanic. Sheep that express the brown eumelanin are incapable of forming black, and so those two colors (brown black) cannot occur together.

Pheomelanic colors do occur in llamas and alpacas. Indeed, these are the most common colors for these species (besides white). The interesting thing here is that the secondaries and primaries are all heavily pigmented in many animals, resulting in the entire range of colors for these species.

The above table may help see which shades are available in which fibers, which need to be documented, and which may be possible through selection.

TABLE 1

	black to grey	*brown to beige*	*red to tan*
Sheep	common	rare	may be impossible
Angora goat	rare	rarest	rarer, but possible
Llama	present	?	common
Alpaca	present	?	present
Angora rabbit	present	present	present
	(Dark shades rare in all pigment types of rabbits)		

Expression of the Badgerface Pattern in New Zealand Longwool-Type Sheep

R. M. W. Sumner
"Landell", R. D. 9,
Frankton, Waikato, New Zealand

Recent reviews on colour inheritance in sheep[2, 6] summarize studies showing clearly that, as in the case of other mammals, patterning in sheep is due to the expression of a series of alleles at the A-locus. The alleles at this locus cause a switching in the type of pigment incorporated into the growing wool fibre between eumelanin (black or moorit) and phaeomelanin (red or tan). Genetically, phaeomelanin production is dominant to eumelanin production on any particular area of the body. Genes at other loci either dilute, reduce or eliminate production of the red or tan pigment resulting in the growth of our ordinary white wool. Thus the top dominant allele in the series produces an all white sheep and the bottom recessive allele a fully pigmented black or moorit sheep.

Eight distinct colour patterns have been described or postulated in non-white sheep as being due to the action of alleles at the A-locus[2] namely:

1. White (White)
2. White body with lightly pigmented belly (Grey badgerface)
3. White body with pigmented belly (Badgerface)
4. Agouti colouring in birth coat with a light-coloured belly (Agouti mouflon)
5. Lightly pigmented body with white belly (Grey mouflon)
6. Pigmented body with white belly (Mouflon or reverse badgerface)
7. Lightly pigmented all over body (Grey)
8. Pigmented body with red or tan patches surrounding eyes (Red-eye)
9. Pigmented all over body (Self-colour)

As well as resulting from the action of single alleles these patterns can also arise from the combined action of two pattern alleles. Thus while the grey mouflon pattern in Icelandic sheep was shown by segregation tests to be due to the action of both a single grey mouflon allele and the combined effect of mouflon and grey alleles the grey badgerface pattern could only be shown to be due to the combined effect of a badgerface and a grey allele[1].

This paper reviews information from segregation tests showing that single alleles for the expression of both the grey badgerface and badgerface pattern exist within New Zealand longwool type sheep breeds.

Source of Badgerface Pattern Sheep

In 1978 the author was offered 2 yearling badgerface type pattern Perendale (interbred Cheviot x Romney) rams and 5 yearling badgerface type pattern Perendale ewes. The sheep were all ½ sibs born in a small flock of pigmented ewes kept especially for handcraft use. The farm, near Te Puke in the North Island ran approximately 2,000 white Perendale ewes from which the pigmented ewes were descended.

Initial matings were carried out to confirm that the badgerface patterning was repeatable in unrelated sheep. In early 1980 S. Adalsteinsson, during a visit to New Zealand, identified the markings as similar to those of the grey badgerface pattern occurring in Iceland as a combination of the badgerface and grey A-locus alleles. A detailed mating programme was then begun to confirm the presence of a grey badgerface pattern allele at the A-locus in these sheep. A known heterozygous badgerface pattern Romney-cross ram was obtained from R. Lundie in the South Island as a comparative control.

Description of Pattern

While it is postulated that both the badgerface patterns may occur on either a black or moorit background, all segregation trials in New Zealand to date have, because of the availability of suitable numbers of sheep of known genotype, been carried out using black (*BB*) background sheep. No moorit (*bb*) background type badgerface sheep have yet been reported to describe the colour shadings which would occur. To minimize ambiguity in comparative descriptions the badgerface pattern is referred to subsequently as "black badgerface".

The areas of pigmentation for both the grey and black badgerface patterns are similar with only the intensity of pigmentation on the belly and body differing (Figures 1 and 2). At birth both patterns have a white to off-white body with

Figure 1 Ventral view of 4-month-old heterozygous grey badgerface ($A^{gb}a$—left; A^{gb} A^{b}—centre) and black badgerface ($A^{b}a$—right) pattern rams.

the grey badgerface's body being lightly pigmented particularly in an area above the cervical and thoracic vertebrae while the black badgerface has virtually no pigmentation on the body. The belly and perianal regions are pigmented in both patterns varying from mid-grey to black in the grey badgerface and black in the black badgerface. The patterns can however be accurately differentiated in males on the basis of the extent of pigmentation on the scrotum and sheath and in either sex on the extent of pigmentation on an area on the medial aspect of the hock. These areas are consistently light grey in the grey badgerface and black in the black badgerface (Figure 1). The lower legs of sheep carrying either pattern are black.

With the exception of the chin, patterning on the head of both pattern types is similar, with a white blaze on the forehead and bridge of the nose, white on upper lip, white around the eyes and white edging to the ears on a black background. The grey badgerface has a white chin and the black badgerface a black chin. The belly becomes lighter with age while other areas of the body remain unchanged.

Figure 2 Posterior view of 4-month-old heterozygous grey badgerface (A^{gb} a—left; $A^{gb}A^{b}$—centre) and black badgerface ($A^{b}a$—right) pattern rams (as for Figure 1).

Confirmation of Patterns Due to A-Locus Allele

Initially the segregation study involved interbreeding the original badgerface pattern ewes with one of the original rams to generate a ram homozygous for the patterning gene. Due to the closeness of the relationship between the original rams only one was used with the other retained as a reserve in case of misadventure. The reserve ram was never used. Both grey badgerface and black self-colour pattern lambs resulted from these matings in approximately a 3:1 ratio as would be expected if both the ram and the ewes were heterozygous for a badgerface pattern allele. The patterning of all the grey badgerface lambs was similar. All resulting grey badgerface rams born over two years were progeny tested as yearlings over 10 black self colour Romney ewes to check for homozygosity of the pattern controlling genes. Two of the tested ram hoggets were believed to be homozygous by generating only grey badgerface lambs. As with the original rams one was held in reserve and the other used for further test matings where he was mated to a group of white Romney ewes which were known from both their pedigree and test matings to be homozygous at the A-locus. All the resultant progeny were white with no distinguishing marks. The biggest of the ram progeny was in turn mated as a yearling to 15 black self colour Romney ewes leaving 12 white and 8 grey badgerface lambs. The resultant ratio between the patterns, which was not significantly different from 1:1, confirms that the grey badgerface pattern carried by the original purchased ewes was due to a single gene at the A-locus and has been called A^{gb} after the adaptation of nomenclature by Adalsteinsson[1]. A mating programme carried out by R. Lundie[5] on his own farm using the ancestors of the black badgerface ram purchased from him confirmed that the badgerface pattern gene carried by the ram was also located at the A-locus and not different from that called A^{b} by Adalsteinsson[1]. Progeny testing of the black badgerface ram obtained from R. Lundie showed him to be heterozygous ($A^{b}a$) at the A-locus.

Dominance Relationships

The relative dominance of the A^{gb} allele was checked by crossings with individuals known to be carrying A^{b} (badgerface) and A^{w} (mouflon) alleles as well as two unrelated rams born with grey body wool, white on muzzle and black legs. The effect of crossing with other alleles at the A-locus could not be checked as the A^{+} (agouti mouflon) and A^{gw} (grey mouflon) have not been identified in New Zealand to date. Also the A^{re} (red eye) pattern in New Zealand[4] appears to be different to that reported in the Corsican breed.

A^{b} Allele

The ram identified by progeny testing as being homozygous for the A^{gb} allele was crossed with a small group of heterozygous black badgerface ($A^{a}a$) Romney cross ewes over four years. All the progeny were visibly similar showing a grey badgerface pattern. A total of 10 ram lambs were retained and each progeny tested as a yearling to between 10 and 15 black self colour (aa) Romney ewes. Five of the tested rams left both grey and black badgerface pattern offspring (Figure 3) in the ratio of 7:9, 8:6, 3:7, 1:2 and 4:5 while the other five left only grey badgerface and self colour pattern offspring in the ratio of 4:7, 6:2, 4:4, 5:8 and 9:5. As the $A^{gb}A^{b}$ and A^{gba} rams were visibly indistinguishable (Figures 1 and 2) it was concluded that the partial production of phaeomelanin at a particular site, in this case the belly, dominates over complete eumelanin production at that site. This supports trends observed with other A-locus patterns[2]. There was no evidence of any consistent sex linked bias in segregation within either of the pattern types.

A^{w} Allele

To check whether there may be any visual difference between sheep carrying $A^{gb}A^{w}$ and $A^{b}A^{w}$ pattern genes a heterozygous ram carrying the A^{w} gene was mated contemporaneously to groups of heterozygousA^{gb} and $A^{b}a$ ewes. Progeny of the two groups of ewes were indistinguishable with an off-white body and belly, head markings similar to the badgerface and black legs (see Figure 4). This also supports trends observed with the other A-locus patterns[2].

Grey Pattern Alleles

Genetically grey sheep are uncommon in New Zealand with the genes involved possibly differing from those observed in Iceland.[4] Two rams possibly carrying a gene(s) for greyness were mated to groups of heterozygous $A^{b}a$ and homozygous aa ewes. Both rams proved to be heterozygous for a grey pattern gene which segregated in approximately a 1:1 ratio when crossed into aa ewes. However, although

Figure 3 Ventral view of male progeny resulting from progeny testing the $A^{gb}A^{b}$ ram in Figures 1 and 2 over aa ewes.

both rams left lambs with a definite black badgerface pattern neither left lambs with a pattern similar to the grey badgerface. The results are thus inconclusive as to whether sheep carrying the A^b allele combined with an allele for greyness are visibly similar to $A^{gb}A^{gb}$ or $A^{gb}a$ sheep.

Figure 4 Adult heterozygous combined pattern ($A^{gb}A^w$—left; A^bA^w—right) rams.

Histological Basis of Greyness

Skin samples were taken from 14 $A^{gb}a$ and A^ba lambs at 4 months of age, sectioned horizontally through the sebaceous glands and examined to determine if a pattern of pigmentation existed between primary and secondary follicles.

Pigmentation was present in the mid-side region of the body for both badgerface pattern types, but while it tended to be primary follicles which were pigmented, the frequency was too low to be of significance. Regions on the belly near the navel and the base of the scrotum on males and anterior to the nipples on females had a secondary to primary follicle ratio similar to that on the mid-side region of approximately 3.4:1. In the case of the black badgerface lambs all the primary follicles and in excess of 80% of the secondary follicles were pigmented while between 70 and 100% of the primary follicles and generally less than 40% of the secondary follicles were pigmented in the grey badgerface lambs. A similar pattern of pigmentation also existed on the medial aspect of the hock where the secondary to primary follicle ratio was approximately 2.0:1. These trends in pigmentation suggest that in young lambs all follicles in pigmented areas of the A^b pattern grow pigmented fibres whereas it is only those follicles initiated during the early growth waves of follicle initiation on the foetus[3] that produce pigmented fibres in the case of the A^{gb} pattern.

Overall Productivity

Data collected over 10 years of segregation trials with A^{gb} and A^balleles suggest they are without any pleiotropic effects on overall wool and lamb production compared with equivalent type *aa* sheep.

The wool, particularly that of the grey badgerface, has a unique place in the handcraft industry where subtle shading changes may be required within an individual fleece for a particular item. Grey and black badgerface skins are also unique as tanned skins for floor rugs.

Both badgerface patterns have been observed in flocks of coloured longwool type sheep in New Zealand, other than those from which the studied sheep were sourced, since the commencement of the segregation trials described in this paper. The frequency of the alleles in the population is however low. Expression of the gene within New Zealand Merino type sheep has not as yet been evaluated.

Summary

Segregation studies have demonstrated the existence in New Zealand long wool sheep breeds of two distinct badgerface pattern genes, the grey and "normal" or black badgerface, within the allelic series located at the A-locus. The two alleles are designated A^{gb} and A^b. Dominance relationships both between the two alleles and between the badgerface alleles and other alleles in the series indicate dominance of phaeomelanin production over eumelanin production as previously reported between other alleles in the series. To date it has not been possible to satisfactorily combine the A^b allele with an allele for greyness from New Zealand sheep in a heterozygote to show a sum effect similar to that of the A^{gb} allele. Greyness within the light badgerface pattern is due to the inhibition of eumelanin production within the later initiated secondary follicles.

References

[1]Adalsteinsson, S. (1970). Colour inheritance in Icelandic sheep and relation between colour, fertility and fertilization. Journal agricultural research, Iceland 2:3–135.

[2]Adalsteinsson, S. (1983). Inheritance of colours, fur characteristics and skin quality traits in North European sheep breeds: A review. Livestock production science *10*:555–567.

[3]Hardy, M. H., and Lyne, A. G. (1956). The pre-natal development of wool follicles in Merino sheep. Australian journal of biological science *9*:423–441.

[4]Lundie, R. S. (1984). Colour patterns in New Zealand sheep. Coloured Sheep and Their Products. Ed. H. T. Blair. New Zealand Black and Coloured Sheep Breeders Association, 128–138.

[5]Lundie, R. S. Unpublished.

[6]Ryder, M. L. (1980). Fleece colour in sheep and its inheritance. Animal breeding abstracts *48*:305–324.

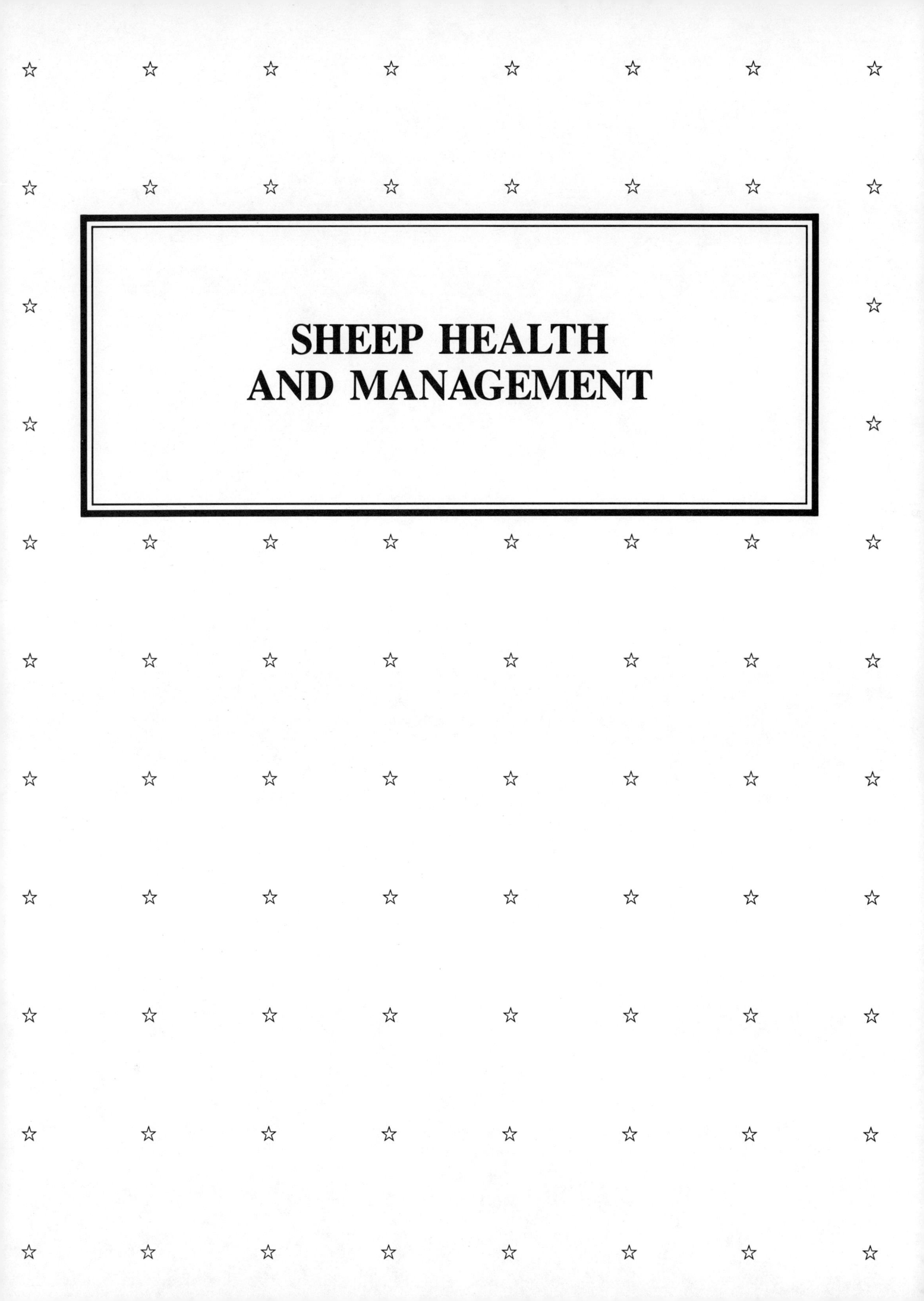
SHEEP HEALTH
AND MANAGEMENT

Sheepcoats . . . A Wrap Up for Clean Wool

Wendy S. Dennis
"Tarndwarncoort"
Warncoort Victoria 3242 Australia

According to *The Australian* newspaper, October 19, 1935, the Phoenicians were skilled woolgrowers and introduced fine wool sheep to Europe. Before the Trojan War, these flocks were the best in the world, producing raw material for Kings' clothing. It is believed that much of the Greek, Roman and Spanish sheep and wool knowledge was handed down by the Phoenician masters who paid special attention to the protection of the growing fleece from dust and dirt. It is believed the Romans went as far as removing the coat and combing the wool, moistening with oil or pure wine and washing with water 3–4 times a year. We may be sure, therefore, that coating of fine wool sheep is as old as history itself.

Coated Polwarth wethers, Tarndwarncoort.

First Australian Trials

In 1934 "a black faced animal of no particular breed, held at Port Adelaide for use as a leader when trucking or shipping sheep" was coated by Mr. Spen. Williams, Manager of the Wool Department, Goldsborough, Mort & Co. The trials that followed were to reduce the dust, dirt and burr contamination in the wool of Australian Merino sheep using jute or light canvas coats.[20]

This was followed by trials on the effect of coating on the milk production of Merino ewes and on the growth of their lambs resulting in no significant increase in either.[18]

The Waite Agricultural Research Institute in South Australia carried out further trials[19] and experiments were carried out in New South Wales.[16] The overall picture was rather contradictory, depending on where the trials were done. Some found body weights increased, an advantage in fat lamb production, a higher yielding wool and more attractive appearance, therefore calling for a higher value. Others didn't.

Further Trials in Australia

The 1960's saw a breakthrough in the material and the cost of sheepcoats: Polyethylene, a similar material to that used in wool bales and about ¼ the cost of the old canvas type coats.[23]

In 1971, Abbott & Leader of the CSIRO Division of Textile Industries together with chemical companies, farmers and State Departments of Agriculture, experimented on 100 different fabrics. These were tested on outdoor racks at Geelong (Victoria), Griffith (N. S. W.) and Barcaldine (Queensland) for 1–2 years. Rapid advances were made in that time for a most suitable material. A white polythylene cloth emerged, which contained a newly developed hindered-amine stabilizer, retaining 97–100% of its strength after 1 year and nearly 90% after 2 years. Various designs of coats made from the different materials were also tested, some of which were at "Tarndwarncoort".

During the various trials, it was discovered that the white material reflects more heat and is up to 8 degrees cooler in the summer.

From 1971 onward, the CSIRO, University of N.S.W. School of Wool and Pastoral Science, Department of Agricultural, the Australian Wool Board and the Bureau of Agricultural economics with Gollin Company Ltd., ran field trials on some 20,000 coated and 20,000 uncoated sheep on 40 properties in New South Wales. The sheep ran under the widest possible range of physical, geographical and climatic conditions possible, and results were closely supervised.

Three years later, field trials were run on 2,500 coated and uncoated sheep in Western Australia in areas subject to severe dust, sand and weed problems. The results of these trials indicated the following:

1. The sheepcoats used insulated against cold, wet and windy conditions and markedly reduced "off-shear" deaths, making winter shearing a more practical proposition.
2. The incidence of fleece rot and staining was reduced.
3. Sand, dust and discolouration were eliminated.
4. Burrs, medics, grass seed and vegetable fault were eliminated.
5. Reduction in body strike.
6. General improvements in the condition of the sheep, being warmer in winter and cooler in summer, giving a more uniform body temperature.
7. Cleaner and faster shearing.
8. Six- to nine-month-old lambs are protected from grass seed infestation and so the export meat and skins are protected.
9. The oxidised or weathered, burnt tip of the wool was reduced.

Abbott (1978)[1] concluded that the use of sheepcoats to protect fleece wools throughout a full growing season produced higher yielding, cleaner wools per sheep. The mean length of fibres in the top was increased and the proportion of noil and vegetable particles reduced.

In extensive field trials in 1956–58 Lipson, Ellingworth and Sinclair reported a typical analytical wool result as follows:[15]

	Content %	
Component	*Rugged*	*Unrugged*
Wax	15.1	10.7
Suint	6.1	6.9
Dirt	3.1	8.9
Burr & Seed	0.9	3.0
Dry wool	59.9	55.5

In 1981, an experiment in "Fading with light and greying with age in the fleece of black Australian Polwarth sheep"[10] was conducted at "Tarndwarncoort". Coated and uncoated wool samples were taken from sheep at 3 month intervals for 12 months. It was concluded that the coats do prevent oxidation or fading of the tip of coloured wool, but cannot prevent any greying of wool which comes with the aging of sheep.

Trials in America

The Laramie Plains north of Laramie, Flag Ranch (Albany County) & Rawlins, Wyoming were the areas used for 6 experiments between 1938 and 1944 by the Wyoming Experiment Station. The object of these experiments was to discover whether the coats were practical for "range sheep", for increased wool, mutton and lamb production.[4]

Based on experimental results and observations, the summary was as follows:

Advantages:

1. Particularly useful on older sheep grazing on cold, windy, sandy ranges.
2. Improve the appearance of the wool, particularly the colour and blockiness of the tip.
3. Decrease the number of burrs in the wool.
4. Decrease the weathering of the wool.
5. Decrease the dirt content and shrinkage of the wool on dry, sandy ranges.
6. Increase the length of staple of the fleeces.
7. Decrease the death loss in older ewes.
8. Decrease the amount of wool pulled out by brush catching the fleece.
9. Protect the sheep from storms after early shearing and during the winter.

Disadvantages:

1. The coat must be properly tailored, designed and adjusted to stay on the sheep.
2. The coats tear on wire fences and snags.
3. The coats need adjustment at least every six months.
4. Coats of No. 10 duck need replacement at least every 2 years.
5. The initial cost of coats, additional help and vigilant herders necessitate additional expenditure.
6. Coats under the conditions of the tests are not practical for 7 months from October to April.
7. It is difficult to obtain a corresponding increase in price for the increased yield and length of wool.
8. Even heavy burlap is not sufficiently durable for sheep coats.

Later Stauder (1955) and McFadden & Stauder (1958) experimented in the New Mexico area and found value in coating sheep for protection against sudden storms, reduced shearing and marketing costs and improved fleece appearance. Oxley & Burns (1959) reported "no death from coyotes" and believed that the sound from the plastic coats or the different colours deterred them.[17]

Trials in New Zealand

In 1974, 3,000 sheepcoats were sent to New Zealand for trials on the South Island by the Department of Game and Fisheries. From these trials, it was reported that because farming conditions are so clean that although there was some improvement in the wool quality, it was not enough for sheepcoats to be a viable proposition. These coats were tested on stronger wool type sheep than those in Australia.

Sheepcoat Management and Fittings

In Australia today, sheepbreeders are using sheepcoats for a variety of reasons. White Merino flocks are grazed in paddocks in sheepcoats for 12 months and the wool sold at auction. Super fine white Merino sheep, sometimes referred to as Sharlea Sheep, are shedded and coated to prevent contamination from the dry feed dust. This practice is at times governed by wool prices. An ultra fine Merino woolgrower in Victoria now has 5,000 sheep under cover in specially designed sheds and wearing sheepcoats.

Young sheep in colder climates are coated to grow them out quicker at a younger age. Some woolgrowers only coat their sheep at strategic times of the year . . . i.e., over the summer months to prevent burr and grass seeds; for a few weeks after shearing to prevent off-shears shock in the case of cold, wet and windy weather; or in the winter in very muddy areas. A number of coloured sheepbreeders use them for super clean wool for the handcraft industry, using coats for the whole 12 months or over the summer months only.

The coats are useful on older-type sheep to help retain condition and prevent sunburnt skin on their backs.

The more elaborate ram rugs are used on show sheep prior to showing.

At Tarndwarncoort, the coats help prevent a burnt tip on the wool of the sheep kept for eventual tanned sheepskins.

There are contrary indications to coating sheep especially coloured wool for handspinning:

1. Old sheep with "doggy" wool.
2. Young growing sheep if careful and regular checking is impossible.
3. Strong wool sheep of approximately 46's (34 microns) and stronger. The wool tends to mat and cross fibre under the coat.

Sheep should be run in paddocks as free as possible from faulty fences or thorny bushes and sheepyards should be checked for protruding objects to prevent the risk of tearing coats. Young growing sheep must be checked every few weeks for tight coats. There is rapid growth in body as well as wool.

Ewes can be mated, lamb and mother the lamb(s) and rams can be joined to ewes in coats. The coats do reduce the incidence of fleece rot, staining and body strike, so sheep can be run in more humid conditions in full wool.

It is possible to crutch sheep in coats, but a better job is done by removing the coat completely. At Tarndwarncoort, where 700 sheep are coated, crutching is done when the coats are changed to a larger size at approximately 8 months wool growth. Standard, non gusset coats are fitted off-shears and changed to a gusset coat without ties, stitching or studs at crutching time.

Sheep can be treated with pour-on body lice insecticides, then coated, but if a plunge or spray dip is used, the sheep must not be coated for approximately 3 weeks after shearing so any cuts may heal. Naturally, with small numbers of sheep, they can be coated before dipping and recoated again afterwards. It is also important not to coat wet sheep.

Whatever the reason to use them, it is most important to make sure the coat fits the sheep. Too big and the back leg slips out of the strap or the front leg goes through the neck hole. Too small and the coat rides up over the rump and the back leg straps will rub.

The best way to coat a sheep is to lightly fill a small pen making it possible to move about among the sheep. Too empty and the sheep will rush around the pen, too full and one cannot move easily with the sheep. Have a range of sheepcoat sizes within handy reach.

Pull the coat over the head through the front opening, stretch the coat over the back, place your hand through the back leg strap, grab the hoof and lead it through the strap; The same for the other leg. In this manner, it is possible to coat 100 shorn sheep in 3/4 hour.[9]

The coats are removed on the same principle. Pick up the back leg, slip it back through the strap, same for the other leg, pull the coat over the back and head. One hundred woolly sheep can be uncoated and classed in one hour by one person.[9]

After coating some sheep will run and jump about with the feel of the coat but they soon settle down. Sheep who have been previously coated take no notice at all.

Coated fleeces are a little more difficult to throw onto a wool table as they do not have the normal weathered tip to hold the wool together.

At Tarndwarncoort, fleeces for the handspinners are skirted first in the usual fashion, removing shorter, discoloured and heavily conditioned wool and fribs. They are then skirted a second time to remove fleece wool not covered by the coat. 35% to 40% of wool is skirted off the finer wooled sheep, less of strong wools.[9] After testing for soundness and classing for quality, the fleeces are rolled up carefully, and baled up or stored on wooden racks in a well ventilated wool room.

Advantages of Coating Sheep

Polwarth & Border Leicester Cross sheep, both white and coloured have been coated at "Tarndwarncoort" over the past 15 years. Without doubt, the most outstanding attribute being the cleanliness of the wool. The Western District of Victoria is regarded as one of the cleanest and finest wool growing areas in the world, yet the coats have enhanced the wool.

Other wool properties of importance especially to the coloured sheep and handspinning wool industry include:

1. Well grown and nourished wool because of a more even all year round body temperature and healthier sheep.
2. No burnt tip giving a softer handling wool.
3. Slightly longer staple length due to both the above.
4. Absence of faded, discoloured, oxidised tip, making it possible to grow an even coloured fleece, especially black.
5. Blocky staples within the fleece, an advantage to handspinners.
6. The wool grows in such a way that wool carding can be eliminated.
7. Wool grown along the back line is unaffected by solar degradation which causes protein breakdown in the fibre when exposed to temperatures of 90 degrees F(32 degrees C) and above.
8. Reduced tender wool due to less fly strike and point number 1.
9. Possibly less skin pieces and second cuts in the fleece due to easier, cleaner shearing.
10. Other advantages are listed under results of trials in Australia & America (pages 2, 3 and 4).

Disadvantages

1. Fitting and frequent checking of coats is time consuming, but must be done for: prevention of rubbed back legs leading to fly strike; disintegration causing wool contamination; falling off during the year; and sheep cast in full wool or off shears; heavily in lamb.
2. The initial cost of the coats 12 months before selling the wool.
3. Depressed neck wool making show preparation difficult. Sometimes a felted band of wool is difficult to shear through.
4. White wool sometimes has a yellowing appearance but will wash out to white.
5. Heavy skirting of uncoated fleece wool.
6. Fine wool occasionally has a greenish tip after prolonged wet weather.
7. Masking wool faults if your breeding sheep are coated.

Present Day Coat Designs, Material & Usage

There are various different coats on the market in Australia today, most having a similar design:

1. The standard coat has a front neck seam running in front of the front legs.
2. Gusset type coats have a gusset sewn into the front seams which lets out with ties, stitching or press studs to accommodate extra wool growth, neck folds or sheep breeds with broader type chests.

3. Some back leg straps are cut and folded material sewn into place while others have a heat sealed cut in the fabric. Others have tapes or ropes sewn to the coat.
4. Neoprene rubber or similar is sewn into the neck opening, along the flank and around the back or hindquarters on some coats.
5. Colours range from white and yellow to blue and green in high density, woven split Polythylene tapes, U.V. Stablised uncoated material.
6. The material will last 12 months or more depending on how and when the coats are used.
(e.g., coats used in the Northern Hemisphere will last longer due to a lower degree of U.V. radiation. Few manufacturers will guarantee a length of time.)
7. Most coats are stitched with waterproof thread.

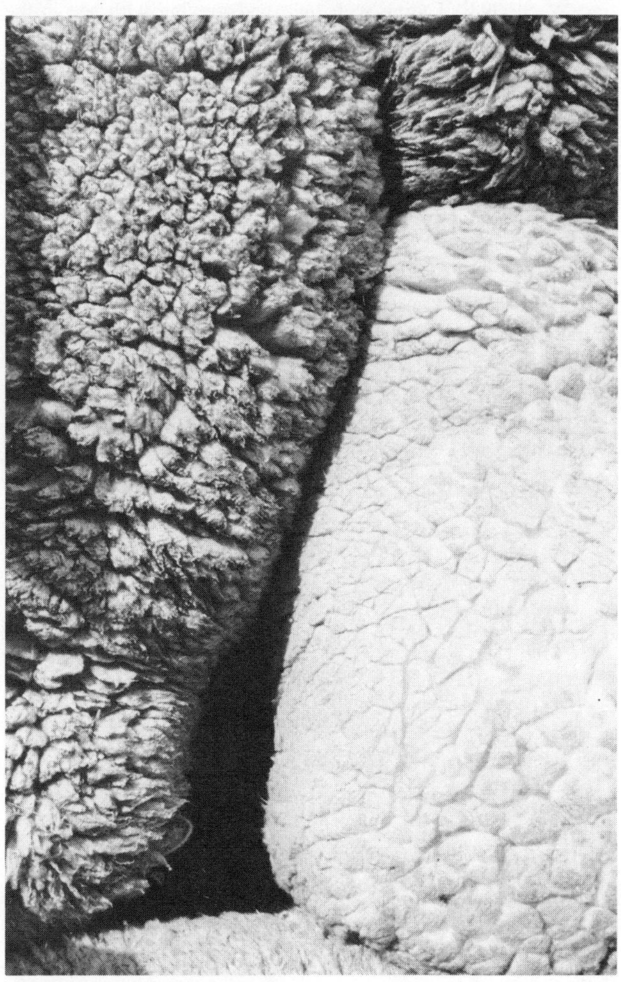

Birds-eye view of coated (R) and uncoated (L) white Polwarth sheep.

Attempts have been made to design a coat which will cover more of the valuable neck wool by extending the neck opening up toward the ears and possibly attaching it to small horns with ties or velcro tape.

The sheepcoat of the future is already being designed for biological wool handling. The CSIRO and Coopers Animal Health are at present experimenting with a net type coat, basically of 6–6 stretch nylon, to support the wool on the sheep for 6 weeks between inoculation and harvesting the wool. The nylon will blend with wool should any contamination occur and the coats may be reusable for 3 to 4 years.

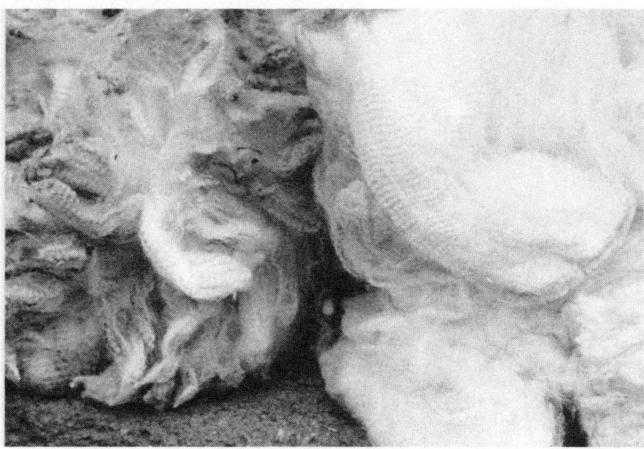

Fleeces from coated (R) and uncoated (L) white Polwarth sheep.

Of the 160 million sheep in Australia today, it is difficult to estimate the total number of coats being used. Information received from a questionnaire sent to all sheepcoat manufacturers and some woolgrowers before writing this paper indicate an increased usage over the past 3 years. New South Wales, a state where high sheep numbers and harsh, dry conditions prevail, uses the greatest number of coats, followed by Tasmania in the South, Victoria, South Australia, Queensland, Western Australia and none in the Northern Territory.

Angora and cashmere goat breeders use sheepcoats to keep the fibre clean and dairy goat breeders find them useful to lift milk production.

Many overseas countries have shown an interest including U.S.A., where Australian sheepcoats are available, New Zealand, Canada, U.K., Russia, China and South Africa.

The following are contact addresses for those Australian sheepcoat manufacturers I know of:

Aust-Ag Sheepcoats,
PO Box 157
Bungendore, NSW 2621

Gollin Sheepcoats
Frank Tope & Partner
PO Box 10
Camperdown, NSW 2050

Hyline Sheepcoats
PO Box 827
Queanbeyan, NSW 2626

OE & DR Pope Pty Ltd.
5 Boundary Road
North Melbourne, Victoria 3051

Nimmity Sheepcoats
69 Fox Valley Road
Wahroonga, NSW 2076

White Polwarth ewes in full wool showing the effectiveness of coating, Tarndwarncoort, Victoria.

Summary

Although time consuming, work intensive, costly and at times, very frustrating, sheepcoats are the answer to ultra clean, attractive, well grown, long, soft, blocky stapled wool without a faded burnt tip, which need not be carded for the creative and caring handspinner.

References

[1]Abbott, G. M. (1978) C.S.I.R.O. Div. Tex. Ind. Report No. G36.

[2]Austin, K. M. (1936) Sheep rugging trials on Therribri, NSW. Pastoral Review 46, 11:1207–1208.

[3]Blummer, C. C. & Cotsell, J. C. (1938) Agr. Gaz. NSW, 4:188–190, 49.

[4]Burns, R. H. & Johnston, A. (1950) A. Wyo, Agr. Exp. Sta. Bull., 298.

[5]Dircks, A. (1956) Wool Tec. Uni. of Tec., NSW, 3, 2:31–33.

[6]Duncan, J. E. (1938) New Zealand Jour. Agr. 56, 409.

[7]Duncan, J. E. (1939) New Zealand Jour. Agr. 58, 315.

[8]Dennis, W. S. (1979) Breeding Coloured Sheep and Using Coloured Wool. Peacock Pub. 100.

[9]Dennis, W. S. (1987) Women and the Family Farm "Black is Beautiful" Marcus Oldham Farm Management College, Geelong (Aust.).

[10]Elliott, E. A. (1939) gric. Gaz. NSW 50:65.

[11]Lauvergne, L. L., Burrill, M. J., Dolling, C. H. S., Dennis, W. S. (1981) Fading with light and greying with age in the fleece of black Aust. Polwarth sheep. Ann. Genet. Sel. anim. 13:93.

[12]Levy, A. L. (1939) Agr. Gaz. NSW, 50, 3:142–161.

[13]Lipson, M. & Black, A. F. (1944) J. & Proc. Royal Soc. NSW 78:84–93.

[14]Lipson, M. & Walls, G. W. (1965). J. Tex. Inst., T104, 56.

[15]Lipson, M. Ellingworth, J. S. Sinclair, J. F. (1970) Text. Inst. Ind. 8:100.

[16]Montgomery, I.W. (1938) Jour. Counc. Sci. & Ind. Res. 11, 3:221–228.

[17]Oxley, J. & Burns, R. H. (1959) Wyo. Agr. Sta. Bull. 361.

[18]Pierce, A. W. (1938) Jour. Counc. Sci. & Ind. Res. 11, 3:229–232.

[19]Pike, K. A. (1939) Jour. Dept. Agr. for So. Aust. 43, 4:315–318.

[20]Scott, R. C. (1935) Jour. Dept. Agr., So. Aust. 39:534.

[21]Williams, S. (1935) Past. Rev. 45, 4:380.

[22]Wheeler, J. L., Hedges, D. A. & Mulcahy, C. (1977) Aust. Jour. Agr. Res. 28:721.

Evolution—Elephants, Sheep, Salt and Selenium

Ronald B. Parker, Ph.D.
Sammen Sheep Farm
Rt 1, Box 153
Henning, MN 56551 USA

Modern living organisms are a product of a lengthy course of evolution extending backward in geologic time for more than a billion years. During that enormous span of time many simple and complex organisms have appeared and later died off to become extinct. A smaller number of species are still remaining.

It is axiomatic that extinct forms disappeared because of their inability to adapt to changing environmental conditions. The kinds of conditions that could give rise to extinctions are legion, ranging from climatic variation to decimation by predation or failure to compete for food with rival organisms.

The Chemical Environment

A central environmental condition that affects all organisms is the chemical makeup of the environment, including the chemical composition of available feed. Organisms show a remarkable ability to thrive in extreme chemical conditions such as the absence of oxygen, or extremes of acidity or alkalinity.

However, the vast majority of life environments are not chemically extreme, and most successful organisms are partly so because they are adapted to average environmental chemistry. Organisms have automatically adapted to chemical elements at the levels of concentration to which they and their ancestors are and were normally exposed.

For example, most Earth organisms are well adapted to the presence of large amounts of a chemically active, one might even say corrosive, gas—oxygen. Such an adaption would seem remarkable to a visitor from another world. In a less dramatic way, organisms are well adapted to other common elements such as iron, calcium, sodium, and a host of others.

The naturally-occurring form of an element is an important factor. Organisms are highly adapted to phosphorous in the form of phosphate, which is in fact an essential nutrient for most. They are ill adapted to pure elemental phosphorous which is a corrosive poison. The reason is simple; phosphorous does not occur in nature in elemental form so there was never an evolutionary advantage in adapting to it.

A comparable situation exists for aluminum, a very abundant element in the outer part of the earth. Aluminum occurs in the life environment almost entirely in geochemically inert compounds such as clays. Thus it is not available in the geochemical environment, so organisms never adapted to its presence. It is as if it was a rare trace element instead of the third most abundant in the outer part of the Earth. One deadly aspect of acid rain is its ability to liberate aluminum into a biologically-available state in which it is toxic to many organisms.

Nutrients, Trace Elements, and Poisons

The heading of this section is the type beloved of those who like things in neat packages. It sets some items apart in a clearly-defined way. It is perfect for writing tables in textbooks or for examination questions.

Yet, it is the classification itself that could do with examination. What distinguishes a nutrient from a trace element or a poison? That question is one that sheep producers should examine with the same interest as academic biologists—perhaps more so.

A nutrient is clear enough. It is some substance that is essential for the healthy growth of the organism. Sheep need some source of protein and they need a source of energy. They also need water. Beyond that what? They need calcium and phosphate. They need sodium. Those are some major nutrients.

Sheep also need a host of so-called trace elements—also called trace minerals. They need iron, manganese, cobalt, iodine, zinc, molybdenum, copper, selenium. The list could go on.

What don't sheep need? If they eat too much phosphate and not enough calcium, they may produce kidney or bladder stones, urinary calculi, and succumb to water belly. If they eat too little phosphate, they will succumb to other health problems. If they ingest too much copper, they will die dramatically when stressed. If they lack selenium, they suffer and die from white muscle disease. If they are oversupplied with selenium they suffer equally lethal consequences.

In the context of the foregoing, I will discuss two common nutrients that are significant for sheep—sodium and selenium. The role of these elements can be understood by an examination of their geochemical occurrence in space and time. Lessons learned from the significance of those elements in sheep health may help us understand a bigger picture.

Sodium and Evolution

Sodium is widely available in large regions of the world. For at least the past billion years there has been a state of equilibrium with the rate of sodium carried to the sea by rivers balanced with the amount returned to the land masses through the air and through marine sediments that become part of the land.

However, there are many parts of the land masses where sodium is in short supply. There are also types of natural food sources that are inherently relatively low in sodium.

If one looks into the ancient ancestry of various lines of descent of animals, it seems that most lines begin with carnivorous types[1]. Because sodium is an important part of the flesh and fluids of animals, carnivores ingest considerable quantities of sodium as a normal part of their diet. For carnivores, sodium is an element in excess. Metabolic processes for carnivores therefore favor excretion of excess sodium to keep body levels to an acceptable minimum.

Successful lines of carnivores soon reduce their food supply to the extent that evolution favors descendants that include plants in their diet. The course of evolution leads to more and more herbivorous animals within any given line of descent.

Plants contain a great deal less sodium than animal tissue[2]. As a result, the herbivorous descendants of carnivores are faced with a problem of sodium retention rather than excretion. That is to say, the herbivores that survive the merciless culling of evolution are those whose metabolic processes allow them to retain an essential nutritional element that is in short supply.

Horses

Horses are probably the best-documented line of terrestrial mammals. One reason that the horse kindred is so well documented is that the line persisted for the last 60 million years or so of Earth history. Whatever the horses did, it must have been right.

Horses appeared on the scene at a time when land vegetation was undergoing rapid change. Modern flowering plants appeared in profusion. Over the course of millions of years, grasses became relatively more important at the expense of trees, shrubs, and forbs. As the horse line evolved, the succeeding types of horses became more and more adapted to the relative increase in grass forage. The change from tree and shrub browsers to grass grazer is recorded forever in the evolutionary change in the teeth of the line of horses.

But, what of sodium? Grasses are very low in sodium relative to woody plants. Therefore, the horses must have adapted to a diet low in sodium. Heinrich Toots and I tested that idea. We studied the sodium content of enamel from fossil horse teeth. We used enamel of teeth because it resists sodium removal more than other skeletal tissue[3].

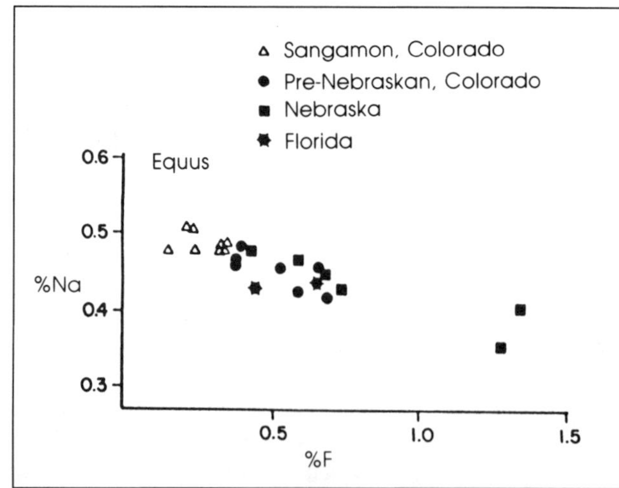

Figure 1 Sodium and fluorine in *Equus* enamel.

Figure 1 shows the enamel sodium levels for a number of fossil specimens of the horse genus *Equus*. The plot is sodium versus fluorine. Fluorine in an element that is added to fossil teeth as sodium is removed by natural processes. We theorized that sodium removal was proportional to fluorine addition. The straight line relationship in Figure 1 supports that theory. The line also passes through a hypothetical composition of an enamel mineral with the maximum possible fluorine added (3.8%), and with 0.1% sodium—the amount in the most highly leached skeletal samples we examined[3,4,5]. We assume that all fossil samples tend toward that hypothetical composition.

Teeth from living mammals contain negligible fluorine (ca. 0.02%). Because of that fact, we can reasonably project the data from fossils to zero fluorine to obtain a normalized sodium value that is presumed to be close to life values. In Figure 1 that value would be just over 0.5% sodium.

One more conclusion can be drawn from Figure 1. Sodium levels in enamel could be a function of either geochemical life environment, or metabolic regulation in the animal. The effect of environment is rejected because the specimens of *Equus* used are from three different regions. Environmental sodium levels in Florida were presumably many times those in Colorado and Nebraska.

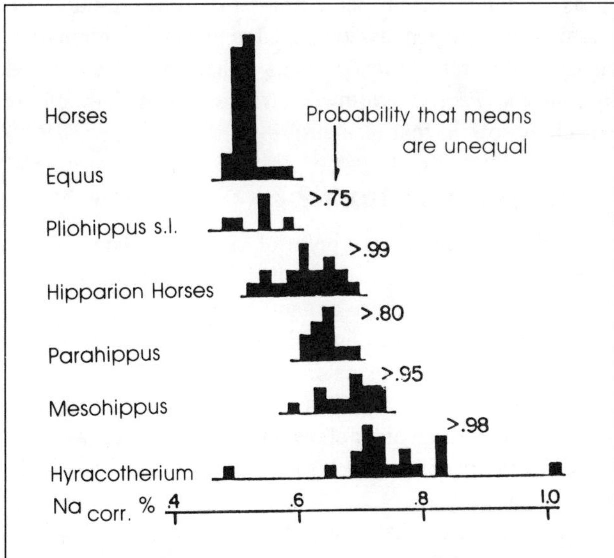

Figure 2 Enamel sodium in the horse lineage.

One would expect normalized sodium levels to be highest in carnivores and least in highly evolved herbivores. Table 1 compares sodium levels for four genera. As predicted by the model, the coyote and bear have the highest sodium. The bear, more omnivorous than the coyote, contains somewhat less. *Subhyracodon*, an early member of the rhinoceros family, has still less. *Equus*, a highly advanced grazing herbivore has the least sodium.

TABLE 1 Sodium in Enamel

	%Na
Canis latrans (coyote), modern	0.86
Ursus americanus (black bear), modern	0.74
Subhyracodon, fossil	0.70
Equus, fossil	0.52

The same trend can be seen in the horse family alone. Figure 2 shows normalized sodium values for the horse lineage from *Hyracotherium* (the official name for *Eohippus*, the dawn horse) through several intermediates to *Equus*. The progressive decline in life (corrected) sodium values is plain.

Other Mammals

Toots and I examined a variety of other mammals. Modern types of Pliocene or younger age ($<$5 million years) grazing animals such as camels, bison, goats, sheep all have sodium levels comparable to *Equus*. *Bison*, the grazer *par excellence*, has even lower sodium than *Equus*. Rodents have the lowest levels we have encountered, so don't turn your nose up at that mouse or rat. It is more advanced than you in at least one way.

The extinct titanotheres, giant herbivorous mammals that were abundant in the Tertiary, have sodium levels comparable to carnivores. Was their extinction caused by their inability to retain sodium? The proboscidians (elephants and their kindred) have high sodium (0.68%). Was that a factor in their extinction over most of the world in the Pleistocene?

The distribution of modern elephants in Africa is closely correlated with environmental sodium,[6] and they depend on food that is rich in sodium[7,8]. Modern elephants are also avid seekers of natural salt accumulations. For a herbivorous animal they have primitive sodium retention.

Sheep

Sheep are among the sodium-retention elite. Some studies of modern sheep give further insights. Studies of sheep and sodium are reported in a phenomenal compilation by Denton[9]. Some highlights will be reported here.

That sheep regulate tissue sodium effectively is shown by studies in which sheep were given excess sodium or deprived of sodium. Bone sodium levels were little affected by either excess or lack.

Also, sheep effectively regulate sodium in their milk irrespective of their dietary sodium input. Sheep deprived of sodium in their diet produce less milk. Sheep with adequate sodium produce more milk. The sodium level of the milk stays constant, but the total milk volume changes. The lesson is clear. If we want our sheep to produce enough milk, we must provide them with adequate salt. Pliny noted in the first century that with adequate salt the sheep's ". . . supply of milk is much more copious."

Numerous experiments have been made in which sheep are deprived of sodium, and then offered saltwater or another sodium source. They consume it eagerly, but through some highly effective feedback mechanism, consume only what is required to restore normal body levels. The experimental sheep were Merinos. Excess salt consumption has been reported for Suffolk and Dorset breeds[10].

Sodium appears to be especially important for reproduction. Sodium intake doubles in early gestation, and remains at the increased level throughout that period. At parturition there is an abrupt, temporary drop in consumption followed by a rise to four times the intake of a non-pregnant sheep. At weaning there is a dramatic drop to gestation levels which is quickly followed by a reduction to normal levels. Female shepherds should understand this pattern inasmuch as they may have experienced the well-known craving for salty foods such as pickles and potato chips during their own pregnancies.

Sheep are mixed grazers and browsers so one might expect that they would not have evolved to be such efficient sodium regulators inasmuch as they would have received plenty of sodium from leaves and bark of trees and shrubs. However, the ancestors of domestic sheep probably evolved in continental interiors where environmental sodium is low. Also, they probably evolved in mountainous regions—once again places with low environmental sodium. Adaptation to their environment meant adaption to sodium conservation.

The lessons are clear for the shepherd.

1. Most sheep seem to limit their intake of salt to appropriate values, so it can be provided free choice.
2. Adequate salt is especially important during gestation and lactation. There are indications that salt levels during breeding are also important.
3. Removal of salt supplements just prior to weaning should aid in reducing milk production.
4. During maintenance periods between weaning and breeding, sheep will maintain adequate sodium levels with minimal supplementation if they are not in a markedly sodium-deficient environment. Barking of trees can indicate a sodium deficiency, and would suggest that supplementation is needed.

Selenium in an Evolutionary Context

Selenium is a classic case of the blurry distinction between nutrients, trace elements, and poisons. Selenium first became known to science as a poison. It was a toxic by-product of the smelting of ores. It was an element to be avoided except as a coloring agent in red glass for automobile tail lights, and the light-sensitive material in early photocells. It now has an essential commercial application as the light-sensitive material in the modern photocopying machines that we can't seem to get along without.

Background

It was in the poison context that O. A. Beath of the College of Agriculture at the University of Wyoming began to study selenium in animal feeds. The university is situated in the Laramie Basin, a region underlain by sedimentary rocks with mountains to the east and west composed of older, crystalline rocks.

Beath found that certain kinds of Cretaceous sedimenatry rocks in the basin produced soils with 2–10 ppm selenium with some as high as 100 ppm. Ordinary soils contain only 0.1–2 ppm. Further, growing in the seleniferous soils are plant species that accumulate selenium. One species of locoweed (*Astralagus*) contains as much as 1.5% selenium[11,12].

Occurrence of seleniferous soils can be highly localized. In the Laramie Basin the soils form on seleniferous shales that may have an exposure width of only a few meters or tens of meters. In other regions selenium can concentrate in low lying regions where drainage from surrounding areas brings in selenium. Beath was fooled by the geological localization of selenium. He purchased some hay for feeding trials, and was careful to obtain one load of hay from a mountain ranch in order to avoid the seleniferous soils of the basin. He later found that the mountain hay was cut from a meadow that was underlain by the same seleniferous rocks as in the basin.[13]

It was not until the late 1950s that selenium was recognized as an essential nutrient[14]. Since that time, research on selenium as a nutrient has expanded rapidly[15]. Selenium deficiency in domestic animals seems to have been recognized first in the Pacific Northwest where irrigated alfalfa is especially low in that element.

Nutrient or Toxin?

From an evolutionary point of view, organisms should have adapted to average levels of all chemical elements in their environment. The average amount of selenium in the outer parts of the Earth is 0.1 ppm. Therefore, diets with 0.1 ppm selenium should be adequate.

As a rule of thumb, deficiencies can arise if an element is present at levels an order of magnitude less than the average. Toxicity can arise if the level is an order of magnitude higher than the average.

In the Pacific Northwest and Florida, selenium levels in forages are 0.02–0.03, and deficiencies in sheep are widespread. Levels of 0.05 ppm in the Great Lakes region produce less severe deficiencies, but are still inadequate.

As Beath and others since have found, selenium levels of 3 ppm and above cause toxic effects with sheep. The rule of thumb works pretty well.

Once again, organisms adapt to their environment. Sheep are adapted to average levels of selenium, and their diet has to fit in a gap between deficiency and toxicity. On the other hand, selenium-concentrating plants have evolved to thrive in soils that are toxic to ordinary plants. Almost every possible environment is occupied by some organisms.

Selenium Status

Figure 3 is a modern version of a widely copied map that purports to indicate the geographic distribution of selenium in forages. Like all generalities, it is useful, but also misleading.

The difficulty with a map like Figure 3 is that it fails to take into account the geological fact that distribution of selenium is highly variable on a small scale. I did a study of my region in west central Minnesota. Blood samples were taken from sheep on several farms within a limited area. Adjacent farms had levels that differed by as much as 100%. The region is all within the selenium-adequate region in Figure 3, yet deficiencies were indicated by both blood selenium levels and clinically diagnosed white muscle disease in lambs.

The region is underlain by glacially transported material. The sources of the material are widely varied, depending on ice movement as well as reworking of previous glacial deposits.

The toxic reputation of selenium lives on in the halls of the conservative Food and Drug Administration. Approval of selenium in feed additives was resisted at first. Finally approved salt mixtures with 20 ppm, and then 30 ppm selenium reached the market. Extensive trials showed that such products had negligible effect on raising body levels in sheep to required levels. Current recommendations are 90 ppm selenium in free-choice salt-mineral mixes. Selenium has finally made it as a required nutrient. (Note: These levels are in a salt mix, not total feed!)

Musings

The ancestry of domestic sheep in geologic time is a determining factor in their requirements of chemical nutrients. Their efficient sodium retention suggests an origin in low environmental sodium retention regions. That their selenium requirements and sensitivity fit well with average values suggests that their region of origin was average in environmental selenium.

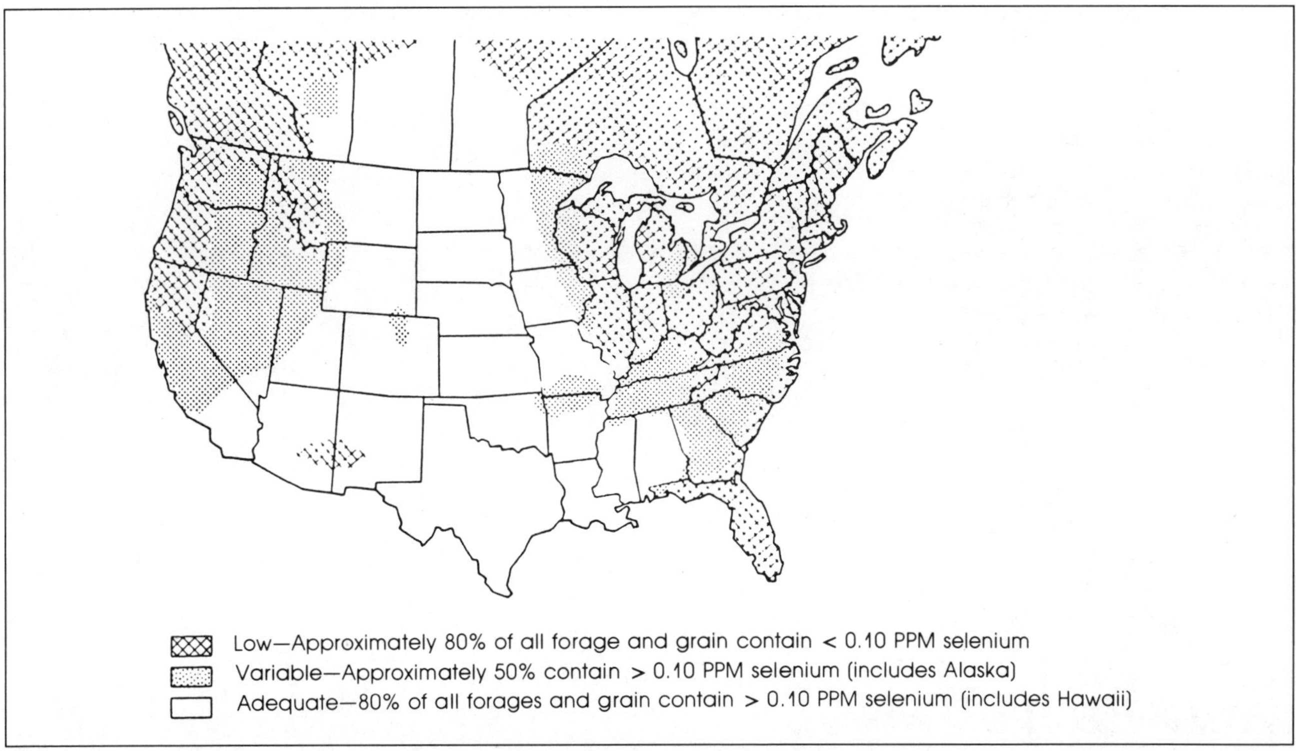

Figure 3 Regional distribution of selenium in forages[15].

The point is that while Figure 3 is of some use, each farm or ranch must be examined on an individual basis. Forage analyses, blood tests, and clinical examination of animals can be combined with feeding trials to determine flock needs.

The extreme toxicity of copper for sheep is notable. Average copper content in the outer earth is about 45 ppm. Yet copper toxicity can arise from diets containing 30 ppm copper. In molybdenum-deficient feeds, as little as 10 ppm copper can be toxic. Domestic sheep must have originated in regions with relatively low environmental copper, or perhaps higher than average molybdenum.

To further the copper puzzle, and to emphasize the importance of ancestry: Copper blood levels and tolerance are different in Finnsheep, Merinos, and the British breeds with the Finns more tolerant than the British breeds, and Merinos in between. Is it the ancestry and the geochemical environment of their places of origin that made the difference? I would like to think so.

References

[1]Bakkar, R. T. 1975. Dinosaur Rennaisance. *Sci. Am.*, 232 (4)58–78.

[2]McDonald, P; Edward, R. A.; and Greenhalgh, J. F. D. 1966. *Animal Nutrition* (2d ed.). Edinburgh: Oliver and Boyd.

[3]Parker, Ronald B.; Toots, Heinrich; and Murphy, J. W. 1974. Leaching of sodium from skeletal parts during fossilization. *Geochem. et Cosmochem. Acta* 381317-21.

[4]Parker, Ronald B., and Toots, Heinrich. 1976. Sodium in fossil vertebrates as a paleobiological tool. *Arizona Acad. Sci. Jour., Proc.* 11:87.

[5]Parker, Ronald, B., and Toots, Heinrich. 1980. Trace elements in bones as paleobiological indicators. In *Fossils in the Making* eds. A. K. Behrensmeyer and A. P. Hill. p. 197–207. Chicago: Univ. Chicago.

[6]Weir, J. S. 1972. Spatial distribution of elephants in a central African national park with relation to environmental sodium. *Oikos* 23:1–13.

[7]Dougall, H. W.; Drysdale, V. M.; and Glover, P. E. 1964. The chemical composition of Kenya browse and pasture herbage. *E. Afr. Wildl. J.* 2:86–121.

[8]Laws, R. M.; Parker, I. S. C.; and Johnstone, R. C. B. 1975. *Elephants and their habitats.* Oxford: Clarendon Press.

[9]Denton, D. 1982. *The Hunger for Salt.* New York: Springer Verlag.

[10]Magee, B. 1984. Personal communication. Some Suffolk and Dorset ewes ate so much salt given free choice that bedding costs became excessive—exceeding costs of hay.

[11]Beath, A. O.; Eppson, H. F.; and Gilbert, C. S. 1934. *Sel- and other toxic minerals in soils and vegetation.* Wyo. Agr. Expt. Sta. Bull. 206, Laramie, Univ. Wyo.

[12]Mason, B. 1958. *Principles of Geochemistry* (2d. ed.). New York: Wiley.

[13]Knight, S. H. 1961. Personal communication. The seleniferous hay was cut from a downfaulted block of basin rocks surrounded by crystalline rocks. Beath, not a geologist, failed to recognize the situation.

[14]Schwartz, K.; and Folz, C. M. 1957. Selenium as an integral part of factor 3 against dietary necrotic liver degeneration. *J. Am. Chem. Soc.* 79:3292–3293.

[15]Nat. Res. Coun. 1983. *Selenium in Nutrition.* Washington: Nat. Acad. Press.

Flock Recording

Leslie Jensen
69 Bell Street
Featherston, New Zealand

There are only two kinds of sheep in the world—those with records and those without. This paper will discuss the reasons, requirements and rewards of recording your sheep. I hope that breeders who use pedigree and performance records as tools in selected matings and flock culling will discover practical suggestions and pertinent information that will assist them in reaching their goals.

Why Record?

Recording sheep has a simple, clearly defined aim—to document information which can be used to make decisions.[1] These decisions are selection decisions that are concerned with the question of which animals to keep to breed future generations, and which animals to cull.[2] It is generally accepted that breeding programs seek to alter an existing flock. This alteration can be directed toward increased production, the quality of wool produced or changing the fleece colour. Breeding objectives are diverse, but there are only two ways the change can come about; through the environment or genetics.

Assuming that the environment is already being skillfully managed for maximum results and that it is basically the same for all the sheep in the flock, any further improvement must come from the influence of genetics. Although environment and genetics work together to produce the animal, the genes cannot be directly affected by the environment. Any genetic improvement must come through the breeding program. Individual records tell the breeder which members of the flock have the traits he wants. Selection decisions are based on this information.

Sheep Identification

Before you go rushing out to the shed or sheepyards, clutching pen and paper in hand, one important requirement must be met. You have to be able to identify the individuals in the flock. There are both temporary and permanent methods of doing this.

Temporary identification is usually used as an aid in short-term flock management. Such tasks as identifying cull ewes, marking ewes at mating time, numbering sale sheep or selecting prime lambs can be made easier with the help of stick-on, clip-on or tie-on labels, sprayed numbers, raddle marks or neck tags.

Permanent identification is with the sheep for life. It is used for pedigree and performance recording and showing ownership, sire group and year of birth. Ear marks, coloured plastic ear tags, numbered metal ear tags and coloured wire twists can be relied on to withstand most of the usual wear and tear in a sheep's life. Examples of permanent methods of identification are shown on the next page.

In flock sheep, the most commonly used identification system is a numbered, metal ear tag, inserted in the leading edge of the ear, at birth. This is a reliable, permanent identification which interferes least with shearing and can be easily read.[4] Shearers should be advised when sheep are wearing metal tags.

throughout its life. Tags are used in sequence from lowest to highest, to make recording easier. It also indicates the early, middle and late lambs, making them conspicuous by their position in the sequence.

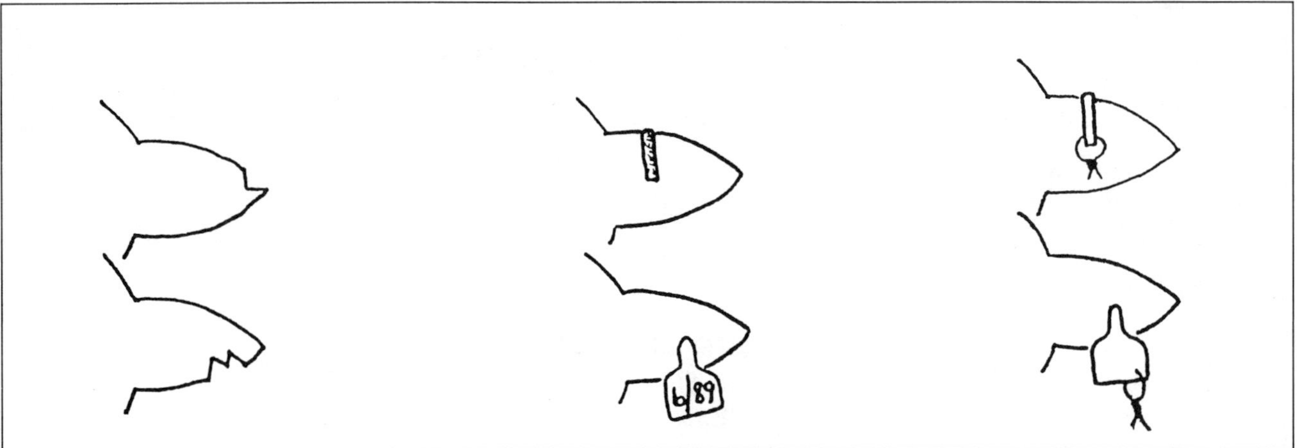

Figure 1 Rim cut ear marks. **Figure 2** Plastic and metal ear tags. **Figure 3** Coloured wire twists.

Large plastic tags are often used in conjunction with metal tags because metal tags are only readable when the animal's head is being held. The plastic tags display the same number as the metal tags and are easily seen without the need to catch the sheep.

This practise of double tagging is justified because it helps prevent loss of identity and the subsequent loss of data that can occur if only one tag is used and that tag is cut, rubbed or broken off. The plastic tags are inserted at the hogget or two-tooth stage, when the ear is big enough to support them. Most types of plastic tags are available in a range of colours. Different colours can be used to denote year of birth or sire group.

A big variety of tags and tagging systems are available. Breeders have to investigate to discover the one best suited to their needs. As soon as the decision of what to use is made, each animal must be given a unique identity. In order to record data collected at mating, shearing, lambing and weaning, a breeder must be able to repeatedly and accurately, distinguish one sheep from another. For this to happen, each animal must be allocated a number.

A unique identification number can be made up from two components.[2] (1) an individual number—its tag number; or (2) the year born—its tag year.

The individual number sequence may start at 1 or 10 or 100 and proceed upward, depending on the size of the flock and the breeder's preference, i.e. 100, 101, 102, 103, etc. The year born is adequately shown by using the final two numbers, i.e. 1989 becomes /89. Therefore, the twenty-third lamb born in the flock in 1989 would be tagged 23/89. This unique identification is given to the animal at birth and remains with it

Whatever method you decided to use, remember that your main aim is to find a simple, repeatable system to identify your sheep.

Figure 4 A sheep properly tagged for recording.

Pedigree and Performance

A sheep's pedigree can be defined as the list of its ancestors.[3] This list provides the breeder with the families or bloodlines that have been mixed or matched to produce a particular individual. They provide information that enables a breeder to repeat successful matings, outcross using new blood to widen the gene pool or line-breed by re-introducing previously used families.

Three factors are necessary for accurate pedigree recording: (1) Individual identification, (2) Ewes correctly assigned to rams at mating time, and (3) Lambs correctly assigned at lambing.

Once sheep have been numbered and tagged, mating records are the next step in putting together pedigree records.

The final aspect of pedigree recording comes with lambing. This is the most critical time in flock recording because now the pedigree is complete and the lamb is written into the flock for the first time. The matching of dam identity with that of her lambs is of the utmost importance: on this one event hangs the accuracy of all future records.

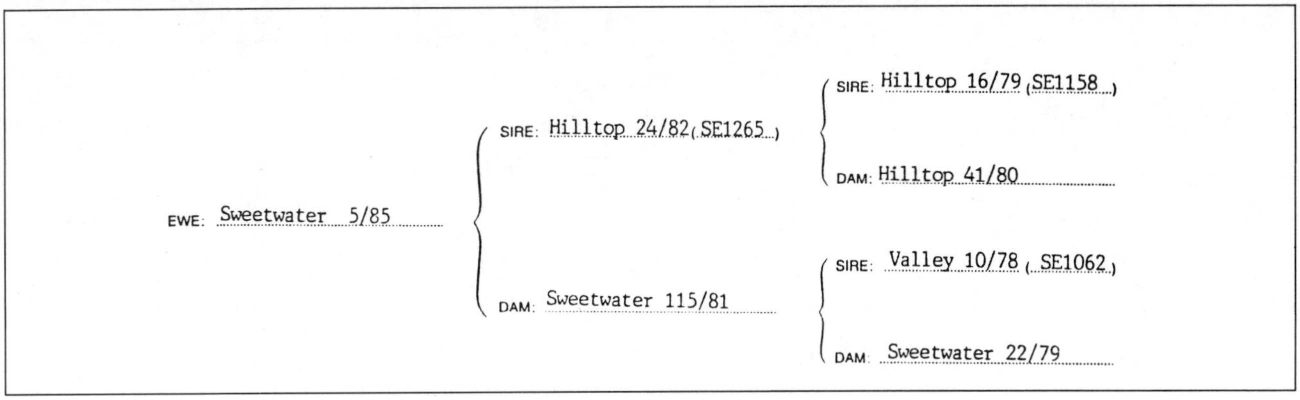

EWE: Sweetwater 5/85

SIRE: Hilltop 24/82 (SE1265)

SIRE: Hilltop 16/79 (SE1158)

DAM: Hilltop 41/80

DAM: Sweetwater 115/81

SIRE: Valley 10/78 (SE1062)

DAM: Sweetwater 22/79

Figure 5 An example of pedigree recording, showing the generations of mating that produced one member of the flock.

It is generally accepted that the sheep production year begins at the joining of the rams with the ewes. The purpose of mating records is to record a successful mating of the ewe by a ram. That is, the sire of the lamb(s) which are eventually born to the ewe.[2] Ewes are 'single sire' mated by dividing the ewe mob and putting one ram out with each group. A mating harness with a coloured crayon can be used on rams to mark ewes mated on different cycles.

The main task at mating is to allocate some ewes to each ram. It is advisable to use some form of checklist from this stage to ensure that animals are not missed or duplicate numbers recorded. The most efficient system is to have an accurate list of all the animals in the flock, in numerical order, within ages prior to mating.[2] When ewes are put through the race, each one can be marked with raddle, a different colour for each ram and later drafted accordingly.

Sires	*12/85*	*16/86*	*18/87*
2-Tooths	5/87 11/87	9/87 14/87	6/87 12/87
4-Tooths	3/86 14/86	11/86 19/86	18/86 25/86
6-Tooths	9/85	4/85	7/85

Figure 6 Suggested mating sheet. Ewes are allocated prior to mating and checked off the list when drafted into groups.

Multiple births increase the chance of errors and this difficulty can be compounded by a high concentration of ewes in a paddock. Lambs only a few hours old are easily caught and tagging as soon as possible after birth helps alleviate these problems. Brass tags are placed in the lamb's ear while its mother is closely protective, removing any doubt of its identity. The record book should be close at hand, to record data as soon as the tags are used. Coloured twists can also be placed to identify sire groups for future evaluation or sex of lamb for easier drafting.

Pedigree records show who an animal is in terms of its ancestors. Performance recording tells what an animal does in respect of wool and lamb production.

Performance recording has often been considered as a separate aspect from pedigree recording but, to be effective, it is essential that there is close association between the two for both administrative and genetic reasons. The recording of pedigree is of course a basic component of any performance recording scheme where the central core is the individual records of the animal.[2]

Unless performance records are used in making better selection decisions, then their value is limited. Selection is defined as allowing some animals to be parents of the next generation while depriving others of the privilege. In practise, selection involves choosing some ewes and rams as parents and disposing of or culling those not required for the breeding program. The main cull or keep decisions in a flock recording system are made on the replacement two-tooths rams and ewes before they enter the flock.[2]

Lamb/Hgt No.	Date	Ewe Ident.	Lamb/Hgt. Fate	Sex					Remarks
1/89	30/8	14/86		R					
2/89	30/8	14/86		R					
3/89	30/8	9/87		E					
4/89	31/8	12/85		R					
5/89	1/9	3/85		E					
6/89	1/9	3/85		E					
7/89									
8/89									
9/89									
10/89									
11/89									

Figure 7 A page from a field notebook. Ear tag numbers are listed and the data is recorded as they are used.

EWE: Sweetwater 5/85 SIRE: Hilltop 24/82 (SE 1265) DAM: Sweetwater 115/81

BREED: Romney COLOUR: Dark Grey RANK: 2 / 2

FLEECE WEIGHTS— Lamb: not shorn Hogget: 4.5 kgs. 2th: 4.1 kgs.

Remarks: Show prospect. Good conformation. Open face.
Even, lustrous fleece. 29 microns.

Year	Shearing Date	Fleece Weight Kgs	Sire of Lamb	Lambing Date	Ram Lamb	Ewe Lamb	Rank	Weights Wean	Weights Hog	Colour	Remarks & Disposal
'87	30/7	4.1	(SE1302) 105/85	5/9	6/87		2/2	40.9	51	dark grey	keep
						7/87	2/2	34.5	43	light grey	sold L. Byron
'88	24/7	4.0	(SE1447) 83/86	11/9		18/88	2/1	35.2	45	grey	keep
						19/88	/				found dead 1 week old.
							/				
							/				
							/				
							/				
							/				
							/				
							/				
							/				
							/				
							/				

Figure 8 An example of performance recording, showing the contribution one member made to annual flock production.

In most flocks, the yearly culling point is governed by the number of replacements required and can alter from one year to the next. Breeders are wise to set and maintain minimum performance levels for fleece weight, growth rate, fertility and points of conformation or behaviour which will disqualify sheep that slip in any one of these areas. This will prevent any overall drifts in areas of flock development. Top producing and top looking sheep are far from synonymous so, in practise, breeders have almost a double culling to do. Conformation faults should not be overlooked just because a sheep produces well and vice versa. A well put together animal should not be retained if it doesn't meet the minimum production standard. Eye appraisal and flock records should work together when sheep are culled.

A breeder should strive to have each succeeding generation of sheep, on average, better than the one preceeding, in both conformation and productivity. This involves selecting parents of the next generation very carefully. Pedigree and performance records provide valuable information to make progressive selection decisions possible.

Making it Work

Three features are essential in order to use flock records as an aid to furthering a breeding program: (1) Establishment of clear objectives for the breeding program, (2) Measuring and recording the performance of the individual animals, and (3) Processing these records to assist in making selection decisions.[5]

Objectives

Breeding objectives are a projection into the future, providing a direction to go in and a goal to work toward. They do not represent a place to start: they are a place to end up, in five or ten years time.

Breeding objectives must be practical and attainable, realistically within the capabilities of the breeder and his farming and financial situation. They help focus attention and enthusiasm on the job in hand. They give a person added reasons for doing a lambing beat and increase interest in the routine work of shearing and drafting.

The range of breeding objectives is as diverse as the breeders themselves, but it is generally accepted that the majority of them are governed by the economic considerations of producing more wool and/or lambs. Coloured sheep breeders often place additional emphasis on fleece colour or wool quality.

Measuring and Recording

It is one thing to decide the main objectives of selection; it is another to specify what records the shepherd must keep to achieve this aim.[1]

When gauging an animal's productivity, the most important factors to measure are fertility, growth rate, wool weight and aspects of wool quality.

Traits must be capable of being measured for recording so that each sheep can be ranked in relation to the others in the flock. For example, a sheep that shears four kgs. of wool ranks higher than one that only shears three kgs. of wool. The breeder must also be aware of what factors influence the trait he is measuring. Is it connected with genetic inheritance or environment? Does the lamb owe its size to good breeding or good feeding?

The degree of heritability must also be a consideration. That is, the amount of a trait that is passed from the parent to the offspring. Heritability values differ quite markedly between traits. For example, traits having to do with reproduction and fertility generally are quite slowly heritable, lying in the 0 to 20 percent range. Those having to do with growth characteristics and efficiency of feed conversion lie normally in the moderate range: 20 to 40 percent. Most wool traits are found in the high range: 40 percent and above.[6]

Fertility

The data necessary for recording fertility can be taken from the lambing record. The lambing date will show how early each ewe cycled and conceived. The sire identification makes it possible to calculate how many lambs each ram sired and the lamb tag numbers and ewe identification shows how many lambs each ewe gave birth to.

Lambing percentage is a flock statistic and can be confusing because of the variety of figures used to calculate it. Some are listed below. Ewes 'joined' refers to the number of ewes that went to the ram.

1. Lambs born/ewes lambing
2. Live lambs born/ewes lambing
3. Lambs born/ewes joined
4. Lambs docked/ewes lambing
5. Lambs docked/ewes joined
6. Lambs weaned/ewes joined

Weight Recording

Weight records are an important part of a performance recording scheme because live weight is an important trait as such and it is also important because of its indirect effect on other traits such as reproduction. Apart from their importance as a genetic trait, weight records are very valuable in assisting with management decisions.[2]

There are two main sources of error in recording live weight of sheep. The first is 'gutfill' which is a common phenomenon of all ruminants. The other is the wetness of the fleece. The ideal is to weigh sheep which have been starved overnight and have dry fleeces (especially bellies).

However, this is not always practical so attempts are usually made to weigh sheep full and as quickly as possible at each weighing. The practical way out is often to accept 'gut-fill' and wet sheep, but make sure that all animals are at the same stage of fullness and wetness when weighing.[2]

Birth weights, if recorded, can be obtained by using a pocket spring balance hooked around the lamb's hind leg. Later and heavier weights (e.g. weaning, hogget and ewe weights) are best recorded on the various types of scales which are available.[2]

Fleece Recording

Fleece recording is concerned with two main aspects of wool production. The first is fleece weight, just as it comes off the sheep. The second is fleece quality which is made up of a number of components.

Fleece weight is a clearly defined trait while fleece quality covers a number of both objective and subjective criteria. The fleece information recorded varies greatly, depending whether it is done for marketing or genetic reasons.

Recording fleece information poses few practical problems as it is usually carried out under cover when sheep are dry. Fleece weights are recorded after the fleece has been picked up from the shearing board, prior to throwing it on the wool table.[2]

The main practical problem is to remove the fleece from the shearing board, with the correct identity of the sheep from which it came clearly recorded in some form. Various techniques are used, from stick-on labels to separate sheets for each shearer. Check lists are often used by breeders or the data can be transcribed directly onto the more permanent record or computer input sheet.[2]

As fleece weight is a moderately to highly heritable characteristic, one record is sufficient to indicate lifetime rank for wool production. The most convenient and accurate time to record fleece weight is at the hogget stage, before the effects of differing lamb drops influence the result.

Wool Quality

There are many different qualities of wool that can be measured. Some have an objective measurement such as staple length, fibre diameter (micron) or clean scoured yield. Others, such as tenderness, colour, cotting or lustre are subjective. These qualities can sometimes be harder to handle because they require an opinion that is translated to a number on a scale, but all can be recorded. Note that if the list of traits to be selected for is too long, the overall progress will be slow.[7]

Processing Records

Records must be kept efficiently if they are to be used effectively to improve the flock and repay the time and expense of keeping them.[1]

Day to day flock management means that data have to be collected in the paddock at lambing time (wind and rain notwithstanding), the shed at shearing time (coupled with the noise and speed) and in the yards (complete with dust and dirt) when drafting or weighing. If you are already fully occupied during these times, don't expect to add recording to the list of existing responsibilities without making proper provision for it. Employ extra labour. If you find that recording and transposing figures quickly and accurately is difficult, don't exacerbate an already demanding situation by trying to do it yourself. Instead, see if you can find a member of the family or a friend able to lend a hand.

Keep your records simple, accurate and easily referred to. A field notebook should be carried so that notes can be made whenever necessary. The glove box of the farm vehicle is a better place to keep it than the drawer in the office desk.

History

Early methods of recording in the ram-breeding flocks of the sheep industry were primarily directed at pedigree information on individual sheep, an aspect which became necessary with the formation of breed associations. Most of these pedigree records were written by hand into books.[5]

The first advance on the above method was the introduction of a card index system of record. The cards were 23 x 13 cm. in size with 84 holes punched around the sides to allow the use of a needle sorting device.[5]

The computer is now the most up-to-date method for storing and sorting data. Animals' tag numbers and collected data can be fed in and, depending on the program, lists will be printed out, giving relevant information.

Technical Developments

A most important aspect is the progressive change in the processing of records. In the early recording schemes, the actual records were used, seldom in a systematic way. The next phase was the use of adjustments for non-genetic effects such as birth and rearing rank, age of dam and age of lamb to increase the accuracy of selection. The third development was the calculation of breeding values [BV is an estimate of the animal's breeding worth based on each productive character recorded and some pedigree information] from the corrected records and the use of selection indexes to put the traits together into a single figure. Thus, the evolution has been from using actual records (confounded by all the environmental differences between animals) to the use of corrections to eliminate the effects of known environmental factors and then calculating breeding values which predict the performance of the progeny of the individual in the flock.[5]

Of the many systems available, there will be one for every breeder to use or adapt to particular needs. Research is needed to find the one best suited to breeders' varying situations.

Conclusion

To maintain enthusiasm for consistent record keeping, breeders must be confident that this addition to the long list of farming duties and expenses is justified. They must accept that the rewards outweigh the disadvantages, both in economic terms and satisfaction with the sheep raised, be they pedigree show champions, commercial flock sheep or a handful of hobby farm ovines. Planning and working, looking always into the future, can be a wearying business, especially if progress is set back by disease, a killing storm at lambing or drought. Join the breed's society. Contact with its members can often provide a much-needed dose of sympathy, understanding or encouragement.

Communication with fellow breeders is essential to ensure that you keep up with the current breed trends. Farmers, no less than doctors or mechanics, need to stay familiar with developments in the industry. Many sources of information and advice are available.

Perhaps most important of all is that you keep your eyes wide open, especially when you appraise your own flock. Analyse and be critical. Keep your breeding program moving forward. You must tag conscientiously, record efficiently, examine and use the results; feed well and cull hard. You will be amply rewarded by the results you achieve.

References

[1]Owen, J. B. 1971. Performance Recording in Sheep. 1976 Sheep Production.

[2]Dalton, D. C. Performance Recording. Introduction to Practical Animal Breeding.

[3]Oxford Universal Dictionary.

[4]Federation of Livestock Breeding Groups, 1975. Proceedings of the Second Conference.

[5]Rae, A. L. Historical Development of Performance Recording in N. Z.

[6]Burrill, M. J. Quality Characteristics of Coloured Sheep.

[7]Dalton, D. C., Callow, C. F., Uljee, A. E. Recording Systems and Their Use for Coloured Sheep.

Genetic and Environmental Factors Affecting Ewe Reproduction Rate

Dr. Howard H. Meyer
Department of Animal Science
Oregon State University, Corvallis, OR 97331 USA

Technical Paper No. 8801, Oregon Agricultural Experiment Stations

Reproduction rate is the most important single factor determining profitability of commercial sheep flocks, particularly in the U.S. where income from lamb far exceeds wool income. It has been estimated that under grazing conditions the net returns from a flock of 5,000 ewes each raising one lamb could be achieved by just 480 ewes each rearing two lambs! This is due to the costs of maintaining a ewe for 12 months in order to achieve one lambing; hence, the first lamb weaned is needed to pay the year-round ewe maintenance costs while the second lamb represents most of the profit.

Under conditions of high winter feeding costs, flocks may have to achieve as high as a 150% lamb crop before being profitable. The U.S. national average in 1987 was only 104% and the preliminary estimate for 1988 is 99%, far below the reproductive potential of sheep flocks but reflecting the many genetic and environmental factors influencing reproductive rate.

I. Biology of Reproduction

Research in recent years has revealed many fascinating mechanisms and features of sheep reproduction. Through integration of biology and management, better practices can be developed to increase reproductive rate.

The basic reproductive pattern in sheep differs little from other species and consists of estrus, ovulation, fertilization, embryo implantation, gestation and parturition. At each step, however, sheep have unique mechanisms operating. I am going to limit this discussion to the period from ovulation through early gestation.

A. Ovulation, Fertilization and Embryo Migration

Ovulation occurs near the end of estrus. This is logical timing since fertilization occurs in the oviducts and sperm deposited in the vagina must pass through the cervix into the uterus, swim the length of the uterus and travel up the oviducts to meet the eggs.

Ovulation is essentially random between the two ovaries (Rhind et al., 1980). Single ovulations have about the same likelihood of originating from either ovary, and the ovary responsible for a single ovulation in one cycle is equally likely to be the source of the egg if the ewe is a single-ovulator again the next cycle.

Multiple ovulations are also random; one-fourth of twin-ovulating ewes produce both eggs from the left ovary, one half produce one egg from each ovary and one-fourth produce both eggs from the right ovary. The same principle holds for triple and quadruple ovulators with some ewes having all eggs released from one ovary and none from the other. The pattern could be just reversed in the following cycle.

The lack of equal distribution of multiple eggs from the two ovaries would at first seem to be a disadvantage in efficient use of both uterine horns. However, ewes somehow manage to distribute ova to both horns following fertilization (Doney et al., 1973). Unlike most other species, the two uterine horns in sheep share sufficient connection at the body of the uterus to allow migration of ova from one horn to the other. While single ovulations result in nearly all fetuses developing in the horn adajacent to the ovulating ovary, nearly all twin-ovulators that produce both eggs from one ovary end up with one lamb in each horn. This migration is clearly not a random event since twin ovulators producing one egg from each ovary virtually never have both lambs developing in one horn.

B. Implantation

The dispersion of ova allows maximal use of uterine "feeding area" for fetal placentas. The uterus has many placental attachment sites known as caruncles. Placental cotyledons develop in contact with the caruncles almost like a fist in a mitten. Each such attachment is a site for nutrient transfer from the uterus to the placenta.

The number of cotyledon attachments influences lamb birth weight (Rhind et al., 1980). During the period of early placental "colonization", multiple embryos compete for caruncles, particularly if there is more than one embryo in a single uterine horn. Embryos which fail to claim enough caruncles have a reduced chance of survival; however, subsequent death of one embryo does not make additional uterine caruncles available to other embryos since placental spreading will have already been completed. This explains why super ovulation or surgical transfer of numerous embryos often results in the birth of only one or two lambs, generally of very light birth weights. Similarly, the fairly common observation of twin lambs born with very different birth weights is probably the end result of a triple implantation with death of one of the two embryos competing in the same uterine horn.

C. Factors Affecting Ovulation Rate

Since ovulation sets the upper limit to litter size (identical twins are rare in sheep), its importance is obvious. Ovulation rate is influenced both genetically and environmentally.

1. Environmental Influences

a. Season

While season influences whether or not ewes are ovulating at all, it also influences their ovulation rate during the mating season. Mean ovulation rate tends to be highest from about the third estrus onward, declining toward the end of the breeding season (Thompson et al., 1985). Hence efforts to breed ewes as early as possible result in fewer lambs born than if ewes were mated later. Likewise ewes such as Dorsets which are less seasonal than most breeds produce a smaller mean litter size when lambing in the autumn than when lambing in spring.

b. Body Condition

Other factors being equal, ewes in good body condition exhibit a higher ovulation rate than those in poor body condition (Thompson et al., 1985). Ewes of condition score (CS) 3.5 or higher on a scale of 0 to 5 are at their maximum ovulation rate and ewes of successively lower body condition drop off in mean ovulation rate accordingly.

The effect of body condition on ovulation rate is modified if ewe body condition is changing during the mating season. Ewes of poor to moderate body condition will ovulate at a higher than expected rate if on good nutrition prior to mating. Conversely, ewes in good condition but in the process of losing weight may ovulate at a lower rate than suggested by their actual body condition.

2. Genetic Influences

a. Variation Between Breeds

Well-documented differences between breeds have been reported in virtually every sheep-producing country. The fairly recent availability of Finnsheep genes in Great Britain and North America have greatly increased the potential reproduction rate of commercial flocks. In general, Finnsheep incorporation into breeds of moderate prolificacy increases mean litter size by about one lamb per 100 ewes for each 1% of Finn breeding. Hence, quarterbred Finns average about 25 more lambs per 100 ewes than contemporary ewes with no Finn breeding.

While most of the increased prolificacy is due to higher ovulation rates, Finn breeding also appears to increase uterine efficiency (see Section D) and perinatal lamb survival (Meyer and Clarke, 1982).

Recent data on the Romanov breed suggest its effect on reproduction in crossbreds is probably similar to the Finnsheep.

The other high fecundity breed becoming available in many countries is the Booroola Merino. The Booroola is unique in having a single gene (designated "F" for fecundity) which dramatically increases ovulation rate (Davis et al., 1982). Heterozygous (Ff) ewes typically average 1.1 to 1.5 more ova than non-carriers and homozygous (FF) ewes increase by about the same amount above heterozygotes. The hormonal basis of gene action has not been determined, and the gene appears to have no effect on other reproductive parameters.

b. Variation Within Breeds

Most producers can recall ewes which consistently produced multiple lambs while others produced only singles. Part of this apparent consistency is due to chance (within a ewe flock with an average litter size of 1.5 there are very likely to be some who have four sets of twins in a row while others have had four singles). Because the heritability of litter size is low ($h^2 = .05 - .10$; Turner, 1969), researchers are investigating the possibilities of selecting for ovulation rate as a means of increasing litter size. Ovulation rate determination with special laparoscopic equipment has become fairly common in research programs and could possibly be applied on farms if results warranted.

D. Embryo Survival/Uterine Efficiency

Embryo mortality is very difficult to study biologically. Early loss of a single embryo is likely to appear only as a delayed return to service. Loss of one embryo following multiple ovulation will not be detectable until lambing and then only if ovulation rate is known. Of course, by this time it is impossible to determine whether the "missing" lamb is due to embryonic mortality or fertilization failure.

Early attempts to determine timing and extent of embryonic death utilized sequential slaughter of large numbers of ewes, with embryo mortality estimates being based on statistical differences in average number of fetuses present at successive stages of pregnancy (Restall et al., 1976; Rhind et al., 1980).

An alternative approach has considered embryonic loss as a maternal trait, comparing breeds or treatment groups based on their success at producing "extra" lambs from "extra" eggs. By observing ovulation rate via abdominal laparoscopy and recognizing that one fetus is required to maintain pregnancy, ewe genotypes can be compared on the basis of litter size resulting from conception to twin or triple ovulations. This has been termed uterine efficiency (UE) (Meyer, 1985).

It has been shown that ewes conceiving to twin ovulations produce a mean litter size of about 1.8 lambs, i.e. the UE of twin ovulators is about .8 lambs from the second egg. Likewise, triple ovulators produce a mean litter size of about 2.4 lambs, or a UE of .6 lambs from the third egg compared to litter size of twin ovulators.

Genotypes have been shown to vary in their uterine efficiency. In one study, litter size of twin ovulators ranged from 1.93 for Finn x Romney crosses down to 1.60 for straightbred Romneys. In another study, twin-ovulating whiteface ewes produced a mean litter size of 1.84 compared to 1.72 for their blackface crossbred contemporaries (Meyer et al., 1988).

Recent studies have incorporated ultrasound scanning from implantation onward in an attempt to determine environmental effects on the number of embryos implanted and the timing of subsequent losses. Preliminary results indicate that both poor body condition at mating and inadequate post-mating nutrition reduce the eventual litter size of multiple-ovulating ewes (West and Meyer, unpublished data).

II. Altering (Improving) Reproductive Efficiency

Understanding the sheep's reproductive biology allows us to manipulate flock reproductive performance through both management and breeding decisions. It has also led to technological developments which offer great potential for increasing reproductive efficiency.

A. Management Applications

1. Ewe Body Condition and Mating Nutrition

By increasing quality and/or quantity of feed to thin ewes two to three weeks prior to mating, we can boost their ovulation rate by simulating better body condition. This practice has long been known as "flushing". As body condition increases, the benefits from flushing decrease. Good condition ewes are already at their peak ovulation rates and unlikely to show a flushing response.

It may be more economical to carry ewes in moderate condition and flush them than to carry ewes in good condition throughout. However, it makes no sense to deliberately reduce body condition in order to get a flushing response since the end result will be no net improvement.

Since thinner ewes show the greatest response to flushing, it is advisable to sort ewes on body condition score and spend flushing feed resources on the group likely to give the best return. It must also be remembered that good conditioned ewes must be maintained since weight loss in this group will result in reduced ovulation rates.

2. Breeding Decisions

a. Prolific Breeds

Producers in many countries are able to dramatically increase flock reproductive rate through incorporation of genetic material from prolific breeds such as the Finnsheep, Romanov or Booroola Merino.

The high prolificacy of the Finnsheep and Romanov breeds is essentially additive so that the first cross is intermediate between the base breed and the prolific breed. Backcrossing to the local breed will reduce the prolificacy half way from the first cross and so on. Accordingly, a breeder can use the amount of Finnsheep or Romanov breeding that gives the desired prolificacy increase. As mentioned earlier, studies have shown an increase of about 1 extra lamb born per 100 ewes for each 1 percent of Finn breeding. The Finnsheep was used as one of four breeds combined (along with Targhee, Rambouillet and Dorset) to produce the Polypay breed. Polypay ewes average about a 200% lamb crop under good feed and management conditions.

Through a combination of crossing to selected prolific rams, backcrossing to the base breed and selection, it is relatively easy to increase reproductive rate by .25 lambs/ewe and have a flock differing little in wool and growth characteristics from the base breed.

High prolificacy in the Booroola Merino is under quite different genetic control than in Finnsheep and accordingly requires different breeding strategies for utilization. Since ewes homozygous (FF) for the gene have mean ovulation rates of about 4 compared to 2.8 in Ff heterozygotes and 1.6 in local breeds, the heterozygote is clearly the most desirable genotype under most commercial conditions.

Unfortunately, it is impossible to produce only heterozygotes in a self-replacing population. Segregation of the F gene means that all three genotypes would be present in varying proportions in each lamb crop with no means of their identification before their own eventual lambing. The mating of two heterozygotes could result in triplet ewe lambs representing each of the three genotypes (FF, Ff and ff) and very different in their reproductive output. Since the only known effect of the F gene is on ovulation rate, a flock of replacement ewe lambs could not be selected until after reaching puberty and probably not until after their first lambing. Until a marker for the gene is identified, the Booroola gene is likely to be of limited use in commercial sheep production, at least in the U.S.

b. Crossbreeding

The crossbreeding necessary for incorporation of genes from prolific breeds often leads to undesirable changes in other traits such as fleece weight or wool type. However, if breeders chose two breeds of acceptable standard for wool-quality (for instance), they would produce a crossbred with superior reproductive capabilities due to its hybrid vigor, yet of acceptable wool qualtiy. Such crossbred ewes would be expected to show as much as a 30% increase in weight of lamb weaned compared to either parental pure breed (Nitter, 1978). The increase is due to a combination of improvements in fertility, litter size, lamb survival and milk production.

Among the wool breeds we have many examples of breeds formed by combining existing breeds. The Targhee and Columbia have been produced in this way in the U.S. and the Corriedale, Coopworth and Perendale are New Zealand products of controlled crossbreeding.

c. Selection Within Breeds

Breeders wishing to improve reproductive performance of an established breed without introducing outside genes are limited to progress from within-breed selection. Since the heritability of litter size has been repeatedly shown to be low, many people assume that genetic change from selection would be very slow. Such is not the case.

Genetic change results from the combined effects of heritability and selection pressure. Although heritability for litter size is small, the genetic variance (and selection differential) of the trait is very large. Accordingly, the rate of change in litter size is comparable to the rate of change expected for more highly heritable traits such as growth rate and fleece weight.

Long-term selection studies for litter size have resulted in an increase of 2 lambs/100 ewes per year of selection (Clarke, 1972). While this rate of change may seem slow, it should be remembered that such changes are cumulative and permanent within the population.

B. New Technical Developments

1. Immunization to Increase Ovulation Rate

Most reproductive steps are caused not by the presence or absence of specific hormones but by the relative levels of one hormone to another. Hormone administration as used in the past for estrus synchronization and superovulation have altered hormone ratios by artificially elevating the level of particular hormones and thus changing hormone ratios.

An alternative approach developed in Australia induces the ewe to alter her own hormone ratios (Cox et al., 1982; Scarzmuzzi et al., 1982). Administered hormones do not normally produce an antibody response in animals. However, a technique has been developed whereby a naturally occurring hormone, androstenedione, is chemically linked to a foreign protein before being injected into ewes. The animals respond by providing antibodies against the foreign protein molecules. In the process of copying the pattern of the foreign protein, the antibodies coincidentally also copy a portion of the androstenedione molecule. When these antibodies circulate through the system they tie up a portion of the available androstenedione, thus altering the ewe's hormone balance. The result is an increase in ovulation rate.

The immunization material has been patented and is now marketed in several countries under the name Fecundin. Reproductive response to Fecundin varies between flocks but averages about 22 extra lambs born per 100 ewes treated (Meyer, 1988). The response to immunization is reported to be limited to the season of treatment with a booster injection required for an ovulatory response the following year. However, the limited U.S. work with Fecundin indicates a significant carryover effect without booster administration (Meyer and Lewis, 1988; Meyer et al., 1988).

Fecundin could be an extremely useful tool for breeders wishing to increase reproductive rate of their flock without introduction of prolific breeds. Its use could be limited to just a portion of the flock, depending on varying feed or other resources. However, unlike genetic increases in reproductive rate, immunization treatment will need to be applied each time that the response is desired.

Likely future availability of Fecundin to producers in the U.S. is dependent upon industry demand and the projected size of the market. Demand, in turn, is linked to anticipated cost. Supplier cost will be influenced by the fact that shelf life of Fecundin is currently limited to a single season.

2. Artificial Insemination/ Estrus Synchronization

The efficient use of artificial insemination (A.I.) requires identification of superior sires, preservation of their semen for future use and extension of semen to allow insemination of larger numbers of ewes. All three of these hurdles are being overcome simultaneously.

Ram semen preservation has been a difficult problem to solve. Even when techniques were developed to allow good apparent sperm health after freezing and thawing, fertilization rates have been very disappointing. It was found that the primary obstacle to success was the ewe cervix itself.

Unlike the cow cervix, the cervix of the ewe is tightly closed and convoluted, making it impossible to manually pass a pipette into the uterus for semen deposit as in cows. However, it was found that by surgically placing the thawed semen directly into the uterus, good fertilization was possible. Hence, freezing did not destroy either sperm motility or fertilizing capacity but rather weakened sperm so they could not both make the arduous trip through the cervix and continue on to fertilize ova.

As a result of these findings, techniques have been developed for direct intrauterine insemination with the aid of laparoscopy. Ewes are placed on their backs in a surgical cradle with their body tilted head downward at a 45° angle. After CO_2 inflation of the abdomen, the endoscope is inserted through the abdominal wall and used to locate the uterus. An insemination pipette with a small hypodermic needle on the tip is also inserted through the abdominal wall. The needle is then used to place half of the semen dosage directly into each uterine horn.

Recently reported results indicate that a single intrauterine insemination with 40 million sperm can be expected to produce a conception rate as high as 75%, comparable to that achieved with intravaginal insemination with fresh semen containing 200 million sperm (Tervit et al., 1986).

Efficient use of A. I. requires good synchronization of estrus among ewes to be inseminated. Controlled intravaginal drug releasing devices (CIDR's) have recently been developed in New Zealand to replace progestagin pessaries (intravaginal sponges). Reports indicate a very low loss rate of CIDR's and sufficiently precise estrus synchronization that ewes can be laparoscopically inseminated on a time basis (Harvey et al., 1986).

Neither CIDR's nor pessaries are currently available to U.S. producers. As with Fecundin, availability will undoubtedly depend upon industry demand. Frozen semen of several sheep breeds is currently available for import from overseas. There is growing interest in using this approach for obtaining new genetic material.

With the recent development of the National Sheep Improvement Program (NSIP) it has become possible to identify genetically superior sires for widespread use. The combination of NSIP with the described recent developments in both laparoscopic A.I. and uterus synchronization give the sheep industry tremendous increased opportunity for genetic progress.

3. Pregnancy Diagnosis/Fetal Counting

Producers have long recognized the advantages of identifying non-pregnant ewes for differential management or culling. Several techniques with varying degrees of accuracy have been used.

The development of real-time ultrasound scanners for use in human medicine has created equipment capable of both diagnosing pregnancy in ewes early in gestation and determining the number of fetuses present. The management opportunities that such information affords producers could be very significant economically.

Ultrasonic fetal counting is now widely available commercially in Great Britain. Studies are just underway in the U.S. to assess the economic returns possible to producers from differentially feeding ewes carrying single versus multiple lambs. The greatest advantages are likely to accrue to commercial producers who have limited resources of supplemental feed. By directing such feed to multiple-pregnancy ewe producers should be able to reduce incidence of pregnancy toxemia (ketosis), increase lamb birth and growth rates and increase perinatal lamb survival.

III. Conclusion

Sheep are probably unique among livestock species insofar as having reproductive potential beyond current producer ability for utilization. Reproductive rate influences not only direct returns, but also determines the selection intensity that can be applied to genetically improving other traits. Through a better understanding of both the biology of reproduction and the biological impact of management decisions, producers can better utilize resources at their disposal and make informed decisions on application of new tools as they become available.

References

Clarke, J. N. 1972. Current levels of performance in the Ruakura Fertility flock of Romney sheep. Proc. N. Z. Soc. Anim. Prod. 32:99.

Cox, R. I., P. A. Wilson, R. J. Scarzmuzzi, R. M. Hoskinson, J. M. George, and B. M. Bindon. 1982. The active immunization of sheep against oestrone, androstenedione or testosterone to increase twinning. Proc. Aust. Soc. Anim. Prod. 14:511.

Davis, G. H., G. W. Montgomery, A. J. Allison, R. W. Kelly and A. R. Bray. 1982. Segregation of a major gene influencing fecundity in progeny of Booroola sheep. N. Z. J. Agric. Res. 25:525.

Doney, J. M., R. G. Gunn, and W. F. Smith. 1973. Transuterine migration and embryo survival in sheep. J. Reprod. Fert. 34:363.

Harvey, J. G., D. L. Johnson, R. L., Baker, B. K. Trust and B. C. Thomson. 1986. Artificial insemination in sheep—comparison of storage time, dose rate and insemination technique. Proc. N. Z. Soc. Anim. Prod. 46:229.

Meyer, H. H. 1985. Breed differences in ovulation rate and uterine efficiency and their contribution to fecundity. In: R. B. Land and D.W. Robinson (Ed.) Genetics of Reproduction in Sheep. pp. 185–191. Butterworths, London.

Meyer, H. H. 1988. Increasing ewe reproductive rate through steroid immunization with fecundin—a review. SID Sheep Res. J. 4:40.

Meyer, H. H. and J. N. Clarke. 1982. Effect of ewe ovulation rate and uterine efficiency on breed and strain variation in litter size. Proc. N. Z. Soc. Anim. Prod. 42:33.

Meyer, H. H. and R. D. Lewis. 1988. Reproductive responses of ewes to immunization with Fecundin (ovandrotone). J. Anim. Sci. 66:2742.

Meyer, H. H., R. D. Lewis and S. L. Berry. 1988. Increasing reproduction of commercial ewes by Fecundin (ovandrotone) immunization. Prof. Anim. Scientist 4:39.

Nitter, G. 1978. Breed utilization for meat production in sheep. Anim. Breed. Abst. 46:131.

Restall, B. J., G. H. Brown, M. de B. Blockey, L. Cahill and R. Kearins. 1976. The assessment of reproduction wastage in sheep. I. Fertilization failure and early embryonic survival. Aust. J. Expt. Agric. Anim. Husbandry 16:329.

Rhind, S. M., J. J. Robinson and I. McDonald. 1980. Relationships among uterine and placental factors in prolific ewes and their relevance to variation in foetal weight. Anim. Prod. 30:115.

Scaramuzzi, R. J., R. I., Cox and R. M. Hoskinson. 1982. The use of an immunological method to increase twinning in sheep. In: R. A. Barton and W. C. Smith (Eds.). Proc. World Cong. on Sheep and Beef Cattle Breeding. Vol. 1, Technical, 359–363. Dunmore Press, Palmerston North.

Tervit, H. R., R. L. Baker, R. Hoff-Jorgenson, S. Lintu Kangas, S. C. MacDiramid and V. Ramis. 1986. Viability of frozen sheep embryos and semen imported from Europe. Proc. N. Z. Soc. Anim. Prod. 46:245.

Thompson, K. F., S. F. Croshie, R. W. Kelly and J. C. McEwan. 1985. The effect of liveweight and liveweight change on ewe ovulation rate in three successive oestrus cycles. N. Z. J. Agric. Res. 28:457.

Turner, H. N. 1969. Genetic improvement of reproduction rate in sheep. Anim. Breed. Abst. 37:545.

New Zealand Coloured Sheep and Cider Vinegar

Rupert Martin
President, Black and Coloured Sheep Breeders Assoc. of New Zealand
Redwood Wool
R. D. 1. Richmond
Nelson, New Zealand

My wife Grace and I have been livestock farming for more than 40 years. We run 1000 natural coloured sheep, 1000 white Romneys and 30 head of cattle at our Redwood Valley farm near Nelson.

We market all our coloured wool, skins and yarns from our farm. All the products from our coloured sheep are sold direct to the consumer—even the meat.

Let me review the national coloured sheep situation before moving on to explain why we began using cider vinegar as a stock health supplement, and how it has helped us.

We've got 65 million sheep in the national flock in New Zealand, and black and coloured sheep make up only a small proportion of these. There are about 150,000 black and coloured sheep of many different breeds, with Romneys the predominant breed.

Other breeds and crosses of them make up the rest, and include Merino, Corriedale, halfbreeds, Perendales, English Leicesters and Coopworths.

Although black and coloured sheep were first brought into New Zealand by the early settlers in the 1820's, few were kept because white sheep were the most valuable for their wool production.

It was only later this century that the value of coloured sheep began to be recognised and the Black and Coloured Sheep Breeders' Association was formed in 1976.

Since then the quality of both the sheep and their wool has improved remarkably.

New Zealand has many craftspeople, who are very talented, and produce a wide range of fibres, garments and other articles. As well there are many commercial users, whose products are a credit to our industry.

A lot of craft wools are sold direct to craftspeople in small parcels and sent by parcel post.

We use only a small proportion of our coloured wools on the domestic market, so much is sold by auction. The commercial wools are sorted into colours and presented in the best possible manner for auction.

Cider Vinegar

I was the manager of a company farm in Nelson which took 5000 acres (2020 hectares) of waste and scrub land to pasture. We went from no stock to running 6000 ewes and replacements, which gave us a flock of 12,000 head to shear. We also farmed 2000 cattle.

With such large stock numbers we had stock health problems, often in a big way, which were difficult to get on top of. The main problem was grass staggers.

I knew cider vinegar was used on horses, but no-one would tell me why. So in desperation one day when I had two young lambs dehydrated and down with grass staggers, I decided to try the cider vinegar on them.

When I told the makers of the vinegar what I had in mind

they said to be careful, and to dilute the vinegar a bit.

I gave the lambs a cupful each and the next day they were up and grazing. So I gave them a bit more for luck.

That was in the February—our summer was very hot and we had drought conditions. Much to our surprise in May these two lambs were in better condition than the rest, except that they had a break in their wool.

This led us to do some trial work. In our first trial we drenched the sheep once a month from weaning in November to shearing the following October.

We had four groups, and kept the wool of each group separate. The wool was all sold by auction, and the wool from the sheep drenched with cider vinegar made NZ$1.43 a head more than the rest.

We were getting quite excited with our find but no-one would believe us. Still, we carried on using more and more of the vinegar.

At this time I was lambing 2600 two-tooth ewes and I believed they were deficient in iodine. I mixed minerals in with the cider vinegar and drenched just before lambing.

During lambing in previous years I was going around the sheep three or four times a day, and assisting up to 14 ewes per round.

The very first time after we had used the minerals mixed in with the cider vinegar we reduced our problems at lambing down to assisting only two ewes per day. The lamb death rate at birth was reduced by a massive 80 per cent.

Well this was good news for us, and for the next 15 years we drenched our sheep three weeks before the rams went out, and then six weeks before lambing. We drenched the ewes again at three weeks before lambing, and found the results were very good.

I was asked to speak on stock health at the local branch meeting of the Black and Coloured Sheep Breeders' Association. I joined the association, and felt I had something to offer.

Stock health problems and marketing of our coloured wools were then the two main problems to deal with.

I had a few coloured sheep, and their wool was given to friends and staff. I started using a coloured ram over the ewes, and found that quality of the stock was a problem too. Although some good fleeces were produced, there were many rejects.

So I decided to drench every month with 20cc of cider vinegar per sheep.

The results were amazing. We shore in May, and sold more wool in a day than we expected to sell in a year from our woolshed operation.

That went on for two and a half days, and sales have been steady ever since.

We found that the cider vinegar seemed to help disperse the grease in the wool right along the fibre, making it softer and easier to shear.

I still couldn't convince people that what I was doing was good, so I bought vinegar and gave it to friends to try. It took a long time to get going, but when the news media took an interest, it just took off.

This spurred me on to do more research. We found grass staggers disappeared altogether in sheep; sleepy sickness was easily cured. Scouring in calves was also easily cured. In fact any disorder the animals had appeared to benefit from the cider vinegar.

When I first started out with the coloured wools, the natural coloured skins had no value. But the first shipment of pelts I sent to be tanned were all stolen. That proved they were worth something, so I kept going. The next shipment got through all right.

They were quite easy to sell so we bought in skins and sheep for slaughter. We found we could produce the skin okay, but had up to 30 per cent of the skins grading out as seconds.

That was too high, with the quality only good to average.

After looking through the tannery and inspecting skins we found that to produce a variation in colours, and to obtain large skins, we had to use skins from older sheep.

Then I discovered the skins which I had bought in were not as good as my own. That led me to believe cider vinegar was playing a part in giving us quality skins.

Now we prefer to condition the sheep on our own farm before slaughter, and reject skins are down to one per cent or less. Our sheep skins just sell themselves.

With the number of skins we were producing, we had to market the meat.

For years friends had been telling us there was "something" about Redwood Valley meat—it was sweeter. No-one knew why they liked it but they did, and our customers just grew and grew.

We found the vinegar disperses the fat, which really is a bonus for the meat producer. First of all it tenderises the

TABLE 1 Suggested Dose Rates for Animals

Method 1— Monthly Dose Procedure		*Method 2— Daily Dose Procedure*	
Stock	*Dose*	*Stock*	*Dose*
Dog	10 ml	Goat	10 ml
Goat	10 ml	Lamb	10 ml
Lamb	5 ml	Ewe	20 ml
Ewe	20 ml	Deer	25 ml
Deer	50 ml	Calf	25 ml
Calf	50 ml	Cattle	50 ml
Cattle	100 ml		

Method 3—Daily Procedure. Added to food or water			
Stock	*Dose*	*Stock*	*Dose*
Calf	5 ml	Deer	20 ml
Cattle	20ml	Poultry	1 ml
Horse	150 ml	Rabbit	1 ml
Pig	10 ml	Fitch, Cat	1 ml
Goat	10 ml		

meat. You have never really tasted the very best quality meat until you have had the pleasure of eating meat from an animal which has been treated with cider vinegar.

The vinegar seems to reduce the smell of mutton cooking, so should appeal to a good many cooks.

We are now at the stage where we can sell the meat faster than we can sell the skins.

To speed up the operation I decided to make garments or anything I could think of out of natural sheep skins.

First up, a very appealing ladies coat was made, then a coat for men. Car seat covers and slippers followed. All these items created extra problems to be dealt with.

One example is that for rugs, the wool length of 6.25cm was a standard length with a little bit of colour variation being the only requirement.

For garments, car seat covers, etc., wool length starts at 1.25cm which means we can take a fleece for the craft trade and soon after harvest the skin and meat. Lamb skins become more useful but have less variation in colour so we will now breed for variation in colour in our lambs.

Due to the increased demand for cider vinegar, scientific research has become a necessity, and the results so far have proven the beneficial nature of the product.

Athough we recommend minimum doses, for stock in poor health we have found from experience that the dosage may be increased up to five times without any side effect.

The Rugging of Sheep in Australia

Judy Baker
"Weddin View", Grenfell
N.S.W. 2810, Australia

Sheep coats can be a useful tool in sheep management. They protect freshly shorn sheep and improve wool quality by preventing weather damage and contamination by dust and vegetable matter. This paper reviews the development and use of sheep coats in Australia and demonstrates their particular benefits for coloured sheep breeders.

History and Development

Australia is a large country with vast areas of land well suited to sheep and wool production. In 1988 the sheep population was estimated at 165 million—10 sheep for every Australian.[1] The state of New South Wales (NSW) has the largest sheep population of any state in Australia—nearly 70 million in 1988—and sheep are raised under wide variations of climate and pastures. Thus any study of sheep in NSW will apply in general to most sheep-raising areas of Australia.

While Australia has an envied world-wide reputation for quality wool, any development which can improve wool quality or sheep management will be closely investigated. The idea of rugging sheep is not new as, in records of early Greek history, instances are quoted of sheep being clothed in a covering of skin to defend the wool from filth and protect it from being torn by hedges. It has long been the practice to rug sheep for wool improvement in preparation for the show ring, but, so far as general sheep husbandry in Australia is concerned, the history of rugging sheep goes back about fifty years[2]. Mr Spen. Williams, Manager of the wool department of Golsbrough, Mort & Co. Ltd. in Adelaide, South Australia, was concerned with the large quantities of inferior wool being offered for sale, the chief causes of damage being dust and dirt. In 1934 he tried rugging a few sheep with coats made from hessian (burlap) bags. Following the success of these trials and press publicity, considerable interest was aroused amongst South Australian sheep owners and by 1935 probably some four to five thousand sheep were being rugged. Most coats were made from used bags as these were readily available to farmers.

Though the primary object of rugging was to protect the wool from dirt and burrs, many claimed additional advantages and many trials were carried out in the 1930's to assess the efficacy of sheep coats. In 1938 Pierce[3] experimented with the effect of year-round rugging on the milk production of Merino ewes and on the growth of their lambs. Also in 1938, Blumer and Cotsell[4] reported on the rugging of aged Merino wethers during winter in an attempt to improve their body condition and sale value in the spring. In 1939 Levy[5] experimented with the rugging of Corriedale ewes in winter and in a separate trial[6] the rugging of weaner Corriedale lambs. In 1939 Montgomery[7,8] reported the results of a two year trial on the rugging of 250 Comeback (Polwarth blood) sheep. The trial was conducted at Albury NSW, an area typical of good wool-growing country, producing reasonably clean wool and representative of a large area of NSW. In 1942 Montgomery and Blumer[9] reported on trials carried out at Armidale NSW

See color photographs on page 150.

(a tablelands area) on the effects of rugging ewes to determine (a) lamb rearing ability and (b) general health and thus, indirectly, on gastro-intestinal parasitism.

All these trials showed that while rugging could reduce dust and burr contamination and the rugged wool appeared more attractive and of better growth than unrugged wool, year-round rugging was unlikely to be an economic proposition because of the cost of the coats and the extra labour involved. Montgomery's trial[8] showed an actual difference of + 2.3% in the clean scoured yield of the rugged fleece. However, in those days, wool value was based on the visual assessment of the wool classer and as the classer's estimate of clean scoured yield was not always accurate (being unfamiliar with rugged wool), the sale prices for rugged wool did not necessarily reflect their true value.

Trials carried out by Lipson, Ellingworth and Sinclair[10] from 1956 to 1958 assessed fleece properties and processing of wool from rugged and unrugged sheep run on three properties in Victoria, chosen to produce wool of high, medium and low burr content. They concluded that rugging greatly improves the subsequent processing performance of wool. Rugged wool from these trials returned an average of 12% more than the unrugged wool when sold at auction. Coats for these trials were constructed from canvas duck, proofed and unproofed, and polyethylene sheet. The rugging materials were considered too expensive for economic widespread use.

In 1968 Panaretto, Hutchinson and Bennett[11] tested the protection afforded by plastic coats to shorn sheep. Freshly shorn Merino ewes, with and without plastic coats, were exposed to cold, wet conditions in a controlled environment. Fifty-five per cent of the sheep without coats died, whilst only five per cent of the sheep with coats died. They concluded that the plastic (0.05mm polythene) coats reduced heat loss considerably and could profitably be used on shorn sheep when heavy losses could be expected because of severe weather conditions.

Development of woven split-film synthetics during the 1970's led to a re-assessment of the economics of sheep rugging. An extensive trial with 34,200 sheep, involving 35 properties throughout NSW was conducted in 1972–1973 by the NSW Department of Agriculture, in conjunction with the Commonwealth Scientific and Industrial Organization (CSIRO) Division of Textile Industry, the University of NSW School of Sheep and Wool Sciences, and a sheep coat manufacturer, Gollin & Co., who supplied the coats, made of woven polyethylene fabric. Gollin & Co. spent $(Aust.)1.25 million on the trials. As new coat designs were developed, coat fabric was improved and other problems were overcome. When the elastic (sewn into the sides of the coat for good fit) deteriorated, sheep were actually fitted with pedometers in a Western Australian trial to determine how many times the coat flexed with walking. This led to the development, with the rubber manufacturer, Dunlop, of a neoprene rubber that could withstand up to twelve months constant flexing without breakdown. Wool from these trials was processed by the CSIRO Division of Textile Industry and results showed that average clean scoured yield for rugged wool was 5.34% higher than for unrugged wool. The Australian Bureau of Agricultural Economics estimated that up to 55% of Australian sheep could benefit from rugging. Notes taken during the trials showed that coats generally caused no interference with mating or lambing though maiden ewes could be a little shy of the rams at first because of the noise of the coats when the rams mounted the ewes. There was no record of flystrike under the coats. Though an average premium of $(Aust.) 0.23/kg was received for the greasy rugged wool, the greatly increased labour inputs was generally considered to make the coats an uneconomic proposition.[12] Nevertheless, these trials led to improved coat materials and better coat designs and sparked much interest in the idea of rugging sheep.

In 1979 Abbott reported on the benefits and economics of rugging sheep with polyolefin fabrics[13,14] and assessed fabric and coat durability.[15] The trials were carried out at six different locations and the results are set out in Table 1.

TABLE 1 Effects of Rugging on the Amount and Quality of Wool Tops

Location (State)	Group	Wool base % at 19% regain	Vegetable matter (%)	Top mfl (mm)	Vegetable Part./100g of top 3–10mm	Vegetable Part./100g of top > 10mm	Value of top+noil/sheep ($A)	Cost benefit ($A)
Birchip	R	65.2	0.4	71	4	0	8.63	0.64
(Victoria)	UR	58.1	0.9	64	2	0	7.99	
Wanbi	R	63.5	4.1	76	90	14	11.50	0.49
(S. Aust.)	UR	54.1	4.7	71	160	11	11.01	
Deniliquin	R	72.9	0.7	79	6	0	11.39	1.25
(NSW) 1972	UR	70.6	2.5	76	20	0	10.14	
Deniliquin	R	77.5	1.4	70	3	0	14.8	0.69
(NSW) 1977	UR	72.2	3.2	69	7	0	14.11	
Sandalwood	R	65.3	4.4	55	16	2	8.84	1.16
(S. Aust.)	UR	55.7	9.7	52	45	8	7.68	

R—Rugged
UR—Unrugged
mfl—mean fibre length

Abbott concluded that the use of sheep coats to protect fleece wools throughout a full growing season produced higher yielding fleece wools and, in most trials, more clean wool per sheep. Rugging also reduced the proportion of noil, increased the mean length of the fibre in the top and reduced the number of vegetable particles in the top. He advised that it was not possible to generalise on the economic viability of sheep coats because the advantages of rugging, and the expected life of the coat itself, will vary from area to area and from season to season.

Since earlier trials had indicated that sheep coats could have definite economic advantages under certain circumstances, several trials have been carried out in recent years by officers of the NSW Department of Agriculture. Messrs John Cahill and Bob Marchant, Sheep and Wool Officers in the Monaro and Southern Highlands regions of NSW, have investigated the use of sheep coats in relation to lambing performance.[16] They conducted environmental chamber tests to prove the effectiveness of polyethylene coats in insulating sheep from heat loss after shearing. Their field observations showed large behavioural differences between rugged and unrugged sheep, especially in the first two weeks after shearing. The rugged sheep grazed more extensively and were untroubled by wind and showers. In trials where ewes were rugged for approximately four weeks between shearing and lambing, rugged ewes produced a significantly higher percentage of viable lambs. Birthweight increases for lambs and improved lactation response from rugged ewes are thought to be responsible for this improvement. Mr Ian Simpson, Special Livestock Officer at Young, NSW, evaluated the strategic use of sheep coats applied for three to four months in late spring/summer, in reducing the vegetable matter contamination of wool in 1984/1985[17] and 1985/1986.[18] Mr Allan Casey, Sheep and Wool Office at Warren, NSW, evaluated the use of 'shearing to shearing' sheep coats in reducing vegetable fault in Western NSW where high vegetable matter contamination can significantly reduce wool value.[19] While Casey found that coats definitely reduce vegetable matter in the fleece, he also noted the problem of contamination of the fleece wool with poly fibre from the coats. Where hot iron cuts are used to form the leg slits in some brands of coat, prolonged wearing of the coats caused the edges to fray and loose poly fibres to become entangled in the fleece wool. Such contamination is a serious problem for wool processors and coats with hot iron cuts definitely cannot be recommended for long-term use. Both Simpson and Casey noted that many coats did not last undamaged for twelve months. Even in Simpson's short-term trials, only 56% of the coats remained totally undamaged after four months' wear.

This experimental work has shown that sheep coats can be of definite benefit. In some circumstances short term rugging may be sufficient. In other conditions long term rugging may be more beneficial.

Short Term Rugging

In Australia, nearly one million sheep die each year during the first thirty days after shearing.[20] Cool, wet, windy weather striking freshly shorn sheep may cause severe or even disastrous losses in a flock. Some owners have lost almost 15% of their entire flock almost overnight. While it may seem sensible to change the time of shearing, that may not be practical. In Australia the timing of the annual shearing may depend on many factors e.g. the availability of shearers; shearers delayed by prolonged wet weather; timing to shear before lambing; timing to fit in with other farming activities on a mixed farming and grazing enterprise.

In 1968 Panaretto et al.[11] tested the protection given by plastic coats. Merino ewes were shorn and half were fitted with 0.05mm polythene coats. All the sheep, both coated and uncoated, were then placed in a controlled environment to simulate cold (3°), wet conditions. Heat loss was measured for coated and uncoated sheep, both when the wool under the coat was dry and when it had been wetted. It was found that the coats decreased heat loss by 31–43% when the wool was dry and by 25–34% when the wool was wet. Fifty five per cent of the sheep without coats died, whilst only five per cent of the sheep with coats died. It was concluded that the plastic coats had considerable protective value and losses could be almost eliminated at an economic cost in high-risk areas. A field trial with the plastic coats showed that the simply designed coats stayed on until removed after two weeks and no evidence was obtained that wearing the coats for this length of time would induce any form of fleece rot. Cahill and Marchant's later work[16] in controlled environment tests with woven polyethylene coats confirmed these findings.

Mr Geoff Cottle, a grazier of Nimmitabel NSW (altitude 1180m), took a keen interest in the 1972/1973 trials with the Gollin coats. After he lost 1300 out of 2600 freshly shorn Merino weaners during a cold March night in 1978, he decided to 'insure' his sheep in future by rugging after shearing.[21] When Gollin coats became available he set about designing his own. The simply designed 'Nimmity' coat evolved and went into commercial production. Following the success of the Nimmity coat and extensive field testing by Cahill and Marchant[16] nearly all graziers in high altitude regions who shear between May and September now use some sort of protection for freshly shorn sheep.[22] Possibly as many as 65% use sheep coats and the remainder use permanent shelter sheds. The coats are left on for 4–6 weeks only after shearing as, after that time, there is sufficient wool growth to insulate the sheep from cold stress. Mr John Cahill, Sheep and Wool Officer in the Monaro, estimates that approximately 300,000 sheep coats are now used annually on freshly shorn sheep in NSW. This practice has enabled shearing to be spread throughout the winter months, facilitating greater flexibility in arranging a suitable shearing date.

Long Term Rugging

Though the work of Abbott[13] and others has shown that sheep coats can definitely improve wool quality, the year-round use of sheep coats has not been widely adopted by large-scale graziers in Australia. While a few consider it worthwhile, most are unaware of the possible benefits or are not prepared to bother with all the extra work involved in rugging, changing coats, regular surveillance etc. One group who do use coats year-round are the Sharlea breeders. The Sharlea is a Merino sheep with wool fibre diameter of 16–17 microns and kept shedded throughout the year. The sheep are fed a specially prepared diet designed to keep the wool fibre fine and of even diameter throughout its growth. The sheep are also rugged to keep the wool perfectly clean and to prevent wool biting. The premium price paid for Sharlea wool makes it economically viable for Sharlea breeders to shed and rug their sheep.

While most experimental work has been carried out some years ago, when wool prices were much lower, it is possible that with current good wool prices in Australia, rugging could significantly enhance returns to woolgrowers. Good, clean 21 micron Merino fleece was making $(Aust.)12/kg at auction in December, 1988.[24] However, similar wool containing up to 3% vegetable matter was liable to a discount of $(Aust.)0.60/kg and with 5% vegetable matter, a discount of $(Aust.)/kg applied. With the average cut per head for Australian Merinos estimated at 4.56 kg in 1988,[1] this difference more than covers the cost of the coat plus labour.

The experimental work has been carried out on white sheep and mostly on Merinos as they are by far the most numerous breed in Australia and the breed whose fine wool has the greatest value to commercial wool processors. It has been noted by Abbott that sheep coats do have particular benefits for coloured sheep breeders.

Rugging Coloured Sheep

Most coloured sheep flocks in Australia are bred to produce fleece for the handcraft market. This means there are important differences in the requirements of the fleece buyers. Commercial wool processors have large machinery to prepare the fleece for spinning and can blend large quantities of fleece, thus evening out differences within a fleece or between individual fleeces. However handcraft fleece buyers are dealing with one fleece at a time and are mostly using simple equipment such as a comb or flicker to prepare the fleece for spinning.

An ideal handspinner's fleece has the following characteristics:

(a) Good length
(b) Cleanliness
(c) Sound tensile strength
(d) Good handle
(e) Lustre
(f) Colour

By rugging the sheep all of these characteristics can be improved, as follows:

(a) Length
Abbott[13] has shown that rugging can increase mean fibre length by preventing tip damage. While this may be only a small percentage increase, rugging reduces weather damage to the tip. Thus a spinner need not reduce the fibre length by cutting off the weather-damaged tip.

(b) Cleanliness
Perhaps the greatest nuisance to handspinners is the need to remove contaminants from the fleece before spinning. While dust may be washed out (before or after spinning), vegetable matter contamination must be laboriously removed. The main contaminants in Australia are burrs, seeds and thistles.[23] Burr medic (Medicago polymorpha), also known as Trefoil, is an excellent pasture but the seed pods have hooked spines which readily adhere to the wool. Many grass species, such as Barley grass (Hordeum leporinum), have long thin seeds which penetrate the fleece and lie parallel to the wool fibres, making them difficult to remove. Thistles, such as Saffron thistle (Carthamus latanus), have leaves bearing sharp prickles on the edge, making them painful to remove. Rugging the sheep can keep all these contaminants out of the rugged portion of the fleece completely.

(c) Soundness
If a sheep is subjected to stress then the wool fibre may be reduced in diameter at that point of growth and will break during the preparation or spinning. Rugging can reduce stress, such as cold stress after shearing.

(d) Handle
The natural grease in wool, formed by the secretions of the sebacious (wax) and sweat glands in the skin of the sheep, aids in the soft-handling characteristics of a fleece and the ease of spinning. Rugging prevents weather damage to the fleece so the natural grease is not washed out by rain. The fibre population density affects the amount of the wool staple exposed to weathering. The general average fibre population for the Australian Merino is about 5,000 wool follicles per square centimetre of skin. This means the fibres are very densely packed, like trees in a tall forest, and only the very tip of the staple is exposed to weathering. However, in stronger-woolled breeds, where the fibre population is not nearly so dense, a much greater percentage of the staple length is exposed. In breeds such as the Romney up to half the fibre length may suffer weather damage. Rugging will eliminate this, producing a well-nourished fleece with grease from base to tip of the staple.

(e) Lustre
If the wool fibre is exposed to weathering as described in (d) then the weathered portion will lose its natural lustre, making it less attractive. This is particularly important to weavers who value the lustrous sheen of wool.

(f) Colour
While colour is a very individual preference in handcraft fleece users, most prefer an even colour from base to tip of each wool fibre. The ultra-violet in sunlight degrades colour in naturally coloured fleeces, causing all shades of black to turn a gingery-brown and all brown (moorit) fleeces to fade a pale beige. If coloured fleeces are protected from sunlight by coats, then there is no ultra-violet degradation and the wool retains its true colour e.g. black stays totally black from base to tip (see accompanying photographs).

Disadvantages of Rugging

Though the advantages of rugging make it worthwhile, there are some disadvantages in rugging sheep.

Putting coats on, taking them off and adjusting them as the fleece grows takes considerable time and effort. It's back-breaking work, the sheep don't seem to appreciate what is being done for them and a typical comment from those who rug their sheep is: "It's a pain!". However, seeing the beautiful wool that comes off the rugged sheep at shearing time makes it all worthwhile!

Coats cost money. Unless the rugged fleece can be sold at a premium price to cover the cost of the coats and the labour, it is not worthwhile to rug sheep year-round. However, it is worth considering the other values of rugging. Coats have an 'insurance' value and save many a grazier sleepness nights when shearing takes place in cold weather.

Coats which are too tight because of sheep or wool growth can cause chafing; coats which are too loose can become entangled in fences or branches etc. Thus rugged sheep must be checked regularly.

A sheep coat only covers about 80% of the total useable fleece. Since the unrugged portion is of a distinctly inferior quality to the rugged portion it is better to sell or use it separately. This unrugged wool can be sorted into several different shades and commercially scoured and carded into sliver, making it then easily saleable to spinners.

Survey of Sheep Coats Available in NSW

Most sheep coats are designed to fit Merino sheep and may not be a good fit on breeds with a broad chest, such as the Romney. Trial and error is necessary to find the coats which fit a particular breed best. Coats with an adjustable gusset in the front seam are much easier to fit correctly and a good fit is essential. The colour of the coat does have some consequences. A darker colour (blue or green) is easy to see in snow and absorbs more heat from sunlight thus aiding in keeping a freshly shorn sheep warm. A lighter colour (white or beige) reflects sunlight better, thus helping to keep a black sheep cool in summer. In summer at Grenfell NSW, where temperatures can reach 40°C, rugged sheep can be observed grazing when the unrugged black sheep are still resting in the shade. Most manufacturers have a wide size range of coats available including large stud ram coats. A summary of the coats is tabled below:

TABLE 2 Sheepcoat Survey

Brand name/maker	*Fabric*	*Colour*	*Leg slits*	*Adjustable*	*Cost $A**
Aust-Ag Industries	UV stabilised woven p.p.	Beige	Hot iron cuts	2 seams in gusset	2.50
Nimmity	UV stabilised woven HD p.e.	Blue or yellow	Sewn	No	3.50
Frank Tope	UV stabilised woven HD p.e.	White	Sewn	Tie-up tapes in gusset	4.20
Hyline	N/A	Green	Hot iron cuts	Press studs in gusset	1.50
Clarente	N/A	Green	Sewn	No	3.50
Pope	N/A	Green	N/A	No	2.00

*Average coat = flock coat for 90 cm size with front gusset if available.
HD = High Density
N/A = No Details available.
p.p. = polypropylene
p.e. = polyethylene

Use of Sheep Coats by Black Sheep Breeders in NSW

All 160 members of the NSW Black and Coloured Sheep Breeders Association were sent a questionnaire on sheep coats. Flock sizes of those that replied to the questionnaire ranged from 11 to 500. Prices obtained for unrugged fleeces ranged from $(Aust.)3 to $(Aust.)8/kg, except for one member who asks $(Aust.)12/kg if the fleece is very clean. The average price for unrugged fleece was $(Aust.)6.50/kg. Prices obtained for rugged fleece ranged from $(Aust.)8.50 to $(Aust.)15/kg with an average of $(Aust.)12/kg.

Some comments from members who did not rug their sheep included: "coats are too costly;" "seed and burr is no problem;" "waste of time in hilly country" (coats get caught in fallen timber etc); "rugged sheep need close supervision."

The comments from members who did use sheep coats included; "fleeces are noticeably cleaner with no weather damage;" "the darkest fleeces really do need coats to stop bleaching;" "have no problems with cold, wet (weather changes after shearing;" "the fleece is clean, no flystrike, and the sheep seem quite happy in them;" "while the coats create quite a lot of extra work, we feel they are well worthwhile." One member at Tabulam NSW (average rainfall 1050mm p.a.) has just started using coats on some of her coloured sheep. The area is not generally considered sheep country, being too wet, but though weather damage and fleece rot has been noted in unrugged sheep, the fleeces of the rugged sheep show no ill-effects caused by rain. Perhaps sheep coats can extend the climatic range over which sheep can be raised to produce quality wool.

In our own flock of 500 coloured sheep (Merinos, Corriedales and Romneys) we rug about 300 a year. We are amongst the few people in Australia who make their living from coloured sheep and we specialise in producing top-quality fleece for handspinners and our own commercially processed, naturally coloured handknitting yarn. Rugging the sheep does make a big difference to the quality from the handspinner's point of view. The absence of weather damage means that the tip of the fleece opens very easily and the natural grease lubricates the wool from the base right up to the tip. The wool is thus very easy to prepare and spin.

Other Uses for Sheep Coats

(a) To keep animals clean for the show ring (sheep, goats, etc.)
(b) For sick animals (reduces loss of body heat)
(c) Insulating effect helps lactation of dairy goats in winter

Conclusion

This paper has shown that the advantages of rugging sheep far outweigh the disadvantages. When sheep are shorn under adverse weather conditions, coats can effectively protect them from cold stress. Coloured sheep breeders particularly can benefit from rugging their sheep. The coats will protect the fleece from dust and vegetable matter contamination, weather damage and ultra-violet degradation of the natural colour. Handspinners will happily pay a premium price for the rugged wool because of the greatly improved quality.

References

[1]Australian Bureau of Statistics.
[2]Scott, R. C. Jour. Dept. Agr., South Aust., 39, 534. (1935).
[3]Pierce, A. W., Jour. Counc. Sci. and Ind. Res., 11, 229. (1938).
[4]Blumer, C. C. and Cotsell, J. C., Agric. Gaz. of NSW 49, 188. (1938).
[5]Levy, A. L., Agric. Gaz. of NSW 50, 142. (1939).
[6]Levy, A. L., Agric. Gaz. of NSW 50, 297. (1939).
[7]Montgomery, I. W., Jour. Counc. Sci. and Ind. Res., 11, 221. (1938).
[8]Montgomery, I. W., Jour. Counc. Sci. and Ind. Res., 12, 169. (1939).
[9]Montgomery, I. W., and Blumer, C. C., Jour. Counc. Sci. and Ind. Res., 15, 10. (1942).
[10]Lipson, M., Ellingworth, J. S., and Sinclair, J. F., Text. Inst. Industr. 8, 100. (1970).
[11]Panaretto, B. A., Hutchinson, J. C. D., and Bennett, J. W., Proc. Aust. Soc. of Animal Prod., 7, 264. (1968).
[12]NSW Dept. of Agric., Summary of Sheep Coat Trials, (unpublished). (1973).
[13]Abbott, G. M., C.S.I.R.O. Division of Textile Industry, Report No. G. 36. (1978).
[14]Abbott, G. M., Wool Tech. and Sheep Breeding, 27, 29. (1979).
[15]Abbott, G. M., Text. Inst. Industr., 213. (1979).
[16]Cahill, J. R., and Marchant, R. S., Sheep Coats in relation to lambing performance. NSW Dept. of Agriculture, Paper 12. (1982).
[17]Simpson, I., Sheep Coat Trials 1984/1985, unpublished data.
[18]Simpson, I., Sheep Coat Trials 1985/1986, unpublished data.
[19]Casey, A., Sheep Coats to Reduce Vegetable Fault in Western NSW. NSW Dept. of Agriculture. (1987).
[20]Hutchinson, J. C. D., Aust. J. of Exper. Ag. and Animal Husb. 8. (1968).
[21]Cottle, G., Private communication.
[22]Cahill, J. R., Private communication.
[23]D'Arcy, J. B., Sheep Management and Wool Technology, 2nd ed. Kensington, New South Wales University Press, 1981.
[24]The Australian Wool Corporation.

Milking the Coloured Flock

Ruth M. Thompson
The Sheep Farm
Pimpama
Queensland, Australia

In Australia folk tend to look aghast when we tell them that we milk our sheep, yet this is not something modern. The milking of sheep and the making of cheese goes back before the dawn of agriculture when nomadic flocks were walked from pasture to pasture. Today there are more than 1,000 million ewes being milked throughout the world, producing in excess of 1½ billion gallons of milk, and sheep milking is not exclusive to white sheep. Flocks of nomadic sheep are still being milked seasonally in exactly the same way as has been done for thousands of years, and a big percentage of these ewes are coloured.

At The Sheep Farm we did not set out to intentionally milk coloured sheep. A few years ago when we thought about it we found that there were no milking strains of sheep in Australia. Research had disclosed a couple of efforts to establish dairies in Victoria a decade ago, milking mainly Border Leicester-Merino cross ewes. All projects had fizzled out. In 1977 we had begun to keep coloured sheep, starting with Merinos over which we crossed a Border Leicester ram. There being few purebred coloured Border ewes available at that time we had obtained white stud ewes and gradually crossed these to obtain our purebred coloured Border flock. Eventually we culled out the original white ewes and retained the heterzyogous ewe progeny, along with the coloureds. We also run registered studs of Wiltshire Horn and English Leicesters. Before the recent intervention of the 'Rare Breeds Survival Trust' both of these breeds had come close to extinction, yet they had once dominated the areas of Britain noted for sheep milking and cheese making prior to "field enclosure" of the 16th century. So it is not surprising to find that these ewes are very milky, although not used in the milking flocks of Britain today. As the costs of importation of specialised dairy sheep are quite prohibitive, we set out to devise the best possible milking ewe available from our collection of sheep breeds.

We had by this time a flock of some 200 coloured Borders and Border crosses in black, grey, moorit and heterzyogous. Wiltshire Horn and English Leicester were crossed over all the heterzyogous and a number of the coloureds giving us still more heterzyogous. However when we used one of our Wiltshire cross rams over our Border ewes, many of the lambs were born with a large red patch on the back of the neck. It resembled red halo hair and disappeared at about 10 weeks. These animals grew up to produce carpet wool fleece and wonderful wool-free udders . . . perfect for the milking machine. The following year when many of these lambs were backcrossed it was elementary that a high percentage would be coloured.

Milking the Sheep

When a goat dairy plant came on the market we made the final decision to enter the sheep milking industry. We purchased a custom made steel stand which holds 12 ewes, 6 each side of a centre well. The electric milking plant being mounted overhead. The goat cups were unsuitable for sheep. Alfa Laval had stainless steel sheep cups available but we were able to find another brand of teat cup made from clear plastic which still fitted our Alfa Laval plant. Although we have had only a short life from some of these cups (cracking around the base) the great benefit of them is the light weight factor and having the teat visible within the cup. Any abnormalities can be noticed immediately, mastitis for example, and the cup quickly withdrawn before the whole milking line is contaminated.

Modern ewe milking parlour.

Our milk is pumped directly into the cheese room, where it is either packaged in 500 ml. raw milk bags or pasteurised before making into cheese, yoghurt and ice cream.

Legalities

There had never before been a sheep dairy in this state, Queensland; so there was no legislation to cover one. We finally had to get approval from no less than five authorities, namely, the Minister for Primary Industries, the Dairy Board, The State Health Dept., the Local Authority Health Dept., and Weights and Measures, plus licences to sell raw milk and manufactured products.

Training the Sheep

The first day is the hardest, with each new ewe being pushed, shoved and almost carried up the ramp and on to the stand. Once they get their head in the feed bin and the headlock is on they settle quickly. Of course there is always the ewe who stamps her feet and tries to kick the cups off. Having nearly finished milking our second lactation, we are now selecting more on wide spread back legs, well suspended udders with downward pointing teats, and udders free of wool. The ewes normally take about three days to adjust to the routine. Udders are washed and thoroughly dried before milking and are treated with teatdip after the cups are removed.

Managing the Flock

The milking flock does receive a bit more attention than our other flocks. They have access to more lucerne (alfalfa) and receive a high protein supplement. Our main problem has been the control of worms. Being situated in the sub tropics only 2 kms from the sea, the environment can be hot, muggy, humid—usually the lot together—and these conditions are relished by internal parasites, especially the resistant Barbers Pole worm. Most of the new drenches used today to control this problem worm have a withdrawal period of up to 28 days. One drench with no milk withholding period is Nilvern, but our experience is that it cannot be relied on to control the resistent Barbers Pole. So although all our other sheep are wormed each month with one of the new drenches, the milkers get Nilvern fortnightly. Faeces are regularly sent to the Animal Research Institute for egg counts. But it also means that newly lactating ewes being shifted into the milking flock must be carefully monitored for their last drenching date, and vice versa. In addition each fortnight a sample of our milk is sent to the Gold Coast Dairy's laboratory to test for coliforms, bacteria, iodine, antibiotics, etc.

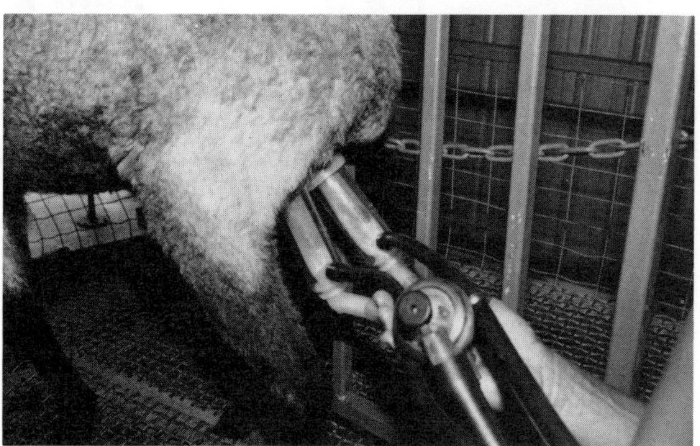

A coloured ewe being machine milked.

	Percentage				
	Fat	*Protein*	*(Casein)*	*Lactose*	*Minerals*
Sheep	7.8	5.6	(4.2)	4.4	0.87
Goat	3.9	3.3	(2.5)	4.4	0.80
Cow	3.5	3.25	(2.6)	4.6	0.75

Wool and Milk

Now, not only can your coloured ewe produce a lamb each year plus a nice spinning or weaving fleece, but could also keep the household in milk. Depending on the breed, ewes can milk from ½ litre to 3 litres a day. Sheeps milk is sweeter than cows milk and really quite delicious. . . THE NATURAL ALTERNATIVE FOR YOUR COLOURED FLOCK, is sheeps milk.

Reference

Mills, O. Practical Sheep Dairying.

Brucellosis in Sheep and the Accredited Free Flock Scheme in New South Wales

Laurie Pryde
"Wavering Downs"
Cumboogle Road MS-3
Dubbo NSW 2830 Australia

Brucellosis is a bacterial disease affecting the reproductive tract of both ewes and rams. Wethers can, for practical purposes, be ignored. While it is possible for wethers to become infected, there are no resultant ill effects on either their health or productivity. Further they are incapable of spreading the disease.

The causal organism is Brucella ovis which in Australia has been isolated only from sheep. Consequently there is no between-species transmission and other species, particularly goats, have never been shown to interfere with an eradication program or to have been thought responsible for the introduction of the disease.

Transmission

Transmission is primarily by venereal means. Infected rams excrete bacteria in their semen passing the disease to ewes at service. These ewes may then pass the infection on to other rams during the time they remain infected.

The effects of Brucellosis can be looked at from both the individual animal and flock perspectives.

Ewe Effects

Upon acquiring the infection susceptible ewes, i.e. those which have not been previously exposed to the disease, develop a mild inflammatory condition of the reproductive tract which normally prevents conception. Recently infected ewes show no outward sign of their condition. Most ewes develop an immunity and throw off the infection after a short period. As a rule of thumb most ewes can be expected to return to pre-exposure fertility levels after three heat cycles.

Overall the effect on ewes is simply a temporary infertility. There have been very few cases of abortion due to brucellosis recorded in Australia. This is not the case in some other countries where abortion is a significant finding.

Ram Effects

Brucella organisms find their way to and localise in the testicles, epididymis and accessory sex organs. In these sites a chronic inflammatory condition results in changes to these organs which impair the sperm production process. As a result the ram suffers a progressive deterioration in fertility as well as intermittantly excreting bacteria in the semen.

Rams never recover from infection nor is there any effective treatment. The degenerative changes resulting from infection often produce palpable abnormalities particularly to the testicles and tail of the epididymis. Up to 50 per cent of infected rams may have palpable abnormalities. Many other infectious, physiological or traumatic causes can produce similar lesions so their presence is not diagnostic of Brucellosis.

Flock Effects

Many factors contribute to the observable effects resulting in a variety of possible presenting problems. Some of the major factors include:

- flock fertility prior to the introduction of disease
- length of joining
- proportion of susceptible ewes
- stage of reproductive cycle when infection occurs
- proportion of ram flock infected
- degree of infertility in individual rams

Usually the first observable effect that concerns producers is a reduction in lamb marking percentage (LM%). It may be seen as a rapid fall in particular mobs, especially maidens, or in other cases as a progressive deterioration over a few years. Where joining is in excess of eight weeks a change in lambing pattern may be noticed with peaks at both the start and end of lambing. This second peak is occasioned by the reversion of fertility as ewes develop immunity.

An additional, but seldom noticed effect is increased ram wastage. Most Australian flocks use 1.5 per cent rams for joining. In infected flocks ram 'power' is reduced contributing to the reduced LM% but also the ram's useful life is shortened. This is translated into a higher annual ram replacement rate.

Testing for Brucellosis

In Australian laboratories the Complement Fixation Test (CFT) is the standard test for Ovine Brucellosis (OB). This test detects antibodies which are produced by sheep only in response to infection with B. ovis. Antibodies appear in the blood about three weeks after initial infection. While the level (titre) of antibodies may fluctuate with time, infected animals generally will always give a positive test.

Only rams are routinely blood tested because they are the reservoir of infection and they never recover. Ewes will recover and over time their titre will fall to undetectable levels.

CFT results are reported as negative, 8, 16, 32, 64 or 128. Under the NSW Accreditation Scheme any ram with a titre of 32 or greater is regarded as positive. Rams with titres of 8 or occasionally 16 are termed inconclusive. Such rams are subjected to additional testing to determine their true status. Reasons for inconclusive results include abnormal test sensitivity, poor sample quality, cross-reactivity with other infections and early cases.

The additional testing required for inclusive reactors includes a repeat blood test, and semen examination and culture. In extreme cases where this additional regime of testing fails to resolve the ram's status, the animal will be slaughtered and the entire reproductive tract subjected to rigorous laboratory investigations.

Eradication

Eradication can be achieved by testing all eligible rams at 60 day intervals until the ram flock returns two successive negative tests.

At any test, any ram returning a non-negative result is immediately removed from the ram flock and isolated. Positive rams are slaughtered and inconclusives retested in isolation as described above. In New South Wales results are usually available within a week of blood collection.

Private veterinary practitioners palpate the testicles and collect blood from all eligible rams. Prior to the commencement of testing the owner gives an undertaking to comply with the conditions of the scheme. Of particular importance are the individual identification of all rams, the maintenance of stock-proof boundary fences and a halt to the introduction of any sheep during the testing program.

Rams eligible for testing include: i) All sires and teasers; ii) Any ram intended for use as a sire; and iii) A sample of all rams 10 months and older.

The sample size is calculated to give a 99 per cent confidence limit of detecting disease if it were present at a 2 per cent flock prevalence.

In addition to the above, all rams over 4 months of age are palpated and any with abnormalities are also blood sampled.

Keeping the Flock Clean

The steps to follow are:

- Maintain stock-proof boundary fences
- Introduce only clean rams
- Only use licensed semen in AI programs
- Isolate introduced ewes for 4 months
- Monitor flock fertility
- Test stray rams

Accredited flocks may introduce rams from other accredited flocks without further testing. Ewes from non-accredited flocks may be introduced provided they are kept in isolation for 4 months after arrival, or if in lamb, until 4 months after lambing is completed. Should a ram stray on or off an accredited holding it should be isolated and tested at an appropriate interval.

Why an Accreditation Scheme?

The main motivation behind the establishment of a voluntary accreditation scheme came from the Australian Society of Breeders of British Sheep (ASBBS). Field experience and surveys had shown the prevalence of infection was much higher in these breeds than in the Merino. The breeds represented by ASBBS are exensively used for cross-breeding with the Merino for prime lamb production. The ASBBS was sufficiently persuasive amongst both its own members and with government to have the scheme implemented. Initially it was restricted to British Breeds but after a few years it was opened to all breeds.

A working party comprising representatives from ASBBS, Agriculture Department, Australian Veterinary Association (AVA) and A Webster Pty Ltd (laboratory) was responsible for drawing up the original guidelines. The scheme commenced in 1981 and now well in excess of half of all stud flocks of all breeds are accredited. An expanded working party meets annually to review the scheme and where necessary modifies the guidelines.

The benefits of the scheme to ram buyers are obvious. Studs progressively found their clients only wanted to buy accredited rams. Initially this resulted in a better price for these rams but now very few studs worthy of the name can survive unless they are accredited.

The promotion of the scheme and its acceptance by stud and commercial breeders has largely been due to the joint efforts of Agriculture Department and the individual breed societies. The former took on the role of both the registration body for accredited studs and provided an advisory program about the disease and the scheme. Breed societies initially encouraged and later pressured their members to belong. Various show societies also became involved with the segregation of rams at shows and sales and finally a decision not to accept non-accredited rams. In the near future some breed societies will require Brucellosis accreditation as a condition of registration of a stud.

As a consequence the number of flocks that have become accredited continues to grow steadily. It is anticipated that within ten years of commencement all stud flocks will be accredited. A major spin off of the accreditation scheme has been the increased producer awareness of Brucellosis and resultant increased testing of commercial flocks. An important part of the advisory program for commercial producers was to make them understand the futility of purchasing accredited rams if they were to be mixed with their own, untested, rams.

Scheme Guidelines

The guidelines under which the scheme is conducted are very comprehensive and only some of the major points are mentioned. Clearly the scheme is voluntary and exists for the benefit of the industry. For it to be successful all participants must at all times be aware of and comply with the conditions.

Veterinarians. Only those approved by the Department of Agriculture may participate.

Properties. The veterinarian must satisfy himself that the necessary standards of management and facilities are in place in order to maintain disease freedom.

Accreditation. Is only granted following the veterinarian's recommendation. Agriculture Department maintains lists of accredited flocks and approved practitioners.

Certificates. Issued by Agriculture Department have a one year validity. Re-accreditation testing is required before re-issue. After three consecutive years of accreditation biennial accreditation replaces annual testing.

Suspension. The detection of disease or the failure to comply with the conditions of the scheme result in suspension. Re-accreditation following suspension may ocur provided the reason for the breakdown is determined and disease eradicated.

Recommendations for Similar Schemes

The involvement of all parties in the development and continued monitoring of disease accreditation schemes is essential. The OB model has been used for a CAE scheme for goats and a soon to be released footrot scheme for sheep and goats.

The disease must be one which is of economic importance to the industry. It also must be one about which we have sufficient knowledge and adequate tools to effectively eradicate.

Promotion of the scheme must take place at many levels and is not the sole responsibility of just one group. Information must be readily available in an easily understood form. Peer pressure as well as economic pressure are important motivating forces.

The studmaster must have a sound understanding of the disease and particularly of the scheme conditions. Ignorance of or failure to observe the conditions has been responsible for the majority of problems to date. An essential component of developing this understanding is a close working relationship between the studmaster and his veterinarian.

Voluntary disease eradication schemes are now seen as the future direction for most animal diseases. If and when an industry determines the cost of a particular disease warrants its suppression or eradication then an accreditation scheme will most likely be the path taken. Governments are increasingly reluctant to commit large amounts of taxpayers funds to undertake compulsory eradication programs, such as our Bovine TB and Brucellosis programs. It may be that such programs will only ever be considered after the flock prevalence has been reduced to minimal levels through accreditation schemes.

Reducing Predator Losses with Livestock Guarding Dogs

Jay R. Lorenz
Extension Wildlife Specialist
Oregon State University
Corvallis, Oregon 97331 USA

Predation on sheep (*Ovis aries*), primarily by coyotes (*Canis latrans*) reduces profits for growers in Oregon and throughout the United States. The most recent survey of losses for Oregon estimated that 2 percent of ewes and 4.7 percent of lambs are taken by predators (deCalesta 1979). Depredations amount to over $1 million and $100 million annually to Oregon and the nation, respectively. A wide variety of lethal and nonlethal tools have been developed to mitigate losses, including livestock guarding dogs.

Old World livestock guarding dogs were virtually unknown in the United States until the late 1970's. Eurasian shepherds used the cultural practice of grazing their flocks under the watchful eyes of 100 lb (45 kg) floppy-eared dogs for centuries. Navajos in America's southwest, with a 200 year tradition of using dogs for protecting livestock, were a notable exception (Black and Green 1985). While accounts of ranching and sheep production in Mexico and America's southwest have been traced to Spain (Kollmorgan 1969, Butzer 1988), accounts of guarding dogs accompanying sheep are missing. Interestingly enough, the Navajo tradition, probably inherited from Spanish settlers, never diffused into Anglo-American sheep husbandry. Most of the U.S. might still be without guarding dogs had a crisis in predators not developed.

Beginning in the late 1960's and through the 1970's, sheep growers faced mounting opposition to traditional methods of predator control. Two special commissions of scientists recommended changes in approaches to controlling predation (Leopold 1964, Cain et al. 1972). The prevailing goal of eliminating all coyotes was changed to focus on problem individuals. Indiscriminate use of poisons such as strychnine and sodium monofluoroacetate (compound 1080), found to be killing many non-target species, was banned by President Nixon in 1972 (Executive Order 11643). Over 360 anti-trap bills were initiated at various levels of government from 1968 to 1986, representing society's disgruntle with leg-hold traps (Gentile 1987). Secretary of Interior Andrus (1979) summarized the status of predator control policies as leaving "neither the livestock industry nor the environmental community . . . satisfied."

The introduction of livestock guarding dogs was linked to the search for a new idea that would satisfy both sheep growers and environmentalists. Livestock guarding dogs were a likely candidate because they had already passed the test of time in Europe. Although they were unfamiliar with sheep *protecting* dogs, sheep growers were familiar with sheep *herding* dogs. The new challenge was one of simultaneously educating ranchers, testing dogs in a new geographical setting, and developing a breeding program. This paper reviews the process of introducing livestock guarding dog programs to Anglo-American sheep growers, focusing on projects initiated at Hampshire College and Oregon State University.

Establishing Programs. The Livestock Guarding Dog Project began at Hampshire College, Amherst, MA in 1977 (Coppinger and Coppinger 1980a). Hampshire's program had three goals: 1) To test dogs in the field and learn how to manage them; 2) To study the basic behavior of the dogs; and 3) To transmit information from field and laboratory observations back to producers. The original intent was to test a population of 100 dogs in the northeast. However, word of early successes and unexpected mortality created a need to expand the program.

By 1984, requests for over 100 dogs annually were received from throughout the country. Meeting that demand from one breeding center in Massachusetts became increasingly difficult. Sources of good dogs were limited, as were regional and readily available educational programs. Diffusion of guarding dogs into American agriculture expanded when Hampshire College, with breeding stock imported from Europe, joined forces with the Oregon State University Extension Service, which had an institutional organization for educating ranchers.

Sheep growers and environmentalists supported a pilot Extension demonstration program in Oregon which began in 1984 with special state and federal funding. Four objectives, derived from Hampshire's program, were established for Oregon: 1) Develop and conduct training programs for livestock growers, county Extension agents, dog breeders and Animal Damage Control personnel; 2) Customize strategies for raising new and replacement dogs; 3) Contribute records of performance to the national database at Hampshire College; and 4) Evaluate the performance of the dogs.

Criteria for Evaluating Performance

Evaluating the performance of dogs in diverse habitats (e.g. mountains, deserts, valleys, varying distances from towns), in different schemes of management (e.g. small and large flocks, pasture and range operations), and unknown densities of predators was a methodological challenge. Criteria for evaluating the dogs had to provide objective information independent of environmental variables and provide an understanding that could be useful for improving the system.

Four approaches were used to measure the performance and adoption of livestock guarding dogs: 1) Behavior of the dog; 2) Estimates of reduction in losses; 3) Survivorship; and 4) Adoption of the technique. Data were collected through an annual questionnaire, breeding records (birth, sex, breed, date and location of placements, survivorship), on-site visitations and telephone conversations. Dogs used were the offspring of working stock (Anatolian shepherd, Maremma, Shar Planinetz) imported from Europe.

Results and Discussion

From 1978–1988, over 1100 livestock guarding dogs were placed in 37 states. Reporting data on the performance of livestock guarding dogs has been an on-going process and was most recently reviewed in Coppinger et al. (1988). Readers who desire details beyond the highlights in this paper are referred to additional trade and technical articles listed in the bibliography.

1. Behavior. Coppinger and Coppinger (1980b) hypothesized that guarding dogs had to have three basic behaviors to be successful: trustworthiness with sheep; attentiveness to sheep; and protectiveness of sheep. These measures were useful because they did not require observations of interactions between predator and dog. They were standards that could be recorded independent of predator density, habitat type, size of flock, and previous history of predation.

Over 80% of Maremmas, Shar Planinetz, and their crosses, and 70% of Anatolians received excellent or good ratings in trustworthiness from their handlers (data summarized over 7 years, Coppinger et al. 1988). Differences between breeds were not significant except in 1981 and 1985.

Scores for attentiveness (7 year average) varied between 49% (Shar Planinetz) and 80% (Shar Planinetz/Maremma crosses). In two years, 1980 and 1986, differences between breeds was significant. Attentiveness was influenced by trustworthiness. Dogs that were not trustworthy were generally removed from the flock, thereby skewing cooperators assessment of attentiveness. There appeared to be a large amount of individual variation in attentiveness. Dogs that were inattentive to sheep generally showed greater attentiveness to people (Coppinger et al. 1983). Attentiveness is the most complicated and least understood of the three basic behaviors.

Nearly 75% of dogs were rated as excellent or good in protectiveness. Protectiveness was closely related to trustworthy and attentive behavior. Nearly 100% of dogs receiving high scores in trustworthy and attentive behaviors received a high score in protectiveness. Protectiveness did not appear as a distinct set of behaviors. Although the name *guarding* dog may imply aggressiveness, many dogs with mild dispositions were successful at reducing losses. Dog-predator interactions that can disrupt the predatory sequence of an intruder include scent-marking, play, dominance, greetings and ritualized aggression. Studies have concluded that attentiveness is the best indicator for predicting protectiveness and reductions in losses.

2. Reductions in Losses. Reductions in losses were the ultimate goal. This could not be the sole measure of successes because data could be confounded by variables such as variations in predator density, abilities of the handler, and other control techniques in use. However, using each farm as its own control, a comparison before and after getting a dog could be made. Changes in predation over time were particularly useful when correlated with behavioral traits. A comparison of losses before and after getting a dog (7 year summary) showed that 20% of dogs were on farms with no predation, 64% on farms with reduced losses, and 16% with no change or increased losses (Coppinger et al. 1988). The number of growers who reduced losses to zero after getting a dog was striking. For example, in 1986, predation was reduced to zero on 53% of farms with a history of losses.

Reductions in losses for Oregon followed the national pattern and showed that dogs were successful in farm flocks as well as in range operations. In 1986, in flocks averaging 105 sheep (range 30–400), losses were reduced from 10 per farm to less than 1 after getting a dog. At the prevailing market price of $67.70/cwt., and assuming 100 lb (45 kg) market lambs, a savings of $626 per ranch or $501 per dog (some ranches had more than one dog) was achieved. On large operations, averaging 644 sheep (range 500–2600), losses were reduced from an average of 31 to 14, amounting to savings of $1151 per ranch or $615 per dog (Coppinger et al. 1988).

3. Survivorship. Survivorship studies indicated the number of dogs displaying minimal standards of acceptable behavior over time. Unsatisfactory dogs eventually were culled. Survivorship also contributed information for calculating replacement needs and economic analyses.

Mortality rate of livestock guarding dogs is high, particularly in the early years. One study showed that half the dogs survive to 19 and 33 months on range operations and pasture operations, respectively (Lorenz et al. 1986). During the first 2.5 years the semi-annual mortality rate was 13% and beyond 2.5 years, it was 5%. The primary cause of death was accidents, followed by culling and disease. Recent review of survivorship data suggests that mortality of older dogs (> 2.5 years) may be greater than estimated by Lorenz et al. (1986).

Survivorship has important implications for cost and management of livestock guarding dogs. Dogs that die before their second birthday may cost more than $1000 per year of useful service compared to an annual rate of $200 (slightly more than the annual cost of dog food) for a 10-year-old dog.

At a the mortality rate calculated by Lorenz et al. (1986), approximately one litter (6 to 8 pups) must be raised annually as replacements for every two dozen adults. More than one litter per two dozen adults would enable the population of guarding dogs to expand. A reservoir of extra juveniles must be kept by sheep and goat producers to replace untimely losses and for expansion.

4. Adoption. Adoption of livestock guarding dogs can be studied on an individual and institutional basis. In the first case, the question is how many, over what period of time, and where. In the second case, changes in policy and programs would be a measure of institutional acceptance of livestock guarding dogs.

The diffusion of livestock guarding dogs appears to be following a pattern common to the diffusion of many other new technologies. Livestock-guarding dogs initially spread locally from two research stations (Hampshire College and U.S. Sheep Experiment Station). In distant states (Oregon, Kentucky, Texas, Minnesota) dogs radiated from the locations of the early adopters. On a smaller scale, clusters of dogs can be found within regions. For example, in Oregon, guarding dogs radiated outward from locations of early placements in central Oregon, the northern Willamette Valley, Jackson County and in the Klamath Falls basin.

Lorenz (1987) noted a high rate of adoption by growers who tried a livestock guarding dog. From 1983 to 1987, 85% (124 of 144) of cooperators in Oregon and Washington had a satisfactory experience with a livestock guarding dog. Satisfied cooperators included growers who had dogs or had good dogs at the time they dropped out (sold sheep) of the program.

Institutional adoption of livestock guarding dogs has lagged behind individual adoption. Beginning in the mid-1970's and continuing through the mid-1980's, the federal government supported several research projects at Hampshire College and the U.S. Sheep Experiment Station. The Oregon State University Extension Service guarding dog project represented the first state-wide educational program. Extension Specialists in at least two other states have since made guarding dogs an integral part of their educational programs. Animal Damage Control, the federal agency in charge of operational control, incorporated livestock guarding dogs into their program beginning in 1988. Incorporating a non-lethal method of predator control in the Animal Damage program represents a significant policy change in an agency whose founding legislation mandated killing predators.

Sheep farmers and dog breeders are also contributing to the infrastructure of livestock guarding dog use. New breed clubs have formed within the past decade to promote and keep records on breeds imported from Europe. Many clubs have predator control committees whose special interest is to support the livestock industry. Sheep growers, most notably in Oregon, have formed their own organizations for the purpose of supplying dogs to one another. Finally, field-testing research and breeding at Hampshire College has evolved into a program for maintaining a national database on the performance of livestock guarding dogs.

Conclusions

Over a period of 10 years, livestock guarding dogs have grown from near obscurity to an accepted form of predator control in the United States. Keys to the successful introduction of livestock guarding dogs included an innovative method for field-testing dogs; educational programs, especially with support from the Extension Service; an infrastructure for supplying good dogs; research supported by private and government sources; and finally adoption by the federal agency in charge of predator control.

References and Bibliography

Andrus, C. D. 1979. Memorandum to Assist. Sec., Fish and Wildl. and Parks, on USDI Predator Control Policy, Nov. 8

Black, H. L. and J. S. Green. 1985. Navajo use of mixed-breed dogs for management of predators. J. Range Mgmt. 38(1):11–15

Butzer, K. W. 1988. Cattle and sheep from Old to New Spain: Historical antecedents. Annals Ame. Assoc. Geog. 78(1):29–56.

Cain, S. A., J. A. Kadlec, D. L. Allen, R. A. Cooley, M. G. Hornacker, A. S. Leopold, and F. H. Wagner. 1972. Predator Control–1971, Report to the C.E.Q., Institute for Environmental Quality, Univ. Mich., Ann Arbor, 207 pp.

Coppinger, L. and R. Coppinger. 1980a. So firm a friendship. Natural History 89(3):12–26

Coppinger, L. and R. Coppinger. 1982. Livestock-guarding dogs that wear sheep's clothing. Smithsonian 13(1):65–73.

Coppinger, R. and L. Coppinger. 1980b. Livestock guarding dogs. Country Journal 7(4):68–77.

Coppinger, R., L. Coppinger, G. Langloh, L. Gettler and J. Lorenz. 1988. A decade of use of livestock guarding dogs. pp. 209–214 in Crabb and Marsh (eds.) Proc. 13th Vert. Pest. Control Conf., Monterey, CA.

Coppinger, R., J. Glendinning, E. Torop, C. Matthay, M. Sutherland, and C. Smith. 1987. Degree of behavioral neoteny differentiates canid polymorphs. Ethology 75:89–108.

Coppinger, R., J. Lorenz, J. Glendinning, P. Pinardi. 1983. Attentiveness of guarding dogs for reducing predation on domestic sheep. J. Range Mgmt. 36(3):275–279.

deCalesta, D. 1979. Documentation of livestock losses to predators in Oregon. Spec. Report 501. O.S.U. Ext. Serv., Corvallis, OR, 20 pp.

Gentile, J. R. 1987. The evolution of antitrapping sentiment in the United States: a review and commentary. Wildl. Soc. Bull. 15(4):490–503.

Green, J. S. and R. Woodruff. 1983. The use of three breeds of dog to protect rangeland sheep from predators. Applied Animal Ethology. 11(2):141–161.

Green, J. and R. A. Woodruff. 1988. Breed comparisons and characteristics of use of livestock guarding dogs. J. Range Mgmt. 41(3):249–251.

Green, J., R. Woodruff, T. Tueller. 1984. Livestock guarding dogs for predator control: costs, benefits, and practicality. Wildl. Soc. Bull. 12(1):44–50.

Kollmorgen, W. M. 1969. The American cattleman—an unwelcome progeny. Annals Ame. Assoc. Geog. 59:215–239.

Leopold, A. S., S. A. Cain, C. M. Cottam, T. N. Gabrielson, and T. L. Kimball. 1964. Predator and rodent control in the United States. Trans. No. Ame. Wildl. and Nat. Res. Conf. 29:27–49.

Linhart, S. A., R. T. Sterner, T. C. Carrigan, and D. R. Henne. 1979. Komondor guard dogs reduce sheep losses to coyotes: a preliminary evaluation. J. Range Mgmt. 32(3):238–241.

Lorenz, J. 1987. Distribute, manage, evaluate the use of livestock-guarding dogs in the states of Oregon and Washington, Year-end Report for FY 87, USDA/APHIS/ADC Coop. Agree. No. 12-16-73-0129, 13 pp.

Lorenz, J. R. and L. Coppinger. 1986 (revised 1989). Raising and training a livestock-guarding dog. Ext. Circ. 1238, Oregon State Univ., Corvallis. 8 pp.

Lorenz, J. R., R. P. Coppinger and M. R. Sutherland. 1986. Causes and economic effects of mortality in livestock-guarding dogs. J. Range Mgmt. 39:293–295.

Sheep Lice Control

Laurie Pryde
"Wavering Downs"
Cumboogle Road MS-3
Dubbo NSW 2830 Australia

In Australia the sheep body louse, *Damalinia ovis,* is second only to the sheep blowfly in terms of the economic loss caused by ectoparasites. The prevalence of lice infested flocks does not vary greatly from year to year or from state to state, generally being in the vicinity of 20 to 30 per cent. Interestingly, in excess of 85 per cent of flocks are regularly treated for lice (regardless of whether they are known to be infested) following the annual shearing.

Given these statistics it could be concluded that successful lice control is difficult to achieve in practice on the farm. However true this may be, we do have the knowledge and resources to eradicate lice from every flock. The primary ingredients so often lacking are motivation, management and co-operative neighbours.

There are many examples of individual flocks in which lice have been eradicated and routine treatment following shearing has ceased, for as much as 25 years, without re-infestation occurring. Potentially every flock can reach this happy state of eradication with its associated increased productivity and reduced costs.

Sheep lice are obligate parasites spending their entire life on the sheep. They reproduce sexually with an approximate life cycle length of 35 days, of which 10 days is required for incubation of the egg until hatching. The environment found at skin level in a fleece of more than 1 centimetre is fairly stable at around 37 degrees centigrade and 75 per cent relative humidity. Lice are adapted to this environment and a pair of lice (male–female ratio in populations is close to 1:1) are considered capable of producing about 1 egg every 1½ days. Longevity of adults is considered to be around 60 days.

The population of lice on untreated sheep increases with wool length, reaching a maximum prior to shearing. Shearing removes up to 50 per cent of both lice and eggs. The resulting skin level environment on freshly shorn sheep causes a further drop in the population for up to a month until wool growth provides an insulating barrier. Approximately 4 months after shearing, lice populations in untreated sheep can recover to their pre-shearing levels. Further significant increases will then occur until the next shearing.

Transmission of lice between sheep only occurs when they are in physical contact. Lice are generally found close to the skin surface where they feed on epidermal debris. Lice dislike sunlight and when a fleece is parted they move back into the wool. Lice are more likely to move away from the skin toward the staple tip when the fleece is shaded. Both a low light intensity and, in summer, a reduction in wool tip temperatures are thought to be involved.

Opportunities for transmission in the paddock usually occur only in sheep 'camps' and between ewes and lambs when suckling. On other occasions management practices may influence the amount of sheep contact, e.g. mustering and yarding, providing restricted access to feed and water troughs and during joining. In any case there are, in the normal course of events, numerous opportunities for the spread of lice throughout the flock during the course of the year.

The presence of lice in a flock can have only three sources:

1. residual infestation
2. introduction of infestation with strays
3. introduction of infestation with purchased sheep

Residual Infestations

Basically we are looking at sheep which missed treatment following shearing or sheep which were ineffectively treated.

Traditionally Australian sheep have been plunge or shower dipped 2 to 6 weeks after shearing. Some flocks are wet dipped 'off the board' but this practice is not usually recommended due to the potential losses that can be associated with infection of shearing cuts. The advantage of dipping off the board is that it saves an extra muster and any sheep that miss treatment can be easily identified (they have not been shorn). Many graziers who dip 2 to 6 weeks after shearing brand their sheep with a coloured scourable dye. Untreated (unbranded) sheep however are still not as easily identified in the paddock as are woolly sheep amongst shorn sheep.

Thus the quality of mustering is critical and must be followed by a period of observation which attempts to detect untreated sheep. The nature of the country, both topography and vegetation, is a major determinant of the success of mustering. One additional factor, often overlooked, is sick sheep. Sheep suffering from, for example, footrot or flystrike are likely to 'plant' and not be noticed by the musterers and his dogs. They may have the opportunity to reinfest the flock after treatment. Thus the post treatment observation is important to find and deal with these sheep. Counts of numbers in mobs cannot be relied upon because frequently there is interchange between mobs, strays from other flocks and deaths to contend with.

The importance of effectively treating 100 per cent of the sheep in the flock at the one time, is well recognised. Unfortunately it is not achieved as frequently as would be desired.

The efficiency of treatment is currently being looked at more critically than ever before. The withdrawal of arsenic as an ectoparasiticide some three years ago has left graziers with only a choice between an organophosphate (OP) or a synthetic pyrethroid (SP) chemical for wet dipping. Fortunately when used at correct concentrations, in properly managed plunge or shower dips, eradication is achievable using chemicals from either of these groups.

There has in recent years been a trend toward the use of SP backline products applied either manually through a gun or automatically as sheep pass through a race. Whilst their cost per treated sheep is high this is outweighed by the ease of application and the avoidance of an additional muster. Until recently all such products wre registered for use within 24 hours following shearing.

It is now apparent that in some flocks these products are incapable of achieving eradication. Their mode of action involves the diffusion of the active ingredient through the lipid layer which covers the entire skin and wool fibre surface. Applied as a small volume of highly concentrated chemical to the back line, the chemical, to be effective, has to spread over the entire woolgrowing skin area and achieve and maintain lethal concentration for at least 10 to 14 days. This period is necessary as eggs are largely unaffected by SP's. If an inadequate dose is given, if the dose is poorly applied or if certain skin conditions are present which impair the diffusion of the chemical, then treatment will be less than 100 per cent—meaning lice will survive. Recent studies have shown the SP's can migrate at the rate of 11 cm per hour from a highly concentrated dose applied to the back line. Total body coverage is possible within 24 hours from application, but the times taken to achieve lethal concentrations at sites remote from the backline are not well known.

Another factor is the selection for resistance by the use of sub-lethal concentrations. This appears to only be associated with SP's used in back line products. The mechanism that occurs is that within a population of lice, their inherent susceptibility to SP's will vary—this can and has been measured. Thus as the dose is increased, an increasing proportion of the lice population will die, until at a certain high dose all lice will be killed.

Some back line SP products were marketed with a recommended dose that was only capable of achieving whole body concentrations just in excess of the required level. Consequently in any sheep where there was underdosage, poor application or impaired dispersion of the chemical, the necessary concentrations were not achieved and some tolerant lice survived. The resultant lice population that developed had a much higher proportion that were tolerant and in successive years this became manifest as poor results and continued infestation. Resistance, as it has become known, is making many graziers reassess their lice control programs.

Concurrently there is renewed interest in dipping or jetting races which I believe is due to a desire to treat sheep 'off the board' using the wet dip method. In wet dipping, usually in plunge or shower dips, the objective is to saturate the entire skin area with dip wash containing a lousicide at the correct concentration. Undoubtedly plunge dipping is the most effective method of wetting sheep. Unfortunately, plunge dips are not well regarded, mostly because of the high labour component usually necessary to force sheep into the dip. More modern designs, particularly of lead-in races, have largely overcome these problems and systems can be constructed which are virtually self-feeding with a minimum of effort necessary.

Shower dips of either the standard or constant replenishment versions can, if properly managed, achieve almost as thorough a wetting as plunge dips. Sheep showers have other features which in some cases make them a more favoured option than plunge dipping—in particular more control over dip stain and infections which result from polluted dip wash. Eradication of lice with either plunge or shower dips is easily achievable.

Unfortunately the same comment cannot be made of the dipping races. In my area of New South Wales there are over a dozen different brands and it appears a new one appears at every field day. The basic construction is a length of race which is fitted out with nozzles designed to coat the sheep with dip wash as they run through the race. There is extreme variation in the length of the spray area, the number and position of nozzles and the volume, pressure and spray pattern they produce. Some are designed to operate continuously, others are activated as the sheep pass through. Some recirculate the wash, with others it is discarded.

Problems with these machines relate to both sheep flow and their inherent ability to wet the entire skin area. Every machine I have seen in operation has problems with sheep flow. The rate at which sheep run through is uncontrollable—consequently dosage is variable. None have a separation mechanism and as a result it is possible for sheep to be so close that one may be shielded by another, preventing complete wetting. Currently none of the machines are recommended by the Agriculture Department and none of the chemical companies recommend their products be used through them. Still, more and more producers are buying and using them. I feel this is because they have become used to the convenience, high throughput and low labour involved with backline products and are unwilling to return to the older and perhaps slower methods.

Strays

It is generally accepted that lice transmission through a stock proof fence is indeed a rare event. Lice are spread from flock to flock by sheep which cross through, over or under a fence. Properly constructed and continuously maintained fences utilising netting or ringlock are an effective barrier. Stock grids, gates and water courses deserve special attention. In much of our sheep raising country fence damage by kangaroos, emus, wombats and feral pigs is a major problem. Too few graziers spend money on fencing and there are a number of understandable economic factors causing this situation. However the recurring annual losses associated with lice infestations ought to be weighed against the costs of fence improvements where strays are the problem.

Various surveys indicate that up to one third of all re-infestations are caused by straying infested sheep. A number of things can be looked at here. First, neighbours should be requested not to return straying sheep by putting them back over the fence. They should be yarded, treated and isolated for an appropriate time before returning to their flock. There are a number of longwool treatments now available, applied as either a backline or by hand jetting. If strays are a continuous problem co-operation with the neighbour working toward close shearing and treatment times on both properties can markedly reduce the problem for both parties.

Introductions

All introductions should be considered by a prudent manager to be potentially infested. Sheep with less than three months wool are very difficult to call lice free by even a detailed visual inspection. These sheep should preferably be treated on arrival in addition to being kept separate from the rest of the mob until next shearing.

Introduced sheep often pose other practical problems in relation to management. If their wool length is markedly different to home bred sheep a decision has to be made about shearing, i.e. either to bring them in line with the rest of the flock or to continue with split shearings. (This means that not all sheep will be shorn at the one time. Some properties, particularly where dealing in sheep is practised, may have three or more shearings each year.) In flocks with split shearings, lice control becomes much more difficult because following the principle of treating all sheep at the one time means some will be treated off shears and others with various wool lengths. Longwool treatments are not considered to be capable of achieving eradication so the infestation is likely to remain with the potential of spread to other mobs. Mostly farmers with split shearings tend to treat each group after shearing which is unsatisfactory from a whole flock perspective.

Protective Period Following Treatment

All lice treatments give, to varying extents, a period of protection from re-infestation following the application of the treatment. The length of this protective period is determined by the chemical group used, the final concentration achieved in the wool grease and rainfall following application.

Because of the unpredictable variability in the length of the protective period, it should not be relied upon and should not be realistically considered as part of the control program.

Backline Treatments

General recommendations concerning the use of these products following shearing are as follows:

- Use within 24 hours of shearing.
- Set dosage to the heaviest sheep in the mob—weigh, do not guess.
- Check accuracy of dose delivery.
- Check accuracy of dose calculation.

- Apply only as a long backline stripe.
- Do not use on ewes due to lamb within eight weeks following treatment.
- Sheep should be clean shorn.
- Do not use on sheep with dermatophilosis.
- Only use where there is no suggestion of resistance.
- Do not use on ewes with lambs at foot unless lambs are also treated.
- The entire flock must be treated at one time.

Management of Resistance

In any flock where resistance to backline SP's is suspected the advice is to shower or plunge dip using an OP after the next shearing. A field test kit is being trialed in New South Wales to give an indicator of resistance. The kit consists of four tubes each containing a pad impregnated with 0, 2.5, 5 and 10 p.p.m. of cypermethyrin. Forty lice are harvested and ten placed in each tube which are then protected from light and kept warm in a polystyrene case. After 24 hours the lice are emptied out and numbers recorded in each category as dead, knockdown and live.

To date a number of resistant 'strains' of lice have been detected in flocks in New South Wales. There is a range of resistance amongst these 'strains' to the SP's applied as backline treatments. Resistance to one SP generally means a degree of resistance to other SP's. Fortunately no resistance has been detected to the same SP's when used in wet dip formulations. Neither has any OP resistance been demonstrated.

Components of an Effective Lice Control Program

1. Motivation—The grazier needs to believe that lice control is economically worthwhile and within his management ability.
2. Shearing—All sheep in the flock are shorn at the one time regardless of wool length of introduced sheep.
3. Treatment—All sheep are thoroughly treated following shearing using an effective lousicide at the correct concentration and in an appropriate manner. Treated sheep are visually identified.
4. Fencing—Stock proof boundary fences are constructed and maintained regularly, especially prior to and during the time sheep are grazing paddocks on the boundary.
5. Strays—A policy is developed with all neighbours whereby any strays are yarded. On return the sheep are inspected, treated and where possible, run separately until next shearing.
6. Introductions—Limit the number of introductions where possible to close to shearing time. Inspect, treat and isolate until shearing and subsequent treatment.
7. Inspections—A representative sample from each mob comprising the flock should be examined minimally twice a year for lice. These inspections ought to be prior to shearing and at crutching. Any sheep with evience of fleece derangement should be examined whenever the opportunity arises.
8. Long wool treatments—If lice are detected the sheep should be treated immediately with a long wool treatment. Investigations need to be undertaken to determine the source of the infestation. If resistance to SP's is suspected, then hand jetting with an OP (as for fly control) is the only alternative.

Lice Control in Small Flocks

Many small flocks do not have access to good facilities, especially dipping facilities. It is here that the use of backline lice treatments or manually applied dip wash is the only practical method of treatment.

Advice to small flock owners with respect to backline treatments would be to use 2 to 3 times the recommended dosage. This will go a long way to ensure that the treatment will be effective and resistance will be much less likely to develop. Obviously the costs associated with treatment will be greater.

Wet dips have been applied to individual sheep by standing them in a tank or bathtub and literally massaging dip wash into their fleece. Other techniques involve the use of knapsack type sprays or even jetting or firefighter plants to wet the sheep. Again the procedure will be relatively more expensive and wasteful of dip wash as well as being slow. Operators should be careful to protect themselves from the chemical. Provided sheep are thoroughly wet, eradication is achievable.

All the other procedures appropriate to larger flocks with respect to fences,introductions and strays are equally appropriate to the small flock. The success rate in achieving and maintaining a lice free flock should be greater in small flocks because of the opportunity of greater attention to detail. One of the major problems in large flocks is the pressure to achieve high rates of throughput with any job—particularly treatment. This often leads to a lack of attention to detail and thoroughness with the consequence of poor results.

An Overview of Sheep Embryo Transfer Technologies

C. R. Youngs[1], L. K. McGinnis,[1] and S. C. Duplantis, Jr.[2]
Animal Science Department
University of Idaho
Moscow, Idaho 83843 USA

Breeders of colored sheep are confronted with the challenging task of combining the appropriate qualitative genes at each of several gene loci to obtain a desired fleece color. A breeding program aimed at producing a certain fleece color may take years to accomplish. However, once the fleece color has been obtained in a particular breeding animal, there are no guarantees that the animal's progeny will also exhibit the desired phenotype since an animal transmits a random ½ of its genes to its offspring. In addition, breeders are also concerned with quantitative genes controlling other fleece characteristics such as staple length, fleece grade, and fleece weight.

Traditional breeding programs have been utilized by colored sheep breeders with varying degrees of success. Perhaps the main drawback to these programs has been the relatively slow rate of progress. Two reproductive technologies which can accelerate the rate of genetic improvement are artificial insemination and embryo transfer. Artificial insemination involves the collection and distribution of semen from genetically valuable males, whereas embryo transfer permits the exploitation of genetically superior females through collection and dissemination of fertilized eggs. The objective of this paper is to provide an overview of embryo transfer technologies which may be of interest to breeders of colored sheep.

[1]Present address: Iowa State University, Department of Animal Science, Ames, Iowa 50011, USA.
[2]Present address: Granada Genetics, Inc., 100 Research Park, College Station, Texas 77840 USA.

Embryo Transfer

Embryo transfer is a series of procedures involving the harvest of genetic information (contained within the egg) from a donor female and the subsequent transfer of this genetic material to a recipient female. The first successful embryo transfer in sheep was reported in 1934.[1]

The entire process of embryo transfer actually encompasses several different procedures including estrus synchronization, superovulation, embryo collection, and embryo transfer. When applicable, other techniques such as embryo splitting and embryo cryopreservation may also be employed.

Estrus Synchronization

Estrus synchronization is the process of regulating the time at which ewes will exhibit heat (or estrus). It is the initial step in embryo transfer and needs to be performed in both donor and recipient ewes. The timing of estrus is important for the production of quality embryos from donor ewes and for the establishment of pregnancy in recipient ewes. The latter is highly dependent upon estrus sychrony among donor and recipient ewes.

The most common method of synchronizing estrus involves the administration of progestagens (progesterone-like compounds). Progestagens are given to induce a hormonal state that mimics what naturally occurs in cycling ewes. Ovulation occurs near the end of estrus in sheep, and a corpus luteum (CL) develops at each site where an ovarian follicle ruptures. The CL secretes progesterone, the hormone responsible for suppression of subsequent estrous cycles, during days 4–14 of the estrous cycle. If a viable embryo is present, the CL (and the pregnancy) will be maintained. If the ewe is not carrying a viable embryo, a signal is sent to destroy the CL, and the ewe exhibits estrus about three days later. Attempts to synchronize estrus involve the administration of exogenous progesterone to override the endogenous hormonal state in an effort to duplicate the progesterone profiles of cyclic ewes. Removal of the progestagen causes the ewe to exhibit estrus 2–3 days later.

Progestagens typically are administered via an intravaginal pessary or sponge.[2] The pessaries may be impregnated with either FGA (fluorogestone acetate, 30 or 40 mg) or MAP (medroxyacetoprogesterone, 60 mg). Sponges are inserted for 12–14 days, and estrus is usually exhibited 24–48 hours following progestagen withdrawal in about 70% of the treated ewes. Administration of progestagens is effective for both estrus synchronization (during the breeding season) and estrus induction (during the non-breeding season). Another progestagen, norgestomet, may also be administered via a subcutaneous implant with similar results.[3] In most cases, ewes are also treated with gonadotropic hormones to achieve ovarian stimulation, a topic which will be addressed in the next section.

A different compound may be utilized for estrus synchronization only during the breeding season.[4] Prostaglandins are compounds whose biological actions include induction of luteolysis, or regression of the CL. Administration of exogenous prostaglandin $F_2\alpha$ will cause destruction of the CL and subsequent estrus in the ewe. The use of prostaglandins is limited to ewes which are cycling, and there have been reports that ewes synchronized with prostaglandin $F_2\alpha$ (or an analogue) will exhibit reduced fertility at the first post-treatment estrus due to changes in the cervical mucous which impair normal sperm transport.[5]

Superovulation

The efficiency of embryo transfer is greatly enhanced when multiple embryos can be recovered from each donor ewe. In order to accomplish this, ewes are given hormones (gonadotropins) which stimulate the ovaries to release a greater than usual number of ova during ovulation. Traditionally, the hormones PMSG (pregnant mare serum gonadotropin) has been given to superovulate donor ewes.[6] However, FSH (follicle stimulating hormone) is now the preferred superovulatory agent,[7] as it results in a more consistent ovarian response with higher ovulation and fertilization rates. Methods of superovulation in sheep have recently been reviewed.[8]

PMSG has a long half-life relative to FSH. Because of this, PMSG may be given as a single injection whereas FSH must be given as a series of injections over a three day period. The dose of PMSG ranges from 500–1500 international units (IU), and the dose of FSH is typically 12–24 milligrams (mg). The appropriate dose to use depends upon the breed and age of the ewe, the season, and the method of estrus synchronization. Most attempts to superovulate ewes result in 8–10 CL and 4–6 embryos of transferrable quality. Keep in mind that a common outcome of superovulation is unfertilized ova and degenerate embryos.

These same compounds are often given to recipient ewes, especially during the non-breeding season, to achieve a mild ovarian stimulation. Lower doses are administered to recipients than are given to donors for superovulation. PMSG doses of 300–500 I.U. appear to work well, but little information is available regarding the use of low doses of FSH in recipient ewes. Limited data from our laboratory indicate that two injections of 2 mg of FSH each may be adequate for treating recipient ewes.

Breeding and Estrus Detection

One of the more important but often overlooked aspects of embryo transfer programs is the mating of donor ewes. More "ram power" is needed when breeding synchronized ewes as compared with naturally cycling ewes. Ewe to ram ratios as high as 10:1 have been suggested for breeding synchronized ewes, but lower ratios may be a safer approach to maximizing fertilization rates.

An alternative to breeding by natural service is artificial insemination (AI), and it has been reported that the surgical deposition of semen results in a higher percentage of fertilized embryos in superovulated ewes.[9] The increased fertilization rate likely occurs as a result of placing the semen past a barrier to sperm transport (i.e., the cervix). A new technique, laparoscopic AI, also places semen directly into the uterus and may similarly prove to be an effective AI method in superovulated ewes. In laparoscopic AI, the ewe is inseminated with the aid of a fiber optics instrument placed into the ewe's abdominal cavity. This permits the technician to deposit the semen into the uterus without performing major abdominal surgery.

Recipient ewes should be synchronized at the same time as donor ewes. However, recipients should not be mated with fertile rams but rather be placed with a sterile "teaser" ram which has undergone a vasectomy or epididectomy. Estrus should be checked at least twice daily to ensure the most accurate possible determination of estrus onset. Many breeders fit their rams with marking harnesses to aid in estrus detection.

Embryo Collection

The principle method of harvesting sheep embryos is via a mid-ventral laparotomy. This procedure is an abdominal surgery where the ewe's reproductive tract is exteriorized to permit flushing of the embryos from the uterus. Isolated reports of non-surgical[10] or laparoscopic[11] embryo collection have also been published. The main drawback to the surgical method of embryo collection is the resultant adhesions which limit the number of surgeries which can be performed on each donor ewe. Field reports suggest that this problem may be overcome by forcing the donor to establish and maintain a normal pregnancy between each embryo collection. This has the obvious disadvantage, however, of limiting the number of embryo collections per year to two, a constraint which may be inappropriate for old, injured, or diseased (non-genetic) donor ewes.

Embryo Transfer

Most embryo transfers are performed surgically using a technique similar to that of embryo collection. A mid-ventral laparotomy is performed, and the embryos are placed into the uterine horn(s) ipsilateral to the CL. Pregnancy rates of 50–60% are often achieved. However, many researchers have also transferred embryos laparoscopically.[12] This procedure is similar to the laparoscopic AI technique, and pregnancy rates rival those achieved with surgical embryo transfer.

One technique which needs further investigation is the use of trophoblastic vesicles. Trophoblastic vesicles (TVs) are small spheres of trophoblast tissue produced by dissecting elongated sheep embryos collected at days 11–12 of gestation. Each piece of embryonic tissue, when placed into culture, will form a TV. A TV can not produce a lamb; however, research in cattle has shown that when TVs are co-transferred with embryos, a significant increase in pregnancy rate occurs.[13] It is suspected that TVs produce one or more proteins which promote embryo development, as evidenced by the ability to culture (in the presence of TVs) fertilized 1–2 cell sheep embryos beyond the 8-cell stage where *in vitro* development normally is blocked.[14]

Embryo Engineering Technologies

Embryo Cryopreservation

From time to time, the number of available embryos exceeds the number of available recipients. Rather than discarding these embryos or transferring them into asynchronous recipients, it may be preferable to cryopreserve ("freeze") the excess embryos. In addition, transporting embryos rather than live animals may be easier, more cost effective, and pose less risk for disease transmission.

The first successful attempt at cryopreserving sheep embryos was reported in 1974.[15] Since that time, a number of reports of successful sheep embryo cryopreservation have been published. Embryos are frozen after being placed into a cryoprotectant, an organic compound which permeates the cells of the embryo and causes dehydration of the cells.[16] By reducing the amount of water inside of the embryo, the cryoprotectant helps prevent damage caused by ice crystal formation during cryopreservation. Commonly utilized cryoprotectants include glycerol, dimethylsulfoxide (DMSO), propylene glycol, and ethylene glycol.

Embryos are placed into standard semen straws for the freezing process which takes between 2 and 3 hours. The computer-controlled embryo freezing machines are programmed to cool the embryos at a specific rate. After reaching a temperature of about −35°C, the embryos are submerged in liquid nitrogen at −196°C where they may may be stored for long periods of time.

To thaw the embryos, straws are submerged in a 37°C waterbath for 20 seconds. The embryo is removed from the straw, as the cryoprotectant must be removed from the embryo prior to transfer. Embryo survival rates following transfer of frozen-thawed embryos have been reported as high as 58%.[17]

Embryo Sexing

Most breeders would be interested in knowing what sex of lamb was being produced from a particular mating. Although not yet perfected, the technique of sexing embryos may offer the opportunity to transfer embryos of a predetermined sex. Certain breeders may wish to produce only ram lambs, whereas others may be interested in propagating only replacement ewe lambs. This technique, however, is not as advantageous as being able to separate male from female sperm prior to inseminaiton (a technique currently being developed). If this were possible, breeders could purchase either "male" or "female" semen and circumvent the need for embryo sexing.

Two methods of sexing embryos are being used, and both of these focus upon the fact that rams have a Y chromosome and produce products unique to males. The mammalian Y chromosome codes for a male-specific antigen called the H-Y antigen. Detection of this antigen can be used for determining the sex of a sheep embryo.[18] There are several variations of the H-Y antigen test currently used for embryonic sex determination. All of these tests utilize intact (whole) embryos, are noninvasive, and are rapid. However, they are only 80–85% accurate.

A second method of embryo sexing being developed is a more sophisticated version of a karyotype, a procedure where the chromosomes are examined to determine if the

Y chromosome is present. This technique, which utilizes a DNA probe, requires the removal of up to 10 cells from the embryo but is 95% accurate.[19] Unfortunately, the DNA probe test requires 24 hours to complete and currently depends upon the use of radioactive isotopes.

Embryo Splitting

Embryo splitting is a method which allows further exploitation of genetically superior ewes by producing twice as many potential offspring as can be obtained through conventional embryo transfer. It also results in the production of identical twins. Early attempts at dividing embryos in half utilized a complicated technique called blastomere separation. This technique was applicable only for very young embryos required a lengthy and complex culture period prior to transfer of the half embryos into recipients. However, simplified techniques have been developed which are better suited to production agriculture.

The simplified embryo splitting techniqe utilizes later stage embryos (morulae and blastocysts) which could be transferred to recipients without a lengthy culture period. This method of bisecting embryos requires the use micromanipulators, equipment which translates the gross movements of the human hand into very precise, fine movements. The micro-manipulators support two microinstruments—one which holds the embryo in place (with gentle suction) and the other (a metal blade or glass knife) which actually bisects the embryo. Several researchers have produced lambs from split embryos with pregnancy rates (per half embryo) ranging from 20–60%.[20–25]

The production of identical twins via embryo splitting enables an interesting application to breeding programs. One ram with very desirable fleece characteristics could be used at farm A while its identical twin could be used as a sire at farm B. This could enable a breeder to more rapidly propagate a valuable genetic line by allowing other breeders to have access to the ram during the same breeding season. Using the same ram at two locations would also permit between farm genetic comparisons of breeding stock.

Nuclear Transplantation

The technique of nuclear transplantation is related to embryo splitting in that its main objective is in the production of genetically identical individuals. This technique is based on the fact that each cell nucleus contains all of the genetic information necessary to create an entirely new animal.

Lambs have been produced by the process of nuclear transplantation.[25] In this procedure, an 8-cell embryo is mechanically divided into the 8 individual cells. The nucleus from each cell is then removed and is fused with an unfertilized egg (from which the nucleus has been removed) using electrofusion techniques. The resulting embryos are cultured in the ligated oviducts of recipient ewes for 4.5 to 5.5 days and then removed and transferred to a second recipient ewe to be carried to term (the same complex "culture" system used in blastomere separation).

Although not yet practical for producers, the attraction with nuclear transplantation is that large numbers of genetically identical individuals can be produced. Theoretically, hundreds of "clones" could be produced by performing serial nuclear transplantation, a process by which "offspring" from the first 8-cell embryo themselves serve as donors for subsequent nuclear transplantations. This is repeated with later generations also serving as nuclei donors. In conjunction with cryopreservation technology, a single genotype possessing desirable fleece characteristics could be cloned and stored for later use.

Gene Injection

One of the newer embryo engineering technologies which has great appeal is the injection of specific genes into embryos. With this technology, it is possible to introduce, for instance, the gene responsible for a particular fleece color without also transferring other undesirable genes (as occurs with conventional crossbreeding programs).

Animals produced by this technology are called transgenic animals. Transgenic sheep have been produced,[26] but success rates of producing transgenic sheep appear to be much lower than with other species. About 1 in 1,000 attempts at producing transgenic sheep is successful. The primary approach to producing transgenic sheep involves direct injection of the gene (DNA) into the pronucleus of the embryo, but other methods such as introduction through retroviral vectors, embryonic stem cells, and liposome carriers are also being utilized.[27]

Most efforts to produce transgenic animals have worked with the growth hormone gene, as it was believed that animals with higher levels of growth hormone would grow faster, be more feed efficient, and produce leaner carcasses. These results, however, have not been observed in sheep, and other genes are being examined. Researchers in Scotland have produced transgenic sheep which produce factor IX and -1-antitrypsin.[28] These compounds are used to treat the human diseases hemophilia B and angioneurotic edema, respectively. Researchers in Australia are presently attempting to increase wool production by inserting wool keratin genes.[29]

In addition, it may become possible to prevent certain diseases caused by enzyme deficiencies by inserting genes coding for production of the enzymes which are lacking.

In Vitro Fertilization

In vitro fertilization (IVF) is the process of producing so-called "test tube" babies. Eggs are harvested from the females and are fertilized in the laboratory with sperm collected from a male. Although a relatively successful procedure in humans, the technique has met with limited success in farm animal species until recently. The IVF procedure may be useful to the sheep industry in supplying multiple lambs from a genetically superior ewe which is unable to establish and maintain a pregnancy (due to age, sickness, or death).

Summary

Many advances in embryo transfer technologies have been made during the 1980's. The next decade promises to be even more exciting as these new technologies are refined and applied to the sheep industry. Breeders of colored sheep will have several options available to them and will no longer depend solely upon conventional breeding programs.

Despite our ability to manipulate various biological processes, many producers will not adopt these practices due to economic constraints. Breeders with limited numbers of animals may not realize a return on their investment unless current success rates are greatly increased.

Literature Cited

[1]Warwick, B. L.; R. O. Berry; W. R. Horlacher. 1934. Proceedings of the American Society for Animal Production 27th Annual Meeting, p. 225.

[2]Robinson, T. J. 1967. The Control of the Ovarian Cycle in the Sheep. Sydney University Press, 258 pp.

[3]Ruttle, J.; S. Lucero; S. Key; M. Daniels; F. Rodriquez; H. S. Yim. 1988. Ovine estrus synchronization and superovulation using norgestomet B and follicle stimulating hormone-pituitary. Theriogenology 30:421–427.

[4]Fukui, Y. and E. M. Roberts. 1977. Fertility of ewes treated with prostaglandin F_2 and artificially inseminated at predetermined intervals thereafter. Australian Journal of Agricultural Research 28:891–897.

[5]Hawk, H. W. and B. S. Cooper. 1977. Sperm transport into the cervix of the ewe after regulation of estrus with prostaglandin or progestogen. Journal of Animal Science 44:638–644.

[6]Hunter, G. L.; C. E. Adams; L. E. Rowson. 1955. Inter-breed ovum transfer in sheep. Journal of Agricultural Science 46:143–149 (plus plates).

[7]Armstrong, D. T. and G. Evans. 1983. Factors influencing success of embryo transfer in sheep and goats. Theriogenology 19:31–42.

[8]Smith, C. L. 1988. Superovulation in sheep. The Compendium on Continuing Education 10:1415–1424.

[9]Killeen, I. D. and G. J. Caffery. 1982. Uterine insemination of ewes with the aid of a laparoscope. Australian Veterinary Journal 59:95.

[10]Coonrod, S. A.; J. Bowen; D. C. Kraemer. 1986. Non-surgical collection of ovine embryos. Proceedings of the 5th Annual Convention of the American Embryo Transfer Association, pp. 83–87.

[11]McKelvey, W. A. C.; J. J. Robinson; R. P. Aitken; I. S. Robertson. 1986. Repeated recoveries of embryos from ewes by laparoscopy. Theriogenology 25:855–865.

[12]McKelvey, W. A. C.; J. J. Robinson; R. P. Aitken. 1985. A simplified technique for the transfer of ovine embryos by laparoscopy. Veterinary Record 117:492–494.

[13]Heyman, Y.; P. Chesne; D. Chupin; Y. Menezo. 1987. Improvement of survival rate of frozen cattle blastocyst after transfer with trophoblastic vesicles. Theriogenology 27:477–484.

[14]Heyman, Y.; Y. Menezo; P. Chesne; S. Camous; V. Garnier. 1987. In vitro cleavage of bovine and ovine early embryos: improved development using coculture with trophoblastic vesicles. Theriogenology 27:59–68.

[15]Willadsen, S. M.; C. Polge; L. E. A. Rowson; R. M. Moor. 1974. Preservation of sheep embryos in liquid nitrogen. Cryobiology 82:560(abstr.).

[16]Leibo, S. P. 1988. Cryopreservation of embryos. Proceedings of the 11th International Congress on Animal Reproduction and Artificial Insemination. Dublin, Ireland. 5:370–377.

[17]Heyman, Y.; C. Vincent; V. Garnier; Y. Cognie. 1987. Transfer of frozen-thawed embryos in sheep. Veterinary Record 120:83–85.

[18]White, K. L.; G. B. Anderson; R. L. Pashen; R. H. BonDurant. 1987. Detection of histocompatibility-Y antigen: identification of sex of pre-implantation ovine embryos. Journal of Reproductive Immunology 10:27–32.

[19]Seidel, G. E., Jr. 1988. Sexing mammalian sperm and embryos. Proceedings of the 11th International Congress on Animal Reproduction and Artificial Insemination. Dublin, Ireland. 5:136–145.

[20]Willadsen, S. M. and R. A. Godke. 1984. A simple procedure for the production of identical sheep twins. Veterinary Record 114:240–243.

[21]Gatica, R.; M. P. Boland; T. F. Crosby; I. Gordon. 1984. Micromanipulation of sheep morulae to produce monozygotic twins. Theriogenology 21:555–560.

[22]Yoshiba, N.; M. Ohtake; Y. Shioya. 1985. Production of identical sheep twins using a razor blade for bisection of hatched blastocysts. Japanese Journal of Animal Reproduction 331:126–129.

[23]Chesne, P.; G. Colas; Y. Cognie; Y. Guerin; C. Sevellec. 1987. Lamb production using superovulation, embryo bisection, and transfer. Theriogenology 27:751–757.

[24]Shelton, J. N. and A. Szell. 1988. Survival of sheep demi-embryos in vivo and in vitro. Theriogenology 30:855–863.

[25]Willadsen, S. M. 1986. Nuclear transplantation in sheep embryos. Nature, London 320:63–65.

[26]Hammer, R. E.; V. G. Pursel; C. E. Rexroad, Jr.; R. J. Wall; D. J. Bolt; K. M. Ebert; R. D. Palmiter; R. L. Brinster. 1985. Production of transgenic rabbits, sheep and pigs by microinjection. Nature, London 315:680–683.

[27]Murray, J. D.; C. D. Nancarrow; K. A. Ward. 1988. Techniques for the transfer of foreign genes into animals. Proceedings of the 11th International Congress on Animal Reproduction and Artificial Insemination. Dublin, Ireland 5:19–27.

[28]Clark, A. J.; A. L. Archibald; H. Bessos; P. Brown; S. Harris; M. McClenaghan; C. Prowse; J. P. Simons; C. B . A. Whitelaw; I. Wilmut. 1988. Molecular manipulation of milk composition. Genome30 (Suppl. 1):45 (abstr.)

[29]Ward, K. A.; I. R. Franklin; J. D. Murray; C. D. Nancarrow; K. A. Raphael; N. W. Rigby; C. R. Byrne; B. W. Wilson; C. L. Hunt. 1986. The direct transfer of DNA by embryo microinjection. Proceedings of the 3rd World Congress on Genetics Applied to Livestock Production 12:6–21.

Healthy Lambs Start with Healthy Ewes

Don E. Bailey, D.V.M.
248 NW Garden Valley Road
Roseburg, Oregon 97470 USA

This is my outline of flock care principles, designed for producers in my area of Oregon, and applicable to most situations with a little correction for season.

A. Ewe Nutrition and Immunology
Important nutrients in sheep are:
1. *Water*—important, sheep are fussy drinkers
2. *Energy*—insufficient energy or carbohydrate intake most important cause of malnutrition
   Lack of energy may cause:
   - reduced fertility
   - slowing and stoppage of growth
   - depletion of body fat and loss of weight
   - premature birth of weak lambs
   - high death loss at birth and after
   - poor milk production of ewe
   - poor wool growth and quality
   - reduced capability to withstand stress and diseases
   - Ketosis
   - do not respond to vaccination

Excessive intake of energy will cause:
   - difficult births
   - more vaginal prolapse
   - reduced fertility (score 5)
3. *Protein*—in sheep the quantity rather than quality is important
   - oil meals = 35–45% crude protein
   - legume hays + 12 = 16% crude protein
   - cereal grain + 8 − 11% crude protein
4. *Minerals*—two groups

Major	*Trace*
calcium	iodine
phosphorus	iron
magnesium	copper
sodium chloride	molybdenum
potassium	cobalt
sulfur	manganese
	zinc
	selenium
	fluorine

   - "Mineral deficiencies are easier to prevent than to treat."
   - Calcium-Phosphorus ratio (Ca=P) grasses and legume hays are an adequate level of calcium, but grains are low in calcium, but just the reverse with phosphorus;—both are required for bone, muscle, and nerve maintenance.
   - symptoms of Ca and P deficiency: slow growth
   - rickets, slow growth, depraved appetite and downer ewes
   - The ratio of Ca to P can vary from 1:1 to 7:1 in sheep as long as the total equals the minimum daily requirements
   - Magnesium—for enzyme systems—will cause grass tetany

- Cobalt—needed for B_{12} production in rumen—cause, unthrifty sheep
- Iodine—for thyroid gland—lack of may cause abortion and goiter in lamb
- Selenium—white muscle disease—lower growth and reproduction
- Copper—usually don't need to supplement sheep—three way relationship with molybdenum and sulfate

Vitamins

Vitamin A—from green forage and hay—important in sheep

Vitamin D—not needed to supplement unless kept indoors

Vitamin E—from green forage and grain—aids in white muscle prevention

Different classes of sheep have different nutritional requirements

1. *Maintenance* of non-pregnant, non lactating ewe and non-breeding ram
2. Maintenance plus *reproduction* for pregnant ewe and breeding ram
3. Maintenance plus growth or *production* for young lambs, feedlot lambs on lactating ewe

   Production
   Reproduction
   Maintenance
   (Compare feeding to filling a bucket; the bottom must be filled first.)

Factors that determine the level of nutrients required by sheep are:

1. Breed—age—size
2. Nature and rate of production desired
3. Stresses of environment
4. Balance of nutrients in the ration
5. Hormonal activities of the animals feed

Feeding ewes—According to their body condition (score), and the stage of their breeding cycle.

a. *Pre-breeding*—feeding—flushing—pasture or concentrate—thin ewe flush better
b. *Post breeding* feeding—wait three weeks after breeding to reduce energy intake
c. *Before lambing* feeding—final six weeks critical —½ to 1½ lb. grain—inadequate energy can result in
   1. pregnancy toxemia
   2. birth of weak or dead lambs
   3. loss of mothering instinct
   4. lower adequate milk production
d. *Lactating ewes—need high level of nutrition—separate ewes with twins and feed separately*
e. *Post weaning*—reduce ration to maintenance

Immunology—Vaccination program for preventable diseases

- Clostridial diseases are included in eight way vaccine
- Abortion disease—Vibrio—EAE and Lepto
- Pneumonia vaccination
- The selenium influence on vaccination
- Timing of vaccination and method used
- Importance of colostrum to the lamb

B. *Lambing Problems*

Losing a lamb means twelve months of ewe cost wasted. One hour of full labor is long enough to wait before examination. When examining the ewe, be *clean* and *gentle*. Some of the variations seen are:

1. The big lamb
2. No feet—head only showing
3. One foreleg turned back
4. Elbow lock
5. Both front feet only
6. Hind feet first
7. Tail only
8. Lying across
9. Twins—together or one backward
10. Ringworm

C. *Grafting Techniques* (not successful after three weeks)

1. Slime grafting—washing lamb, tie legs
2. Skinning graft
3. Tying ewe up
4. Ewe Stanchion
5. "Odorize"—Kerosene, vanilla, milk, deodorants, etc.—apply to ewes' nose and lamb's body
6. Blindfold ewe
7. Tie dog close by
8. Grafting second lamb—keep both away and return together—repeat
9. Vaginal—cervical stimulation

D. *Lamb Incubator and Hypothermia in Newborn Lambs (Low Body Temperature)*

1. High risk lambs
   - lambs under 5–7 pounds birth weight
   - lambs from thin or old ewes
   - born during and in bad weather
   - premature lambs
   - extra large lambs after difficult birth
2. High risk period
   - *Birth to 5 hours*—inability for wet newborn to regulate body temperature along with excessive heat loss
   - *Ten hours to three days*—due to starvation and complicated by heat loss due to environment
   - Thermometer is essential in detecting hypothermia
     - Normal temperature—102–103°F
     - *Above* suspect infection
     - *Below* suspect hypothermia

With lamb under five hours old and weak:

Step 1—Take temperature; if over 98°, dry lamb and feed. If returns to normal (102–103°), it will probably be all right

—If temperature is under 98°, incubator should be used with a temperature of 104–112°. Place the lamb in the incubator for 30 minutes. Recheck the temperature. If it is still under 98°, give 4cc/pound 20% dextrose IP (intraperitoneal). Return to the incubator for 30 minutes and recheck the temperature. When over 98°, remove from the incubator and feed with a stomach tube 100–200 cc colostrum milk. Give 1cc pen strep. Continue to monitor. Return to mother or feed 6oz three times daily. Use oral antibiotics.

With lamb over five hours old.

Take temperature. Feel stomach to check for milk. Feed 20cc/pound body weight three times daily. Check ewe for milk supply, mis-mothering, mastitis, etc.

- Use follow up antibiotics
- Pen/Strep, Dexymiycin, etc.
- return to mother or graft when temperature over 100° and lamb shows improvement.

Lamb incubators can be made of bales of straw with plastic cover or use a plywood box 4′ x 4′ with wire mesh floor and heat (electric) blower keeping the temperature at 104–112°.

E. *Lamb Health Care Standards*

—Checklist of lambing barn needs

____ 1. Propylene glycol for ketosis
____ 2. Heat lamps
____ 3. Iodine for navel dipping (7%)
____ 4. Lamb puller and leg snare
____ 5. Antibiotics along with syringe and needles
____ 6. Uterine boluses—for use after OB
____ 7. Frozen colostrum (ewe, cow, or goat)
____ 8. Bottles, nipples—tube feeder
____ 9. Dextrose for injection (5–20%)
____ 10. Thermometer
____ 11. Plastic OB gloves
____ 12. Lubricant (KY-soap)
____ 13. Mastitis ointment and treatment tube
____ 14. Record sheets, ear tags, marking chalk
____ 15. Slack lime for disinfecting lambing pens
____ 16. Clean OB bucket and disinfectant
____ 17. Ewe halter
____ 18. Oxygen tank with valve and tube (optional)
____ 19. Towels
____ 20. Eye ointment and scissors for entropion eyes
____ 21. Scour medicine and pepto-bismol

Management influences losses and is most critical period of the sheep production year.

Types of lambing operations

a. Open pastures—hope for good weather.
b. Shed-off systems and separate newborn by driving through gates on daily basis.
c. Modified confinement where twins and trouble lambs are housed, singles stay out on pasture.
d. Confinement or shed lambing with drop pens where every ewe goes through; plus singles only stay long enough to dry off and fill up; twins stay 2–3 days depending on the weather

Biggest losses at lambing occur during first *3 days following birth*

1. Difficult birth
2. Weak lamb
3. Too small or too big
4. Abortion
5. Starvation
6. Not claimed

Then first two week losses due most commonly to:

1. Starvation
2. Pneumonia
3. Scours
4. Injuries

After two weeks the most common are:

1. Pneumonia
2. White muscle
3. Tetanus
4. Injuries
5. Enterotoxemia
6. Coccidiosis
7. Navel infection

Rules on confinement lambing include proper disinfecting the navel:

- clip navel to about 1 inch from the stomach wall
- dip—immerse the stump of the navel in a bottle containing 7% iodine and roll lamb on its back to hold bottle tight against stomach for 30 seconds
- strip—open ewes' teats and check for mastitis
- sip—see that lamb gets colostrum as soon as possible after birth
- water—ewe needs 1–2 gallons of water soon after lambing
- withhold concentrate feed for 24 hours after lambing to reduce udder congestion
- turn singles out as soon as dry and full of milk and weather is good
- keep twins 2–3 days in pens and then into mixing pens—paint brand or ear tags will help identify trouble.

Some medical supplies that are good to have on hand:

1. Pepto-bismol for weak lambs, simple stomach ache
2. Scour medicine for lambs with scours: Kaopectate, Biosol M small animal tablets (1 every 12 hours), Tribrisses injectable and tablets (½ every 24 hours or 1 tablet every 12 hours)
3. Dextrose—5% for giving underskin (2–4cc/lb) peritoneal injection (4cc/lb) 20% dextrose
4. Misc. 5 grain aspirin, eye ointments for entropion (turned in eyelids), tube feeder, frozen colostrum

Management Calendar for Sheep Production in Western Oregon

A. *Pre-breeding* (July-August)
1. Cull—body condition, teeth, udder, age
2. Condition score ewes and rams—divide according to score and start feeding 4–6 weeks before breeding (all ewe and rams should score 3–3.5 by breeding time)
3. Worm and fluke
4. Trim feet and foot bath
5. Vaccinate for abortion diseases (when indicated—vibrio, EAE, lepto)
6. Sample flock on selenium blood tests

B. *Breeding* (August and September)
1. Start flushing 14 days before breeding—turn in teaser rams at same time
2. Breeding soundness examination on rams—especially check for symptoms of epididymitis—vaccinate if present
3. All new rams should have semen sample taken
4. Watch for breeding activity—where possible use marking harness to adjust ram availability to modify lambing per day capabilities
5. Restrict breeding to 5–7 weeks (2–3 heat cycles), good flushing program along with the use of teaser rams should result in short breeding season
6. Use teaser buck with marking harness to mark open ewes after breeding season
7. Start bucks on recovery program after breeding season. Increase nutrition, worm, low level antibiotics in feed

C. *Pre-Lambing* (November-December)
1. Shearing or tagging
2. Worm and fluke
3. Condition, score, and divide and feed according to score (score 3–3.5; score by lambing)
4. Booster vaccination C and D or 8 way Vibrio—Lepto maybe
5. Spray—dip or pour treatment for lice
6. Sample flock by blood test for calcium, phosphorus, magnesium, and slelenium
7. Prepare lambing facilities and equipment
8. Start coccidiosis medicine in trace mineral and salt 30 days before lambing

D. *Lambing* (January-March)
1. Be ready—have medicine and equipment organized
2. Iodine all navels and check milk on all ewes as soon as possible after lamb is born
3. Be ready to treat weak and hypothermic lamb
4. Watch lambs carefully for "filling up"; pick up and feel milk in stomach
5. Have dead lambs necropsied
6. Keep count on losses and what the cause is
7. If footrot is a problem, use zinc sulfate foot bath going into lambing barn—8 pounds to 10 gallons of water—to prevent footrot and especially foot scald
8. If did not worm ewes at pre-lambing, then worm ewes as they go through lambing
9. Consider use of footrot vaccine

E. *Post lambing and pasture season* (April-June)
1. Worm ewes and any lambs over 30 pounds around the end of March—continue to worm lamb every 30 days until weaning in June
2. Trim feet—use zinc foot baths on 7 day intervals if footrot and scald is a problem
3. Castrate and dock—use method preferred
4. Vaccinate lambs with Type D toxoid at 7–14 days and then repeat two weeks later (include tetanus anti-toxin if ewes are not hypered with 8 way vaccine twice the first year)
5. Set up creep feeding system (away from barn if possible)
6. Watch for sore mouth and white muscle symptoms—be ready to vaccinate
7. Have death losses necropsied—run fecal tests for worms and coccidia
8. Treat for liver flukes and keds, and shearing
9. Booster ewes with 8 way vaccine at shearing in chronic fluke areas
10. Cull dry ewes at shearing—also mark mastitis ewes for culling

F. *Weaning* (July)
1. Vaccinate replacement ewe lambs with 8 way Clostridial Vaccine
2. Vaccinate feeder lambs with Type D Toxoid
3. Reduce nutrition on ewes
4. Cull ewes—age, condition, production, mastitis, teeth, etc.
5. Trim feet—best time to find chronic footrot carriers and sell

G. *Finalize Year End Health and Production Records for Sheep Operation*

____ Total ewes bred
____ Number of productive ewes for this year
____ Ewes lost
____ Total lambs born
____ Total lambs lost
____ Total live lambs
____ Death lost percentage
____ Lambing percentage
____ Total lambs weaned
____ Weaning percentage
____ Pounds of lamb/ewe
____ Pound of lamb produced/acre

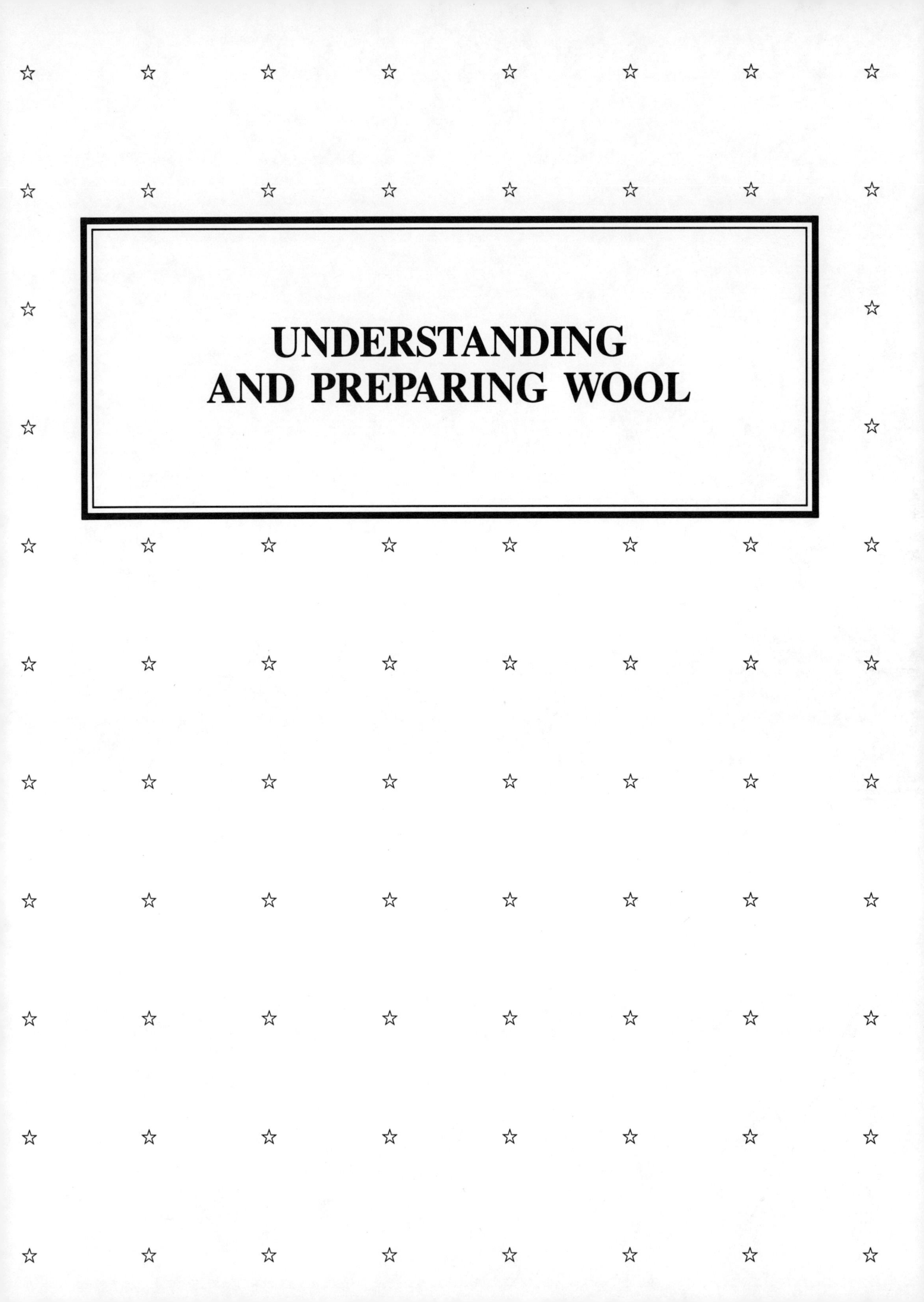

UNDERSTANDING AND PREPARING WOOL

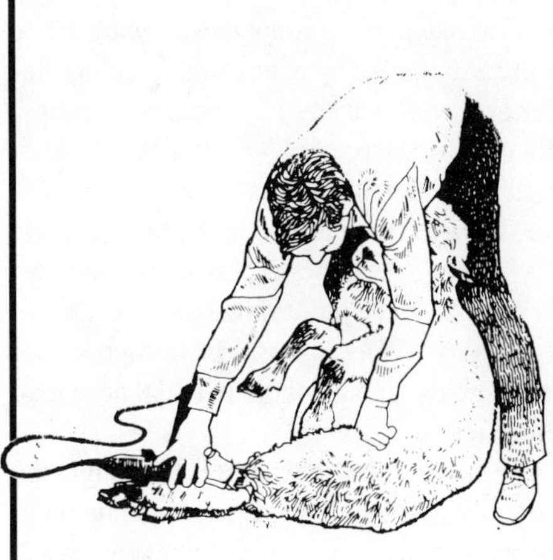

Wool— The Outstanding Textile Fiber

Rosalie R. King, Ph.D.
Home Economics Dept.
Old Main 560
Bellingham, WA 98225 USA

Wool continues to be the most important natural fiber today. Its unique and desirable characteristics are as much in demand now as they have been for centuries. What are the characteristics that contribute to the success of this fiber even in times when hundreds of variations of man made fibers are readily available?

Wool primarily comes from the hair covering of sheep. There are many breeds which vary in physical appearance, wool characteristics and geographical distribution. The term "wool" also includes hair of the angora goat, camel, alpaca, llama and vicuna.

The use of wool dates before recorded history. We generally accept that wool was discovered about ten thousand years ago in central Asia when the sheep was found to be a two product animal, providing fiber for protective covering as well as meat and milk for food. The Egyptians, Babylonians, and the Romans all used wool in both household items and in clothing. Later it was used in Italy, England, and Spain and in the mid 1600's the woolen industry came to the colonies in America. America's wool industry slowly increased its production while competing with England. The Merino sheep were eventually imported, which started a large supply of fine staple wool and created a greater interest in fine woolens. Between 1840 and 1870, the speed and overall improvement of looms was increased with more machinery for worsted processes.

In more recent years, wool consumption has decreased because of the introduction of man-made fibers, which in turn has caused a decline in sheep population. However, the wool fiber is still in great demand in the United States for a wide range of end uses.

Physical Properties

The wool fiber consists of the cuticle (the plate-like cells on the outside), the cortex (inside the cuticle) and sometimes the medulla (the core made of air filled cell or sometimes just a hollow tube). In finer wools or those fibers closer to the body, the medulla will not be present. Fibers exhibit a wide range of variation. The outside scales determine the luster of wool; luster can vary from a high reflection of the straight, smooth fibers to a subtle luster of the highly crimped or waved fibers such as those of the Merino sheep. The fibers are relatively light with a specific gravity of 1.34. They range anywhere from 1–8 inches (2.5–20 cm) long.

The differences in type of fiber can also determine the hand or the tactile feel of the fabric. Wool can have a soft and delicate feeling to it if the fiber is fine and highly crimped, or a more crisp hand if the fiber is less fine in diameter. Resiliency, the ability to resist deformation, can also vary but with all wools it is high. Wool retains the body heat of the wearer. This has been important to people who live in colder climates or those who hike in a wet environment as well as population groups in warm regions. Body heat is trapped in the waviness or crimp of the fibers. The fibers on the outside of the garment give added insulation providing another layer of still air between the garment surface and the skin.

The insulation properties even hold when the wool is wet, and wool fabric has a unique surface tension which resists water until the tension is broken. It has excellent moisture absorption properties as its moisture regain is 16%. Of all the natural fibers, wool is the most water absorbing, and it absorbs moisture away from the body so the wearer does not feel cold and clammy.

Wool is self-extinguishing when the flame source is removed. However, it will support combustion and leave a black crusty material. When heat is applied to the wool fibers, they will become dry and harsh feeling at 212 degrees F (100 degree C) and will scorch at 400 degrees F. Overdrying of wool fiber should be avoided as permanent drying will occur.

Wool is prone to shrinkage; the outer scales appear to shorten. Many shrink resistant treatments have been developed to counteract the undesirable quality. Shrinkproofing treatments can be applied to wool at almost any stage of processing after scouring. Two recent treatments have been a chlorination treatment and interfacial polymerization, a process that leaves the fiber coated with a thin layer of man-made polymer such as nylon.

Wool fibers compact and swell when in moisture, heat, and alkaline conditions leading to felting. When felting occurs, the fibers become entangled with each other and the effect is irreversible. This has sometimes been a desirable trait—as in the production of melton fabrics where a firm dense cloth, impentrable to windy climate conditions is required. Shrinkage accompanies the felting process.

Chemical Properties

Wool is a protein fiber partially composed of a keratin and it is related to hair, horn, and feathers. Approximately nineteen amino acids join in peptide linkages that are repeated many times, giving an extremely long chain of molecules. Wool also has cross links that join the peptide chains. These cross-links (called cystine links) give the fiber crimp and resiliency, unique to this fiber. Various factors may break these cross linkages thus the fiber becomes less efficient. Such factors include sunlight which will begin to break down the linkages and the fiber will weaken. Wool is not damaged by dilute acids, however it will be damaged by concentrated acids. Because of the number and wide variety of functional groups, wool is probably the most reactive of all textile fibers to dyestuff of many different formulations. A disadvantage of wool is the susceptibility to damage from moths and other insects.

Wool grease is purified to lanolin, an important by-product of the wool industry. The amount varies on different parts of the sheep's body. Wool grease is composed of nitrogen, phosphorous, acids, and alcohols in small amounts, fat, iodine, and cholesterol.

Sulfur is distinct in wool and hair fibers although the proportions will vary. Wool is not a homogeneous chemical compound but is composed of chemically distinct substances. The wool molecule is characterized by flexibility to assume many folded or spiral configurations.

Although wool is the predominant animal fiber in the apparel and interior design industry, other lesser known but more expensive fibers are also used alone or in combination with wool to create increased lustre, strength, or variety.

Reference

King, Rosalie Rosso. *Textile Identification, Conservation, and Preservation.* Noyes Publication: Park Ridge, New Jersey. 1985. pp. 48–54.

TABLE 1 Properties of Wool (Natural Protein)

I. Form	*II. Physical*	*III. Chemical*
1. Length	1. Density (specific gravity 1.34)	1. Structure—helical protein
2. Cross section shape-elliptical	2. Moisture a. moisture regain 16% b. Swelling phenomena	2. Composition—50%C, 24%O, 16%N, 7%H, 3%S
3. Crimp—yes	3. Stress-strain a. strength 1.0 to 1.7 g/d b. elongation 35% c. stiffness low d. toughness low e. elasticity very good f. resilience very good	3. Behavior toward a. acids: resists mild acids b. alkalies: destroy c. dyes: high affinity d. solvents: resistant
4. Surface character scale (varies) slightly with species of animal	4. Color white, gray, brown, & black	
	5. Luster high in low grades 6. Scorching or melting point 300°C. 7. Static electricity only when extremely dry	

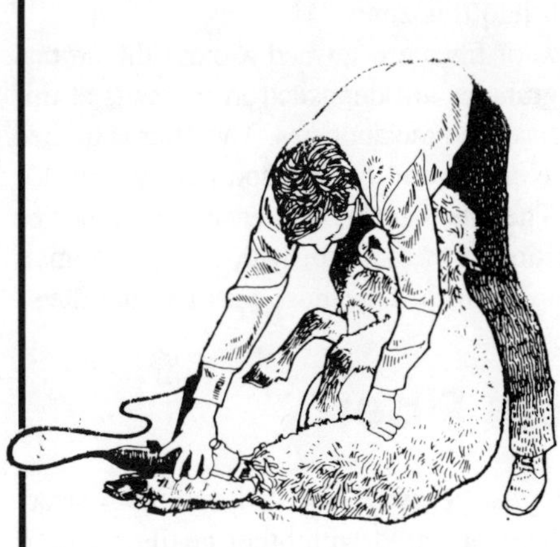

Wool Structure and Color Formation in the Sheep

Melinda J. Burrill, Ph.D.
Department of Animal Science
California State Polytechnic University
Pomona, CA 91768 USA

For those interested in the production and use of wool, an understanding of the basic structure of the fiber and the way it is produced by the sheep may be helpful. Through a knowledge of these, management, care and breeding of the animal may be designed so as to result in the most superior product possible.

Wool fibers are similar to the hairy fibers produced by all mammals. They are composed of several cell types which are "glued" together and which originate from special structures, called follicles, situated in the skin. Breeds of sheep that have been developed for their wool have fibers that grow continuously. However, some breeds of sheep have fibers that are much more akin to the hair produced by cattle. These breeds are classed as hair sheep. Other breeds are rather intermediate in that they have a wool-like fiber which is, however, shed annually. Some of the mountain longwools of Great Britain fall into this category.

The Wool Follicle

Wool fibers are classified into two types; the separation between the two being predicated upon the time at which the fibers are formed in the lamb along with the presence or absence of several structures associated with the follicle. The primary follicle develops rather early in fetal life (between the thirty-fifth and fortieth day of gestation) and may be distinguished by the presence of the following accessory structures:

1. the sebaceous or wax gland
2. the sweat or suint gland
3. the arrector pili muscle

The secondary follicle appears later in the life of the lamb, with some perhaps being formed after birth. These follicles generally lack sweat glands and arrector pili muscles. Occasionally, follicles are found that have both gland types but lack the arrector pili. These are referred to as primo-secondary follicles.

Primary follicles, which tend to be found in clusters of three (trios), may often produce a coarser fiber which contains a medulla (see Figure 1). In some cases kemp is also produced by this type of follicle. The former type of fiber, though always fewer in number, are often more apparent in fleeces especially in the young lamb. Kemp fibers are unusually thick, short in comparison to the rest of the fleece, and often a chalky white color, even in colored fleeces. They have poor spinning and dyeing qualities and are quite undesirable.

Secondary follicles are formed after the primary follicle trios, normally in the late pre-natal development of the lamb. The sebaceous gland associated with these follicles is often smaller than that associated with the primary follicle. Follicles grow in clusters around the primaries and the average number of follicles in each cluster varies from one part of the body to another, and from one breed to another.

The smallest clusters are located in the cod or udder region where often only primaries are found; the largest are found on the back near the neck. Clusters of intermediate size are located on the sides of the sheep. Breed differences can be striking with respect to cluster size with Merinos having the largest (50 or more secondaries per primary) and breeds like the Lincoln having small clusters. Most breeds fall in the range of from between 3 and 50 secondaries per primary follicle. Thus breeds and animals within a breed are often characterized by the ratio of secondary to primary follicles, called the S/P ratio.

Along with the S/P ratio, the primary follicle density (the number per unit area of skin) and the total surface area of skin determine the total number of follicles on a sheep. Adding in the factors of fiber thickness and length, these together determine the total weight of the clean fleece.

A simplified diagram of a primary follicle is illustrated in Figure 1. At the base of the follicle is the bulb into which the dermal papilla is invaginated. The papilla has a blood capillary to provide nourishment to the dividing cells surrounding its interior surface. Both cells that will become the wool fiber proper and those that produce pigment granules are found there.

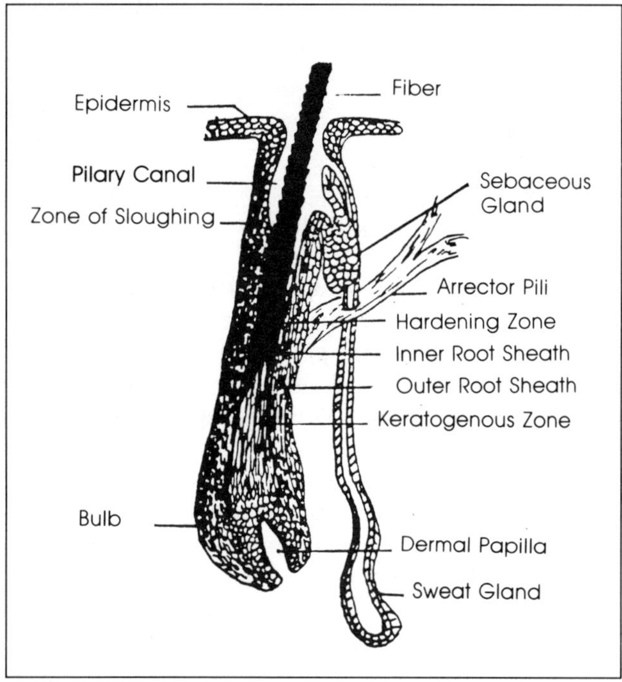

Figure 1 A primary follicle.

Above the bulb is the region of the follicle called the keratogenous zone. Proteins that form the fiber and inner root sheath are formed there. It is in this region that cell elongation also occurs.

Progressing upward, one finds the region in which hardening of the fiber occurs. It is also in this region that cells of the inner root sheath begin to degrade and be resorbed.

The zone of sloughing is found below the opening duct of the sebaceous gland. In this zone, the degraded inner root sheath cells slough off along with some of the outer root sheath cells.

The pilary canal is located above the zone of sloughing and extends to the surface of the skin. The sebaceous and sweat glands open into this zone.

Cells of the wool fiber are formed around the dermal papilla. Pigment granules are deposited in the cells at this time by special cells called melanocytes. The fiber is pushed upward, the cells elongate and harden and finally protrude from the surface. The rate that this sequence occurs may be related to the nutrition of the animal, with well fed animals showing faster wool growth than those on marginal diets.

The Wool Fiber

Under the microscope, wool reveals its complex structure (Figure 2) when compared with other textile fibers. It is composed of a very large number of cells arranged in two (sometimes three) primary layers. The larger central layer, the cortex, is made up of relatively round, spindle-shaped cells that taper at the ends. These cells are quite closely packed. The cortical cells contain pigment granules in colored wools.

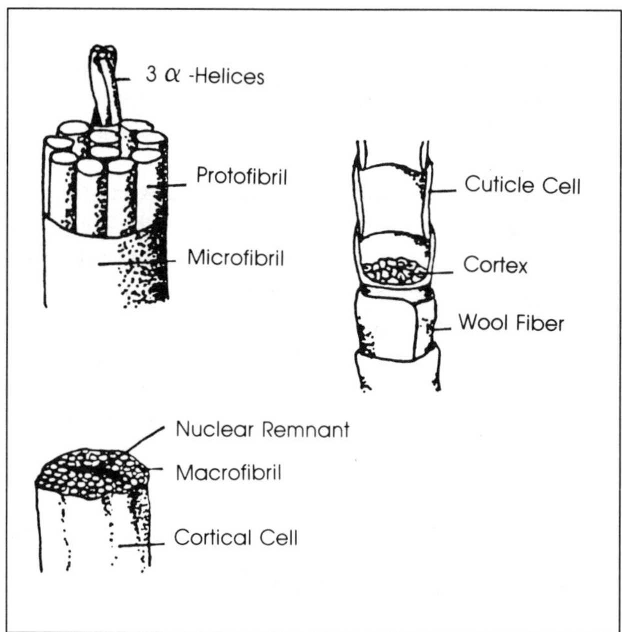

Figure 2 Components of the wool fiber (after D'Arcy, 1981).

In fine, highly crimped wool, two types of cortical layers are readily apparent. The orthocortex runs on the outside of the crimp curvature and the paracortex runs along the inside.

In coarser wools, the distinctions between these two layers is less visible.

The cortical cells consist of bundles of filaments called macrofibrils. Inside each of these are found smaller filaments called microfibrils surrounded by a non-fibrous protein matrix which acts as a glue. The structure of the microfibrils consists of bundles of protofibrils. Each protofibril is composed of three alpha-helical protein chains wound around each other. The protein making up both the glue and the microfibrils is called keratin. The keratin of the glue is quite high in sulfur while that of the microfibrils is low in sulfur.

Diets which are insufficient in protein have been shown to cause reduced wool growth and fiber weakness. However, feeding excessive protein will not stimulate wool growth above the genetic capacity of the animal to produce wool. As dietary protein is expensive, excess feeding to promote wool growth is not advised. Protein in the diet may be provided by feeding high quality legume hays, such as alfalfa, or by feeding linseed, cottonseed or soybean meals.

Surrounding the cortex are the cuticle cells. These cells are flattened and rather square shaped and fold around the outside of the wool shaft. They also overlap each other like shingles on a roof with the "free" end pointing toward the tip of the fiber. This configuration of cuticle cells can be felt by running one's fingers up and down a fiber. Running up toward the tip, the fiber feels smooth while running toward the root, the fibers feel rough. Over the exposed surface of the cuticle there is a chemically resistant membrane called the epicuticle which serves to protect the cells from various environmental hazards.

A third inner layer of cells, the medulla, may be found in medium-coarse to coarse wool fibers. This layer lies centrally in the fiber and is surrounded by the cortex. The cells of the medulla are box-shaped and hollow. Pigment granules are also found in these cells, though in smaller numbers. When it is present, the medulla may be continuous along the length of the fiber as in the coarsest wools or interrupted, as in the medium-coarse wools. The presence of a medulla results in less favorable spinning and dyeing qualities.

The wool follicle requires large amounts of energy since it is the site of rapidly dividing cells. In animals where the energy in the diet has been insufficient to sustain all the bodily functions of the animal, such as growth, pregnancy and lactation, the wool fibers will grow more slowly and may be substantially thinner than in the well fed animal. The length of the fiber that was growing at the time will be weakened. As wool in the same area grows at approximately the same rate, when this weakened wool grows above the skin surface, it may be grabbed and pulled out in large amounts. The wool fibers will all break at the weakened site. This defect is called tender wool. It may also be caused by sickness. Energy in the diet may be provided by various grains. It is advisable to feed some grain (1 lb. per day is usually sufficient) during late gestation and lactation when quality wool is required.

Color Formation

As stated earlier, the pigment in colored wool derives from the presence of color granules or melanosomes in the cortical and medullary cells of the fiber. The melanosomes are "injected" into these from specialized cells, melanocytes, which are found on the dermal papilla, Figure 3.

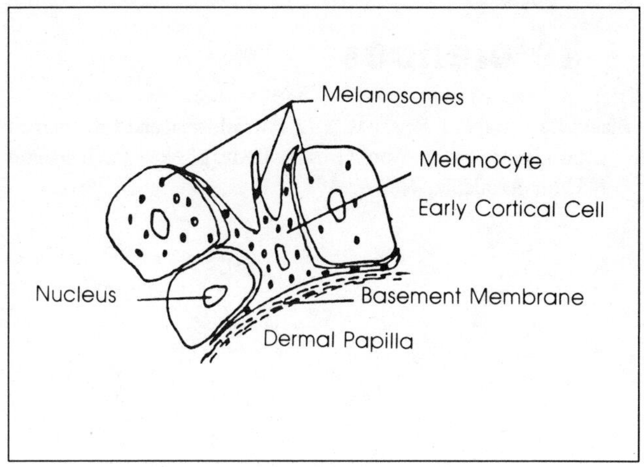

Figure 3 Situation of the melanocyte in relation to early cortical cells at the dermal papilla. (After Rogers, 1979)

Melanocytes themselves are first formed in the neural crest of the early developing embryo and are dendritic or star-shaped cells similar in outward structure to other neural cells. Within these cells, the special structures, the melanosomes, are formed and within the melanosomes, the pigments, eumelanin (black and brown) and phaeomelanin (red and tan) are synthesized.

The biochemistry of synthesis of eumelanin has been elucidated for several species and is virtually identical in all of them so there is no reason to think that the biosynthesis is different for the sheep. The essential amino acid, tyrosine is the basic precursor which through several steps is modified and polymerized to form eumelanin. The biosynthesis of phaeomelanin is similar but requires another amino acid, cysteine. Both melanins are large polymers.

The initial step in both biosynthetic pathways is catalyzed by the enzyme, tyrosinase. If the gene for this enzyme is defective, albinism results.

Some forms of white spotting in various species such as the mouse and the horse have been found to be due to the failure of the melanocytes to migrate from the neural crest to the epidermis (skin) during early fetal development. This has not been thoroughly investigated in the sheep but may be the cause of one or more forms of spotting.

Graying of the wool may be due to lack of tyrosinase production, the cessation of melanosome formation in the melanocytes or to death of the melanocytes themselves. The structural and/or biochemical causes of graying in the sheep have not been totally determined.

In summary, wool is a complex structure and its properties depend very much on this structure. It can be modified by external environmental factors, such as nutrition. By understanding the structure and formation of the fiber, better environments can be provided to the sheep which will maximize the production of a quality product.

References

Black, J. L., and P. J. Reis, eds. 1979. Physiological and Environmental Limitations to Wool Growth. Univ. of New England Pub. Unit. Armidale, N.S.W.

Botkin, M. P., R. A. Field and C. L. Johnson. 1988. Sheep and Wool Science, Production and Management. Prentice Hall, Englewood Cliffs, N.J.

D'Arcy, J. B. 1981. Sheep Management and Wool Technology. New South Wales Univ. Press. Kensington, N.S.W.

Kammlade, W. G. and W. G. Kammlade, Jr. 1955. Sheep Science. J. B. Lippincott. New York.

Rogers, G. E. The cytological and biochemical basis of pigmentation in hair and wool. In: Breeding Coloured Sheep and Using Coloured Wool. 1979. Peacock Publications. Hyde Park. S. A.

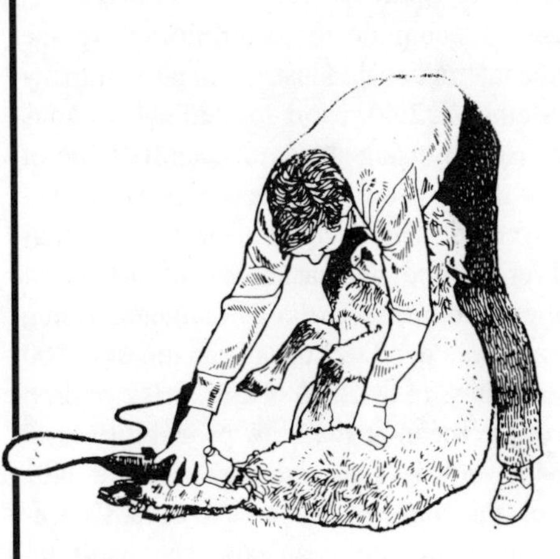

Developments in Objective Fiber Measurements to Express Quality and Improve Marketability of Wool, Mohair and Cashmere

C. J. Lupton
Texas Agricultural Experiment Station
7887 North Highway 87
San Angelo, Texas 76901 USA

An objective measurement is an assessment made without distortion by personal feelings or prejudice. It is the opposite of subjective assessment which is an evaluation made by human judgement using the senses of sight and touch. By necessity, therefore, an objective measurement is obtained using a machine or instrument.

Most transactions including the transfer of ownership of animal fibers involve some form of appraisal by the purchaser. Why should objective measurements be used to assist with such appraisals? In a nutshell, because the human senses of sight and touch cannot be calibrated as accurately as some instruments and cannot consistently produce assessments with the required high degree of accuracy.

Numerous measures of quality are common to wool, mohair and cashmere. In contrast, some quality characteristics are unique to a specific fiber. The importance of individual characteristics is dependent to a large extent on the end use for which a particular fiber is intended. Thus, so far as the commercial worsted manufacturer is concerned, fiber, staple and fleece characteristics of white greasy wool have the significance illustrated in Table 1. Undoubtedly, this order would be changed somewhat for handspinners. Similarly, so far as woolen system manufacturers are concerned, the order and priority of some of these properties would also be changed. In addition, physical properties such as bulk and fiber length distribution after carding would appear prominently on a woolen spinner's list. When considering mohair quality, these same properties are again important but luster, medulated fiber content and mohair style and character are additional characteristics to be assessed. Similarly, the other luxurious goat fiber, cashmere, has other important characteristics that are determined for this unique bicomponent fleece. Again, style and character are important since all fine (13 to 19 microns) goat hair is not cashmere. Mechanical yield of cashmere from the whole fleece and fourthly, the difference in average fiber diameter between true cashmere fibers and guard hairs are other important parameters.

Recognizing that the importance of individual quality characteristics is different for selected end uses, objective methods in current use or those being developed will be reviewed generically. Subsequently, advantages of objective measurements to each segment of our industry will be outlined.

Review of Objective Measurements

Sampling for Yield and Diameter

A high proportion of wool is sold in the U.S. after being tested for yield and fineness. In contrast, only small proportions of mohair and cashmere are sold on the basis of objective measurements, although this situation is liable to change very soon. Unfortunately, much of this precise information is not fed back to the producer because it was obtained for large lots that were produced by combining numerous smaller batches. Advantages of selling wool in this manner have been reviewed elsewhere in detail.[7] This form of testing begins with random sampling of the lot, often referred to as "coring." Detailed procedures for sampling lots of different sizes and types of wool are summarized in ASTM D1060, Core Sampling of Raw Wool in Packages for Determination of Percentage of Clean Wool Fiber Present.[1]

TABLE 1 Significance of Greasy White Wool Characteristics to Worsted System Processors*

Major	Yield (including quantity and type of vegetable matter) Average fiber diameter Average staple length Staple strength and position of break Color/colored fibers
Secondary	Variability of fiber diameter Variability of staple length Cotted wool (felted fibers) Crimp (resistance to compression)
Minor	Weathered tips Age/breed/type Style/character/handle

*Original source Australian Wool Corporation.[11]

Portable, electric coring tools having two-inch or half-inch diameter tubes are probably the most commonly used in the U.S. today. Typically, two-inch coring tools are used to sample bags or bales of wool at the warehouse and mill. Subsequently, the two-inch cores are normally subsampled using a coring machine fitted with half inch tubes. This same machine is often used to sample single fleeces. The half-inch cores are then used in yield and diameter determinations.

Many overseas storage complexes and wool textile mills throughout the world are now using specially designed coring machines for sampling wool bales.

Yield Measurement

The standard method for determining yield (ASTM D584) has changed little since its introduction. Basically, the method involves scouring core samples in hot, soapy water followed by determinations of residual grease, inorganic ash and vegetable matter content on the dried, scoured fiber. Subsequently, "Wool (mohair or cashmere) Base" (or pure, oven-dry wool) is converted to a value known as "Clean Wool Fiber Present" by dividing with a factor of 0.86. This is the factor required to adjust the wool base to a moisture content of 12%, an alcohol-extractives content of 1.5% and a mineral matter content of 0.5%.

Numerous attempts to speed up or circumvent this tedious method using modern technology have been described in the literature. One technique with some promise is near infrared reflectance spectroscopy.[3] To date, a suitable replacement has not been established.

Diameter

A microprojection technique for determining average diameter has been the international industry standard for many years. In ASTM Method D2140, short logitudinal sections are projected onto a screen using standard magnification of 500X. The widths of the projected images are measured using a standard wedge card. This method allows for calculation of both an average and a measure of variability of diameter, that is the standard deviation of diameter. Using the wedge card technique, a good technician can measure 200 fibers in about 20 minutes. In order to obtain ± 0.2 micron confidence limits of the mean at the 95% probability level when measuring 64's wool, it is necessary to measure 2401 fibers. For 50's wool, this number increases to 6146. Consequently, obtaining data in this way is an expensive and relatively slow proposition.

Several U.S. institutions are experimenting with digitizing devices that could replace the wedge card and allow for more rapid measurement. At this time, none of the innovative techniques have been incorporated into standard methods.

Ultrasound and image analysis techniques have been used to measure the diameter of wool with varying degrees of success. ASTM Test Method D3515 described a technique utilizing a PiMc Particle Measurement Computer System. Unfortunately, this instrument is now obsolete and the method was discontinued. Nevertheless, several research groups, including our own and the Commonwealth Scientific and Industrial Research Organization (CSIRO) in Australia are further developing image analysis techniques.[8] This technology shows distinct promise for enhancing the speed with which fiber diameter can be measured and is predicted to be commercially available in the (relative) short term.

For many purposes, a knowledge of the variability of diameter is not required. In such cases, air flow instruments have made a tremendous contribution (ASTM D1282). Three versions that have been used in the U.S. include the Micronaire, Port-Ar and WIRA instruments. These instruments require sample sizes of 4.8, 12.5 and 2.5 g, respectively, of scoured, well-blended, conditioned wool. The actual measurement of average diameter takes less than a minute. Sample preparation takes longer.

For many years, there has been a need for an instrument capable of rapid and accurate measurement of fiber diameter and distribution. This need has been met to a large extent by the introduction of the CSIRO Fiber Fineness Distribution Analyzer and its commercial counterpart, the Peyer Texlab FDA200 System. These instruments undoubtedly represent the most innovative concept for determining animal fiber fineness parameters since the introduction of projection microscopy for this purpose. The electro-optical technique is capable of measuring 2000 fibers, calculating a mean, standard deviation and coefficient of variation and printing this information together with a histogram all in the space of three minutes. Probably because of the high price ($47,500) only three of these instruments are present in the U.S., one of them being in the Texas Agricultural Experiment Station Wool and Mohair Research Lab in San Angelo. Approximately 30 of these instruments are operating throughout the world. The instrument is expected to make a major contribution, particularly in those research and quality control applications where a knowledge of diameter distribution is necessary e.g., selective breeding studies with sheep and worsted spinning of fine yarns.

Length

Commercially, the length of animal fiber determines primarily which system may be used to spin the fibers into yarn, i.e., the worsted, woolen and short-staple (or cotton) systems. A highly significant linear relationship exists between the staple length of sound wools and mean fiber length in top. Fiber length in top has a major influence on spinning speeds, yarn count and yarn quality. Thus, mean staple length is another important parameter to measure. The technique recommended by ASTM (D1234) for shorn fibers is simple, accurate and slow and requires only a rule for measurement and a pencil for recording. Similarly, a good estimate of staple length can be obtained directly on the live animal.

By 1964, researchers with the U.S. Department of Agriculture had managed to speed up the process using a semi-automatic "Staple Length Recorder," a device in which each staple was fed through a slot onto a moving belt and past a light beam.[9] When this beam was interrupted by a staple, electronic pulses in units of 1/10 inch of staple length were counted. The device was not fast but it contained the basic concept found in present designs, none of which have yet managed to qualify for entry into an ASTM standard although several types of instruments are being used routinely overseas.

Staple Strength and Position of Break

Wool fiber strength is a major factor in the strength of yarns. It also has an important effect on the percentage of noils formed and mean fiber length in top-making. This is particularly true if there is a weakness in the staple due to nutritional or health factors. The position of the weakness along the staple is also important. ASTM methods of tensile testing grease wool do not indicate position of break.

The current state-of-the-art staple length/strength/position-of-break instrument is the CSIRO ATLAS (Automatic Tester for Length and Strength). Several of these instruments are currently being used in Australian Wool Testing Authority labs providing mainly presale data for wool buyers. In addition, researchers from the South African Wool and Textile Research Institute (SAWTRI) have developed a length/strength tester which provides information on the profile, length and strength of raw wool staples both at the maximum practical gauge length and also at a short gauge.[2] In our own lab, a relatively inexpensive instrument is used that produces results that are highly correlated to those obtained from the ATLAS instrument.

Color/Colored Fibers

For white wool, it has been demonstrated that scoured color is a reproducible, measurable characteristic. The New Zealand Standards Organization has a published standard and Australia is considering one. Of course, purchasers of white wool are actually more concerned with lack of color or whiteness. Nevertheless, the same instrument, a colorimeter, can be used to accurately specify the color of colored wool, mohair and cashmere. A word of warning, however. In the case of white wool, the whiteness (or yellowness) of the greasy product is not a very good indicator of the whiteness of the scoured product. Similarly, more accurate specification of colored fleeces is obtained when the measurements are made on scoured material.

So far as colored fiber contamination of white wool (or vice versa) is concerned, it is generally appreciated that direct measurement of dark fibers in grease wool is not attainable because of the sampling problem.[5] In contrast, developments in metrology, particularly in image analysis, means we should be able to standardize and automate testing for detection of dark fibers in top in the near future.

Variabilities of Fiber Diameter and Staple Length

When required, variabilities of fiber diameter and staple length can be obtained concurrently with determinations of the means.

Felted (Cotted) Fleeces

There is no objective test for this condition. However, it is readily detected and can easily be discriminated against. All things being equal, the mean fiber length of a scoured and carded cotted fleece is significantly shorter than that of a non-felted fleece. Cotting is often associated with subnormal levels of crimp, particularly in fine wools.

Crimp

No current ASTM method addresses the measurement of crimp. Visual appraisal of staple crimp frequency and definition still forms an important though often misleading part of buying strategy. Image analysis techniques have been developed to provide this information objectively.[6] An indication of crimp level can also be obtained using a measure of fiber resistance to compression or bulk.

Weathered Tips

This fault which results in reduced mean fiber length after carding and uneven dyeing can be detected visually and manually with varying degrees of difficulty depending upon the actual degree of weathering. Reduction in strength due to this condition can be quantified using a staple strength tester.

Age/Breed/Type

If this information is not known for a particular fleece, it cannot be objectively measured. When the information is known, it can be useful to categorize a fleece or a lot of fiber within some broad parameters. For example, knowing that a particular 12-month fleece was produced by a 3-year-old Rambouillet ewe immediately implies the fleece characteristics will probably fall within the following ranges that are typical of the breed:

Yield (%): 40–60
Diameter/microns: 19–25
Staple length (in.): 2–5
Color: white
Grease fleece weight (lb): 9–14

Perhaps for certain end uses, this information would be adequate. Objective measurement of individual characteristics obviously produces more precise information.

Style/Character/Handle

I am not aware of any attempts to develop instrument methodology for directly assessing the handle of greasy animal fibers. Nevertheless, the interrelationships between handle and yield, fiber diameter, strength, crimp and bulk are well documented. Style and character have been subjectively assessed characteristics of somewhat vague definition. Certainly they are important characteristics affecting price. In the case of cashmere, the absence of correct style and character relegates the fiber to fine goat hair. The ringlet, web and flat lock styles of mohair are diminishingly prized due to the claimed increase in variability of average fiber length. Image analysis is currently being evaluated in the labs of CSIRO's Division of Wool Technology to provide an objective measurement of style and character.

Medullated Fiber Content

Medullated fibers in wool, mohair and cashmere cause difficulties when attempting to dye solid, heavy shades due to their apparent inability to absorb dye. Of course, if their presence is planned, kemps can be used to provide pleasing aesthetic effects. Methodology for determining medullated fibers is described in ASTM Test Method D2968. Fibers are prepared in exactly the same manner as for diameter determination. However, only a cursory inspection of 1000 projected fibers is required in order to designate each fiber into the unmedullated, med (less than 60% medullated) or kemp (more than 60% medullated) category.

An alternative method of identifying the specific proportion of fibers capable of causing dyeing problems is currently being investigated by the International Mohair Association. The new method will involve dyeing a scoured sample of fiber with a mixture of dyes followed by a visual evaluation and accurate count of undyed fibers.

Luster

Luster is routinely, subjectively assessed particularly in the marketing of mohair and luster wools. There is no approved objective method for determining luster on greasy animal fibers. However, it appears likely that colorimetric techniques can be developed for estimating this elusive characteristic. As for color, luster is also affected by the amounts of grease and dirt surrounding the fiber. Thus, scoured fibers are required for accurate determination of luster. As unlikely as it may seem, the bulk or resistance to compression of luster wools (and perhaps mohair) is related to luster.[4]

Mechanical Yield of Cashmere

The proportion of a bicomponent goat fleece comprising cashmere is called the mechanical yield. In contrast to the scoured yield, this yield reflects the proportion of fine cashmere fibers that can be mechanically separated from coarse guard hairs. This value can be accurately assessed using tweezers to separate the two types of fibers in a sample followed by weighing of the separated portions using an analytical balance. This very slow procedure has been replaced in several labs by the Shirley Trash Separator that uses an aerodynamic principle to separate guard hair from cashmere in a relatively short time. Although no standard method includes this procedure, it is quite widely used by the cashmere trade.

Relative Diameter of Cashmere and Guard Hairs

If the means of these two populations of fibers are not separated by a significant amount (e.g. 30 microns) physical separation of cashmere from the raw fleece is difficult or impossible. Accurate determination of average diameter and variability of diameter of raw cashmere can be obtained using any of the techniques discussed earlier in the diameter section.

Discussion

Benefits of objectively measured wool, mohair and cashmere characteristics accrue to producers, fiber dealers and processors. These benefits are summarized as follows:

General

- Potential to provide a common language in fiber discussions throughout the world
- Improved definition of fiber quality characteristics
- More specific determination of the value of individual lots
- More accurate feed-back of processors requirements to producers and/or dealers
- More precise matching of fleeces.

Processor

- More efficient design of blends
- More accurate prediction of processing performance
- More efficient processing and improved quality control
- More accurate prediction of yarn and fabric properties.

Producer

- Greater equity in prices (particularly for growers of uncharacteristically fine or high yielding fleeces)
- More accurate assessments of the effects on fiber growth from seasonal, nutritional and health factors
- Improved strategies for shearing, lambing, kidding, and other management and breeding practices
- Potential for improved, more standardized preparation of fleeces for marketing

Dealers

- Potential for increased efficiency and cost reduction in marketing
- Reduced risk when buying and selling; less necessity for hedging.

It is interesting to note that in the last 15 years, sale of wool by sample and objective measurement has increased significantly in Australia, South Africa and New Zealand. Of these systems, the Australian practice is probably the most advanced and may be summarized as follows. Last season, 99% of auction wools were tested presale and sold by sample with the following measurements made on core samples: fiber diameter, yield and vegetable matter. A proportion of these wools was also tested for staple length, strength, position of weakness and clean color. These additional measurements are seen as part of the continuing trend in the international industry toward the ultimate system of selling by description alone. Sale by description implies that there is no recourse to physical inspection of the sample by buyers who will, as a result, determine the value of the wool (and their prices) on the basis of measurements and guaranteed information. It has the potential for major savings in wool marketing costs.[10]

With the leadership of the Objective Measurements Task Force that was appointed following the U.S. Secretary of Agriculture's Conference on Quality and Competitiveness in June 1988, the U.S. wool, mohair and cashmere industries are now moving toward the adoption of a system of marketing based on objectively measured fiber characteristics. It would seem very appropriate that the colored segment of this industry also adopt such a system in order to obtain the documented advantages.

References

[1]American Society for Testing and Materials. 1988. Annual Book of ASTM Standards. Section 7, Vol 07.02.

[2]Cizek, J. and D. W. F. Turpie. 1985. The Performance and Application of the SAWTRI Length/Strength Tester for Raw Wool. Proceedings, 7th International Wool Textile Research Conference, Tokyo. Vol. II. 137–146.

[3]Connell, J. P. 1983. Predicting Wool Base of Greasy Wool by Near Infrared Reflective Spectroscopy. Textile Research Journal. 53:651–655.

[4]Elliott, K. H. 1986. The Ability of the WRONZ Automatic Bulkometer to Rank Wools for Lustre. WRONZ Report R 135. I–11.

[5]Foulds, R. A. 1985. Studies of Dark-fibre Contamination from Fleece to Fabric. Proceedings, 7th International Wool Textile Research Conference, Tokyo. Vol. II. pp. 65–74.

[6]Higgerson, G. J. and K. J. Whitely. 1985. Measurement of the Crimp Characteristics of Greasy Wool Staples. Proceedings, 7th International Wool Textile Research Conference, Toyko. Vol. II, pp. 157–166.

[7]Lupton, C. J. 1987. The Role of Objective Measurements in Wool Research and Marketing. SID Research Digest. 3, 3:21–25.

[8]Marler, J. W. and H. McNally. 1987. Fidam—A New Method of Wool Measurement by Image Analysis. Wool Technology and Sheep Breeding. Dec Jan: 194–196.

[9]Ray, H. D., H. C. Reals, D. D. Johnston and E. M. Pohle. 1964. Measuring Wool by Staple Length Recorder. Marketing Research Report 668, USDA.

[10]Rottenburg, R. A. and M. W. Andrews. 1985. Recent Developments in Raw Wool Specification and the Ramifications for the Processor and Wool Grower. Proceedings, 7th International Wool Textile Research Conference, Toyko. Vol II. pp. 117–126.

[11]Specialist Working Group on Sale by Description. 1978. Report to the Australian Wool Corporation.

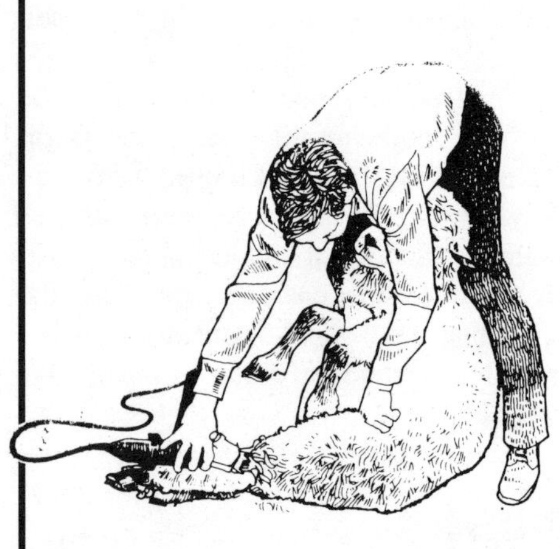

Clip Preparation of Australian Black and Coloured Wools in Victoria

Helen E. Wright
R.M.B. 4163 Elmhurst
Victoria, 3469, Australia

Australia is known world-wide for its production of white wool, in particular white Merino wool, although many wool types from Superfine Merino to Lincoln and Carpet Wools are produced.

Victoria is the southern-most mainland state of Australia and wool of many types is produced across the state.

A Code of Practice for Preparation of Australian Wool Clips[1] has been prepared for members of all sections of the wool industry to ensure a correct approach is taken to clip preparation. The aims of clip preparation are to prepare wool for sale in such a manner as to provide a textile fibre which processors may use with confidence and to maximize the net return to the woolgrower. This Code of Practice is prepared in regard to white wool because to this important Australian industry, black and coloured (pigmented) wool fibre is a serious contaminant or fault. Australian white wool is respected by processors for its relative freedom from dark fibre contamination, less than 5% (percent). Therefore, within the Code, particular emphasis is placed on the rejection of all pigmented fibre from white wool at all times, with the encouragement of rigorous culling of all coloured and spotted sheep. All sheepbreeders have a responsibility to protect and maintain the processing reputation of our white wool industry.

The production of Black and Coloured Wools is a specialist industry and developed in response to the demand of the handcraft home-industry. Not all the wool produced is suitable for this specialist use, therefore in Victoria we have sought to develop an outlet for growers' non-handcraft wools through the commercial market. This development of two market avenues has necessitated clip preparation of slightly varied requirements whilst still being controlled by the same principles and standards.

Handcraft specialists generally desire small quantities of quality wool—well grown, prepared and presented. Commercial users require large quantities of wool for economic viability and can utilise a wider range of wool types than the handcraft specialists. To maintain and hopefully increase their confidence in the performance of our wool, we must also ensure that wools purchased by them are well prepared and presented. A high standard of preparation and presentation is essential to the successful promotion of natural coloured wool, just as it is vital to the Australian White Wool Industry; therefore the principles of the Code of Practice are applicable to our industry. Firstly there are some basic management practices vital to successful clip preparation.

Sheep Management

This factor must come first as the wool characteristics which are important in processing are determined by breeding, nutrition and husbandry.

Breeding: Select and breed from the best animals of your chosen breed or type, which are available to you—why waste money selecting animals growing inferior wools, especially as most wool characteristics are highly heritable.

Nutrition: Aim to have uniformity of feed for all animals to maintain maximum body growth and even fibre growth.

Husbandry: Healthy animals will give you their maximum fibre return.

Presentation for Shearing

The aim is to present sheep for shearing so that contamination of the wool will not occur during shearing or subsequent wool handling. Users of wool want to buy wool fibre, not useless contaminants.

(a) Sheep should be crutched to remove dung and urine stained wool;
(b) Yards should be as free of dust and mud as possible;
(c) Draft sheep to separate—different breeds, lambs and/or hoggetts from grown sheep, different colours—specifically blacks/greys from brown, sick and diseased sheep including flyblown sheep;
(d) Sheep with wool too wet for pressing (packaging) should not be shorn.

Shed Management

Ensure that the shed is cleaned before shearing to remove all contaminants. Hay-baling twine is a serious contaminant to processing. Do not leave any rubbish or equipment around where it could easily be mixed with wool. Have good light, with adequate space and working facilities.

Preparation Management

Who can prepare the wool? Individual growers can prepare their own wool for private sales. If they are new to the industry it will be to their advantage to undertake some basic wool knowledge studies. All wool destined for sale by auction must be prepared by a Registered Woolclasser and all bales must be branded with the stencil denoting the Woolclasser's Registered Number. The woolclasser shall prepare the wool in accordance with the Code of Practice and with the directions and orders of the grower or grower's representative. In addition to classing, the woolclasser is responsible for the supervision of shed staff involved in the preparation of the wool.

The woolclasser may prepare the wool in the shearing shed simultaneous with shearing; or, as is the case with most coloured woolgrowers in Victoria who run only small flocks, the grower may undertake the basic preparation and then bale/package the wool and transport it to a Wool Broker who will employ a Woolclasser to prepare the wool.

Clip Preparation Standards

Shearing Shed Preparation. How do we prepare our wool? Having already paid attention to the basic management practices, we are ready to commence. Familiarity with the terminology for the parts of the fleece is necessary to facilitate easy preparation.

The sheep is shorn. The belly wool will have been segregated and should have been checked for any urine (pizzle), water or heavy mud stains; which should have been removed from the clean portion and placed in a stain line. Pick up the fleece and throw it onto the woolrolling table, spreading it out fully, ready for skirting. Sweep up the locks from the shearing board and place in a pack after removing any long wool (pieces), stains and dags. The object of skirting a fleece is to remove all those parts which vary from the bulk of the fleece, leaving it as uniform as possible; thus all fleeces must be skirted carefully so that only inferior wool is removed and all good fleece wool remains intact.

Inferior wools removed are:

- all stains and all dags; to be placed in a separate bag
- all sweat points and fribs; removed and place in pieces (PCS) line
- long topknots; remove and place in PCS
- short crutch wool, topknots and shankings; remove and place in Locks (LKS) line
- jaw (jowl) pieces and cotted edges; remove and place in PCS, unless heavily matted and seedy in which case they should be kept separate.

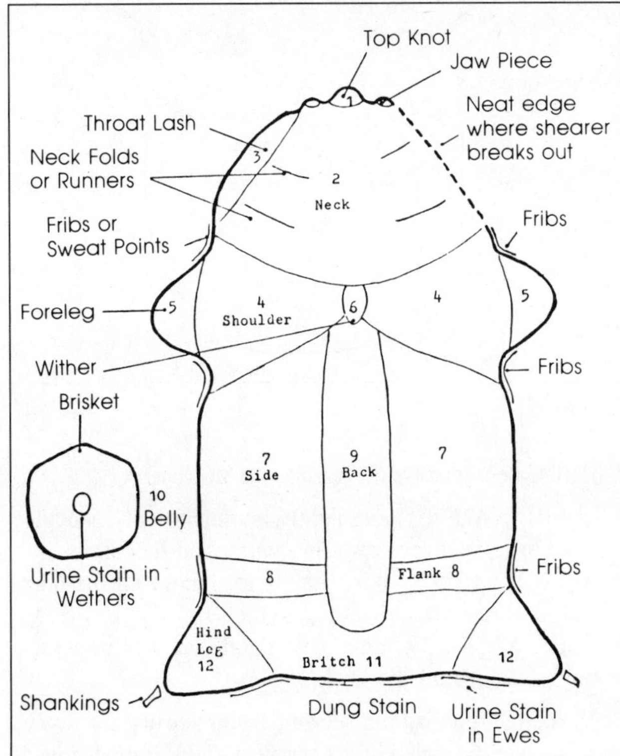

Figure 1 The parts of the fleece.

- clump vetetable matter; remove and keep separate
- hairy britch wool; remove and place in a separate PCS line, since it generally contains medullated fibres. Breeders should be eliminating this fault by culling animals from their flocks

Figure 1 is a diagram illustrating these parts.

- necks; should only be removed if heavily contaminated with vegetable matter or water-stained, in comparison with the rest of the fleece
- backs; should only be removed if very dusty, wasty and tender in comparison with the rest of the fleece and kept separate since they cause processing problems. If time permits the skin may be cut off the wool with hand-shears.
- skin pieces; must always must be removed

The following is a diagram showing light skirting for fleeces which are uniform in quality and free of vegetable fault.

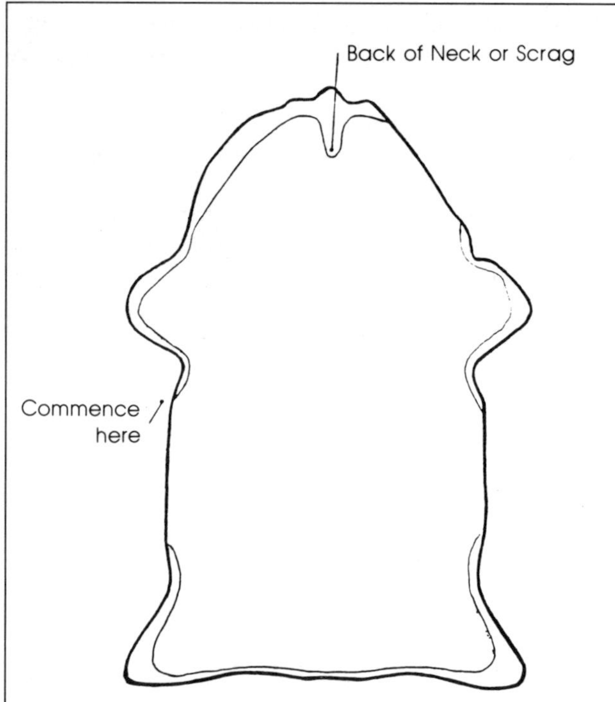

Figure 2 Light skirting.

Different degrees of skirting may be necessary:

- In FREE WOOL, skirt lightly, removing only enough wool to leave fleece even in quality and length;
- In LIGHT BURR, skirt heavily to remove all the burr and leave the fleece free;
- In HEAVY BURR, only skirt lightly, as it is impossible to remove all the burr.

Figure 3 is a diagram showing heavy skirting for fleeces showing quality variation, or carrying a moderate degree of vegetable fault.

Once skirting is completed, then fold and roll the fleece to present it as a tidy, easily handled bundle. Figure 4 illustrates this.

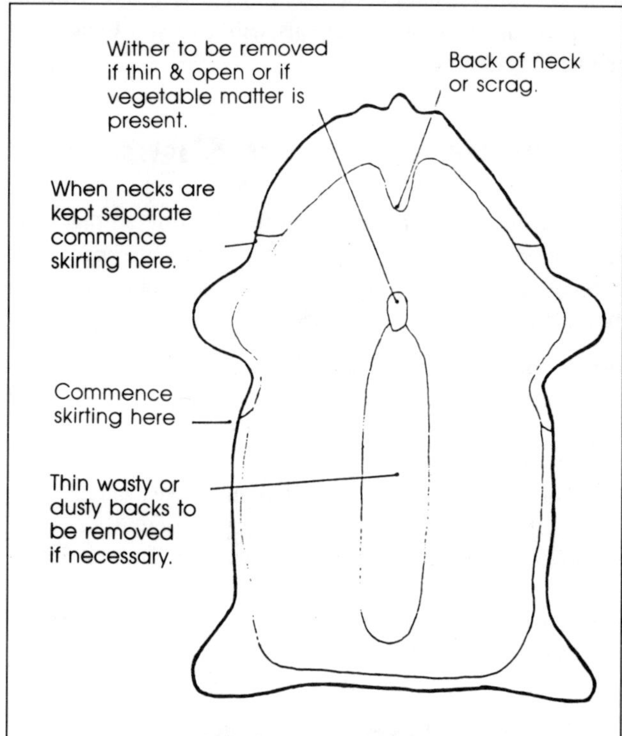

Figure 3 Heavy skirting.

If preparing the fleece for direct sale to the handspinner or weaver, pay particular attention to skirting, maintaining long, even staple length and freedom from vegetable matter as the degree of variation able to be handled by the handspinner is far less than that which can be handled in various forms by commercial mills. Having skirted the fleece for your private customer, do not forget to test its tensile strength, it must be sound. If the fleece is large, possibly greater than 2.5 kg, then it may be better to divide it in half (down centre back from neck to britch) and pack into two (2) bags. Most Victorian Coloured Woolgrowers prepare their best fleeces for direct sale to the handcraft industry and package such wools into clear plastic bags, suitably labelled to indicate the grower's name and address, the breed of sheep or wool type, the weight of wool and the sale price. The Handcraft's person will be willing to pay premiums for your very best wool.

But what about the inferior wool removed from those premium fleeces or the fleeces that were not good enough?

Quality lines must be made in pieces (PCS), particularly in mixed clips.

A PCS BLK— Merino pieces
CBK PCS BLK— Comeback pieces, 58's–60's, from Polwarth, Bond, Merino-based crossbreeds
FXB PCS BLK— Fine and Medium Crossbreed pieces, 56's–54's, Corriedale, Perendale, cross-

breeds of British Longwools and Merino or Comeback

XB PCS BLK— All pieces showing a tendency to lustre, 46's and stronger, Border Leicester, English Leicester, Romney Marsh, crossbreeds of same.

LINC PCS BLK— Lincoln pieces.

Similar quality lines should be made for fleeces, with tender fleeces being kept separate from sound fleeces. Also pay attention to length and style, especially if wool quantities will allow. Very short fleeces should be segregated from good length wools, and very plain fleece, i.e., wool lacking distinct crimp and staple formation, should also be segregated. The lack of good length and attractive style will reduce processing performance.

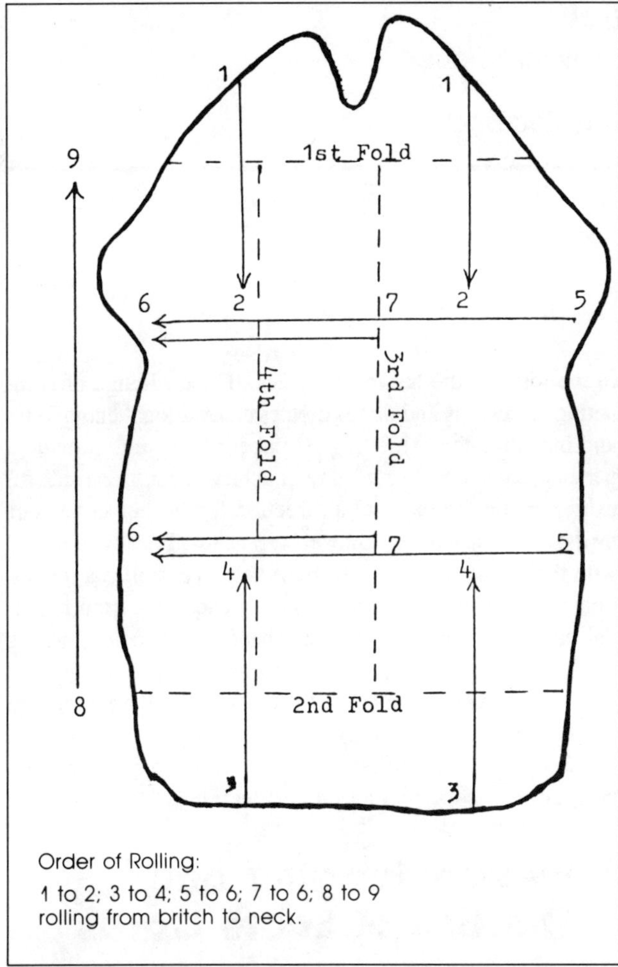

Figure 4 The correct method of rolling a fleece.

Lambswool must be handled separate to the wool from adult sheep. Remove hairy-tipped wool, skin pieces, stain, dag and seedy wool from lambswool:

- If 4 cm or less, no further picking is necessary, except for quality as for the pieces lines;
- If 5 cm or more, sort to a separate line, plus for quality as above;
- If of 10–12 months growth (with lambstip/hoggett type), may be treated as fleece.

For wool to go direct from the grower's shearing shed to the Woolbroker's Showroom floor and Auction, the grower must have one or more bales of the separate lines as designated above. A bale must weigh a minimum of 110 kgs, but be no more than 204 kgs, and be branded with the owner's property brand, the woolclasser's registered stencil number, the wool description, and the bale number. The woolclasser's specification, giving all relevant details of ownership, the descriptions of all bale contents and the suggested sale lotting of bale lines to guide the woolbroker in his sale cataloguing, must accompany all wool to the woolstore.

The production of black and coloured wool fibre is a specialist industry with a relatively limited market highly reactive to fashion trends; thus we must relate to the processor's needs. To maintain the confidence of all processors of our wool—the handcraft industry and the commercial manufacturing industry—we must ensure continuing achievement of high standards of clip preparation. If we maintain these then the successful promotion of our specialist product will continue in the future.

References

[1]Code of Practice for Australian Wool Clips—July 1986, Australian Wool Corporation.

[2]Australian Wool Selling Regulations: Joint Wool Selling Organisation—Sub-committee, Australian Wool Corporation.

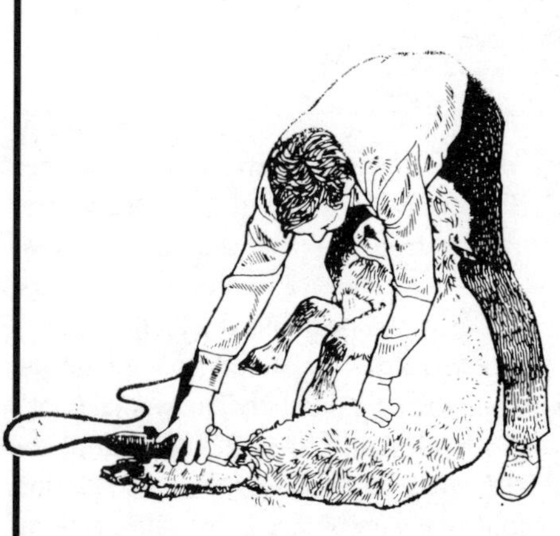

Judging of Black and Coloured Wool

B. M. Tinnock
Department of Wool Science
Lincoln College
Canterbury, New Zealand

Exhibiting and judging of coloured wools in fleece competitions are comparatively new innovations in the predominant white wool producing countries of Australia and New Zealand. In those countries, they have followed the formation during the 1970's of breed societies by a number of people whose aim was to establish flocks of coloured sheep, primarily for the production of coloured wool as a specialty product suitable for handcrafts. At the same time, similar events were occurring across the world in the United States of America and more recently in the United Kingdom.

Breeders of black and coloured sheep commonly arrange local exhibitions of sheep and wool at their meetings and field days. Some breed societies arrange annual coloured fleece competitions on a nationwide basis as well as at a local level. In most wool growing countries, fleece wool competitions are also held at agricultural shows and fairs. Fleece classes at these gatherings probably comprise the greater bulk of coloured wools exhibited for competition and judging. Fleece entries at a show might number from below twenty to over two hundred. This paper focuses on the judging of handcraft potential of coloured shorn fleeces.

Fleece Competitions

Fleece wool competitions serve several useful purposes. For wool producers, they provide a means of exhibiting the results of their sheep breeding and management programmes in the form of the harvested fleece. The challenge of competing for awards and prizes creates an additional enthusiasm and incentive for exhibiting. Competition results provide a personal satisfaction or a sense of achievement for successful exhibitors on the day and a direction for further endeavour by those whose entries achieve lower status. Entries displayed with their prize awards or ribbons help to establish a reputation for breeders. This may result in enquiries from potential purchasers of rams, or from handcraft persons seeking a reliable source of a particular wool type. For the general public, different fleece classes provide a visual appreciation of the variety of raw material from which wool products are made. Fleece competitions thus provide an arena for exhibition, comparison, promotion and learning.

Fleece Judging and The Use of Score Cards

Judging is a major element of fleece wool competitions, whether at local, county, provincial, state or national level. In the case of coloured wools, it is usually done by invitation accorded an experienced sheepbreeder, craftsperson, woolclasser or other skilled person in the wool trade. The role of a judge is to assist a breed or show society as it aims to encourage excellence in coloured wool production.

Fleece judging has long established traditions in many countries where white wool production is an important part of the economy. Its extension to coloured wool classes in recent years has required examination of traditional ideas, since the differences between commercial or industry requirements for white wool and the requirements by the handcrafts fraternity for coloured wools have become more widely recognised.

One result of this has been the re-design or creation of new points or score cards suitable for use in jduging coloured wools. Not all breed or show societies require fleece judges to use score cards. Some require judges to award a points score for each wool character listed in their official judging card, with the fleeces scoring the highest points being awarded the prizes. Some discussion on these two different judging approaches might be appropriate before consideration is given to the fleece factors involved. Since points cards involve more time, more work and generally more stewards to assist the judge, they will be accorded particular attention in the following observations.

Wool has fascinated many who observe its initial appraisal in fleece form and its later manipulation by hand, from raw material to creative products of aesthetic delight. Fleece in the hands of an experienced crafter—its instinctive appraisal by discerning eye and sensitive finger, the collective mental appraisal of its colour, length, handle, bulk and lustre, the assessment of the potential ease of fibre manipulation and of the combined effect of characters in the end product for which the creator is selecting—surely are some of the most basic and fundamentally satisfying of natural or acquired human skills.

By comparison, the separate consideration by a wool competition judge of each fleece character in turn, the allocation of points for each on a judging card with subsequent addition to establish a total point, seems a very artificial and cumbersome means of appreciating the worth or quality of the fleece as a whole. To apply points for various wool characteristics in this manner incorporates an additional mental factor which is quite at variance with the overall practical assessment and appreciation directly resulting from the natural combination of sight and touch.

Furthermore, in a points system of judging, we are relying quite heavily on the assumption that the maximum points provided for each character are correctly "weighted" and that the potential for best results in terms of finished product can be expected from the fleece with the highest total points. In fact, it is possible when using points cards, for a fleece which lacks desirable factors for handcrafts, to be placed above a highly desirable fleece by virtue of the heavier clean fleece weight of the former and the higher maximum points allowed for this factor.

When points are added, it is common to find in large classes that several of the better fleeces have achieved the same total. A judge must then review these fleeces and make adjustments to the recorded list of points so that the fleeces are ranked in order of merit. This is an additional but often necessary exercise resulting from the use of points cards. It is seemingly a superfluous task, for the final ranking of fleeces would probably have been the same (given due credit for clean fleece weight) if the judging cards had not been used at all. When the original time taken to record, add and check points is taken into account, there are grounds to question the contribution of points cards to the overall efficiency of the fleece judging procedure.

One has to accept that since the human factor is involved in subjective appraisal of fleeces, there are bound to be differences in points allocation between judges, a fact easily borne out by people who enter the same fleece at different shows or competitions. This may be a reflection of:

- the degree of human inability to distinguish the worth of particular wool characteristics by subjective award of points and/or
- a personal bias of an individual judge towards particular characteristics and their worth for an end product.

It should further be considered that results essentially reflect the opinions, standards and decisions of a judge on the day and that concern regarding inability of judges to agree is generally unwarranted. It should be realised too, that any experienced appraiser is bound to be less than one hundred percent consistent in his/her own points allocation for each character of the same fleece, when presented with a large number of fleeces to assess for a second time. This human frailty stands to be acknowledged by both judge and competitors alike.

Some Alternatives To Using Score Cards

Notwithstanding all the above, many growers prefer (and even demand) that a list of points scored be provided for each fleece in order that they might compare the judge's observations on their own wool with those of other competitors in the same class. This requires a major time consuming exercise of subjective assessment, points allocation, points recording and tedious addition and checking of points, which in a large entry provides literally thousands of numbers for competitors and viewers to digest.

Because viewers are not usually permitted to handle fleece exhibits, these points are possibly useful where the distinction between the best and second best fleece is a very fine one, eg. slight tender staple, which may be impossible to observe by visual means only. But in the main, the large number of lists of points are usually accorded only a cursory perusal by the general viewer.

As an alternative to using points cards for individual coloured fleeces, a notice could be well displayed for exhibitors and viewers with the following statement—THESE FLEECES ARE JUDGED ON AN OVERALL BALANCE OF THE FOLLOWING CHARACTERISTICS: fleece weight, fibre fineness, staple length, staple strength, character, handle, spinability, freedom from inherited or acquired faults.

This would probably be a sufficient guide for most general viewers. But growers—not viewers—make the competition and it is therefore desirable that some form of grower education should provide the necessary knowledge for exhibitors to appreciate the reasons for the judge's ranking of prize winning fleeces.

A more expedient judging method than the points system is advocated. This would consist of displaying individual cards containing only the judge's written comments, on the award winning fleeces in each class. These comment cards would identify the finer differences between the best entries which, in a close contest, might not be so obvious to viewers. They would also serve as benchmarks by which unsuccessful exhibitors could identify shortcomings in their own individual entries.

A further step in providing grower education without points cards, is to announce a particular time when the wool judge will comment upon the range of fleeces he/she has inspected and will give his/her verbal reasons for having placed the top fleeces in their prize winning order. This is a method already practised with sheep at some major shows, where the top animals in each class are lined up at the conclusion of judging and the judge comments upon the conformation and fleece of each award winner, outlining the reasons for his/her placings. We might well ask why points cards are required for fleece judging where their use is not seen to be necessary for sheep judging.

Structure of Score Cards

As judging cards are regarded by many as being an essential requirement for coloured wool competitions, it is appropriate that we should consider the various combinations of judging factors and associated points currently used on some cards.

A recent perusal of several coloured wool judging cards, including some from Australia, New Zealand and USA, has revealed forty differently named factors for which points are awarded in various coloured fleece competitions. Some of these names or phrases represent slightly different means of expressing the same concept, with most judging cards comprising approximately eight to ten factors. In various ways, they are also designed to allow other recognised factors to be associated with, or incorported in, assessments made under these principal headings. The forty factors noted above are incorporated in the nine fleece characteristics discussed later in this paper.

Another feature of the various coloured wool judging cards is the wide diversity of points allocated for similarly designated characteristics. For example, five selected factors were found to have a wide range of maximum points (each factor contributing to a possible total of 100 points): staple length 5–10; soundness 6–15; handle 5–15; evenness 6–20; clean fleece weight 25–40. Whilst it is not valid to quote points out of their original context, it does raise the matter of the importance or otherwise of the relative value for each characteristic when formulating judging cards.

Rather than attempt to establish definitive relative values for each characteristic, it seems more practical and just as effective to allocate maximum points that facilitate the judging operation. It is more expedient to have a card in which maximum points for most factors listed are similar and which a judge can easily memorise, instead of a steward having to recall in turn a diversity of maximum points for the judge, as he/she evaluates each fleece. By the former system, every fleece still has the chance of obtaining top points for each characteristic, most of which, in fact, are likely to be considered of similar priority or value according to the individual preference of various crafters. This concept is exemplified in the B.C.S.B.A. of New Zealand card, in which seven factors have maximum points of 10, clean fleece weight 25 and staple length 5. The lower point for staple length recognises that most competition fleeces are of good length and it allows sufficient margin to suitably penalise any overgrown or short fleeces.

The relatively high points allocated for fleece weight on most judging cards is a bone of contention with some coloured wool producers, who claim that actual fleece weight has low priority in the eyes of their woolcrafter clients. While this may be true, the fact remains that clean fleece weight is a measure of animal productivity. The high points allocated for this factor recognise the desirability of good breeding and feeding. Producing coloured wool which features crafter requirements and optimum fleece weight, reflects expertise in both sheep selection and management. Without fleece weight standards, the event could resort to merely a part fleece competition, in which only the very best portions of fleeces are displayed. This might be entirely satisfactory from a crafter's point of view, but it does not adequately demonstrate a standard achieved in coloured sheep breeding.

Fleece Preparation

It is appropriate to briefly review fleece preparation before consideration of some fleece characteristics used in judging. Where coloured wool is purported to be grown as a specialty raw material for handcrafts, the preparation of each fleece for either show or sale should be of a high standard.

For optimum preparation, the shorn fleece should be spread out with staple tips uppermost on a wool table for skirting. This means removing the following inferior portions, with the objective of upgrading the fleece as a whole: matted or short neck wool; heavy conditioned edge portions; seed and vegetable matter; inferior back wools; impurities such as brands, stains and dags. Coarser or hairy britch portions, if present, should also be removed as part of the objective of achieving evenness of fineness and staple length throughout the prepared fleece.

Ideally, the skirted fleece is folded lengthwise to a convenient width, then neatly rolled in a form which facilitates easy unrolling and unfolding for inspection. However, many exhibitors prefer to fold fleeces in a form which best displays the more attractive shoulder wool. Contrary to the custom in some countries, the tying of fleeces is not practised in New Zealand or Australia where it is regarded as being unnecessary and inconvenient.

Fleece Characteristics and Their Significance as Judging Factors

When judging cards are being used and where high points have been awarded for each factor, they should add up to or represent a clean, sound, attractive fleece of good weight, requiring minimal (if any) preparation before spinnng and with characteristics that are desirable in an appropriate end product. A similar result should be evident if the same fleece is judged without the use of points cards.

Most classes in coloured fleece competitions provide for wools of twelve months growth and this must be the general basis for allocation of points for *staple length.* Average staple length for a twelve months growth period will vary between sheep breeds and may also be associated with differences in average fibre fineness between those breeds. With handcraft requirements in mind, points should be deducted for those fleeces which are considered to be short and for those which are overgrown. Overlong fleeces are more difficult to handle and a greater degree of fibre breakage is likely to be incurred if they are carded prior to spinning. Short fleeces may also be inconvenient to handle and will result in a shorter average fibre length in the yarn. This is not conducive to optimum yarn strength.

Staple strength or *soundness* must always be considered in conjunction with staple length. In some cases, what may appear to be good apparent staple length may prove to be poor effective fibre length if the staple breaks when stretched. While tender wools can be put to good use, a high degree of soundness is required by most hand spinners. Tenderness or break in the staple are features indicative of a reduction in feed intake during the period of fibre growth. Sound fleeces can only be produced with good management involving an adequate feed supply throughout the year.

Staple formation varies between breeds and between young and mature sheep within breeds. The breed type within a particular class must be taken into account when assessing staple formation. In general, handcrafters prefer staples that are free, full handling, even throughout the fleece and without excessive tip in the case of the column or pencil shaped staples of the coarser woolled breeds. Fine wool breeds feature staples with a blockier shape and flat tip, different degrees of staple separation being associated with various breed types within this category.

Well defined, regularly spaced *crimp* certainly adds much to the aesthetic appeal of a fleece. It is regarded by handcrafters as an indication of potential ease of manipulation. A well defined crimp is usually associated with good staple separation and a minimum of cross fibres. A pleasing degree of elasticity and good draping qualities are also envisaged in end products made from well crimped wools. Crimp type varies widely between different sheep breeds. The degree of expression of the crimp pattern characteristic of a particular breed must be taken into account when allocating points for this factor.

Handle is a most important attribute when considering wool for handcrafts. The ultimate softness or harshness of handle of an end product will be closely related to the handle of the raw material from which it has been produced. Differences in handle can be found both within and between different sheep breeds. A close correlation exists between handle and average fibre diameter and in general, coarser wools cannot be expected to handle as softly as finer wools. Handle can be influenced by the amount of grease present in a fleece. Where some grease has been washed from the fleece in areas of high rainfall, a slightly harsher or drier handle may result. Merino fleeces contain a higher proportion of wool waxes than other breeds. The changes in fluidity of this wax when the fleece is standing in either very cold or very warm conditions can influence its handle. The presence of dust can also affect the handle of greasy wool.

Evenness of fineness, staple length and character are expected of the higher grade fleeces. Evenness is a factor which appears under various headings on a number of judging cards. It is a requirement which crafters associate with spinability and with ultimate consistency of the spun yarn. Fine wools are inherently more even in staple length and fineness than medium to strong wools.

Colour is, ironically, a topic of much debate when formulating judging standards with a points system for natural coloured fleece competitions. While many crafters prefer a fleece to be of similar colour throughout, there is no doubt that individual fleeces of variegated colour also have their devotees. Some shows accommodate this fact by providing special classes for variegated wools. Where such classes are not provided, a fleece featuring colour throughout might generally be preferred where all other factors are equal. Separate classes for fleeces from breeds such as the Jacob, where multi-colour is expected, should be provided in countries where these breeds are found. With the variety of preferences for various coloured wools and since each coloured fleece is a unique entity, it seems prudent *not* to include colour as a judging factor in a points card system for coloured wools. However, it must still be taken into account by a judge. He/she may discount points (under Faults) where a fleece lacks a good proportion of colour.

Inherited or acquired faults or impurities in the fleece, together with the presence of inferior fleece portions as a result of insufficient fleece preparation, can be conveniently considered under the general heading of *Freedom from Fault.* This will automatically accommodate the major handcrafter requirement of cleanliness, which is the first selection priority of many spinners. Noting the judge's observation of specific faults on the back of the points card (or the fleece identification card where points are not used), is recommended as being helpful to exhibitors.

Clean fleece weight is important to recognise, even though fleeces for handcrafts are more commonly sold on the basis of greasy weight. Clean fleece weight is related to greasy fleece weight and where score cards are used, the use of appropriate conversion charts is employed for subsequent allocation of points, taking into account the judge's assessment of average fibre diameter and yield. Conversion charts incorporate maximum fleece weights to ensure that "overgrown" fleeces of more than twelve months growth are not advantaged.

Where score cards are not used, the significance of fleece weight should still be incorporated in the overall assessment of each fleece. This can be achieved by establishing minimum weights for entry and by weighing and recording the greasy fleece weight of each entry. A judge can then take these weights into account when deciding final placings, together with his/her observations on fleece yields. Where other factors are equal, the higher yielding fleeces will receive higher placings.

The use of scales for weighing and the recording of greasy fleece weights should constitute an essential part of fleece competitions, regardless of whether or not points cards are used.

Summary and Conclusions

Broadly speaking, wool judging is an endeavor to rank shorn fleeces in order of merit, taking into account the variations which exist in individual fleece characteristics. It is essentially a balancing act, both within and between fleeces. Because of the subjectivity involved in estimation of the worth of various factors, it is the relative ranking of a fleece with the others in its class that is ultimately the most important thing.

The use of score cards to indicate the order of importance of the ranking is considered by many to be worthwhile. This is mainly because the cards are seen as fulfilling an educational requirement. It is contended, however, that this function, together with the personal and promotional aspects of fleece competitions, can still be realised with an alternative and more straightforward approach to fleece judging.

This would firstly involve judging without the use of score cards, resulting in more effective use of time normally spent in a points system. Secondly, the essential educational requirement would be met by establishing the more direct method of public comment on the overall and winning entries in each class as an integral part of a judge's duties. Provision of comment cards for display on award winning fleeces only, is proposed as a necessary adjunct.

These alternatives would have particular advantages at smaller, one day shows or fairs, where the use of score cards in judging a large wool entry can result in the exercise extending well into the afternoon, allowing little time for the public to view the final results.

At all times, the decisions of a judge on the day must be regarded as final. Not everyone will agree with a judge's decisions, but appointed times at which judges publicly state the reasons for their placings should not be taken as a forum for questions and answers. Most judges, however, are always willing to discuss their judging standards informally with individual exhibitors and other interested persons.

Judging is an important element in coloured wool education and improvement through fleece competitions. Methods by which laborious judging systems may be improved upon, while still retaining or improving their educational role, should be adopted. This will more efficiently and effectively contribute to the aims of the breed and show societies which organise and conduct these very worthwhile events.

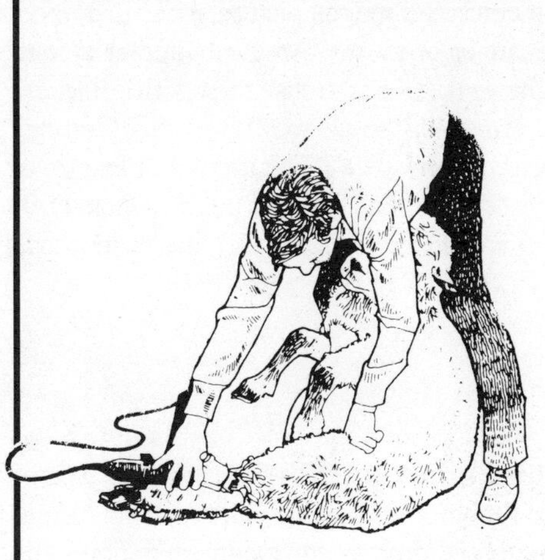

Procedures for Handling Natural Coloured Wools for Craft and Commercial Use in New Zealand

R. M. W. Sumner
Whatawhata Research Centre
Ministry of Agriculture and Fisheries
Private Bag, Hamilton, New Zealand

W. R. Regnault
Wool Department
Massey University
Palmerston North, New Zealand

J. H. Hutchinson
New Zealand Wool Board
P. O. Box 401
Cambridge, New Zealand

B. M. Tinnock
Wool Science Department
Lincoln College
Canterbury, New Zealand

New Zealand has a sheep population of approximately 65 million or 20 sheep per man, woman and child. Although the majority of these are white there are many flocks of black and coloured sheep. These are kept principally to produce wool for craft use, in small flocks of generally less than 100 sheep. The larger flocks of up to 1,000 coloured sheep are associated directly with commercial ventures. Although the exact population of coloured sheep in New Zealand is not known it has been estimated to be about 200,000 producing approximately 800 tonnes of greasy wool annually. White wool production is nearly 350,000 tonnes greasy annually.

With New Zealand being the principal exporter of coarser type wools to markets all over the world, sophisticated wool harvesting, handling and selling procedures have been developed. Against this background many of the procedures developed for white wools have been directly applied to coloured wools. This has, however, caused some problems for coloured wool producers where colour shading and individual fleeces are more important to craft people than large blended lines of wool with specified diameter, length and whiteness values. Because the coloured wool industry is small, with the producer frequently selling direct to the user, there are good lines of communication for the grower to meet the requirements of the user. In the last 10 years there has been an increase in the number of entrepreneurs involved with products containing coloured wool. This has resulted in aspects of bulk handling being introduced into the production and sale of coloured wool, as used for white wool.

Why Handle Wool

Shorn wool is regarded by the commercial processor as essentially valueless until converted to a useable product. For conversion to particular types of yarn, felt or other products it must possess particular attributes. Coloured wool, like white wool, differs over the body of the sheep, tending to be finer on the neck and shoulder than on the britch or hind leg, lower yielding and compressed on the belly and with possibly an area of weathered wool on the back. Wool from the lower body may also show yellow discolouration. As most handcrafters wish to spin yarn straight from the fleece with minimal blending, it is important that the wool be as uniform as possible. Each fleece must therefore receive the utmost individual attention in preparation to enhance its value as a specialty product for the discerning handcraft market. This is achieved most efficiently by sorting or skirting the fleece immediately following shearing to remove faults that would affect the subsequent hand manipulation of the wool. It is important to the grower to retain as much weight as possible in the resultant upgraded fleece, as skirtings are worth considerably less than fleece wool. Overall, the aim of wool growers should be to shear when the wool is most suitable for its potential end use and containing least fault, so that handling and preparation are also minimal.

Skirted coloured fleeces are readily traded by growers and craft outlets in New Zealand and overseas. While colour is a major factor in the appeal of a fleece to a potential buyer, the mystical "feel" or "handle" that is so hard to describe is critical in visualizing and utilizing the natural attributes of a fleece to its best advantage.

How We Describe Wool

As with any product, we need a language to describe wool in order to communicate between growers, traders and users. Craft wool, with all its subtle nuances, presents a real challenge. Until relatively recent times with the advent of objective means of measuring some of the attributes of wool, our descriptions were subjective and essentially craft based. Even today many of these characteristics cannot be measured objectively. The characteristics we consider of greatest importance in the New Zealand context are:

Yield

Defined as the percentage by weight of clean fibre in a fleece or parcel of wool. Since wool is traded by weight, yield is of great importance in pricing wool sold in the greasy state, as the grease is of no direct use to the processor. He is only interested in the amount of clean fibre present. Commercially sold wool is traded with a certificate indicating the yield as measured in a testing laboratory.

In the case of individual craft fleeces, "yield" is often looked at somewhat differently, in that while purchasers prefer a high yielding fleece as described above because it is easier to handle, they also think in terms that the fleece will yield two sweaters and a woolly hat.

Fineness

The single most important parameter of wool affecting the lightness and/or softness of the resultant product. Wool fibres are so fine that only recently have sophisticated laboratory methods been developed to measure wool effectively. Traditionally, however, fineness has been described on the basis of the predicted output of yarn that could be spun from a standard weight of fibre. Assessment was based on crimp frequency, lustre and softness. This learned skill was very subjective and open to dispute for trading purposes, often requiring arbitration.

In British countries the standard was based on the number of standard hanks (560 yards) of worsted yarn that could be spun from a pound of weight of prepared wool. Thus the finer the wool, the more yarn could be spun and the higher the "count" number. America adopted a similar system and we spoke of 46's and 48's for our Romney wool and 64's, 70's and 80's for our Merino wool. Other countries used different names for their grades so a 56's in New Zealand was a "cruza fina 2" in South America; a "croisee II" in Belgium and a "C2" in Germany. Now in New Zealand all commercially sold wool is measured objectively in millionths of a metre called microns (μm). Thus Romney wool maybe 37μm and Merino wool 20μm.

In the case of our coloured craft wools, the earlier system still prevails—it seems more descriptive. Advertising a 50's fleece, rather than a 33 μm fleece for sale, allows the purchaser to form a much better mental picture, as 33 μm wools can come in all manner of guises. An even simpler system seems to serve quite well. This is to state the relative fineness for the breed and age status, so we may have "fine Romney hoggett" or "medium Corriedale". The important feature of these subjective systems is that both parties to a conversation or transaction are at least roughly on "the same wavelength".

Style

A system indicating freedom from fault based on a complex blend of discolouration, vegetable matter, staple formation, tensile strength and dust or earth contamination. The New Zealand style grade system uses three different description systems: descriptive terms like "super, good, average" between traders and growers; a letter system used by warehouses and professional classers; and increasingly a numerical system for easy use with computers.

For coloured craft wools, Regnault (one of the authors) advocated a system with three grades.

Good— free grown, sound, free from vegetable fault, suitable for hand spinning direct from the fleece with only minimal combing out of tips.

Average— may have slight tensile weakness, poor staple formation, free or nearly free of vegetable matter, requires carding before spinning.

Poor— requires scouring and machine carding.

This was never taken up nor any alternative system devised for coloured wools outside the mainstream commercial field where they received very cavalier treatment. It seems that there is no need, as craft wools are satisfactorily traded without it, probably because most wool offered fits into the "good" category.

Tensile Strength

Many aspects of the sheep's life can affect the tensile strength of the wool produced. It can vary from absolutely "sound" to very "tender" or may have a "break"—a point of extreme weakness. Unsound wools have more fibre breakage during carding than sound wools and so produce a shorter fibre which is less desirable for spinners.

Craft people are generally very particular about soundness and commonly overdo the testing of fleeces they may consider using. A simple test with their usual method of preparation will indicate the extent of wasteage. Too many excellent wools are unnecessarily rejected on this score.

Bulk

A measure of the "filling power" of wool and related to a characteristic called "resistance to compression" in Australia. This is an exciting characteristic, long known to craft people and described by terms such as "springiness", "resilience", "bounce" and even "oomph"! All these terms are very subjective and self explanatory. This characteristic is now also being increasingly recognised by commercial users with the development of an objective test to measure specific volume under a defined pressure.

High bulk wools tend to have low lustre while low bulk wools tend to be lustrous with the bulkiness being due to the presence of a marked "fibre" crimp rather than a defined "staple" crimp. It is this attribute which gives bulky wool its filling power improving its insulating qualities and its shrink resistance as well as helping the garment to hold its shape with wear. Bulk contributes a feeling of luxuriance underfoot in carpets. This does not mean that high bulk is automatically good and low bulk bad. Rather it is a case of matching the wool to the end product.

Down breeds produce the highest bulk wools with measurements of over $30cm^3/g$ while Leicester-type sheep have low bulk wools below $20cm^3/g$. Other breeds fit in between. It is generally considered that the effects of using higher bulk wools in yarns start to show for wools above approximately $27cm^3/g$.

Where to With Description

In the mainstreaming of commercial wool trading we are firmly on the path to describe our wool objectively and completely in numerical terms. Increasingly, rectified specifications accompany the wool as it enters the mill where it is used to put together processing blends or to formulate dye liquors using computer technology. In the case of our coloured wools however, beauty is usually of more importance than technical precision, so that the romantic terminology may well be with us for a long time yet.

Pre-Shearing

Wool growers with large flocks of white sheep who sort their wool into saleable lines are required to initially sort their sheep into groups based on breed, age and wool length to assist wool handlers. In small flocks shorn by one shearer, or where all fleeces are bagged individually this is unnecessary. Regardless of the colour of wool, growers are advised to remove all faecal contaminated wool (dags) and to allow the yarded sheep to stand for two hours to "empty out" before being penned for shearing. This reduces faecal staining of the body wool which can affect subsequent dyeing performance.

Shearing

While commercial flocks in New Zealand are shorn in purpose-built sheds where the sheep can be held on a slatted grating floor to keep the wool clean, the same facilities are often not available on small blocks where coloured sheep tend to be farmed. Consequently coloured sheep may be shorn in multi-purpose sheds housing vehicles, workshops and general storage. In some cases the sheep are even shorn outside in yards. Care is taken to keep the wool clean.

Most sheep in New Zealand are shorn with a mechanical handpiece driven by an overhead electric motor through a downtube carrying the semi-flexible drive shaft. The shearer works on a wooden floor for safety, with the woolhandler or rouseabout working with the shearer as each sheep is shorn. The woolhandler keeps the wool clear of the shearer and does some preliminary sorting of the fleece. The board is then swept by the woolhandler while the shearer is catching and dragging the next sheep from the catching pen. Shearers can attain daily tallies of between 200 and 400 sheep each in a 9 hour day, shearing large commercial flocks. There are commonly three to four shearers per shed. The small size of most coloured sheep flocks enables them to be shorn by their owners or friends, where tallies are unimportant.

Fleeces from sheep with more than 8 months wool growth tend to hang together, enabling them to be thrown onto a slatted table for skirting. The fleece is picked up by holding the hind leg wool and drawing the fleece against the handler's body. It is then carried to the wool table and thrown whilst holding the hind leg wool. The fleece should end up laid out on the table, shorn side down, with clear access to all parts of the fleece. Lamb and short fleeces, which fall apart, must be fully sorted on the floor during shearing so that the wool types do not become mixed. Skirting and sorting of white wool into types is a skilled job with each type being sold separately. As a general rule in the case of coloured wools the belly wool is removed along with the short very greasy wool around the edge of the fleece and coarse or stained wool from the crutch area. Areas of bacterial discolouration, matting or cotting and vegetable matter contamination are also removed. Off sorts from coloured fleeces are collected together in a suitable container or woolpack for sale through commercial outlets. The use of a slatted table for skirting allows very short fibres to fall clear of the fleece onto the floor. This is equally important for both commercial and craft wools in minimizing pilling or rubbing in the finished product.

When fleeces are either sold separately or to be rehandled, it is normal to carefully roll them by turning in the sides toward the backbone, folding in half lengthwise and then rolling from britch to shoulder. The rolled fleece is not tied in any way, so there is no knotting or contamination of the wool.

Skirting short greasy edge pieces from fleece on slatted wool table.

Post-Shearing

Several methods of selling coloured wool are available to growers in New Zealand, either as individual fleeces or bulk lots. Individual fleeces sold privately do not incur a levy payment to the New Zealand Wool Board. Commercial coloured wools are levied at the point of sale. This levy supports the activities of the Wool Board including price protection at auction, wool research, and grower, shearer and woolhandler education.

Individual Fleeces

Visually appealing fleeces suitably prepared for craft use are generally sold "in the grease", packaged in plastic, paper bags or cardboard boxes. The fleeces may be sold on the farm where they are grown, at meetings of craft people, at field days, in shops, by postal order within New Zealand or exported. These fleeces are generally sold on a whole fleece basis with the price calculated from a per kg value set by the grower, which tends to be higher than commercial lines of equivalent-type white wool. Some people prefer to buy "a jerseys worth" rather than a whole fleece. However, because of difficulties in matching up the left overs, growers are generally reluctant to split fleeces. Less visually appealing fleeces may be carded commercially and sold as carded sliver.

Bulk Lots

Bulk lots may be sorted on fineness, length and colour, for sale to entrepreneurs on contract or prepared to the requirements of a particular business for a spot sale. Returns from a contract sale are likely to be higher than those from a spot sale where the back-stop is the price of coloured wool at auction. These wools are used to produce a range of commercial products from carded sliver, yarns and cloths to rugs and clothing.

Smaller lots, mixed lots, off-sorts and oddments are often sold to either a wool broker or private wool merchant whose principal business is trading white wool. These firms may sort or blend the coloured wool before selling it, either at auction or by negotiated sale, to a manufacturer or exporter. Growers are charged approximately 22c/kg for sorting or binning by wool brokers.

Valuing a sample of a commercial lot prior to auction. Test result certificate displayed behind sample.

Coloured wool sold at auction is packaged similar to white wool in jute or polypropylene packs. The bales measure 1250 x 660 x 660 mm and are pressed to a weight of up to 200kg. A sub-sample is taken prior to auction and displayed in boxes for valuation by buyers. Each lot carries a test result certificate following objective measurement of yield (clean wool content) and mean fibre diameter. Auction prices for coloured wool are generally below those for equivalent type white wools.

Potential fibre contamination between pigmented and white wool is of particular concern to the New Zealand wool trade. Consequently considerable care is taken in brokers and buyers stores, with the display boxes and in scours and mills to minimize potential contamination of white wool with pigmented wool.

Conclusion

Black and coloured sheep breeders in New Zealand are proud of their coloured wool, with many putting considerable effort into preparing it for sale to craft people. The commercial trade does not require such stringent preparation for coloured wool, where it may be sold by negotiation following binning, or at auction with objective measurements for yield and fineness.

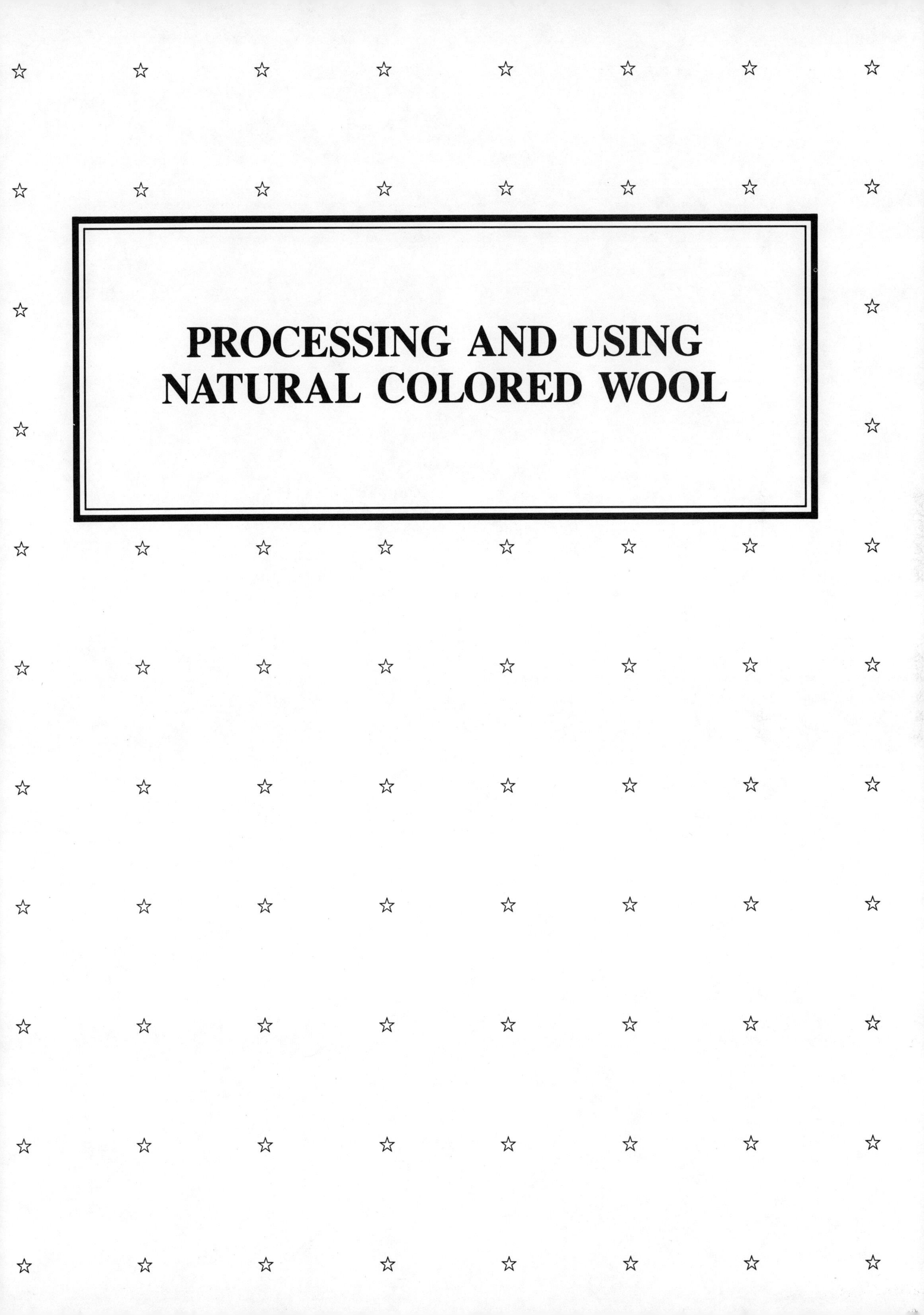

PROCESSING AND USING NATURAL COLORED WOOL

Wool Scouring Technology for Small Scale Operations

Robert Donnelly
P. O. Box 327
Arroyo Hondo, NM 87513 USA

The scouring (washing) of raw (greasy) wool is an important and potentially difficult first step in the converting of fleece wool into a finished product. For the small scale, local resource wool processing industry in the United States, scouring has been limited to washing wool in basins, bath tubs, or small washing machines normally used to clean clothes.

Experiments conducted at Utah State University in Logan, Utah[1] have resulted in wool test results comparable to wools washed in large-scale commercial scouring equipment. This paper presents a comparison of USU's "Aeration Agitation" system with standard industrial scouring technologies, and gives application suggestions for the small wool processing mill. (The term "mill" meaning any facility which processes wool for commercial interests, e.g., to make money.) Major technical departures used in the USU study were:

1) use of air bubbles to agitate the wool in the washing and rinsing bowls
2) extracting water from the wool by centrifuging.

Today (1989), there is at least one new small scale installation under evaluation using the USU technologies.

Fleece Contaminants

"All the fleece is *not* wool!"

First, let's find out what we must get rid of in the scouring process. The non-wool components of a fleece, called contaminants, found in varying percentages and relationships are:

1. *Suint,* which is the "sweat" from the animal, deposits on the wool as water soluble salts, mostly potassium. These salts have a profound effect on the pH of the scouring solution. For example, "the pH of the water extract of greasy wools shows considerable variability, being usually slightly acid or near neutral for Merino, and alkaline for coarse crossbred wools."[2] I might add that to my knowledge, there have been no specific studies to determine the pH of colored wools.

2. *Grease,* or wool-fat which is secreted by special glands in the sheeps' skin, contains a percentage of lanolin. This can be regarded as a mixture of two fractions on the fiber, oxidized and unoxidized. "The unoxidized fraction is easier to remove than the oxidized."[2]

3. *Dirt,* in the form of soil, sand and dust which enters the fleece in various ways and is held onto the wool fiber by the grease. The removal of dirt in the scouring process is dependent on the melting of the grease, solubility of the protein contaminant layer, openness of the fiber tufts, the agitation the fibers receive and the efficiency of removal of the water from the wool.

4. *Other contaminants,* such as hay, burrs, polypropylene twine, paint, etc., are the result of management practices. These must be removed in post-scouring processes, e.g.: carbonizing or combing. They are not soluble in water and are entrapped or are similar in physical nature to wool fibers so as not to "fall out" during the scouring process. Removal of these types of contaminants could be the topic for another paper. The simplest procedure for controlling these contaminants is to not allow them in the fleece at all!

Comparison of "Classical" Scouring and USU Technology

"Scouring is a process carried out on the individual fiber."

Simply stated, the scouring process cleans the individual fiber. The purpose of scouring is to separate and remove the nonfibrous part of the fleece from the fibrous. In commercial scouring operations this is conducted on billions of fibers simultaneously in machinery of large capacities (1,400 kg./hr.). Years ago, these machines were so huge they were called "Leviathans." Modern systems, such as the Wronz Mini-Bowls are smaller than their predecessors, but could be 1.5 meters wide, and up to 40 meters long overall. The USU "Aeration Agitation" scouring system is approximately 0.77 meters wide and 4.5 meters long, not including ancillary equipment. It is a series of standard sized stainless steel dairy sinks!

For the majority of the wool processing industry throughout the world, wool is scoured using hot water containing a soap or detergent and alkali (usually sodium carbonate) when needed, in a series of wash tanks or "bowls." These bowls are equipped with moving "rakes" which enter the wool/water slurry to open the fiber mass, allowing the exit of dirt particles. These rakes also help move the wool through the bowls. At the exit end of each bowl are high pressure squeeze rolls which separate the water from the wool.

The USU scouring system replaces the rakes with air bubbles which accomplish all the purposes except moving the wool through the bowls. It is not necessary for the wool to move from one end of the bowl to the other in the USU system. This is because of the substitution of the squeeze rolls by a centrifuge for separating the water from the wool.

Opening and Blending

"The purpose of opening wool is to improve scouring efficiency."

The first step is opening and blending the wool fleeces. Wool is blended to minimize variations in the finished product caused by the natural differences in wool (within a fleece and between animals). Blending has no significant impact on the scouring process, however the machinery which performs the blending also "opens" the wool in preparation for scouring.

Openers are made in many styles. An opener is a large (90 cm. diameter x 1 meter wide), heavy (100 kg. or more), cylinder fitted with teeth approximately 10 cm. long, spaced 8–10 cm. apart in a staggered pattern over its outside surface. The mass of this main cylinder is important because opening is a very energy intensive process and it is extremely helpful to have a high amount of kinetic energy available.

Wool is presented to the opening cylinder by two (or more) feed rolls, about 10 cm. diameter and equal in width to the main cylinder. These rolls are covered with coarse (about 1 point per sq. cm.) metallic wire for gripping the wool. An apron made of wooden slats attached to a canvas belt, and moving at the same surface speed, carries the unopened wool to the feed rolls.

After the wool passes through the feed rolls, it encounters the tips of the teeth on the opening cylinder. The gap between the teeth of this cylinder and those of the feed rolls is about 1–2 cm. The cylinder surface speed is 200–500 times the surface speed of the feed rolls. The teeth on the main cylinder dig into the wool (this is a relative term), opening the large tufts into smaller "tuftlets," or individual fibers under ideal conditions.

The opening action takes place within an enclosure which has an aperture opposite the feed rolls. Within this enclosure are grid bars or screens which help clean the wool as it is carried around by the teeth. Centrifugal force and air movement generated by the opening cylinder fling the opened wool out the aperture into some appropriate storage container to await the next step.

So the objective of opening the wool prior to scouring is to improve the efficiency of the process. This is realized by the separation of the fiber tufts which have existed in the fleece for a whole growing season. Tufts are created by the grease which coats each fiber and acts like an adhesive, sticking them together. Grease can retard penetration of the scouring solution and prevent scouring of the wools in the center of the tufts. Remember that scouring is carried out on the individual fibers. By allowing each fiber access to the scouring solution, you can realize the best scouring efficiency.

Another effect of opening is to remove (to varying degrees) some dirt and some grease. Sometimes other contaminants like vegetable matter are removed, but opening cannot be relied upon to totally clean the wool. Likewise the scouring process cannot be expected to remove fibrous-like contaminants.

The USU study was made using wools that had been opened at a commercial wool scouring plant. No variables were introduced by using a different opening method.

Washing Wool with Hot Water

"Get the grease off and keep it off."

Some systems use what is called a "Suint Bowl." This is merely a cold water steeping of the wool prior to immersion in the first hot scouring bowl. The suint bowl softens the contaminants and dissolves the salts on the wool to begin the scouring. There has been, and I suppose will continue to be, a controversy regarding the efficacy of the suint bowl. The USU study used a suint bowl as one of five scouring and rinsing bowls to simulate the commercial scouring plant process.

Hot water is necessary to melt the grease on the wool. Generally this melt down occurs at about 40 degrees Celsius (105 degrees F.). Scouring is carried out above this temperature to speed up the swelling of the soil complex or protein contaminant layer as well as to melt the grease, so the water temperature of the scouring bowls is kept at about 55–60 degrees C. (131–140 degrees F.) as a safety factor.

Heating can be accomplished in many ways; direct fire under the bowls, heat exchangers, steam injection, hot oil immersion coils, etc. Bear in mind that commercial sized bowls are filled with hundreds of liters of water, so there is a huge energy input required to heat it. The scouring bowls in the USU study were heated with electric immersion heaters to duplicate the temperatures used in the commercial scouring train. Temperatures in the USU study were: Bowl No. 1 (suint bowl), room tempreature, approximately 25°C; Bowls No. 2 and 3 (wash bowls), 62°C.; Bowl No. 4 (first rinse) 49°C.; Bowl No. 5 (final rinse) 42°C.

The other important ingredient in aqueous scouring is the addition of a detergent or surfactant. There are two major reasons for the addition of surfactant; one is to penetrate the contaminated layer, and the other is to prevent the re-deposition of grease and dirt on the fiber surface. The detergent used in the USU study was Tergitol, NP-9, a product of Union Carbide at 0.033% solution.

It is important to note that the scouring should be carried out at pH levels on the basic side. This promotes the self-detergency of wools by converting the potassium salts (from the suint) to soaps. It also explains why Merino wools are more difficult to scour. (Remember the pH of the water extract of Merino wools is usually neutral or acidic.) Crossbred wools are usually scoured without added alkali. No alkali was used or control of liquor pH was made in the USU experiment. The wools were all crossbred types.

Wool Movement within the Bowls

"Perfect movement removes dirt but doesn't felt wool."

In commercial scouring, with few exceptions, the wool is moved by two media. First, it is moved smoothly through the bowl by the flow of the water itself. Water usually enters the scouring bowl at the same end as the wool does, and exits at the opposite end of the bowl, just like the wool. This is known as the "flow-through" phenomenon. There was no water flow-through used in the USU study.

The other movement medium is the rake or harrow, which is a mechanical device having vertically arranged tines which penetrate the wool/water slurry vertically from the top, then move forward, while still in the wool, toward the exit side of the bowl. This forward movement is variable within the constraints of the equipment, and is usually about twenty-five centimeters.

Once the rake has completed its forward movement, it raises vertically out of the wool and proceeds rearward to its original position to begin the cycle again. This complicated movement is carried out by a series of gears, levers and compound motion joints.

In addition to moving the wool through the bowl, there is another benefit of this mechanical raking motion. That is the penetration or "opening" of the wool to allow dirt particles to fall out toward the bottom of the bowl. Without this action the dirt particles would remain entrapped within the wool slurry and be carried through the squeeze rollers staining the wool under pressure. The dirt thus remaining in the wool will produce manufacturing difficulties such as a discolored finished product, unpredictable processing, and unhappy customers.

Movement of the wool fibers relative to each other (this could be called opening) is vital to maximize the efficiency of the cleaning process. But movement of wool fibers relative to each other in hot soapy water can cause wool to felt together. (For those of you who don't understand the felting phenomenon, let me say it is complex, dynamic, and generally unpredictable.) It is well known that felting can be minimized if not prevented in the scouring process by taking all the above described actions very slowly and gently. What exists is a "trade-off" situation, where you want the maximum cleaning (allowing the dirt particles to drop out) with the minimum of felting. Wool scouring managers dedicate their careers to perfecting this knowledge.

This action of moving the wool fibers relative to each other seems to have been accomplished very well using air bubbles. Air pressure and volume changes in the air supply had little effect on the scouring efficiency, within the parameters of the USU study. No felting was apparent in any of the tests.

The theory is that the air bubbles separate the fibers so that the intimate surface to surface contact which results in felting is never allowed. It is conceivable that the entrainment of the air bubbles is so complete that each fiber is surrounded by air and is totally isolated from its neighboring fibers. So far it has been impossible to record this phenomenon in any way to show what is actually going on. We only know that in the USU study it was impossible to felt any of the wools with aeration agitation.

Separating the Wool from the Scouring Medium

"Here is where the scouring really happens."

At the exit of commercial scouring (and rinse) bowls lurk the dreaded "squeeze rolls." Aside from their apparent danger to human hands, they are subject to a wide variety of potential problems which have significant effects on the quality of the scoured wool.

Some examples of the problems: they can get slick from the wool grease; they can get wrapped with the wool; pressures vary with time, temperature, and with the load of wool passing through. Also, if the slurry of wool passing through the bowls is not even, it will lump at the squeezing zone (nip) causing fiber damage in the densest part of the lump, and preventing sufficient "squeeze" at the thinner parts.

However, despite all that has just been said regarding their shortcomings, squeeze rolls are the best method of removing water (the scouring medium) from wool in large scale, continuous wool scouring operations. Squeeze rolls have evolved over the years with special elastomeric coverings, pressure equalizers, and uniform fiber feeding systems, and are considered "state-of-the-art" in wool scouring operations around the world.

The USU aeration agitation scouring system employs a centrifuge for removal of the scouring medium (water) from the wool. "More water can usually be removed from loose stock (wool) by centrifuging than by squeeze rollers, and continuous centrifuges have been investigated or used in place of the squeeze press (rolls) from time to time. This is an area where further machinery development may yet take place."[2]

Another advantage of centrifuging is the relatively lower capital cost of a used machine compared with building or adapting moving rakes and squeeze rolls onto tanks, tubs, etc. In the USU scouring system, the movement of wool through the system is discontinuous, and the process is labor intensive as a result. However, it will always be the case for small scale wool washing operations that a higher per pound labor input will be required than large scale industrial systems. On the positive side, this presents an opportunity for improved quality control.

Drying the Scoured Wool

"The final step!"

No attempt was made to simulate the drying conditions of the commercial wool scourer in the USU test. All the test batches were dried at room temperature in the USU wool laboratory.

Much can be learned from the commercial wool scouring industry regarding drying scoured wool. This information can be applied to the small scale wool scouring processors.

First, you will want to reduce the water content of the wool out of the last rinse bowl. This minimizes the amount of water that must be removed by the drying process. As stated earlier, centrifuging is superior to squeezing for removal of water. Also, studies have shown that higher temperatures in the final rinse reduce the amount of heat required to dry wool by allowing more water to be expressed from the wool. "The major reason was probably the reduced viscosity of the water at higher temperatures."[2]

Next, uniformity of feed into the dryer is important. Regardless of whether your drying system is continuous or intermittent, you will want as uniform and open a batt of wool as possible. It is very helpful to open wet wool (gently) by means of a feed hopper. These are available on the used machinery market, or they can be made from parts available from suppliers. They are made with a bottom feed belt in a hopper which carries the wool forward to a vertical spiked apron which lifts the wool past a metering beater, then over the top to another beater which flicks the wool off the spiked apron and onto the drying screen. This could also be done by hand for small quantities.

Effluent Disposal

"A hidden asset!"

Another paper could (and should) be written about this topic, but I want to just touch on one point that may be helpful to those of you with a specific problem. If you are in a rural location, consider using the scouring effluent as irrigation water instead of dumping into your septic system.

Campbell et al.[3] have described long-term monitoring of pastures irrigated with effluent from a commercial woolscour in New Zealand. They concluded that land irrigation of a freely draining soil type affords a satisfactory means of disposing of woolscour wastes. I suggest you get a copy of their paper and study it if you intend to take this approach. In any case, you'll need to check with the local water quality officials about using this nutrient rich by-product for irrigation.

Review of Local Resource Small Scale Wool Scouring

"An industry capabilities survey."

In October of 1988 I conducted a survey of the small scale wool processors in the United States that I knew about. I sent a letter to each of them requesting their assistance by answering several simple questions and sending me some pictures.

A total of 45 questionnaires were sent out. Of these, 8 are not in business any longer: that's 18%. There were 16 (36%) who did not respond for reasons known only to themselves. No photographs or slides were submitted by anyone.

Of the 21 (47%) who did respond, the results are very interesting. Three respondents I would classify as large mills gave information, but I have not included their figures in this summary. Two respondents gave no figures, and the rest averaged nearly 1,900 kilograms (4,200 lbs.) scoured per year. The lowest was 103 kgs. (250 lbs.). The highest was 7,250 kgs. (16,000 lbs.)

There was a wide variety of techniques reported; concrete laundry tubs, old bath tubs, and modified washing machines were most popular. In conclusion I can say that there is no standard technique for wool scouring in use by small scale wool processors in the United States. Every respondent was interested in learning about the new technique we've been reporting here, although there were some qualifying statements about cost and chemicals.

Generally, reported immersion times in scouring solutions were far longer than necessary. Provided the temperature of the scouring bath is at or above the critical temperatures discussed in "Washing Wool with Hot Water," immersion times need only be approximately ten minutes. Some processors used Soda Ash on everything they washed, and I'd guess that few (if any) checked the pH of their scouring solutions.

Setting Up or Modifiying a Small Scale Wool Scouring System

"The heart of the matter."

We are now ready to improve (where possible) the techniques of raw wool scouring. If you are scouring already, these suggestions could help. If you are planning to get into scouring, this could save you a lot of time and trouble.

First on the list of improvements is an opener. I neglected to ask if opening was done on my questionnaire, and it was not generally reported in the process descriptions. So I can assume that opening is *not* done by most of the respondents. I do know of one of the processors who just installed a Clark mixing picker (which is an opener), and have experienced a marked improvement in their scoured product. So buy one, or if you're really clever, build an opener. It will greatly improve your scouring results.

Next you'll need to get an air compressor that will deliver about 0.3 cu. mtrs./min. (10 scfm), with a storage tank of 110 liters (30 gal.) capacity. Also get a pressure regulator for controlling operating air pressures up to 7 kg./cm. (100 psi).

You'll need to construct piping that will sit in the bottom of your bowls (tubs). It's a good idea to mount a screen over these pipes to prevent wool from snagging on them. The pipes should have holes drilled in the top side so that air can exit. The smaller the hole, the smaller the bubbles generated. At this time there is not enough data on hand to specify the optimum hole size or population. The USU study used 1.6 mm (1/16 inch) holes on 2.54 cm. (1 inch) centers in 6.35 mm (1/4 inch) iron pipe. It is quite possible that considerably smaller holes (less than 0.5 mm) might do quite well. Possibly a porous material like the plastic blocks you see in fish tanks would work well, provided it could withstand the temperatures.

The method used for heating your scouring liquor is optional. The USU study used electric immersion heaters, which worked quite satisfactorily. I've seen propane burners used, and I wouldn't have been surprised if someone reported a wood fired system. The main thing to remember is you're going to have to use *hot water* for best results.

Wool handling, from bowl to centrifuge, etc., can be done in a number of ways depending on your particular set of conditions. One method would be to construct bags to hold the wool. They should be made in dimensions which fit your tubs inside. The material for constructing the bags should be a very strong and open mesh, like that used for commercial laundry bags. You will probably want to devise some mechanical method of lifting the wool out of the bowls, because *wet wool is heavy!*

A commercial laundry centrifuge is the best device for extracting water. A home washer doesn't spin fast enough so the wool retains too much water. Reduced scouring efficiency and high drier loading are the results of inefficient water removal.

For drying the wool, remember several things: (1) get as much water out of the wool after the final rinse as you can; (2) the wool does *not* have to be "bone-dry," but should have about 16% moisture in it; (3) move as much hot air as you can, with fans, through the wool batt to increase drying capacity; (4) "fluff" the wool as it's drying so that the outside becomes the inside of the batt.

A simple pH tester (or papers) will be valuable to know the pH of your scouring liquor, especially if you do different types of wool. Develop a consistent, efficient procedure so that your product is the highest quality.

Unfortunately, testing the scoured wool for residual grease, dirt, staple length, whiteness, etc., is time consuming and costly. The best test still is a satisfied customer.

References

[1]Aeration as an Agitating Mechanism in Raw Wool Scouring, Dr. Lyle G. McNeal, Ph.D., Utah State University. Logan, UT.

[2]Wool Scouring and Allied Technology, R. G. Stewart, 1983 Wool Research Organisation of New Zealand.

[3]Some Effects on Soil Properties, Herbage Composition, and Blood Chemistry of Sheep, Arising from the Irrigation of Pastures with woolscour Effluent, A. S. Campbell, K. M. McKenzie, N. Z. Journal of Agricultural Research 23 (1980).

The Green Mountain Spinnery as a Small Scale Producer—from Fleece to Yarn

David Ritchie
Green Mountain Spinnery
Box 568
Putney, Vermont 05346 USA

Green Mountain Spinnery in Putney, Vermont is made up of a diverse group of workers who are spinning wool and mohair into yarn and at the same time are trying to weave a business that is both part of Vermont's agricultural fabric and a financial success. In 1988, the Spinnery—housed in a former gas station—produced about 23,000 lbs. of wool and Mountain Mohair in 45 colors and a variety of weights. During the period from 1984 to 1986 our production doubled. We employ about 14 full and part-time people and sell our products—yarn, Vermont Designer Patterns and Knit-Kits—to nearly 1300 shops nationwide and to about 3000 mail order customers.

Meet the people behind our beautiful Green Mountain Spinnery yarns.

In the late winter of 1982, just a few months after we began producing yarn, Mary Guild came to us with the wool from her small flock of Border Leicester and asked if we would spin it for her. Mary and her family live about half an hour away. She designs and hand-knits sweaters to sell at fairs, and takes custom orders. This was a big moment for us. All of the yarn we had made so far was from fleece we had purchased from New England sheep breeders—yarn for us to market on our own. We were still novices. Some days the machines did what we wanted and some days they didn't, and when they didn't there wasn't always someone who knew what to do about it! But spinning yarn for sheepbreeders out of their own wool was one of our reasons for starting the Spinnery, and we had to begin somewhere. Mary was willing to be our first custom customer. She had some white wool and some colored wool, and we made a bulky yarn for her. It worked. She's been back for more almost every year.

By now we've made hundreds of lots for individuals who want yarn from their own sheep. Custom processing has grown to be about 38% of our production. We're very proud that yarn we have spun has won prizes at some of the large sheep festivals, e.g. "best of show" at the Maryland festival a few years ago. Our small scouring facility, using mostly locally available, recycled equipment adapted to our needs, makes it possible to produce a batch of yarn from as little as 100 lbs. of unscoured materials.

Our custom spinning has been a source of challenges and growth for us, not only because growers from all over the country are requesting our services but also because they want us to use a variety of fibers, and often request yarn with specifications different from our Green Mountain Spinnery yarn. Here are a few examples:

Liz Hall and her son, Sandy, began raising Romneys about the same time we began the Spinnery. They attribute the fine quality of their wool to the special care they give the sheep, "enabling (the sheep) to concentrate their energies on growing long, lustrous fleeces." These fibers behave differently in our machines than do the blend of Dorset, Corriedale, and other fleeces in our GMS yarns. We've had to learn to adjust the machinery to work successfully with their wool. (We are fortunate to have Ray Phillips on our staff, as he has worked with machinery such as ours for about 40 years. He trains our less experienced workers.) We spin worsted weight and bulky yarns for the Halls. Liz does hand-dyeing—"a rainbow of colors", and their farm shop, she says, is "filled with the harvest from our sheep. Fiber for knitting and weaving, wonderful handknit sweaters, hand woven apparel and handsome, sturdy wool rag rugs." We now process about 1000 lbs of yarn annually for the Halls.

Our Mountain Mohair Yarn.

Seddon Wylde is a designer and producer of elegant clothing from handloomed fabrics. She came to us knowing the exact specifications of the yarns she wanted—how many yards per pound, how many twists per inch in the singles yarn, and how many twists per inch in the plied yarn. Her requirements are for yarns finer than our GMS yarns. She provides us with small amounts of brilliant dyed fleeces to blend with her colored wools, and the results are rich, beautiful tones. Wylde Weaves' products are marketed through craft fairs and by mail order.

Linda MacMillan turned her back on city life and came to Putney to raise the animals she loves—sheep, angora goats, and angora rabbits. Her shop on the main street of the village features handspun yarns, augmented by Spinnery-spun blends of Merino and angora. This was another first for us. We succeeded, using 18% angora. The dark Merino had enough "grab" to hold on to the angora as it travelled through the machines, and there was enough angora to give a very special texture to the finished yarn. Linda also does beautiful hand-dyeing, and has a mail order catalog.

Now in 1989 we have over 250 customers sending us their fleece from all over the country. In many ways the custom processing is the most exciting part of our business because it puts us in touch with so many breeders who want to creatively market their fiber products. There is much still to be learned. We want to be able to process both a wider variety and higher percentage of precious fibers. We therefore are continually researching ways to adapt our machinery to a variety of fiber characteristics.

Card producing mohair blended with wool.

How We Began

Four of us are founders of the Spinnery. Libby Mills and Claire Wilson are weavers who were inspired by the beautiful yarns brought back by a friend from a tiny mill in Sweden. David Ritchie, Diana Wahle and Claire were part of a study group examining how what we do locally relates to global issues. We were impressed by the reasoning behind E. F. Schumacher's "Small Is Beautiful," and the advocacy of intermediate technology. A study of Vermont indicated that our home state was similar to a Third World Country: out-of-state ownership of land and businesses cause profits to leave this region; many products purchased in the area are produced elsewhere; raw materials from the state are shipped long distances for processing. Based on our own interests, concerns, and abilities, what could we do that would make sense? Many of our attributes combined to get us off to a positive start. Claire, as a journeyman weaver, knew about finished yarns and the use of color, as did Libby, a dean at Putney School and originator of the weaving program there. Diana had marketing experience as an administrator for the Community College of Vermont, and David felt confident that he could work effectively in a business setting because of a degree in economics and because his work in education and social service had involved listening and responding to people's needs.

None of us had been involved in running a business; none of us had worked with heavy machinery. Even so, the idea of a small scale spinnery was very appealing to us. We knew there were sheep on the hills of New England in increasing numbers; that the winters are cold and dressing warmly is high priority; that people here like to buy locally produced goods when possible.

A statement of purpose was developed which became the mainstay of our business teamwork. We developed it as part of our original business plan and included it in our loan applications. It created guidelines for the types of products we wanted to promote, parameters for the effect we wanted to have on our environment and local economy, and our aims for a better workplace. Periodically we update it.

STATEMENT OF PURPOSE

The purpose of the Green Mountain Spinnery can be summarized in the four following areas:

1) To produce and sell the finest American wool yarn from New England sheep for use by handweavers and knitters. We want to process a wide range of natural fibers which are locally available and to enhance them with outstanding colors. We also would like to produce lanolin.
2) To scale our production (approximately 600 lbs. per week) in ways that
   —minimize adverse effects on the environment
   —minimize the need for shipping raw materials and finished yarn over long distances.
3) To work toward self-reliance in our geographical region by
   —providing specialty yarns to area crafts people
   —providing a service to sheepbreeders not otherwise available, through custom processing and fleece acquisition, and thus encouraging diversified farming.
   —using regionally available raw materials to make a product needed in the region.
   —being a model small business geared to meet a regional need; encouraging others in similar ventures.
   —working toward using a locally available renewable energy resource to heat our building, heat our water, and power our a machinery. Converting to that source as soon as possible.
4) To create a vital work environment where workers are challenged to make use of their skills. To evolve into a worker-owned business.

We began visiting mills, surveying yarn shop owners, talking with sheep breeders, and looking at old machinery. We visited our potential competitors and, where possible, toured their mills. A neighboring mill, Harrisville Design, even invited us to observe their production for a week. We learned that the Vermont Industrial Development Authority (VIDA) sometimes gave loans at very low interest rates to new

The Spinning Frames.

businesses, and that our using a local agricultural product for manufacturing might be of particular interest. Through her work at Community College of Vermont, Diana found students at Dartmouth's Tuck School of business who were excited about our plans and willing to help us develop a complicated loan application. We bid on a recently abandoned gas station right at the exit of the interstate highway in Putney; one of us was able to make the downpayment, and the bank provided a commerial loan for the balance. The approved VIDA loan provided means for expanding the building to accomodate the used machinery we were purchasing from New England mills (which in turn were acquiring the latest in electronically controlled machinery from abroad.) About 30 of our personal friends came forward with loans for machinery, for moving machinery, for purchasing raw materials, and for the many predictable and unpredictable expenses for getting started. These loans were also for low interest rates and extended pay-back periods. This very generous local support was a tremendous boost to our morale. Another loan came from the Coop Fund of New England, which hitherto had funded only food coops.

The Mondragon system of worker-owned coops in the Basque area of Spain was an inspiration for us in planning how we wanted the business to be owned and managed. We learned about this approach to worker ownership through our membership in the Industrial Cooperative Association located in Somerville, Massachusetts. With the guidance of the I.C.A., we organized our company to be owned and operated by its workers. It was our intention to follow our statement of purpose and aim for shared control, and also shared responsibility.

It didn't take long for us to realize that not everyone on our team would feel ready to take on an equal role in ownership, particularly knowing what risks we had taken in capital investment and long term commitment. Most of our workers were committed to other priorities in their lives besides the Spinnery—sheepbreeding, farming, professional writing are but a few examples—and could not take on further responsibilities beyond their work hours at the mill. Thus we modified our original intent and established a corporation with the original founders as co-owners. At the same time during these initial years, we have sought to involve all workers in decision making as much as possible. We accomplish this through our weekly production meetings, special problem-solving groups, and now through monthly all-staff potlucks. This form of operation is not everyone's style, but for us has resulted in a challenging teamwork spirit at the Spinnery.

Our backgrounds in teaching and human services left gaps in technical and business skills. We borrowed skilled help in evaluating used machinery, and for putting the machines in operation once they were installed. For the first six months of operation we "scrounged" technical help by apprenticing in other mills and hiring skilled workers on a very intermittent basis. When Ray was able to begin a four-day week for us in 1982 we turned a big corner! It is still difficult to hire full time qualified workers, but we are fortunate in having dedicated staff who have been trained on the job. Two of them are sheepbreeders themselves.

Janet Lum working at the skeining machine.

Making the Product

According to the Wool Bureau, the Green Mountain Spinnery is the country's smallest woolen mill. Modelled after spinning mills in Sweden, Ireland and Wales, we process only New England grown wool and mohair for our own yarn production.

As at larger spinneries, the first step in the spinning process is evaluating the fleece for staple length, grade, and cleanliness. In selecting the best New England fleece for our own GMS yarn, David selects fine fleece used mostly in our Mountain Mohair yarns (Rambouillet and Targee) and medium grade fleece in our 100% wool yarns (Romney, Corriedale, Dorset and others).

Scouring, or washing the fleece comes next. The fleece is steeped in successive baths of hot, soapy water and a rinse to remove dirt and excess grease. A water softener is used. After the last bath, the fleece is spun in a centrifuge and then tumble-dried in an industrial laundry dryer. We scour virtually all of our custom work and mohair. Because of limitations of space and water, wool for most of the spinnery's own yarn is sent to a company in Adamstown, Pennsylvania. Dyeing of the fleece then takes place in Woonsocket, Rhode Island.

Here in Putney, the scoured and dyed material is put through a picking machine in preparation for carding. The Spinnery's Davis and Furber card untangles and blends the fibers into a "web," which is then divided into separate strands of roving in preparation for spinning. In contrast to larger mills, we have two instead of three or four main cylinders on the card, allowing for the homespun look of our final product.

This building houses our manufacturing plant and retail shop.

The process of spinning is the drawing out or "drafting" and twisting of fibers. Spools of roving are brought from the card condensor and placed on the spinning frame (a Whitin Model E). The ends of the roving are threaded singly, in pairs or triples depending on the kind of yarn to be spun. The yarn is wound onto bobbins as it is spun. If the yarn is to be plied, another machine receives the bobbins to complete that process.

While still on the bobbins, the yarn is steamed to set the twist and reduce kinks. From there the yarn is either transferred onto cones for weavers and machine knitters or wound into skeins by a skein winder, a machine that can wind a dozen skeins at a time to uniform measurements. Our Mountain Mohair skeins are washed and hung to dry before they are labelled. The skeins are then tied and twisted by hand before they are labelled and ready for sale.

Challenges for the Future

In reviewing our accomplishments over the last eight years, we realize how much we have achieved. We have used our energy and convictions to produce a beautiful product that's indigenous to Vermont and enables more sheepbreeders to have yarn from their own flocks. As we look ahead and develop our projections, our greatest challenges will be in the following areas:

1. *Space:* we need to expand our manufacturing, administrative and retail space to meet our increasing orders.

2. *Pricing:* how to pay a fair price for the fleece and fibers we use, pay ourselves a fair wage, and still protect our customers from continually spiralling prices.

3. *Production Research and Development:* we will focus on ways to improve our efficiency, especially in the scouring department. We want to make scouring a less difficult task and to create a system for doing all our scouring more locally. Other changes will reflect our concern for our health and the environment: we want to reduce noise problems, improve our air quality and make better use of our waste in production.

4. *Human Resource Development:* managing our business to keep the priority high on professional support and creativity, and low on burn-out and too much stress. Continuing to organize innovative team building, problem-solving and long-term planning including all workers.

Cleaning the Clip at the Wool Scouring Co-op

Jane Deamer
President, Wool Scouring Co-op
9 East Main Street
Winters, California, 95694 USA

The Wool Scouring Co-op was founded in 1981 by the New Franklin Society (N.F.S.) of Yolo County to explore the feasibility of a small woolen mill as a business lending itself to appropriate technology. With the help of a grant from the U.S. Department of Energy, the USDA Western Research Laboratory and Storm Engineering of Davis, designs for a small scale, energy efficient scouring machine were prepared. This batch machine was to be capable of scouring up to 20 lb loads of wool, allowing the small wool producer and user to obtain low volume custom scouring at a reasonable cost. After construction and testing at the laboratory of Dr. Howard Needles in the Textile Department of the University of California, Davis, the now-operational scouring machine was turned over by the N.F.S. to the Wool Scouring Co-op in 1982. The Co-op was formed and continues to operate under the mangement of Mary Major and myself, Jane Deamer, of Davis.

Mary Major has been active in the application of appropriate technologies for 30 years with a special interest in land, water and local resource conservation. I had been raising and showing Columbia Sheep for 3 years and had become painfully aware of the difficulties in co-ordinating the specialty wool production with the consumers. We teamed up to help provide the small wool producer with a more accessible and marketable product. We chose a location in Winters, California, (a small farming community near Davis) owing to its proximity to "The Wool Warehouse," a wool carding operation, and the obvious advantages the complementary nature of our businesses would offer. We set up as a growers co-op partly because the nature of wool growers, spinners and weavers seemed to be compatible with a cooperative; and partly because the nature of our group suggested they had more talent than money.

We now have 117 members. Our original plan was to place several machines in different localities, each doing their own scouring and to develop alternative energy systems for operation, i.e. solar energy, water recycling, pollution control, etc. Most of these have been developed on paper, but were put on the back burner due to other commitments demanding our attention: Kids, jobs, husbands. . . . With $5,000 and a washing machine to our name we decided to go with the first priority. The warehouse we rented had no electrical or plumbing hook-ups but the rent was low. Fortunately for us, Charlie Welsh and wife Ellen had come to town with their weaving business; "The Winters Textile Mill." Charlie, a talented weaver, electrician, plumber and carpenter, was able to help us for a price we could affort—nothing up to next to nothing—so he installed the power, plumbing, scouring machine, and water heater (which we intercepted on its way to the salvage yard). He has subsequently constructed our dryer, installed a used extractor, lights for the building, a sink salvaged from a remodeling job . . . the list could go on and on. In short: Without Charlie and Ellie (a super wool scourer when called upon) we couldn't have gone into operation. They have since moved their business to Yuba City, but we occasionally drag them back for emergencies such as installing a new Clark mixing picker, or helping to unload 90,000 lbs of spinning equipment from Connecticut, our latest capital investment.

During our 6 years in operation, we have ironed out the majority of our difficulties with the scouring process. We no longer make exotic felted sculptures unless we want to. The machine has broad capabilities, yet we've tested its limits. It won't magically restore barn sweepings or wool stored for 6 years. However its action closely resembles the effective "scouring train" process, in that we reverse the flow of water through the wool every 16 seconds, simulating the train's action of moving the wool against a counter current. The gentle cross-current action thoroughly cleans while minimizing fiber damage over a wide range of wool types and qualities.

The Scouring Process

We normally use a 4 cycle process for each batch. A brief description follows:

1. Prerinse—adding 0.1% soda ash by volume. This removes the major dirt, loosens the lanolin and raises the pH for better cleaning.
2. Wash-adding 0.1% Trycol NP9, a commercial non-ionic detergent.
3. Rinse—adding 1 cup of 5% acetic acid to help remove the soap and restore the fiber to a neutral pH.
4. Rinse—with plain water and perhaps adding an antistatic agent to facilitate carding and spinning.

Water temperature of around 55 °C is maintained throughout. The water is drained between cycles and finally the wool is placed in the high speed extractor to remove most of the water. Since clean water is used for each cycle we are able to use less water and soap per wool weight than in a traditional train method. We do leave an average of 4% lanolin in medium to fine wools which allows the wool to be spun by hand easily and is acceptable for carding by machine. After scouring, the wool is dried and in the future will be picked, then sent out for carding.

The Co-op has recognized the importance of finding an outlet for our wool products. Marti Franco, our first employee, started a communications service called the Northwest Woolen Network. This was discontinued when Marti moved to the East Coast. Mary and I have made a stab at the problem by forming Yolo Wool Products; a retail co-enterprise functioning as a consignor for sliver, batting and yarns at shows and through mail order. We will be increasing our efforts in this area as the 1989 wool harvest begins to move.

The problems with American wool have long been recognized, namely dirt, burrs and lack of consistency. We have noticed a great improvement in our customers' wool over the years, but after reconnaissance of the wool processing operations in Australia, New Zealand and England, we have concluded that there are some basic differences in our countries' respective attitudes and therefore in our solutions. Probably our best course of action is to promote the variety of wools which we do produce. The varying climates in the U.S. limit the practicality of producing primarily one kind of wool—as in Australia with the Merino or in England with the coarse wools. The dirt and vegetable matter could be helped by planting controlled pastures, pasture rotation, covering the sheep and shearing before weeds go to seed, as well as keeping the floor on which the sheep are sheared swept. These things are done in Australia and New Zealand where their wool receives higher prices accordingly.

The low return on our wool may reduce the incentive to improve it, but we hope that in a small way the Wool Scouring Co-op can help. Our recently purchased mixing picker is one more step toward a cleaner and better product. After much encouragement from our members, we have invested in a spinning frame with carding machinery to produce roving and yarns. We hope, and are reasonably confident through feasibility studies, that this service will greatly increase the market for our specialty wools. We will make every effort to do this in a co-operative spirit with other mills and with our friends in other countries. By working together we stand a better chance of success with promoting our wool and, besides, it's a lot more fun.

Producing Natural Coloured Sheepskins

Wendy Lynch
The Patch
RD4, Rotorua, New Zealand

Why coloured skins? Producing natural coloured sheepskins can be a useful and economic addition to growing wool.

What are the traps? Producing skins is not easy, and not always possible, especially if there are no facilities for killing the sheep in your district or on the farm, or the distance to the tannery is too great. Home tanning is OK for one or two skins but a line of 50 to 100 would be a major undertaking, as well as very time consuming.

As with black and coloured wool, there is a certain type of buyer who prefers their sheepkins natural, not dyed, bleached or boring natural white.

Growing Sheep Suitable for Sheepskins

Animal Age

In most farming circles one should cull 20% of ewes per year, and if farmed properly these culls make excellent large skins. After weaning their lambs these ewes should be shorn and shifted onto good pasture to allow growth of good wool length and cover of meat.

Two tooth hoggets, wethers, rams, or ewes not suitable for breeding, make good big skins, with the wool being finer and softer than older sheep. Lamb skins are generally too small, unless breeding special meat breeds, and depending on the tanning charge, are normally not worth doing.

Wool Length

Depending on the sheep breed, the wool takes from 8 to 16 weeks to reach the minimum of 50mm, but 75mm is the preferred length of wool. Anything over 75mm can look long and luxurious, but the skin is prone to damage in the machines at the tannery, because of wool cotting and pulling.

Shearlings (wool length under 25mm) are unable to be sold at the prices fetched by long wool sheepskins. Under our present tanning charges, shearlings are thus not worth doing, although there could be a market for them in the clothing trade for jackets, hats, vests, etc.

There are 17 processes the skins go through at the tanneries, mostly mechanical, and if the wool is too long, it can easily get caught up and tangled or ripped. Most tanneries mechanically damage about 10% of a batch of skins. These losses of first grade skins must be calculated in your costings as generally you will only just recover outgoings on the resulting seconds.

Slaughtering Sheep

The skins from sheep that are to be tanned must be in perfect condition, before sending off to the tanneries. There must be no knife marks, rips, torn off legs, excessive fat or holes.

Selecting Sheep

The sheep itself must be in excellent health.

Not too skinny—the skin when tanned becomes papery and brittle.

Not too fat—the skin retains some fat in the skin, and even with several degreasings with cleaning fluid, the excessive fat will not completely come out. These skins will be greasy on the back, and cannot be put onto carpets, as they would leave a greasy stain. They are OK on wooden floors.

No flystrike—the skin where the sheep was struck, is of different thickness and will not tan well. The wool also tends to grow a different colour.

Dipped back—the wool in a dipped back tends to be shorter and coarser than the rest of the wool cover and makes the wool on the sheepskin uneven.

Hairy britch—excessively hairy britch does not look attractive

A good sheep for slaughter must have a good cover of meat down the backbone, if you can feel it all knobbly and hard, it is too skinny, and if spongy and soft, too fat.

Slaughtering the Sheep

To slaughter the sheep cut the throat high up under the chin. Cut small strips of skin down the inside of each leg and remove the skin from each leg. Hang the carcass up by the back leg tendons on a hook. Continue to remove the skin from the back legs cutting carefully around the tail and anus.

Slit the belly from the inside tail straight down the centre to the throat. Punch the skin with the fist around the backside and pull it off like a banana, from the tail right to the ears on the head.

Use a knife only to cut the skin off under the ears. The skins must then be allowed to cool off. Cool the skin by either washing in cold water, spreading out on the ground, or draping over a fence, washing the blood off from the neck area if required.

Do not pile the skins up on top of each other at this stage, as heat will build up and ultimately the wool will fall out in patches. This effect is called woolpull.

After cooling and drying, fold into four and freeze. Alternatively place on a wooden pallet 100 mm off the ground, and salt each skin with at least 2 kilos of coarse plain grade 22 salt. They can be piled on top of each other folded lengthwise at this stage, as long as the skins are cool and well salted into the corners of each skin.

The skins can then be sent off to the tannery either frozen or salted. Do not tie up the skins with twine to freeze. When the skin thaws it is very difficult to remove this twine and can result in wool being ripped out.

Freezing Works or Abbatoirs

When sending sheep to the freezing works or the local abbatoir, it is essential to liaise with the management and workers of these factories to explain exactly your requirements for skinning. It can be heartbreaking after growing your sheep for skins to have them ruined at this first stage.

Tanning Process

The most successful tanning is chrome tan, which leaves a slightly green tinge on the skin side. The leather comes up soft and supple. The tanning process has little effect on the woolly side.

There are many home tanning recipes, the easiest is the salt and alum method. The bulk of the work is in fleshing the skin and softening it after the tanning process, all of which is very time consuming. In the long run home tanning is not economical especially in terms of time. The salt and alum method does not produce a permanently fixed rot resistant skin.

Trimming and Presentation for Sale

It is essential to trim and comb up the skins after tanning. Always try to get the best shape with an even rounded neck and four distinctive legs. A rounded shape looks better than a squared off shape of neck and legs. See Figure 1.

Have a standard measurement of skin to determine your sizes and price structure. Generally the length of wool does not dictate the price of the skin but more likely the area measurement or length and breadth.

Preferences for Coloured Skins

From my experience, customers prefer badger pattern, saddle pattern, light grey, beige and multi-coloured sheepskins. The black and dark brown skins are slower to sell, even though they look more luxurious. Canadians and Germans seem to prefer the dark skins over the light, while Americans and Japanese prefer the lighter shades, and patterns. New Zealanders, surrounded by 65 million sheep from cradle to grave, tend to buy sheepskins for overseas gifts rather than for their own consumption.

Conclusions

Producing skins can be a rewarding sideline to your mainstream sheep operation. It also allows heavier culling part of a breeding programme. However, to be financially viable, close attention must be paid to quality control, tanning, freight charges, and marketing.

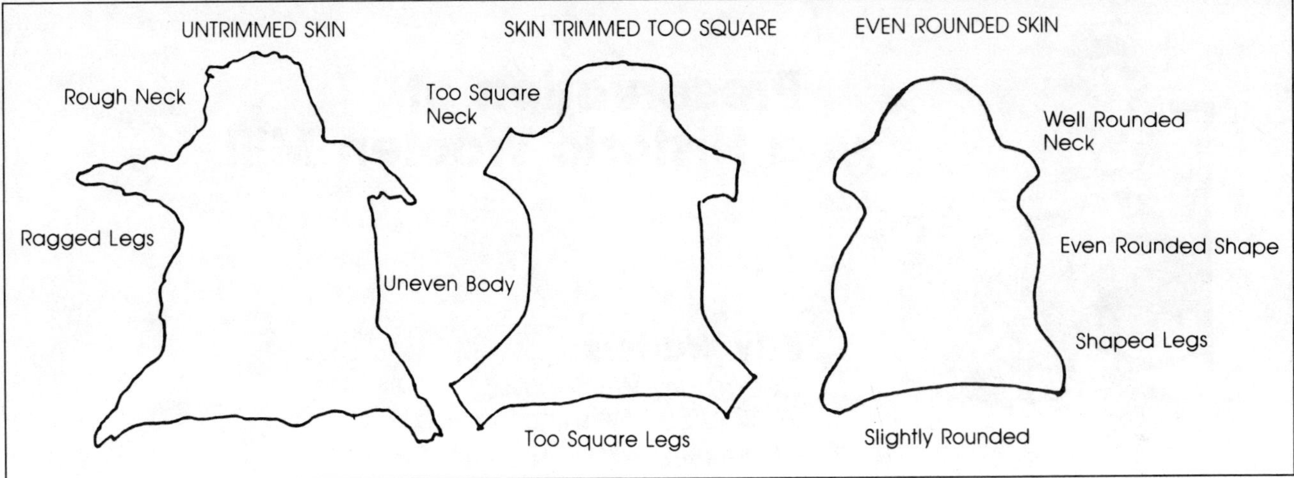

Figure 1.

Preservation of a Historic Woolen Mill

Kay Walters
Cedarburg Woolen Mill
W 62 N 580 Washington Ave.
Cedarburg, WI 53012 USA

Historic Cedarburg is a community of about 10,000 people twenty-five minutes north of Milwaukee. Settled in the 1840's, it was based on an abundance of flowing water for power and proximity to the well known Green Bay Road—a primary trade route from Milwaukee to Green Bay. Within a three mile radius there developed two flour mills, two grist mills, a nail factory, and two woolen mills, each with dams for their source of power.

The Hilgen-Wittenberg Woolen Mill, completed in 1865, was incorporated in 1782 as the Cedarburg Woolen Mill. In 1875 Mr. Hilgen and Mr. Wittenberg purchased the Patzer Woolen Mill in West Bend, some fifteen miles away and consolidated it with their mill in Cedarburg. By 1880, they baled and shipped $17,000 worth of blankets and woolen goods annually. Business was so brisk they decided to build another mill in Grafton called the Grafton Worsted Mill.[1]

In 1883, the Cedarburg Mill added a large three story addition to the existing facility and became known as one of the largest and best mills in the State of Wisconsin.

"Correspondingly, we find that in 1850, the Department of Agriculture, Trade and Consumer Protection recorded 124,896 sheep, mostly in southeastern Wisconsin—center of the population. By 1868, the State had nearly 1,600,000 head and ranked 2nd in the nation. More wool was needed during the Civil War because cotton supplies shrank. The high sheep population prevailed for about twenty years after the Civil War. After 1900, there was a rapid downward trend until World War I. The population then fluctuated in a narrow range between 300,000–400,000 head until the late 1940s. Then began the decline of sheep numbers to present day of 78,000 head (1988). Decline has been steady until the past five years where it apparently has now stabilized."[2]

"The mill in Cedarburg also reflected the decline and closed in 1929. The company's labor force, which had turned out some $100,000 worth of business annually, were added to the list of 'unemployed.' The silent woolen mill was, however, only slumbering. It was reopened by brothers Carl and Fred Wittenberg and operated as Wittenberg Mills. They updated their machinery to process the new nylon and orlon synthetics, but closed their doors permanently in 1968. The building was saved and recycled into a cluster of unique shops, restaurants and a winery."[3]

The Grafton Worsted Mill was changed to Badger Worsted Mills in 1902. It continued operating, producing fine quality worsted yarns, until it was sold and moved to Georgia in 1982. Our machinery was part of their operation until one of the brothers, Howard Roebken, broke off from the family business to start his own carding operation and retail yarn sales.

Ken and I grew up in Cedarburg, with its German heritage. We met in high school and married after college. I graduated from the University of Wisconsin with a double major in Clothing and Textiles, and Home Economics Education. I'm an active spinner, weaver, and fiber artist, selling my work in two juried folk art shows every year: Christmas in the Country and Spirit America. Ken attended Marquette, the University of Wisconsin, the Graduate School of Banking at the University of Wisconsin, as well as the Senior Bank Officer program at Harvard. He has degrees in Finance, Real Estate, and Banking. After 18 years of banking and being Chief Executive Officer of two banks, a securities analyst, investment real estate analyst, and licensed for real estate brokerage and insurance, he joined me in the mill operations. In addition, he developed the 16,000 sq. ft. complex that houses our business.

We purchased our first registered Hampshires in 1976. My interest in sheep and wool was the result of a childhood dream to have a farm and animals, and a natural outcome of my background in textiles. I began acquiring black sheep, mainly Corriedales, which were available in the area. I also liked their medium-fine wool for use in clothing. My coal black ram lamb even won a blue ribbon at the Wisconsin Black Sheep Gathering in 1982. I was totally "hooked" on sheep and wool.

In 1979, when the Grafton Wool Carding business was for sale, I felt it was a unique opportunity for me to further my interest in the industry. First, I had a professional interest in it as a textile major and home economist. Secondly, Ken and I have an interest in preserving the past. We've restored five homes, our present one is an 1885 stone farmhouse.

Man has always valued that which is unique. In this case, I saw value in preserving a wool carding business that in the past was commonplace. This truly was an oppoortunity to preserve and educate the public on the historic and real value of what used to be and what I thought could be again, a thriving sheep and wool business in this area. However small, this last remnant of the once thriving woolen mill should be saved.

It took me two years to convince Ken that it wasn't too great a risk, as the machinery was vintage—1860 patent dates. In the meantime, while we were deciding, Howard Roebken, the owner, did retire and closed the carding and retail yarn store. In fact, he closed completely six months before we decided to buy the carding business. It meant renting space, as we were not interested in purchasing the building. It had been a going busienss for over 25 years in that location. We felt the inaccessibility of the building off the main traffic route was an important consideration. At that point in time this business was part-time and we intended to employ our sons and help them through college. I was to be there on a limited basis. The retail yarn store upstairs was purchased by a former classmate who agreed to continue to take in the wool carding when I wasn't there, as I was selling real estate.

As people realized we were carding wool the business began to grow. Soon the yarn store owner upstairs said she could no longer take in wool for us as it was taking up too much of their time. I believe I was paying her $1.00 per batt for the service. That decision forced us to take another look at the practicality of the business. How viable was it? Could I make as much selling real estate? Would our sons really work here? Do people still appreciate wool for batting? Had polyester or down effectively replaced wool in the marketplace? Was it really important to keep this business alive or purely sentimental? Ken was concerned over the extreme age of the equipment. What happens when something breaks? We hadn't spent our entire life working in a woolen mill as the four "Roebken boys" had. He called two toolmaker friends, making sure they would help keep it going if needed. They said OK. We had to consider all these questions again because it meant I would have to leave real estate and build this business to produce income for my family. It now became more than a novelty and reqired a firm commitment. I felt we could make it a strong, viable entity.

The questions we had were endless, but they all had logical answers which had to be searched out. We spent two months working with the former owner's son Jon who taught us the basics of picking and carding. We toured almost every carding mill from Minneapolis to Maine. We talked and gathered information from libraries; the Wool Bureau; from Bill Mogayzel from Howard Bros.; from Emory Benson from Portage, Wisconsin; from the Museum of American Textile History, North Andover, Massachusetts; from Guy Webster and Eli, Crescent Woolen Mill, Two Rivers, Wisconsin; and many other wonderful people who shared names, ideas and knowledge. There is still much to learn, but it is challenging and rewarding.

After purchasing our business, one of the first things I did was to start scouring grease wool. I had a good supply of it at home in the barn. Howard warned me against it—said it was too much work. Well it is, but I feel there is a need for it. The nearest scouring facility to Wisconsin is Faribault, Minnesota, an eight-hour trip. They would do it for me but it is a long haul—time and also it necessitates a very large stock trailer full of wool to make it worthwhile. As a result, we are doing it ourselves on a small basis.

We scour one fleece per laundry tub. We ask customers to skirt their fleeces. Clean sheep make clean wool. There is no machine that can remove all the vegetable matter from a fleece. Midwestern sheep are especially full of foreign matter as we feedlot our sheep. Few of us have acres and acres where we can pasture our sheep. A good part of our farms are producing cropland. We need those fields to produce two good cuttings of hay every summer to feed our sheep through the long winters.

To chemically remove the vegetable matter involves shipping our wool to the East Coast to a carbonizing plant. That is economically not feasible for us. So we encourage farmers to keep their sheep clean.

To scour, we use hot water (approximately 140–145°F) and a detergent to cut the grease. As part of our original machinery purchase we have a large extractor where we rinse and remove the water. The extractor removes so much of the moisture that it takes only 5–8 minutes to get an entire fleece dry. We use a commercial dryer with an accurate heat control and a cleanout in the bottom. A lot of dirt falls out in this procedure. Sometimes the wool must be scoured more than once.

Everyone's wool is kept separate in our processing. It is all custom work, so you always get back your own wool. We even have some boxes labeled, "Priscilla's—Sue Jones." Even the sheep are identified!

Our picker is a wonderful cast iron "Mestizo Burring Picker" made by Curtis and Marble, Worcester, Mass. Patent: August 30, 1865. It's original color is forest green with red highlights. It has oak lids with solid brass hinges. In our trip to the Museum of American Textiles in Massachusetts, we found the original book that listed our picker with diagrams and instructions. The following is a reprint of that information—I felt it would be interesting to you.

GODDARD BURR PICKER

19. The Goddard, or Curtis and Marble, burr picker is favorably known to the trade and is found in some of the best mills. It is shown in Fig. 8 and in section in Fig. 9. The machine is constructed with the same objects in view as those previously described; namely, to remove burrs, shives, dust, and other foreign matter from the wool, and also to open the stock and make it lofty for the cards without breaking or injuring the fiber.

The principle of this machine is somewhat different from the burr pickers previously described, but still the same general features are present; i.e., opening the wool with a picking cylinder and delivering it to burr cylinders, from which it is stripped by a rotating brush, the burrs being removed by burr guards.

Figure 8 Goddard, or Curtis and Marble, Burr Picker.

20. Construction.—The feed-rolls are set with cockspur teeth, which hold the wool securely while the teeth of the picking cylinder thoroughly open it. The bottom feed-roll is stripped by the picking cylinder, and in order to prevent all chance of the stock winding around the top feed-roll, this roll is driven faster than the bottom roll; thus, the wool is cleaned from the back of the cockspur teeth on the top feed-roll by the points of the teeth on the bottom. The feed-apron and feed-rolls in this machine may be stopped by means of a lever without stopping the rest of the machine, as described in the case of the other burr pickers.

21. The picking cylinder is composed of sixteen crossbars, which are attached to spiders fastened to the main shaft of the machine. The cross-bars are filled with round-pointed teeth of steel, which are placed staggered, so that at each revolution of the cylinder the teeth cover every sixteenth inch of the width of the machine. Above the picker cylinder is a perforated brass screen fastened to an iron frme and held in place by buttons, so that it can be readily removed for cleaning. This screen has perforations of sufficient size to allow fine dust and dirt to be removed from the wool through it. The screen is removable to give access to the picker cylinder for cleaning or other purposes.

Beneath the picker cylinder is a grate, or grid, made of angular iron bars. This grate is made especially firm and will not bend out of shape; thus, the spaces between the bars of the grate are always the same. The grate is made fine enough so that loss of wool is avoided, but the heavy dirt, shives, etc., are allowed to drop through it. The grate is hinged at the rear and may be lowered in order to clean it and the picking cylinder between different lots of wool.

At the rear of the picking cylinder, two burr cyliners are built up on central shafts with alternate steel-toothed rings and solid packing rings. These burr rolls should be made of the right-sized toothed rings, or burr wire, to suit the class of wool that is being operated on, and should be spaced fine or coarse to suit the same. The burr cylinders in the Curtis and Marble pickers are not of the same size, as they are in the machines previously described, but one is smaller and works over the larger one—running in the opposite direction. The larger burr cylinder is provided with a burr guard that knocks the burrs from the surface of the cylinder. Working in conjunction with this burr guard is a smaller one, which prevents the burrs and other refuse from being carried over on to the brush cover. On this machine a rotating brush, with six cross-bars filled with bristles, strips the stock from the burr cylinders and delivers it to the gauze room through a suitable spout or pipe.

22. Others.—In Figure 8, a square tank will be seen placed over the spout where the wool leaves the machine. This is a device for oiling the wool as it leaves the picker

so that it will be ready for carding. This may be attached to the picker when desired. The oiler consists of a revolving brush that throws the oil supplied by a tank over the wool as it passes from the picker. Oiling wool by mechanical devices, either at the picker spout or elsewhere, is not generally approved by the trade. Oilers afford an easy means of lubricating cheap stock, but are seldom used on fine stock, as will be explained in another Section.

23. Operation.—In operation, the stock is fed to the traveling apron either by hand or by a self-feed. The apron *a*, Fig. 9, carries the wool to the cockspur feed-rolls *b* of the machine, where it is loosely held while being combed and opened out by the picker cylinder *c*. The wool is thus opened out at the start and the burrs and dirt loosened. The picker cylinder revolves upwards past the feed-rolls and carries the wool to the large burr cylinder *d*. The exhaust fan, or blower, *e* on the top of the machine creates a strong draft of air, which removes loose and small particles of dirt through the perforated brass screen *f*, and also lifts the fibers of wool from the picker cylinder, so that they are more readily caught by the teeth of the burr cylinder *d*. This cylinder takes the stock from the picker cylinder, receiving the fibers in the spaces between its toothed rings, while the burrs and other matter being larger than the fibers of wool remain on the surface of the cylinder and are removed by the rapidly revolving burr guard *g*. Working in connection with this burr guard is a smaller roll *h*, which prevents the burrs and refuse from being thrown on to the brush cover. The burrs thrown out by the burr guards are carried downwards by the picker cylinder, and such as do not drop through the grate *m* under the cylinder are carried forwards and thrown into a burr box through an opening beneath the feed-rolls of the machine. A small burr, or cotter, cylinder *j* operates in conjunction with the large burr cylinder, revolving slowly in the opposite direction and opening out cotted lumps of wool. The revolving brush *k* cleans the wool from both burr cylinders and delivers it through the spout *l* to the gauze room.

24. The Goddard picker, as shown in Fig. 8, requires a countershaft. The picking cylinder, fan, and revolving brush are driven direct from the countershaft, as is also the oiler, if used. The other parts of the machine are

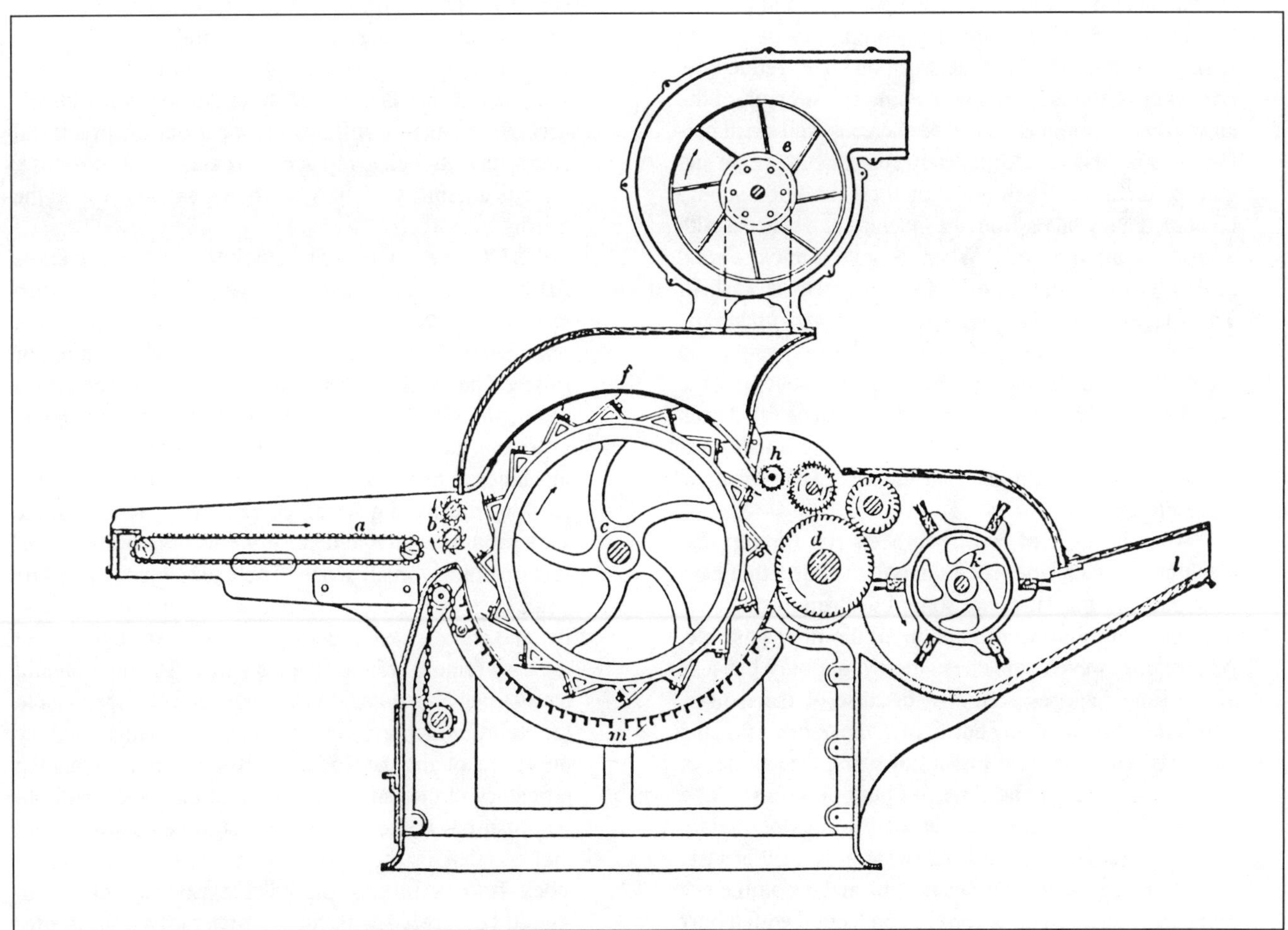

Figure 9.

driven from the main picker shaft. The following table shows the capacity of the different widths of these machines, the amount of work turned off varying, of course, under different conditions.

Size of Picker Inches	*Diameter of Picking Cylinder Inches*	*Capacity per Day Pounds*
24	24	600 to 1,200
30	30	1,200 to 2,400
40	30	1,600 to 3,200
46	36	2,200 to 5,000
48	36	3,000 to 7,000

MANAGEMENT OF BURR PICKERS

25. In regard to the management and proper care of burr pickers, it may be said that one of the most important points is to clean periodically the various parts of the machine. All perforated, or wire, screens, grates, conducting pipes, etc. should be kept clean and clear of dirt and grease. If they are clogged up, they hinder the removal of the dirt from the wool and also reduce the efficiency of the fan, thus weakening the strength of the air-currents through the machine. Screens with small perforations will become completely coated with grease and gummy dirt, and when found in such condition should be immediately taken from the machine and washed with a strong solution of soda. When the gum is thick, a good deal of it can be scraped off before the screen is washed. The spaces under the picking cylinder and under the beaters of a Parkhurst picker should be frequently and regularly cleaned out. On the Parkhurst machine the swing-back bonnet is so arranged that it can be lifted back after the belt has been thrown off from the brush, thus allowing the burr cylinder and interior of the machine to be cleaned.

After a batch of wool has been run through the machine, if much fiber is clinging to the burrs that have been cast out the burrs are sometimes run through the machine again in order to obtain all the fiber possible. As a rule, however, this does not pay, not only because of the time required, but also because of the danger, where there are so many burrs, of some of them passing forwards with the wool instead of being thrown out of the machine the second time. As burr pickers run at a high speed, it is essential that all bearings of rapidly rotating parts be oiled at least twice a day; otherwise, there is danger of the journals heating and becoming fast in the bearings. All belts used in connection with a burr picker should be laced, as there is danger of accidents if belt hooks are used, since the belts are in exposed places and it is necessary in many cases for the operator to work in close proximity to them. The brushes of burr pickers wear out rapidly, and they should be set up closer to the burr cylinders from time to time and ultimately replaced.

26. Setting.—In regard to the setting of the working parts of a burr picker, it may be stated that this depends largely on the character of the stock being run through the picker. For a coarse, long-staple wool, the burr cylinders may be set farther from the picker cylinder than for a finer and shorter wool, when they may be set up as close as possible without any contact. The burr guards should be set as close as possible to the burr cylinders without knocking out the wool as well as the burrs. If set too far from the burr cylinders, many burrs will escape them; while if set too near, they will pull the wool from the cylinder. Great care should be taken not to allow either the picking cylinder or the burr guards to touch the burr cylinders, since, if this is the case, the burr cylinders will be damaged and very soon ruined. The picking cylinder does not require to run very close to the burr cylinders, since the centrifugal force due to its rapid rotation will throw the stock from the picking cylinder into the burr cylinders. The brush should be set to strike into the burr cylinders slightly, so as to clear them thoroughly of wool. Care should be taken, however, that the brush does not strike the burr roll with any great force, as this will quickly wear out the brush and necessitate its being replaced. Again, if set too close, there is a liability of the stock being carried around the brush.

27. Gauze Room.—The stock from a burr picker is delivered by the current of air generated by the brush to a gauze room. This is usually a wooden compartment provided with openings covered with wire screening, or gauze. The stock is blown from the burr picker to this room, the object of the gauze-covered openings being to let out the surplus air, but to retain the wool. Such machines as burr and mixing pickers, which deliver the stock by means of a strong current of air, require some such room in which to deposit the wool in order to collect it within a small space. The gauze room should be made large enough to accommodate a batch of wool easily, and should have a door of sufficient size to permit the easy removal of the cleaned stock. The door should have strong and suitable fastenings, and the openings in the room, which are covered with wire gauze to allow the egress of air, should be of sufficient area so that the efficiency of the current of air from the picker will not be impaired. If the gauze room is to be used for wool that is oiled in the burr picker, or for receiving the stock from a mixing picker, the floor of the room should be covered with tin, or preferably zinc, to prevent the oil from soaking into the floor. Care should be taken to prevent fire in a burr picker, or any similar

machine that depends on currents of air for its operation, as nothing causes fire to spread so quickly as an air-current.

Our card picker is an E. C. Cleveland, Worcester, Massachusetts, patented November 30, 1869. It also is green with red. It is a small card by today's standards with 4 workers and 4 strippers being about 45 inches wide. It has a "Bramwell Automatic Weigher and Feeder"—patented August 1864–79 at the front, into which you can place your clean picked wool. The spiked apron will pick up the wool out of this large bin and feed it into a hopper which accumulates it until the right weight automatically releases it onto the apron feed of the card.

Approximately 50% of our business is washing and recarding someone's old comforter batt. These need to be carded only once to produce a good batt, ready to be covered with fabric for a comforter.

Processing someone's grease wool into a product suitable for spinning is a different matter. This may need carding several times, and even then very fine wools need finer gauge clothing wire on them to get a product free of second cuts. All North American carding mills like ours, that I am aware of, have only one size wire on them. Consequently, there is a limit to what the carding can do on some wools, especially very fine wools.

After carding, the wool is released onto a conveyor which transfers the thin web onto a lowered moving platform. This platform moves back and forth under the conveyor at precisely set intervals, making any width we choose from a 45-inch crib to an 84-inch queen. After an initial buildup of this thin web of wool on the platform, we can make a conveyor apron on the platform move, causing the web to build up in layers on a 90-inch diameter cylinder. Cutting off this cylinder gives a 90-inch length for a comforter.

It is a marvelous piece of machinery that gives us the width and length. No need for stretching the batt, which will give you thick and thin spots where the cold can sneak through.

In our travels, we have never seen a picker or card like ours. No one can make sizes. All others have to stretch batts to get different widths and lengths.

Because of this fact, we have developed a whole new market which never before was addressed—that is the handquilting market. People normally considered wool as only used for winter weight, "fat batts" that had to be tied. The thick batts and tying are not because someone is too lazy to handquilt. Thick batts are necessary for Wisconsin and other-northern winters and the only way to handle the thickness is to tie the three layers together at regular intervals.

Handquilting necessitates a very thin batt suitable for summer weight. We are capable of producing a very thin woolbatt on this card to exact width and length. Because the batt is thin, the wool must be a uniform staple length to insure a good product that will last. Handquilters are very particular, and use natural fiber (100% cotton) fabrics. They must be made aware that a natural fiber batting—100% wool—is available. It has a longer staple which allows the quilting stitches to be further apart than cotton. It is soft and supple and quilts like "butter". Moreover, it breathes, making garments or bed coverings more comfortable for any season of the year.

One criticism is that wool "beards": that is, it comes through the outer fabric coverings. I've found that a few simple gentle handwashings will cure that problem when it does occur. In the eight years we've been in business, we've only had two quilts brought back with "bearding." We've washed them and sent them away to a satisfied customer. Microscopically, wool has scales on the fiber. It is these scales combined with water and a small amount of hand agitation that cause the wool to cling to itself, and it ceases to migrate through the fabric. The ends of the fiber that already protrude must be pulled back into the inside, in the case of a tied comforter, by pulling the fabric away from the batting, or in the case of a handquilted quilt by cutting or shaving them off. Pulling or brushing them only hooks more scales of other fibers and pulls more through.

Other markets we've worked at developing are as follows:

Baste wool into cheescloth covers—make duvet covers that can be removed and laundered.

New, pieced handquilted quilts by Wisconsin Amish using wool batting

Custom processing of fleeces for handspinners

Custom patchwork comforters

Purchase wool from local sheep producers—all breeds, all colors—to provide wool for handspinners

Buy, sell and renovate antique quilts

Enlarge our retail store to include the following: supplies and classes for spinning, weaving, dyeing, quilting, basketry. Natural fiber textiles for clothing and home use.

Our main emphasis of course is wool. Wonderful, soft, comfortable wool. Man cannot reproduce all the unique qualities God gave us in our sheep.

References

[1]Wendt, Alice Schimmelpfennig, *Hilgen Heirs* (Cedarburg, Wis.: Schroeder Printing Co., Inc. 1988), p. 68

[2]Pope, Arthur L., *A Look at the Past and Future of Wisconsin's Sheep Business* (a paper presented at the Wisconsin Sheep Industry Conference, March 25–26, 1988), p. 4.

[3]Edquist, Rita, *Cedarburg History, Legend and Lore* (Cedarburg, Wis.: Schroeder Printing Co., Inc., 1976), pp. 58, 60.

How and Why to Dye Coloured Wool

Kerstin Gustafsson
Skrekarhyttan
S-713 92 Gyttorp, Sweden

Since at least 1000 B.C. dyeing has been used to colour wool. At that time China had a Department of Dyeing which was organized in the "Heavenly Ministry", and around the world dyeing of textiles probably started to be a known and used technique.

But the use of dyeing was not well accepted everywhere. In Sparta, for example, it was said that dyeing "destroyed the whiteness of the wool". This opinion is understandable if one considers all the breeding work which was made in order to get sheep with white wool.

But even without dyeing, people made patterns in their textiles. Textile finds in Denmark from the Bronze Age (1500 B.C.) up to the Iron Age (Christ Birth Year) give an interesting and instructive picture of the wool textile technique development. A nice example of textile made of natural coloured wool is the Skydstrup Woman's cloth from 1500 B.C. in light brown with a white belt. The white belt can be seen as an ornament and indicates that white wool was very valuable.

A find, called the Ekteved Girl, from about the same time, also seems to prove that white wool was an object of value. Beside the young woman and her baby child, among some jewelry, was a lock of white wool.

Another woman from about 600 B.C., found in Borre Mosse, wore a dress woven in twill. The earlier finds were made in tabby. The yarn was made of a more homogeneous white wool containing only 6% coloured fibres. The cloth is light brown, but the brown colour originates, in all probability, from the dyeing effect of the water in the peat-moss.

The "Blue Girl" from about the birth year of Christ found in Lönnehede, had a plain blue dress, probably dyed with "Vejde" (Isatis tinctoria) on white wool.

From about the same time, two dresses were found in Huldre Mosse. They are check patterned, woven of yarn dyed in eleven shades from brown to red passing over yellow and green.

It is fascinating to see how well people from ancient days had mastered the textile techniques.

White wool breeding is the result of the availability of good dyes, and this is the reason why almost all wool is white today. People in the old days seem also to have dyed coloured wool.

Of course you can see the brightness of the dye on white wool better but also dyed coloured wool has a nice tint. I think people in the old days saw and valued the colours better than we who are used to seeing the bright colours of all the modern materials which surround us.

Let us use the Scandinavian traditions as an example of how coloured wool was used during the last 4 centuries.

Perhaps "the farmers grey wadmal" is the first thing one will think about. The grey wadmal was "in the beginning" made of grey wool, but over the years this kind of sheep died out. Up to the first part of this century the mixing of white and black wool was the main method used to get grey yarn because there were no grey sheep. But probably when people wanted whole cloth in dark blue or black, they used black wool to dye.

Another typical Scandinavian wool textile is the "lusekofta" (lice jersey). This type of sweater is made in patterned knitting with one black and one white thread. The technique was used for two reasons: first, in order to be extra thick and warm, due to the loose threads, which waiting to be knitted on the backside of the textile, make it double the thickness; secondly, for the nice pattern.

Over the years the pattern itself has been more and more developed and is today the main thing when knitting that special type of jersey. It is still an important part of the Scandinavian traditions forming our design of today.

Sometimes you can see this kind of jersey not only in black and white but also, for example, in red and black. During the 18th and 19th centuries, before it was common or possible to buy industrially produced and dyed yarn, this kind of knitted textile was made of a white and black thread. The way to make it, for example, red, was to dye the whole cloth in red after the knitting was finished. The reasons were, among others, that when you had to spin all the yarn yourself it was easier to have just two different types of yarn, black and white. You saved yarn because you did not get a remainder stock of yarn in different colours and you saved dyes when you just used as much as was needed for the cloth. Dyeing was something expensive and hard, especially if it was red, blue or green. Of course, dyeing patterned textiles in the piece was also used for textiles in other techniques, such as crochet.

Another reason to dye coloured wool was that it was difficult to get a really black colour on white wool by dyeing. There is no natural dye which gives black colour. You have to use a lot of chemical mordant, such as iron. The disadvantage of these black dyes is that after some time they start to erode the fibres. This can be noticed on old tapestries where the weft threads in some parts that have been black are almost gone.

Even today dyeing coloured wool is a useful technique. Natural black and brown wool will become discoloured in sunlight. If you dye black wool with red or blue, the black colour will be better preserved and become more brilliant and thereby give a deep and nice contrast to the red or blue colour.

Knotted carpets from small villages in the Konya region in Anatolian Turkey, are even today made of naturally coloured wool. The reason seems to be the lack of enough white wool. The yarn is made of greyish brown, brown or beige wool, and that used for the knots is dyed in different colours. The warp is not dyed and that is one of the reasons why it is easy to see the original colour of the wool.

There are three different ways to dye textiles. Dyes can be applied to the wool, to yarns or to the textile after it has been crafted.

There are two basic types of dyes, natural and synthetic. The natural dyes can be applied directly on the fibre or require an additional chemical, a mordant, to attach to the fibre.

All these things work in different ways, and can be combined in a lot of variations. There are some things to take into consideration to get a good result.

Basically, there is no difference between dyeing free wool, yarn or textiles. The dye penetrates easier into each fibre if they are as free as possible. But you must, in this case, take into consideration the special property of wool: felting. When dying free wool you really have to be careful, because of the risk of felting. Work very slowly and do not "touch" the wool more than is absolutely necessary.

Dyeing yarn is the easiest because, compared to textile and free wool, it has a wider area, which gives a good result. Even if the fibres are attached to each other, the threads absorb the dye better and more equally than a woven or knitted textile.

To dye a textile in the piece can be difficult, especially if it is a big one. The trouble is to get a uniform colour. The dye has to be applied at the same time and equally all over the fabric. It is also important to expose all parts of the textile equally in the dye during the whole dyeing process. To get a good result, it is recommended to use an exceptionally large dyepot and a bigger dyebath than needed for the theoretical weight of the textile.

There are some differences to take into consideration when choosing between natural and synthetic dyeing. Natural dyestuff gives "nice" and soft colours, with no hard contrasts. They suit very well together, and form very nice patterns. But natural dyestuffs need mordants, which means that the material has to be processed in at least two different baths. Very often the natural dyestuff is basic, and the material has to be very well rinsed in several baths, and finally rinsed in a solution of vinegar and water.

Among the different types of synthetic dyes for wool, there is one type which works with vinegar, that is the easiest and the best one to dye with. You only have to put the material in the bath with vinegar for a moment, take it up, put in the dyestuff dissolved in some water, and put the material back into the bath. Since the bath is acid there is no need to rinse it, if it is not over dyed. But there are also some disadvantages. The dyestuff is very strong; very little gives very brilliant colours and because of that it can be a little difficult to use. It is important to learn how to mix the colours. You have to know a lot about colouring to be able to make nice patterns. Now, how to combine these different methods?

Free wool is easy to dye in synthetic dyestuff. It is just one bath and you do not have to handle it so much, and that means that the risk of felting is minimal. Yarn works well in both systems.

Dyeing textiles in the piece works best in synthetic dye, unless you can hang the material in the bath in order to get it equally exposed to the dyeing solution. If you cannot, you have to move it all the time to get it uniformly dyed, and that can be too much if you have to handle it in several baths too, provided the material is not already felted.

There are some other things to think about when choosing colour and dyeing method.

Free wool needs a strong bath. The fibres seem to fade in colour when the wool is carded, especially if you mix different colours during the carding.

To dye coloured wool works as if you mix one light and one dark colour. The dark fibers mixed to the light ones, make the light fibers look not so brilliant.

Dyeing black wool gives another blackness, but it can be hard to see the difference if you are not used to it.

To dye brown wool will often give very undefinable colours, but can be very nice and useful, especially in tapestries.

To dye grey wool works as if you were to dye mixed black and white wool fibres. All the fibres take the dye well, but you will hardly see the basic colour, because of the black fibres, and that will make the color dark.

If you want the wool very lightly dyed, you must work in a special way. Of course you can use a thin bath for dyeing, but with such a bath the wool might very well turn pale. To avoid this, it is better to mix dyed fibres with white ones when carding. The fibres that have a rather strong colour can stand the light better, but because of the mixing, the yarn will look light coloured.

To work with coloured wool in all these ways makes the work very exciting and gives you a lot of possibilities. You are free to make all kinds of variations if you just learn the basic demands. Fine handicraft demands good knowledge and experience.

References

Dandanell, B. & Danielsson, U. Tväändsstickat. Stockholm: LT 1984.

Gustaffson, K. Gamla textila tekniker i ull. Stockholm: LT 1988.

Hald, M. Olddanske tekstilier. Kopenhagen: Nordisk forlag 1950.

Nielsen, K-H. Kvindedragten fra Skrydstrup. Haderslev Musem 1979.

Sandberg, G. Växtfärgning. Stockholm: Norstedts 1980.

Sandberg, G. Indigo. Stockholm; Norstedts 1986.

DETAIL OF THE TAPESTRY FROM OVERHOGDAL. 1100 A.D. (Courtesy Jamtlands Lans Museum Ostersund, Sweden)

THE FEATHER—SYMBOL FOR THE SOUL. Tapestry, designed and woven by Kerstin Gustafsson, Nora, Sweden

SWEATER SLEEVES MADE IN PATTERNED KNITTING. Made of natural coloured wool, the lower one dyed in the piece. (Courtesy Dalarnas Museum Falun, Sweden)

MAKING WADMAL BY FOOT-FULLING.

RECONSTRUCTION OF A FULLING MILL. 1988 outside Nora, Sweden.

TAMPERS. In the Fulling Mill, Vemhan, Sweden.

AXLE WITH TAPPETS. In the Fulling Mill, Eksharad, Sweden

Wadmal (Coarse Cloth, Shag, Rough Woolen Cloth)

Kerstin Gustafsson
Skrekarhyttan
S-713 92 Gyttorp, Sweden

What Is Wadmal?

Wadmal is the material that helped the human being survive in the cold nordic climate.

It is windproof, water-repellant and long-lasting, because of its structure, and the quality of the wool. Those things made the Wadmal material meet all the demands and needs for hundreds of years. Ordinary men did not have many changes of clothes. The Wadmal clothes could resist a whole days work out in woods and fields, even in rain and snow, without making the wearer too cold or wet.

What is it that makes this material so marvelous?

The first quality is that the wool can be fulled. There are three things needed for this to happen: water, warmth (heat) and working (mechanical processing). This way to process the wool has been very common in almost every textile technique where one wants a thick and strong product. Today we usually think of felting as a disadvantage, but we need to bring back old knowledge.

How to Define Wadmal?

We must look at some different woolen materials to understand the differences.

The very simplest form is a wool material that has not been processed in any way after it is woven, except being pressed.

Then we have materials for coats of plaid or similar materials. They are prepared in a special workshop. Washing makes them shrink a little and after that they are napped. They then become soft and nappy, looking very attractive. They are not very long-lasting, nor windproof.

Then we have Broadcloth. This is also a woven material but made much thinner, both in the choice of wool and weaving and technique. It is fulled, napped and pruned, often a couple of times.

Materials for Wadmal must go through a much more strenuous process. Wadmal has to be washed and fulled but not napped. It should be fulled to such a grade that one can't pull out a thread of a cut edge.

To summarize: Wadmal is woolen material woven of a relatively rough wool. It is thoroughly fulled and not napped or pruned.

To understand all the good qualities of Wadmal you have to know a little about wool as a textile material. Wool possesses two qualities that seem to be contradictory. It has a very high capacity to hold moisture but is water repellent. The reason is that moisture in the form of steam can be absorbed by the wool fibre itself, while the water drops are shed by the hair.

Wool also has the capacity for felting. This means that by influence of moisture, warmth and working, the fibres creep in between each other and make the fabric, into a compact textile.

After this process, there are more fibres per unit of area and this makes the material stronger, more dense and gives a higher capacity for absorbing moisture.

How to Process a Wool Material to Get Wadmal

While fulling, material should be processed in a way that makes the fibres crawl in between each other and not toward the surface of the fabric.

That's why one cannot process a wool material to Wadmal at high temperature in a washing machine. The material will shrink due to the temperature and the working by the washing machine, but the fibres crawl out on the surface and the material looks napped and becomes flimsy.

The Wadmal material has to be rubbed in a way which forces the fibres to mix. The most common way to do this has been to full it with ones feet in a tub.

There are wall paintings in Pompeii showing how it is done. We also know that in England many fullers became unemployed when the waterpowered fuller mills began. The very first one in Europe existed in an abbey in France during the 12th century. It seems to have been spread from there to England which was a big producer of Wadmal and broadcloth during the Middle Ages.

In Scandinavia it seems that the first water powered fuller mills were built after the 16th century. But the country people probably foot-fulled their Wadmal for hundreds of years before that and this was a common method until the beginning of the 19th century, especially in families living in places where there were no industrial fullers.

Fulling was a work which two or more people did together. It took quite a long time to full a batch of fabrics, but there were certainly a lot of smart constructions to be found for making walking on the material more comfortable.

It was probably not before the end of the 18th century or the beginning of the 19th that fulling mills in the countryside became more common. People of that area could leave their woven cloth to get it fulled. It was often the dyers who also had fulling mills.

These mills were usually used during the high rainfall periods in the spring and autumn.

How Did the Water-Powered Fuller Mills Work?

There have been a couple of different types. The one which lasted longest, used in some places until the 1940s, had an axle from a water wheel passing behind the tampers.

The axle (shaft) has tappets lifting each tamper. The tampers are wooden stocks vertically placed in wooden tubs. After each tamper is lifted by the tappet it falls freely down in the tub where the fabric is laying folded in hot water. There are normally six tampers in a mill working two by two.

The tamper is step-shaped and this together with the round inner shape of the tub forces the fabric to turn slowly in the tub. The tamper edge also has to be bevelled in such a way that the fabric is forced toward the center of the tub. Otherwise the selvedges will be insufficiently shrunk and become corrugated.

Even though the cloth is rotated a little during the process it is important to stop and refold the fabric or it will not be processed uniformly.

After this refolding one pours on new hot water so that the material remains warm. There is not meant to be a lot of water in the tub. The rule is the same as in other types of fulling: the material is supposed to be warm and well wetted but not floating in water.

The time for the fulling process can vary from about 1 hour for a dress-cloth in half-wool up to 12 hours for a heavy-duty working cloth.

In the old days one could ask for whole, half or quarter fulling.

One tub can contain about 20 meters (60 feet) of fabric and depending on how much it shrinks during the process and how long the fabric initially was, it is also important to know how to combine different materials, taking into consideration both quality and color as well as how the fabric is woven.

In the end of the fulling process the length and the width have to be measured in order to see how much it has shrunk and by this be able to check if the fabric is ready.

After the fulling is finished the cloth has to be rinsed because quite a lot of fluff has loosened during the process.

Then it is time to dry the fabric. Perhaps the long drying time for wool material can be seen as a disadvantage, but in this case it gives you time to stretch and form the fabric.

It is important to stretch the material as well as possible. For this purpose a lot of different frames and other facilities have been used. The most common way is to stretch the fabric during winding on a roller. Such a roller construction looks almost like a warp-frame with extra crossbars on which the material goes back and forth before it winds up.

In EkshaR rad (a small community in the west part of Sweden) another type is to be found. Here the fabric is hung up on logs and down in the folds are other similar logs laid so the fabric is stretched while it is drying.

How Much Does the Material Shrink and What Qualities Can One Get by Fulling the Fabric

Quality and shrinking depends on a lot of different things:

1. Wool type
2. The spinning method

3. If the threads are S or Z spun
4. If there are one or two threads in the yarn
5. Which binding is used in the weaving
6. The reed grade
7. The length of the fulling process.
   The material may shrink from 15–50%

1. Different types of wool have different capacity for fulling. Let it become a habit always to make a fulling test with a piece of the wool before spinning it.

The Swedish country breeds have very good qualities for fulling while other types of wool imported to Sweden often are difficult to full.

We have today three different countrybred-wool sheep:

- Fine-wool. Can be used for soft cloths.
- Fur-wool. "Tapestry-wool".
- Long-wool. Good for Long-Pile Rugs.

Types 2 and 3 contain both wool and hair.

The hairs are needed to give the yarn firmness but also to give clothes "outdoor qualities".

The hair must not be too rough or too long, because they will not be able to crawl fully into the yarn. They will instead, during the fulling, penetrate on the surface of the fabric as boucle-loops. This can be very beautiful and good looking as coat material for example, but Wadmal has traditionally never been like that.

2. In a simplified way one can say when spinning a rather long and rough wool on a Spinning Wheel or with a Hanging Spindle that the weight of the spindle stretches the wool fibres and forces them to lay parallel in the thread, giving a strong yarn for warp.

One can spin fine wool on a Walking-Wheel (Great Wheel) or a "Spindle Shaft" which looks like a Hanging Spindle without whorl and hook or slot in the top end.

This technique does not stretch the fibres so much and the fibres will lay more disorganized, giving the yarn a higher grade of plumpness and softness.

Such a yarn can preferably be used for weft in the weave.

To spin with the "Spindle Shaft" technique has almost been forgotten in Sweden during the last 100 years.

3. The direction of the twist of the thread influences the final quality of the Wadmal.

As long as the yarn was "home made" the yarn designed for warp was Z-twisted, the Spinning Wheel running clockwise, and the yarn for weft was S-twisted, the Spinning Wheel running counter-clockwise.

Why it was spun in this way, we will probably never know. The tradition seems to go back as far as to the Bronze Age.

Perhaps when spinning long wool on a Hanging Spindle it is easiest to speed it up by twisting the Spindle in a clockwise direction. Spinning short wool with a "Spindle Shaft" on one's knee is easy to do counter-clockwise.

Besides, it is also a fact that a fabric with the warp and the weft twisted in the opposite directions, will be more pliant and adaptable. Such a fabric is also faster and easier to full. That is because in an S-twisted yarn the "back side" becomes a "Z-side". With the S-back side (Z) facing the Z-front side (Z) the adjoining fibres from the two threads will not become parallel. Therefore the two threads move easily against each other and the fibres easily penetrate in between each other during the fulling process.

4. When looking at old cloths of Wadmal one cannot find any made of yarn with two threads. The fibres in a single thread yarn are not as tight as in a yarn made of two threads. Again, this gives the threads and fibres the possibility to move more easily during the fulling.

5. When it comes to the binding of the fabric, we can choose between tabby and twill. The twill will, in percent, crimp more than the tabby, because of its looser binding.

We can also make the choice between different types of twill. The binding is usually not visible after the fulling, which means that there is no need to think of such effects of the binding as structure or pattern. It is the use of the material which is important when making the choice.

We may for example look at the clothing of the "Bocksten Man", a finding of a man from the 14th century with all his clothes, found in a bog. The warp in the cloak is Z-twisted. The weft is made of S-twisted wool yarn. The fabric is woven in "3-shaft" twill, which makes one side warp-faced and the other one weft-faced. This will give the cloak an outside mostly containing water-repellant hair and a warm, soft and fluffy inside. The cloth is slightly fulled in order to get the best quality, and is also windproof.

This is an illustration of how the knowledge of wool, spinning- and weaving technique has created an almost perfect coordination of all the good qualities.

6. How close the fabric shall be woven depends on how thick and tight the final fulled material shall be.

A comparatively loose material shrinks more than a close one. The fibres are not attached as hard to each other. It does not seem that the materials from older times were made as close.

A sample which I have from an old man is woven in a 35 dent reed with 2 threads in a dent. It seems to be a common quality. I have woven materials in about the same or looser qualities, and they have functioned just fine.

Of course it also depends on, among other things, the quality of the yarn.

7. The fulling time is important. Longer processing time gives a harder and tighter material.

Can We Make a Good Wadmal Today?

This completely depends on the type and quality of yarn available. In Sweden, yarns from the bigger spinning mill companies often contain a large part of shorter imported wool which may be difficult to full.

But the small spinning mill companies have started to make different types of yarns of pure Swedish Breed, which are easy to full. They also offer the opportunity to order both S- and Z-twisted yarns.

If you find a yarn with the right wool hair for warp, you must check that the yarn is spun hard enough. If you have the problem that the warp threads break, you can make them stronger by just painting them with boiled milk or paraffin mixed with some wheat-flour.

Yarn for weft you could probably spin yourself. Fabrics woven from home-spun yarn will get a very beautiful structure. It is the very small variations in the hand-spun yarns which give the material the special look.

To spin yarn for weft is not so difficult. The weft is best single threaded and this gives the possibility to wind up the yarn on the bobbin directly from the spinning spool. There is no need for washing the yarn, as it will be automatically washed during the fulling!

To get a good yarn you have to be very careful with the mixing of the wool before you comb it. Comb the wool two times and mix it in between. This is important because you cannot have one weft bobbin containing wool from one part of the sheep and another bobbin with wool from another part of different type and quality. In such a case the fabric will become banded and the edges undulated.

How to Get a Fabric Fulled Today

You may try to full it by yourself in some sort of a tub. It is best to use a tub made of wood. A tub of plastic or metal is too slippery and gives poor friction. Besides that, the water cools down too fast in such tubs.

If the fabric is made of machine spun yarn, you may need some soft soap to help the fulling process.

Start by thoroughly wetting the cloth in warm water of about 45 degrees Celsius (110–115° Fahrenheit).

Squeeze out the water. Then fold the material and put it in the empty tub, pour in the hot wash (soft soap water). The temperature of the water shall be about 50–60°C (130–140°F). It will cool down quite fast.

Now it is time to step into the tub and tread on the cloth while trying to turn it over with your feet. You may wear a pair of rubber-boots if you think it is too hot, but you will feel best what to do with bare feet.

When the water is cool, lower than about 30°C (85°F), you just pour it out, take up the fabric, refold it and put it back in the tub. Then pour in some new hot wash.

Take up and refold the fabric every time you change wash. At the last two or three times you change it, you can use pure clean hot water.

Take your time and do not hesitate to take a break now and then because it will do no harm for the fulling. Of course if there are two of you, you can replace each other and the fulling will be done sooner.

To full Wadmal in this way will take quite a time, normally 6 to 8 hours to get a well-fulled fabric.

Wadmal cloth is still as marvelous a quality product as it always has been. We must remember the old forgotten knowledge again. Unfortunately it is true that when a technique is retrograding, the refinements are the first to be forgotten. The technique itself usually survives for some time in a simplified shape. By studying old products and techniques we can bring back and use a lot of good knowledge gained by long practical experience.

We have today knowledge of the synthetic fibres. We should take advantage of them and use them for what they really are good for, and complement them with other materials where they are better.

We have today, for example, very good raincoats and rubber boots to use in really bad weather. But for a mountain tour, during mushroom hunting or other outdoor activities in chilly and foggy weather, as well as in a blizzard, the Wadmal protects the body better and in a more comfortable way than many of the overrated synthetic materials.

The Wadmal material is a very good insulator and at the same time it can take care of humidity and dampness, both from outside and from the inside.

We can learn to fashion old knowledge and experience into new high quality cloths with a nice-looking design.

References

Grenander-Nyberg, G. Lanthemmens vävstolar. Stockholm: Nordiska Museet 1975.

Gustafsson, K. & Waller, A. Ull - hemligheter - möjligheter- färdigheter. Stockholm: LT 1987.

Hoffman, M. Om dugmagere og töymagere og redskaperne deres. Olso: Ur By og bydg 1945.

Hoffmann, M. Rog og spinning i tukt og manufaktur husene. Oslo: Ur By og Bydg 1942.

Hoffman, M. The Warp-Weighted Loom. Oslo: 1982.

Nockert, M. Bockstensmannen oc hans dräkt. Varberg: Stiftelsen hallands Museer 1985.

Ryder, M. L. Sheep & Man. London: Gerald Duckworth & Co. 1983.

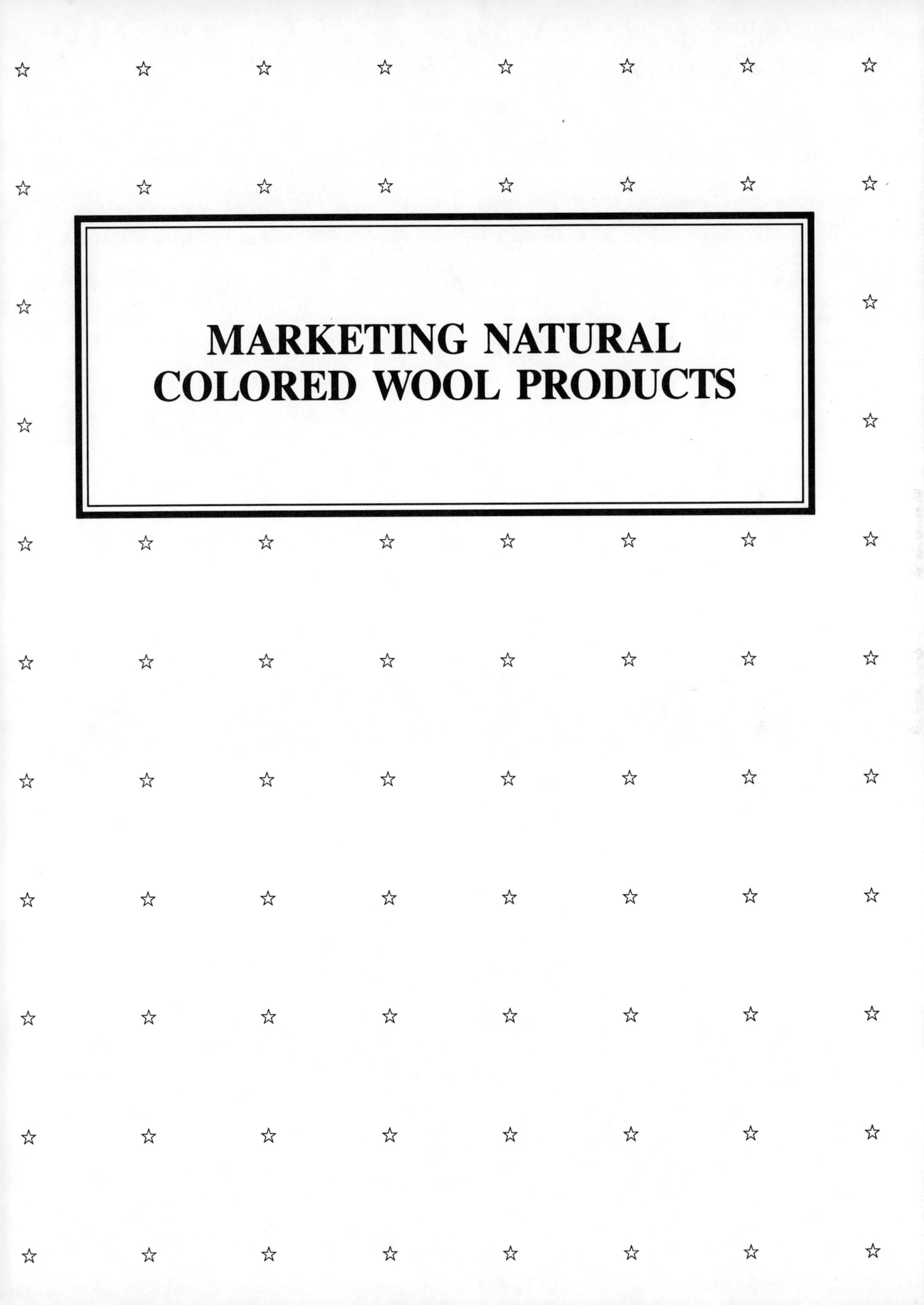

MARKETING NATURAL COLORED WOOL PRODUCTS

What? Colored Wools! Why?

Stanley Bulbach, Ph.D.
239 W. 15th Street
New York, NY 10011 USA

In the ancient Fertile Crescent, rivers in the midst of arid environments created unique conditions that accelerated the normal rate of human development. Here was the opportunity to exploit the waters for irrigated agriculture and for transport. These opportunities encouraged the development of management skills. After the last Ice Age ended in 10,000 B.C., it was there that the seeds of the first cities and their civilizations began to grow. By about 3,200 B.C., these peoples developed writing and history began.

The Mesopotamian cultures, like most of the Near East, used wool for the majority of their textile needs. Their health, comfort, protection from the elements, wealth, art, etc., all depended upon wool. As they learned to organize labor to attend the irrigation needs of agriculture, so too they organized labor and materials efficiently to produce more woolen goods than they needed for their basic personal wants. They were able then to produce non-essential luxury goods and a surplus to export. This was the foundation of a rapidly developing trade economy that profited these early cities and connected them together in a primitive international trade network.

The preeminence of the wool economy in the Near East and the Mediterranean continued with the early Greek shipping expansion through the Roman and Hellenistic Period for similar reasons. The wool industries were among the economic pillars of the Renaissance. Later, wool's primacy in Britain led the Industrial Revolution which transformed Europe and North America, and then the rest of the world.

The spin offs from the wool industries are many, including the dye industries which brought us to the modern photochemical industries. The evermore intricate looms brought us to complex mechanics, to Jacquard looms automated by hole punch cards, and to modern business machines which were the prototypes of today's computers.

However, in recent years, the public has forgotten the role played in history by the woolen textile industries. And most important for us as producers is that the public has lost most of its knowledge of wool. Today, there is almost no public awareness of variations in characteristics and qualities that differentiate wools. The public is provided little idea what to look for and what to guard against in judging wools.

I speak from first hand experience. In the mid 70's when I began to focus seriously on weaving, my first carpets disappointed me. They seemed to lack a physical vitality. Although I had done a considerable amount of research before I had begun, none of the basic publications ever informed me about the complex nature of wools. As I had no idea that wools varied as significantly as woods, I did not suspect that the yarns I bought might be inappropriate for my work's specifications.

At that time I was working on my doctorate in Near Eastern Studies. In examining ethnic carpets from that region, I saw unfamiliar wools with a very bright luster and large fiber diameter. This led me to visit growers of premium wools for handspinners, where I was introduced to Lincoln and

Romney wools. While I was distressed to add the work of selecting and spinning my own yarns to the daunting tasks of natural dyeing and handweaving, the processes of handspinning did provide me the special wool yarns that reproduced the qualities of the ethnic treasures that are so admirable.

As to the rich silver, black, and brown yarns and the muted dyed shades in the ethnic weavings, these of course were the wools from the colored sheep. In the mid-70's, their availability in the U.S. was greatly limited, and despite crop and market growth, they are still relatively rare. It has been most difficult to find information on these, for these colored wools are virtually unknown to most people. These colored wools have become integral to my art work, and I find that the public is very strongly attracted to them. And yet, when I talk with the public about these wools, their response is inevitably, "Colored wools! What?"

We all should note that the public is attracted to these wonderful colored wools. Now we must begin to tell them why they should purchase these wools. The almost total lack of product information for this potential market is the major challenge confronting us all today. Growing colored sheep has changed over the past decade or so. It was originally a very modest undertaking, a hobby with pets as it were, supported basically by personal dedication. The point has now been reached where basic business rules must be applied. Now as growers ambitiously attempt to increase flock sizes and to bring back lost breeds, these projects will need the support of major economic forces.

The ancient Greek origin of the word "economy" is revealing. The first part is from "oikos," the word for "home." The second part is from "nomos," from the word for "law." "Economy" is literally, "the law of the home," or as they say in gambling establishments, "the house rules." Depending upon how growers of premium wools position themselves, economic forces will either propel them or impede them. The future growth of the colored sheep industry needs an economic base to support it. What is that economic base? How is it being developed?

Craftspeople and artists are the prime market for the wools of colored sheep. And we seem to be reaching a point where today's spinners, knitters, and weavers are not buying up enough of these wools at a high enough price to support this small, but growing industry. Why? Because handspun premium wools remain a secret while no supportive market is being developed for this commodity's products.

Growers of premium wools and craftspeople who work with their wools are all part of the same production pipeline. Thus growers have a vital interest in not one but two markets: first, the craftspeople who buy their fleece; and second, the craft buying public. It is this craft buying public which now determines the additional volume and price of fleece that craftspeople can purchase from growers. That public could demand manyfold more products of premium wools, but growers and craftspeople will have to focus cooperatively upon developing this commonly shared market of potential growth.

What does the buying public know about these special wools? Very little. Today the public has almost no positive information on wool, and retains strong prejudices against it. Some of the bad image of this commodity comes from the past promotion campaigns by the synthetic fiber industry. Some of it comes from the common visibility of poor quality woolens, and the rare visibility of top quality woolens. If one talks with the public about wools, one usually hears that all wools are the same: itchy, mothprone, unwashable, changeable in size, and expensive to boot.

Here is the parmount task confronting growers and crafspeople who want to develop their market: The absence of adequate market education on wools. Growers! Why are your colored wools so special? And craftspeople! Why are your products made of these wools so special? The public must be informed.

The Renaissance of handspinning in the 60's occurred after much of the public had lost most of its knowledge of good woolens. Unfortunately, handspinners reintroduced wool to the public predominantly in ideological and visual terms, and not in technical terms. It was "back to nature," "wholesome," "self sufficient," "traditional," etc. It parodied itself by featuring yarns with wild variations in structure and with slubs. These structural defects, seen as ugly by a good many people, were prized largely because they imparted a "handmade look" to the yarns. The 60's craftworld developed a marketing message that has haunted us handspinners ever since: "Handspun yarns are special only because they look homespun." We must begin to tell the public about superior physical properties instead.

Today, we are in the midst of a period where there are vibrant markets in most consumer fields for "gourmet" products. Consumers know that products have common grades and better grades which are more expensive and luxurious. The public is also more curious than ever before, and will stop and listen to such information. And finally, the public is more willing nowadays to pay for those better quality commodities. It is the failure to address this economic reality adequately that obstructs market growth both for premium wool growers and craftspeople alike.

Growers and craftspeople can easily begin to develop a much larger market for their products with very little additional effort and investment. Since craft and market development are both educational fields, special market development supports are available in the vibrant educational network of fiber guilds, conferences, publications, etc.

Growers should participate in this network with a far more critical eye to educational information. It may come as a surprise, but most fiber craft journals do not impart adequate information on wools or on the significance of wool selection and handspinning to the final textile product. And yet, most editors I know would be happy to expand that focus for their readership. Growers who are advertisers should work more closely with the publications to enhance wool information in them. A heightened editorial focus on the integration of fiber used with the finished piece would benefit everyone in our field. It would provide market education for more craftspeople and it would provide craftspeople a curriculum for them in turn to take to their market: The public consumer.

In addition, premium wool growers' organizations should interact with the fiber crafts' many publications, guilds, and conferences, as an ongoing comprehensive program. Growers' organizations should be interacting with editors, curators, writers, etc., and commenting on published information on wool. For example, the developments forthcoming from this ambitious Congress should be interpreted for fiber craftspeople and shared in all their media over the upcoming year. This would not be an expensive undertaking, and it is fundamental to normal market development.

Similarly, growers and their organizations should interact with the major fiber craft organizations to their mutual benefit. For example, here the Handweavers Guild of America offers Certificates of Excellence in Handspinning; however a review of past volumes of their guild publication reveals that HGA has not yet integrated much fiber information into its membership's focus on weaving. The two related concerns, the finished textile and the fibers from which it is made, remain journalistically segregated on parallel but separated tracks. Again, it would be to everyone's benefit if growers asserted their educational priorities for their craftspeople who are organization members and for their public markets.

Growers' organizations should also think about collaborating with craftspeople to develop a specific product which will not only begin to educate the public, but will also be profitably marketable. With an eye especially to knitted goods, if the public knew about the superior qualities of handspun premium wools, they would clamber to buy such products, if the prices are not discouragingly high. What is so costly about knitted clothing of good handspun premium wools? Isn't it the manual labor in the handspinning and the handknitting? Then shouldn't we begin to inspect the difference in qualities between products of handspun that are handknitted and those that are machine knitted? Wouldn't machine knitted products of good handspun yarns retain most of the treasurable qualities we extol, and yet be far more marketable and profitable than handknitted ones?

Currently the craft media focus on art clothing and special fashion, which are unique products for which there is almost no market. And yet the market today is strongest for professionals and business people who need work clothing that is elegant, serviceable, comfortable, sturdy, and medium priced. The craft media focus primarily on female fashion, but the real market is at least half male. The craft media focus primarily on white and dyed yarns, while the market is strongly predisposed toward browns, blacks, and greys; that is, toward the exact spectrum of colored wools. The liason between colored wool growing, handspinning, and machine knitting would find a very strong public market today, and yet this whole area of cooperative activity has thus far been overlooked.

I would encourage the colored sheep organizations to support simple research here. Some growers' organization should plan a cooperative project with a major craft organization, publisher, or teaching institution. They should schedule a workshop where growers, handspinners, and machine knitters will meet to learn about each other's work and to publish their results in cooperating magazines.

When I have traveled to conferences where handspinners congregate, I have always been greatly impressed at the quality of the work they exhibit. This is work that is invariably ignored by "contemporary craft" institutions and the art related media. I find that this work fascinates the public and attracts far more interest than most fiber art does. It is high time for a major public exhibition of handspun wool projects based on the mastery of traditional techniques. Again, growers organizations should ally themselves with a major craft organization, publisher, or educational institution to put together a museum exhibition that could travel widely to fascinate and educate the public.

Growers of handspinning wools and handspinners confront significant economic hardships. And yet, it is amazing to see how far the field has advanced in recent memory. Today growers and craftspeople have it within their grasp to open up a vibrant market. Today, the supports in our fields and the potential public interest are there .

Thus far one major constituency is absent from this equation. What about the major sheep and wool growing organizations? What about the major wool industries? Complaints have been made that many of these have hesitated to support the handspinning wool movement.

Actually, any such hesitations at this point would seem natural, albeit ill advised. Newcomers on the block always disturb the establishment, and premium wool growers and craftspeople can easily be misidentified as distractions or threats. In approaching these organizations for help in my projects, I too have confronted my fair share of slammed doors before finding help. These rejections were very troubling in light of the strong support other crafts enjoy from the major industries using their materials. The wood, paper, ceramic, dye, glass, precious metal and other industries, have a history of generously supporting craft for the public. Those benefactors understand the importance to their market development of art and craft.

Again, craft is education. Corporate support for craft provides corporate industry advertising benefits that are far, far less expensive than standard Madison Avenue costs. Consider that a page of advertising can cost way above $14,000 in a newspaper and many times that amount in a magazine. A $14,000 grant that helps sponsor an exhibition may reach fewer people, but they all pay full attention to the message: "This material is on exhibition because it is important." Furthermore, that craft usually then receives free publicity in many of the same newspapers and magazines.

We should emphasize to the wool industries the extent to which they can benefit from supporting our public projects. Personally, I have found strong support from the wool industry when I focus on concrete business realities and clarify exactly where I think we can help each other by working together. Our common interest is bringing the importance of wool to the attention of the public.

I would first encourage premium wool growers and craftspeople to expect slammed doors and to navigate them as diplomatically as possible. Design the projects to include goals of interest to potential benefactors. This usually means wool information for the public and a serious effort to ensure some publicity. Above all, avoid presentations or appeals based upon cuteness, "homespun" amateurism, etc. Emphasize information and public outreach, and try to detail how effective and inexpensive such a market development project is compared with regular advertising costs. Emphasize how powerfully the public is drawn to exhibited craftwork, and how fine craft lends prestige to their industrial produce.

Often it is wiser first to solicit help indirectly. Such would be the opportunity if a growers' organization successfully interested a host institution in producing a travelling exhibition of crafted wools. With its experience in developing corporate support, that sponsoring museum or gallery would then make the appeal to the wool industries on behalf of the growers' organization.

In recent years the surprising growth in the field of colored wools has suddenly revealed where supports are needed to continue progress. Market development and public consumer education are essential priorities. The general public is our most important market, but the public simply does not have enough product information on why they should seek out and buy products made of colored wools. And since the public will not be buying most of its colored wools directly from the growers, but from craftspeople or their representatives, growers and craftspeople must plan together for their market development needs.

"What! Colored wools?" The public always asks this. We must answer them immediately with specific information "Why"—why they should buy one of history's most important and treasurable aesthetic luxuries: Fine quality, premium wools of beautiful, rich, natural colors.

Integrating Technology with the Romance of Local Wool Processing

Robert Donnelly
P. O. Box 327
Arroyo Hondo, New Mexico 87513 USA

This paper presents a comparison of "romantic" intuitive, creative thinking versus "classical," practical, technological thinking. The need for integrating these two thought processes into an holistic approach for successful wool enterprises is addressed, and we shall discuss examples of how this integration has worked in the world of wool today.

The problem here is that there seems to be a dichotomy. On the one hand are the "romantics" who see the big picture as it is (or could be if only the classical thinking technocrats would open their eyes and their hearts and get out of their narrow technological ruts).

Romantic: 2. fanciful; unpractical; unrealistic: 3. imbued with or dominated by idealism, a desire for adventure, chivalry, etc.
—*Random House Dictionary,* 1981

Then on the other hand are the "classics" who see technology as presenting the capabilities to do anything, if only the romantics would get practical about it and acknowledge its limitations as they exist today.

Classic: 10. accepted as standard and authoritative, as distinguished from novel or experimental. —*Ibid.*

These selected definitions are used to illustrate the basic differences in the motivational impulses which drive those types of individuals. There are probably other definitions that can be used, but these will suffice for this paper. Robert M. Pirsig discussed this apparent dichotomy throughout his book *Zen and the Art of Motorcycle Maintenance* in great detail.

Romantic vs. Classic in Local Resource Wool Processing

In terms of the "Local Resource Wool Processing Industry," we are discussing the people who want to convert their (or their neighbor's) wool into saleable merchandise, but know only the hand processes of doing so. These are the "romantics", they know in their hearts that it could be done, but are unable or unwilling to access the technical information to do so. The beauty, wonderful texture and comfort of the wool products are the paramount qualities they wish to exploit. The romantics "know" intuitively that what they want to do *can* be done, they just don't know how.

Also in this industry or near to it are the "classics". They may not necessarily want to disrupt or prevent the emergence of an alternative wool processing industry, however, they do want to keep the status quo intact and to represent technology as unfathomable and unchangeable in how it works. The classic thinkers also believe that the "romantics" are wishing for an impossible world.

The industrial revolution was concentrated in the textile industry. Furthermore, the wool textile industry has been an integral part of the United States' economy since before its birth as a nation. It has long been the "classic" way of doing things which adds to the value of wool so that its market price at the retail store (in sweater or suit form) is ten to twenty times higher than at the farm gate.

We can visualize our two local resource wool processing characters standing back-to-back, with the romantic thinker looking into the future, and the classical thinker looking into the past, both believing that the present time (now) is immutably fixed and unwilling to acknowledge the viability of the other's viewpoint. This is not a very encouraging picture, BUT by simply introducing the intangible element of "consciousness" we can change the situation from static (no movement) to dynamic, and the process of integration can begin.

The Difference *Can* Be Resolved

The differences, although opposed, are not mutually exclusive when examined in the arena of consciousness. Those differences can be minimized to acceptable and workable levels when both types of people integrate their thinking with that of the other person. This integration is not as difficult as it may seem. Any individual or group of people who have the desire to get involved in wool processing can do it! Here's how.

There needs to be, first and above all, a dedication to the overall objectives of the enterprise. What exactly is it you wish to do? Statements like, "providing quality merchandise at a reasonable price;" "employing local work-force;" "filling a market niche" should dominate the objectives. The objectives themselves will become a blend of the romantic and the classical.

I strongly suggest that "making money" should be understood as a given. If you're *not* planning to make money at your chosen enterprise, it matters little whether it is a success or a failure. Making money in today's world is as necessary as breathing. You could, however, embellish the idea of making money and state that you wish to be "handsomely rewarded" for your efforts.

It is *not* noble to be impoverished, and frankly, it greatly impedes the success of any enterprise. This industry can service a desire for quality products, demand and receive a reasonable return, and provide "Right Livelihood" for all involved. Some words of caution: *Beware of Greed!*

Secondly, the integration process needs an openness to the "other side". What that means is, allow the notion that your way is *not* "The Only Way" some room in your consciousness. Sometimes it is difficult for one to erase the habitually held beliefs that one has acquired, or inherited from one's ancestors, but this kind of dogmatic thinking process will doom your enterprise at a most critical time.

It simply is a question of *responsibility.* It is irresponsible to close your mind to the other viewpoint. You have committed yourself and your associates to be responsible to your customers and to the vendors who sell to you.

You are embarking on a "new" adventure, that seems to be heading through uncharted territory. This reality is very true for you. However, to do this adventure responsibly, you should remember that there has been local resource wool processing going on for thousands of years by spinners, weavers, felters and merchants in many different lands on the face of this planet. The knowledge of the technologies necessary to process wool in the appropriate manner is available to everyone. There are books and other sources which can be tapped to gather whatever information is needed.

If you can't open your consciousness to these two ideas (i.e.: 1. dedication to the objectives of the enterprise; 2. opening to the other viewpoint), take that as a strong warning that any success in the marketplace is going to be a struggle for you. When you as a "romantic" can accept the "classic", and you as the "classic" can begin to open up to the "romantic" viewpoint, the result will be more "wholeness" for you as an individual and an ability to work in both realms. More importantly, from the standpoint of the enterprise, the results of integrated thinking will be a smoother operation, and ultimate success.

Recognizing the Potential Problems

In 1980, there was an article done in *National Wool Grower* magazine, the publication of the National Wool Growers Association, in which I was interviewed by the editor. At that time, the movement to process wools locally had just begun to stir in the United States, and there were a lot of romantic ideas flying about regarding the emergence of this industry. Response forms printed with the article were returned to me with statements that amounted to saying, "Wouldn't it be wonderful if the fiber artists and sheep people could get their wools processed into yarns, fabrics, etc., by some magical method or machinery, at a price and place that is workable?"

There existed at that time (in the United States) the "mystique" of the wool processing industry which was mostly on the East Coast and kept information seekers, who lived mostly in the Western U.S., outside the mill environs.

There existed also the wool buyers who went out into the countryside to buy wool from the growers and to convince them that the market was "soft" and prices were depressed. "Why complain?", they'd say, "the govenment is willing to make up the difference between what you're getting for your wool and what you should be getting." these buyers didn't know (so they said) what was going on in the textile mills, only that the market price for wool was low and was going to stay that way.

In 1985, six years after the magazine interview, I prepared and presented a paper for the Wool on a Small Scale Conference in Logan, Utah. The paper was quite technical in nature, and I suggest you read it. The information is still current, and you'll get some insight into the "classical" approach to wool processing.

At the time I wrote the paper I was thinking in the "classical" mode with but a small amount of sympathy for the "romantics". I knew what they wanted and I was hoping that they would "get the message" and embrace technology as something worthwhile; not rejecting it out of hand. I have spent my entire career working with technology in many forms, and am convinced that it is a wonderful tool that we can use to make our work simple and effortless as well as handsomely rewarding.

Technology is not anybody's property. Patents have been issued to provide a "legal" claim to inventions, and as such, patents and copyrights give the classical impression of "ownership". But the truth of the matter is that ideas are free floating entities which find expression *through* the inventive mind. Inventive and creative people are really "channelers" for the information that is available to all.

These inventive minds can be found in both romantic and classic thinkers. Many early day inventors could be categorized as romantics. Cyrus McCormack, Thomas Edison, etc., were more imbued with the possibilities of the idea that had come to them than with the probability they would get rich. Altruism existed to a great extent in the birth of many of today's greatest technological advances, even television.

Jose Arguelles in his very dense and substantive book *Earth Ascending,* writes about the "psi bank" which he says exists in the radiation belts surrounding the Earth. In the psi bank, he explains, resides all the knowledge that there is and he gives rather detailed examples of how access to this gigantic data bank has been accomplished by civilizations in the past.

Today, information networking is producing some of the same results on a more mundane level. There are many methods of access available, as simple as dialing a telephone. They seem simple to us today, but can still be considered a "breakthrough" in expanding our access to knowledge. Computers are just the "tools" for the consciousness to use (or toys to play with) in the process of creating today's realities. It is possible, with a little imagination, to predict the future will be filled with people who have direct conscious contact with all the information stored in the psi-bank.

So the dichotomy has remained more or less intact for another decade. Our two characters are still standing back to back, stuck, as it were in the immutable present, not looking or moving into the other thinkers territory. Still, there are those who feel in their hearts that something *must* be done. Those "romantics" are still at it. Why can't they see that the "system" has been a long time established and it is not going to change or adapt to their perceived needs? To suggest that an alternative wool manufacturing and marketing system could work is viewed by some classical thinkers as nearly heretical, and certainly hysterical.

I'm pleased to say that in 1989 the integration process has begun, and in some instances is well under way! There are some fine examples of small scale wool processing enterprises throughout this country. I suspect if you'd look at them closely you'll probably find both a romantic dreamer and a classic technologist working together. On rare occasions, this could be one person.

Also, I must say that in some cases the integration process did not "take" and the enterprise is no longer in operation. (Although the enterprise is no longer in operation, its leftover equipment is being put to use by another person or group of people.) So even in the case of failures, there should be no grief, for the technology lives on, as well as the dream.

A Dramatic Change

Wool prices (especially for the finer wools) went up dramatically in the past few years. Why? Many people have many "reasons" for the increase, but it really has to do with demand. Demand by the wool consuming countries who buy large quantities from the Australians and other global wool producing countries has been increasing. As demand increases and supplies remain relatively stable, the price goes up. (Classic supply and demand economics!)

So the "romantic" says, "It's easy to see why people want wool. It's a natural fiber, it feels good, it 'breathes' and has all the other 'well known' attributes of wool." But how did these attributes get so well known? This is not the kind of information that parents pass on to their children in Chicago, Illinois or Tokyo, Japan on a day to day basis.

The awareness can be credited to the Woolmark Program of the International Wool Secretariat. IWS is funded by the wool producers in the Southern Hemisphere nations of Australia, New Zealand, South Africa, and Uruguay. Woolmark is a well financed and professionally managed *merchandising* campaign, that is further supported by another well financed product development and technology program.

There is also a United States wool promotion program from the American Sheep Producers Council, but its mark (Made in America with American Wool) is not nearly so well known, and the technology program is but a fraction of that of the IWS. I personally feel that without the IWS, demand for wool would have been muffled to a mere whimper. Wool producers of the world would be far fewer in number and there would be severe financial stress for the wool industry.

There was another seemingly "magic" and invisible ingredient in the phenomenon of good wool prices and that is the Floor Price Scheme which maintains the price of all wools at a reasonable level, especially when demand is low. Briefly, in this plan the Australian Wool Corporation (similar organizations can be found in other countries, except the United States) establishes a "minimum" price for wool. If auction prices are below this minimum the corporation buys from the producer and holds the wool for future sale. There is a lot of *faith* that underlies this kind of a system, but it has proven to be worthwhile in maintaining reasonable returns to the grower. I am told that there was a great deal of resistance to organizing the AWC for purposes of marketing wool, and later against the Floor Price Scheme.

How the IWS Blends the Classical with the Romantic

1. IWS uses its influence in every possible way to get research programs into various departments of agriculture and other scientific establishments. (Basic research is "romantic" in that its major objective is to go on researching without a particular goal.)

2. IWS also sponsors work in the product development and processing technology fields, either at independent institutions or in its own facilities. (This type of work is "classical" in its pragmatic approach to the objective of maintaining wool as a desirable fiber in the mills and the marketplace.)

3. IWS has developed world-wide recognition of the Woolmark. (The Woolmark is "romantic" in that it appeals to the sensitivities of the ultimate customer. The mark expresses quality graphically and simply. (It is also "classical" by the way it is supported in the marketplace with hard cash promotions and quality control standards.)

4. IWS member nations all have a price support program to maintain the price of wool. (This is "classical" in its results of producing a reasonable return to the wool producer.)

While there appears to be a predominance of "classical" minded people in the AWC and IWS, there is also the "romantic" notion that sheep farming is an honorable profession and the work is good work. (Buddhists call it right livelihood.)

The Australians (and I might add, most sheep producers in other countries, too) exhibit a unique blend of the two thought processes (romantic and classical). How does this work? I think it has a lot to do with sheep, and with shepherding as work inspired by a higher ideal than simply "making money". Many of you are in the sheep business and know intuitively what that statement means. It isn't something that can be entered into the balance sheet for the operation, but is something that is indispensable for its success.

So there was stiff opposition to the ideas presented above, but the persistence of the "romantics" in the organizing group endured the nay-sayers and the results are something to look to as a model.

Can There Be a Coloured Wool Organization Similar to the IWS?

The "romantic" in me says, "There certainly can." The "classical" in me says, "Not in your or my lifetime." This paper is not designed to establish the International Coloured Wool Secretariat, but merely to point out to those who will listen and observe, what has come to pass in the models of the IWS and AWC.

On a less grand scale than international, can anything be done to facilitate a better situation for the manufacturing and the marketing/merchandising of wool? Why not? Surveys, business plans and all the other paraphernalia that go with starting up and managing a small business *are* necessary, but unless you get to do those two simple steps described earlier in this paper, (1. dedication to the objectives of the enterprise; 2. opening to the other viewpoint) it all becomes an exercise to pass the time.

Aphorisms to Help You Integrate

- Transcend your limitations. Know that there is and always will be a market for quality products.
- Know that technology is nobody's property, and it can be accessed for whatever you need, if you open your mind up to the possibility, and don't be dissuaded by the rutted technocrats.
- Establish your enterprise on the highest possible foundation. Good work and right livelihood are the noblest things that can be done by anyone.
- Be honest and impeccable in whatever you do.

References

Pirsig, Robert M. *Zen and the Art of Motorcycle Maintenance,* Bantam Books, New York, NY, 1985.

National Wool Grower, November 1980, pp. 23–27.

Current Status of Technology Available for Processing Wool into Finished Products. Proceedings of the Wool on a Small Scale Conference, June 23–26, 1985, pp. 142–157.

Arguelles, Jose. *Earth Ascending,* Shambala Publications, Inc., Boulder, CO, 1984.

Wool as a Cottage Industry

Paula Simmons
Pat Green Carders Ltd.
48793 Chilliwack Lake Rd.
Sardis, B.C., V2R 2P1, Canada

I consider a wool-based cottage industry as the key to the good life; enjoyment of healthy living and simple pleasures, while working profitably at home. This is something that can involve a whole family, without going off the property to work. It can provide wholesome employment for growing children and teach them valuable work habits, and can be started with a modest investment. There is the option of working as hard and as long as you want, to expand the business quickly or to have your home-business as a part-time occupation, and grow at a more leisurely pace.

Regarding financing, I wouldn't encourage anyone to go into debt to start, better to start small and grow with the business. Less impressive but safe—and you do get a chance to be experienced before having to handle a large number of sheep or a massive volume of business. The really foolproof way is part-time while you have a job to pay the bills. When I started, I had a salary job in town, which I was able to cut down gradually a day at a time as my business grew. With less panic about finances, you can be more concerned with quality and more selective about the items you offer.

Priorities

One real need is to establish priorities for your time and energy. Some sacrifices are usually required in order to achieve any worthwhile goal. It helps to be realistic about the demands of complete commitment, and realize that you must not allow other pursuits to distract you from what you decide is your real priority. Persistence, enthusiasm and a total dedication to your goal are far more important than talent, training or financing. For long-term profitability, the business must come first.

The following thoughts are directed to people with the entrepreneurial spirit, who have the drive and initiative to get things done.

Options

The many potentials of a wool-related cottage industry can be roughly broken down into three phases or categories:

THE BASIC INDUSTRY—farming, raising sheep and selling the raw wool and lambs and breeding stock, plus imaginative use of by-products.

THE HANDCRAFT APPROACH—raising wool, hand processing, selling yarn or knitted or woven garments in addition to sheep and wool.

THE SERVICE LEVEL—usually supplementing income from sheep and from some amount of handcraft. This could include custom shearing, custom dyeing or spinning, custom carding and marketing processed wool.

The demarcation between these three levels is only on paper, in actuality they can overlap considerably as a business grows and diversifies.

Level I. The Basic Industry

Even in "just farming," there are opportunities to enhance your income by related projects that are not normally part of standard sheep raising: Tanned pelts, for instance. Anyone raising sheep which have an unusually attractive fleece, such as your naturally colored sheep, should consider taking advantage of tanned pelt sales. Home tanning is a possibility, but only if someone in the family has the time to develop the skill, and time to do the work involved. Sending pelts away for tanning is the usual route in the United States, with the profit from the sale of an attractive pelt making it well worth the investment in tanning and the cost of shipping. With my dark sheep, I found it convenient to shear any animals headed for slaughter, then wait until enough wool had grown out to make a good shearling length, and ask for the return of the pelt from the slaughter house, for tanning.

Teaching at Prairie Mary's Acres in Iowa, I went crazy over one of her brown Coopworth pelts. It was so lovely, price didn't matter, nor the fact that I didn't need another pelt. Actually, I didn't stop at one, the Jacob pelts were so fascinating. There was one that had been damaged at the tannery and cut in half to sell, so I got one half to use on a loom bench. These pelt sales prove a very good addition to Mary's sheep and wool and yarn sales.

One question always to consider; are you making the very best return possible for your products? You with colored sheep have a great advantage over those with only white sheep. The wool—if it is clean of vegetation and carefully sheared and skirted—should bring a premium price marketed direct to handspinners. The condition of the wool is the main factor, not just the degree of color or the breed of sheep. Fleeces that have tender breaks, excessively weathered tips, a profusion of second cuts and contamination of vegetation, are not worth a premium price and will not promote repeat sales. Many growers are turning to sheepcovers to protect fleeces from both weathering and vegetation such as weed seeds, burrs, hay and bedding. These covers do not protect the neck area or the belly wool, but in prime fleeces these portions are removed by generous skirting.

To sell raw wool profitably takes knowledge and understanding of the market and your product. Whether you are raising wool or buying it for resale, it is essential to have a thorough technical understanding—you must know more about wool and its uses than either your supplier or your customers. Without this knowledge, it is difficult to speak with believable confidence.[1]

While the spinners' market is a large and viable one, many growers do not take it seriously. They still think that the amount of wool used by spinners is just "a drop in the bucket." The larger growers can afford to take that attitude, but small-flock owners need to get the very best return per sheep in order to have a profitable situation.

If you wonder about the potential market, it is estimated that there are at least 200,000 people in the United States who are engaged in spinning and/or weaving. There are many cottage industry wool enterprises that are buying or importing large quantities of raw wool for processing. These are your potential customers. In addition, there are well over 500 local and regional weaving and spinning guilds for easy marketing access.

By taking woolgrowing seriously, and selling at the best possible price, you can make your sheepraising more profitable.

Prime Quality Fleece

Try to learn how a prime fleece should look. It is most important to recognize the value of prime quality fleece, how good it must be, and be able to easily differentiate between the very best and just good average wool. They must be priced accordingly.

It is a waste of your very best wool to send it to a wool pool or a commercial market, where you'll receive a price that is being paid for the average quality. The fleeces you have that are obviously well above average in cleanliness, color and general attractiveness should be marketed in such a way as to obtain a price in keeping with that quality.

Cleanliness from vegetation is not the only criterion. It takes a sheep a year to grow a good fleece, and it is a waste to have its value downgraded as a result of poorly managed shearing. Spinners do not want their black fleeces contaminated with white snips or white fleeces containing dark wool or dark snips. A shearing tarp or plywood floor should be swept between each sheep sheared, and holding areas should be coarse gravel rather than sawdust or wood chips which can contaminate the wool. The shearer should understand what you want and be willing to cooperate, keeping the head and leg wool out of the fleece, and avoiding second cuts. It is better to have the sheep look a little ragged than to have its appearance tidied up by overlapping strokes. For spinners, overly-generous skirting is necessary, as well as removing any other parts of the fleece that are badly contaminated with grain, hay or burrs.[2]

Continue to care for the wool after it is sheared and skirted. Damp fleeces should not be packed into bags, and bags not stored on damp cement or the dirt floor of a barn. Never toss unwrapped fleeces on top of hay bales where they can get littered with hay leaves. Once you have a good reputation for the production of "prime" fleeces, you will not have any long-storage problems.

A separate room or building can be used to store wool that you will sell right from your farm. Priscilla Blosser Rainey of The River Farm has a separate building in which she displays all her fleeces for sale, on slatted wooden shelves. Just before shearing each year, she reduces the price on any fleeces left over from the previous year. Then, she cleans out her shed and moth-sprays it, in preparation for the new clip.

Once you are familiar with the desires of discriminating spinners, you will see that not all fleeces are spinning quality. Sort out the best, and the lower quality could still sell if priced appropriately. It will be the middle-of-the-road product, for the middle price. Even a systematic program of washing and storing the lower quality wool can be profitable. As washed wool, it may be sufficiently attractive to bring a good price from spinners or felters. Machine carding can make it even more attractive and easy to use. Some growers save all of their skirtings and wash them to send away for custom carding. The price of dark wool depends on quality and/or on value-added techniques.

Even if you plan to sell all your wool in the grease, it helps to show some washed, to illustrate how beautiful it can be. To please all customers, you could keep a supply of washed wool on hand to sell, washing more as needed from your supply of grease fleeces.

Packaging can contribute to the attractiveness of wool. If you have a small showroom at your farm, find a variety of ways to put up wool for sale. In addition to whole fleeces, try one-pound packages, both grease and washed wool. Have some "beginner spinner kits" that contain a pound of fleece and a simple drop spindle, for instance.

Dyeing is another excellent option to consider. It can improve the attractiveness of white wool and is really stunning when pale gray wool is used. After washing and dyeing, it can be picked to blend two or more colors. Packaging for sale could be in bags of one color, or of several pounds in assorted harmonizing colors.

Familiarity with wool crafts will be of advantage to anyone who is involved in the sheep or wool business. I am astonished at the number of sheep raisers who have no knowledge of spinning, and show no interest in it. Are they the same ones who do not eat lamb and mutton? Appreciation of your product as well as knowledge about it, is a great asset in selling. Not only is your enthusiasm contagious, but you can give good advice and help to the customer, always useful in promoting repeat sales. The person who sells locker lamb will benefit from providing good recipes for the use of lamb and mutton, and the person who sells wool to handspinners should be knowledgeable about this use, in order to guarantee a satisfied customer. There is no better way to convince a sheepraiser about the spinner's requirements of wool that is clean of vegetation, than to get that grower to spin the wool. A high degree of proficiency is not a critical element for wool selling, but knowledge how to spin *well* has advantages over just a casual skill, for it offers other options. Teaching, for instance. Giving spinning lessons can create a new crop of spinners who are automatically wool users. Students will buy fleece to take home to spin, and thus become steady customers—especially if your wool is the high quality that makes it a pleasure to spin.

Meat

In the marketing of your meat crop, it is also possible to increase your income by concentrating on quality, and on direct marketing. The best quality here also commands the best price, as it should. This quality will need to be promoted in merchandising and advertising, to reach potential customers. You may have read the article on the Jamieson Farms.[3] Their promotion of their lean milk-fed, expertly cut and trimmed Cryovac packaged lamb has resulted in a prosperous mail-order business direct to the consumers, appreciative and demanding individuals and gourmet restaurants. Shipping by Federal Express gets their product to the customer quickly, to almost every state in the continental U.S. Their advertising budget is high and other expenses include a toll-free 800 number, a computer to keep mailing lists and track customer purchases, printed recipes for the use of lamb, brochures, logos and a very specialized cutting and trimming. They find that their customers are willing to pay a premium price for extra quality and special service. The Jamiesons like to eat lamb and are always trying a new recipe or method of cooking, so customer satisfaction is assured because special recipes have been furnished to them.

To realize more profit on cull ewes, and have fun in the process, try turning them into sausage.[4] Home sausage making requires very little investment; a very good recipe and instruction book, grinder, mixing pans, measuring spoons, accurate thermometer and a sausage stuffer. Commercial kits are available that contain everything needed to make a particular variety of sausage, including instructions, pre-mixed spice blend, cures, etc. Once sausage has been successfully made from a standard recipe, individual alterations can produce a very personal and unique product. A home sausage making hobby can develop into a nice little sideline business.

Level II. The Craft Connection

The crafts of spinning and weaving have been respected professions throughout the ages, usually involving whole families in the various processes. After the Industrial Revolution there was a total acceptance of machine-made products, with spinning and weaving becoming "lost arts." Now the trend has reversed, with consumers wanting real fibers and handcrafted high quality. "Mass produced" is no longer a recommendation. Plastic and polyester and pollution do not enhance the quality of life, while organically grown crops and high quality handmade yarn and garments are among the good things of life—highly desired and commanding a premium price.

It is honorable and acceptable to be self-sufficient, to grow your food and make your clothes. And, most envied of all situations, to make your living without going to work as a "servant" in someone else's business.

When I started spinning 35 years ago, it was with the serious intent of earning a living. I could gather no encouragement from others who were doing it, for I knew of none. Even so, it worked into a comfortable business based on a small dark flock to provide a nucleus of my wool needs. The rest of the fleece I bought, averaging about 1000 pounds a year, all turned into handspun yarn and handwoven blankets and garments. Now, an unbelievable number of spinners are making a partial or complete income from selling handspun yarn. There is a fantastic array of handspun available, many textures, colors, fiber blends, plies and yarn sizes. In general, most find it of advantage to zero-in on a particular type of yarn, a recognizable style, once they have decided what they enjoy spinning and what seems to sell best for them.[5] However, a person who sells their handspun should know how to knit and weave, to be able to advise the customer about successful yarn use, and know enough about these crafts to properly spin to order.

Even if you are primarily interested in selling wool, the spinning of some amount of yarn for sale has advantages—you can use whatever wool you do not sell as raw or processed wool, your yarn can actually be a "promotion" for your grease wool by vividly demonstrating how it looks when it is spun. Spinning is something that can be done seasonally, at a time of year when you are not busy gardening or canning or with seasonal sheep care. A really proficient spinner can bring in a substantial income.

To do a handspun yarn business by mail, it is necessary to have sample cards, either separate cards for weavers (more textured yarns) and one for knitters, or one card for both. Charge enough for the cards to make it worth your while to make them, which takes yarn and time, as well as the cost for envelopes, printed price lists, and stamps.

One way to sell handspun yarn (or finished products) is at arts and crafts fairs. These are often seasonal, two or three-day weekend affairs that charge entry fees for space and collect from 20% to 40% in commissions on sales. With most, there is advance screening of entrants, and the well-juried fairs usually sell best.

If you have attended any of these, you will know how much depends on how things are displayed. Avoid confusion in display, or too elaborate a presentation, since it appeals to the esthetic nature rather than the desire to buy, and will draw more compliments than sales. Be sure everything is plainly priced. Have brochures or catalogs to hand out, and plenty of merchandise, with items in a wide price range—something for every pocketbook. Your first fair appearance may not be profitable, but if you observe reactions and listen to comments and see how other craftspeople handle their sales, it could be a valuable learning experience and show you how to manage the next exposure.

The way you present your products, even by mail, will have a great deal to do with their success. You need to pay special attention to efficient packing and wrapping, neat and well packaged. Printed shipping labels, which are not expensive, are a good investment. You need a typewriter for legible replies and a simple brochure or even a one-page price list about your products. In answering letters, your reply needs to be professional in appearance, but friendly and personal, to make a truly confident and quality impression. Try to answer all letters as quickly as possible, not just for the sake of the customer, but also because, if letters start to pile up, it becomes depressing and hard to tackle. It saves time and postage to answer as fully as you can the first time, even trying to anticipate future questions. The next letter to you may then be an order rather than a request for more information. It is good practice to acknowledge all orders received by mail if shipment will be delayed or not from stock. A post card is sufficient. This is doubly important if the payment is sent with the order, so the customer will know it has arrived. Give an estimated delivery time. Even though this may have been stated on your sample card, people do not read and remember every word.

In selling either wool or yarn, give generous weight. Check your scales, then add a little extra measure. If your customer weighs it, the overage will give a pleasant surprise. If you had given slightly under-weight, even though inadvertantly, you would have lost a customer and even all the potential ones who have any contact with that person.

Tissue paper wrapping for yarn is only proper and expected. Even a specially careful wrapping of raw fleece is well appreciated. You want your product to look valuable when it arrives.

Keep a mailing list of all your customers, to be able to contact them when you have a new product. A Christmas open-house at your studio is a good occasion for sending out a mailing. You can display an array of yarn and finished items, all priced, Serve coffee and tea (I served homemade wine) and show people around your studio, explaining processes and equipment.

A newspaper announcement, as distinct from a direct mailing, could be seen by many more people, but it would not necessarily reach the ones who are interested in you and what you are selling. It is the *personal* interest that produces the most results.

While it is possible to earn a living by spinning alone, life is more interesting with some diversification. The development of a product made from your handspun, knitted or woven or crocheted, can lend variety to your routine as well as adding interest to your business and your bank account. Combining two different skills in a product is one way to achieve a distinctive effect. This could be a crocheted edging on knitted or woven garments, macrame edging on woven hangings, or knitted portions in felted wool garments.

One handcrafted product that sells well in cold climates is a handspun knitted mitten with a felted lining. The process is much simpler than it looks in the finished article; it is just plain knitting, but with a slender strip of carded unspun wool carried along on the inside and bound into the stitches as they are knit. When the mittens are done, they are turned inside out and the unspun wool rubbed with thick soap mixture until it is lightly felted. A most impressive effect, with a small amount of work. Made of handspun yarn, with real wool felt lining, they command a price far in excess of the amount that could be charged for the yarn and the time expended in knitting.

While handspun yarn, and its uses, appear to realize the highest return per ounce or pound of wool used, there are even "craft" uses of wool that do not involve the skills of spinning, weaving or knitting. These would include feltmaking with raw wool, the larger garment pieces being the most profitable but requiring good design and sewing to justify the necessary price.

Locker Hooking, with either raw fleece locks or carded wool, can be used to make such products as horse blankets, rugs, heavy jackets and wall hangings. It is an easily learned craft, and does not require any substantial investment or bulky equipment.[6,7]

Level III. Service Activities

Custom services can be related to the craft level, usually, in that they include custom spinning or weaving, and often start with the crafts before progressing into a service area.

The safe way to start a custom service is to start small, with minimal equipment. The smaller tools are not wasted when you decide to get larger machinery, since they have resale value, or can be kept for small sample tasks. While hand tools are very labor intensive, they give the opportunity to measure the demand, get a feeling as to which way you really want to go, and work up a unique product or specialty on which you can expand.

Probably the most interesting and profitable of the many service possibilities is that of wool processing. Many who already process wool for their own projects, find that there is a demand for the specific service of processing. Washing wool and picking and carding is something that many spinners either do not have the time or facilities or the desire to do—they want to spend their available time spinning. Even the ones who spin as a profession, often find that their time is best spent completely on the spinning, not in preparing the fibers, so they will gladly farm out the preliminary processes in order to concentrate on the finished products. Anyone who already has, or wants to have, the equipment to do this work for their own products, can increase their income by taking on custom jobs.

The decision to do custom processing can evolve from the need to obtain a better return per pound of wool than could be realized by selling it in the grease. By taking in custom carding in addition to the work on their own fleeces, the cost of equipment is paid off quicker. There is a relatively low cost in setting up a carding business. Few other enterprises can recoup their capital within the first year, as has been substantiated by many businesses equipped with our "Pat Green Cottage Industry carders."

Most of the owners of our cottage carders do a wide range of services at the beginning—processing their own wool to multiply its value, custom carding for other growers and for spinners, selling carded wool to spinners and to shops, marketing quilt batts and felting wool, dyeing wool and carding it into Rainbow Batts and variegated color rovings (sliver), and using their carded wool for their own craft creations including yarn, weaving, quilted mattress underlays or comforters, plant dusters, etc. Purchasing a small amount of exotic fibers such as alpaca, silk, bunny and mohair, and blending them with wool, can give a far greater return for the time invested. When it comes to custom-blending for others, this is normally charged at a higher rate per quantity than for plain wool.[8]

We have noticed that after the first year, many businesses will settle down and specialize, the choice of the specialty being a balance of what seems most profitable and what they most enjoy doing. The making of this choice would sound as if it limits or restricts their income, but just the contrary. Having a specialty seems to promote more business. There are several reasons for this: specialization confers a certain air of status, you are more easily remembered, and the fact that you are exclusive about what you do gives the impression that you know more about it than do others. In practice, this becomes a reality because it has absorbed all your time and interest. Often a specialty will evolve itself from repeated customer demands for a service not otherwise available.

To establish a custom carding business, get set up to do an efficient job of wool washing and drying. To be machine carded, the wool must first be free of grease. Insist on offering a re-washing service if needed. Poor material, especially heavily contaminated, can damage your equipment. Heavy burrs can bend carding cloth teeth, as well as dispersing into the carded wool to make it appear more contaminated than before. Let customers know that carding does not remove vegetation and second cuts.

Our Cottage Industry carders have the option of a "bump winder" that automatically compresses and draws out the roving (sliver), then winds it up into a center-pull ball of up to 1½ pounds (750 g). It can also make a thick two yard (2 meter) batt. This versatility allows the best wool to be made into roving for easy shipment to spinners or shops, while the lower quality can still be suitable, in large batts, for quilts and for garment feltmaking. Another special feature of the carders is their ability to handle a great variety of fibers, including the long wools and mohair.

Customers who have previously sent wool away to larger facilities will discover that the small cottage operation can do a much more personal and specialized service, keeping each fleece or portion of a fleece separate for maximum variety and interest, or blending in other fibers for exotic effects.

Rainbow Batts, in many layers of dyed or natural color, are particularly popular, as are two-color and three-color rovings, which come out in lush ribbons of color and spin up into a yarn spiralled with these colors. Skillful work, combined with thoughtful and personal service can build a thriving business with loyal repeat customers. Actually, the easy availability of a small processing service can spur spinners into an increased production and wool use by alleviating the burden of pre-spinning chores. In other words, whatever potential there appears to be, may be an underestimation.

Conclusion

The foregoing is not a figment of my imagination, nor is it based on book learning or a collection of other people's experiences. It is really a condensation of my own enterprise, over quite a long successful period.

We ran a flock of about 30 dark sheep, none purebred. Our breeding program was not consciously based on genetics, but depended on visually developing a crossbreed with suitable fleece for the kind of yarn that I marketed. We didn't know much about breeding then, but wanted healthy productive animals, so we selected our rams from the best of our stock, and seldom brought in any outside blood. This policy seemed to have acted as an effective disease control. With careful culling and selection for fiber length, softness and color, the resulting fleeces were spun into a soft medium-weight one-ply yarn that sold well to spinners and knitters. While I spun some finer yarn for my own use, it was not economic to offer it or to offer two-ply, because either would have necessitated a higher price than was acceptable in those days.

We sold lambs locally, ate the cull ewes, did custom-shearing for the wool, and sold handspun yarn by mail. Weaving and garment sales added a wider-based income. A powered carding machine made it possible to double the yarn/garment output the first year that it was used.

The logical result of years of experience initiated some other diversifications: I wrote books and taught workshops

Pat Green's Extended Cottage Industry carder with automatic bump-winder as well as six-foot (2 meter) batt making capacity.

around the country, usually three week trips, flying every night and teaching every day. For my cross-country trips, I had to mail ahead some of my nicely carded wool to each place to assure that classes had the well-prepared fibers needed to learn new techniques. I noted that there was an obvious lack of cottage industry size carders. This observation eventually led to my involvement with the design and manufacture of wool processing equipment. My new endeavor gives me the opportunity to be of assistance to others who are ready to take the plunge and pursue the cottage industry life.[9]

References

[1]Simmons, P. (1976, 18th printing, new edition 1989) Raising Sheep the Modern Way, Storey Publications.

[2]Simmons, P. "Prosperity with a Small Flock," Sheep! Magazine, January-February 1987.

[3]"Jamieson Farms goes Direct to the Consumer," Sheep! Magazine, October-November 1988.

[4]Salsbury, D. L., "Home Sausage Making," Southern Sheep Producer, September 1988, Southeastern Sheepman, October 1988.

[5]Simmons, P. (1979) Handspinner's Guide to Selling, Pacific Search Press.

[6]Rough, J. Z. (1982) Australian Locker Hooking, Fox Hollow Fibers.

[7]Nickerson, S. (1988) Australian Locker Hooking, The Crafty Ewe.

[8]Simmons, P. (1985) Turning Wool into a Cottage Industry, Madrona Press.

[9]Simmons, P. (1977) Spinning and Weaving With Wool, Globe Pequot Press. (1982) Spinning for Softness and Speed, Graphicom.

[10]Simmons, P. Pat Green Carders Limited, (604) 858-6020.

Developments in the Coloured Wool Industries in the Australian States

C. H. S. Dolling
Sheep Breeding Consultant
P. O. Box 74, McLaren Vale,
South Australia 5171 Australia

Since World War II, the coloured wool sold at auction in Australia has shown three trends.

The first period, from the wool selling season of 1946–47 to that of 1977–78, saw virtually no change in the number of farm bales of Merino wool sold at auction (Hayman and Cooper, 1965; Dolling, 1984a). And since 1971–72 at least, the same obtained for crossbred wools (Figure 1).

The source of the coloured wools, termed 'Black and Grey' in the Australian Wool Corporation Type List, was from sheep born in white woolled flocks and shorn before they were slaughtered.

The second period, from 1978–79 to 1983–84, saw an increase from 2,576 bales to 9,843 bales. Assuming that coloured sheep born in white flocks continued to produce about 2,500 bales each year, one concludes that the additional 7,000 and more bales were produced from flocks reared especially to grow coloured wools, these flocks having grown rapidly during the five-year period.

The third period to 1987–88 has seen a gradual decline for four years down to 8,449 bales which probably reflects a plateau or slight reduction in numbers of sheep in flocks of coloured sheep.

The beginning of the second period coincided with the first Congress on Breeding Coloured Sheep in 1979, the beginning of the third period with the second Congress in 1984. The recipe for change in coloured wool production could well begin "Firstly, find your Congress . . . "!

Merino Wools and Crossbred Wools

The Type List of Australian wools, used in their domestic and international marketing, contains a total of 727 types, differentiated on average fibre diameter, staple length, and all of the individual characteristics which contribute to style. There are 348 Merino types, 354 crossbred types, 9 downs wool types and 16 carpet wool types (Australian Wool Corporation, 1988)

Merino wools are of 24.5 or fewer microns in average fibre diameter. Most crossbred wools have higher fibre diameters, but those which do not are classified as crossbred by virtue of the absence of the style characteristics of Merino wool. There is not a precise one-for-one relationship between breed of sheep and type of wool. A very coarse fleece from a Merino sheep would be typed as a crossbred wool; and a fleece with all the characeristics of Merino wool, but from a non-Merino sheep, would by typed as a Merino wool. And crossbred wools don't all come from breed crosses—they include the wools from British longwool breeds and several breeds developed in Australia and New Zealand.

There are 7 types which cover black and grey wool—3 Merino and 4 crossbred. The types are shown in Table 1. The letters F, B, C, D, Y, and K are six of the ten secondary classifications based on vegetable matter content which may be superimposed over the numbered types.

TABLE 1 Black and Grey

Merino Fleece		PCS/BLS	Crossbred Fleece			PCS/BLS
Good	Average	Lms. etc.	26.5 up	26.6-30.5	30.6 down	Lms. etc.
24.5 up	24.5 up	24.5 up				
320	321	322	606	607	608	609
—F-B-C-D—		B-C-D-Y-K	—F-B-C-D—			B-C-D-Y-K

sheep or Sp^{sp}/Sp^{sp} black spots (or piebald) sheep (Dolling 1989) and that most coloured crossbred wool is grown on sheep in the coloured sheep flocks, not in white flocks.

Avenues for Sale of Coloured Wools

There are four avenues of sale of coloured wools for Australian growers, namely

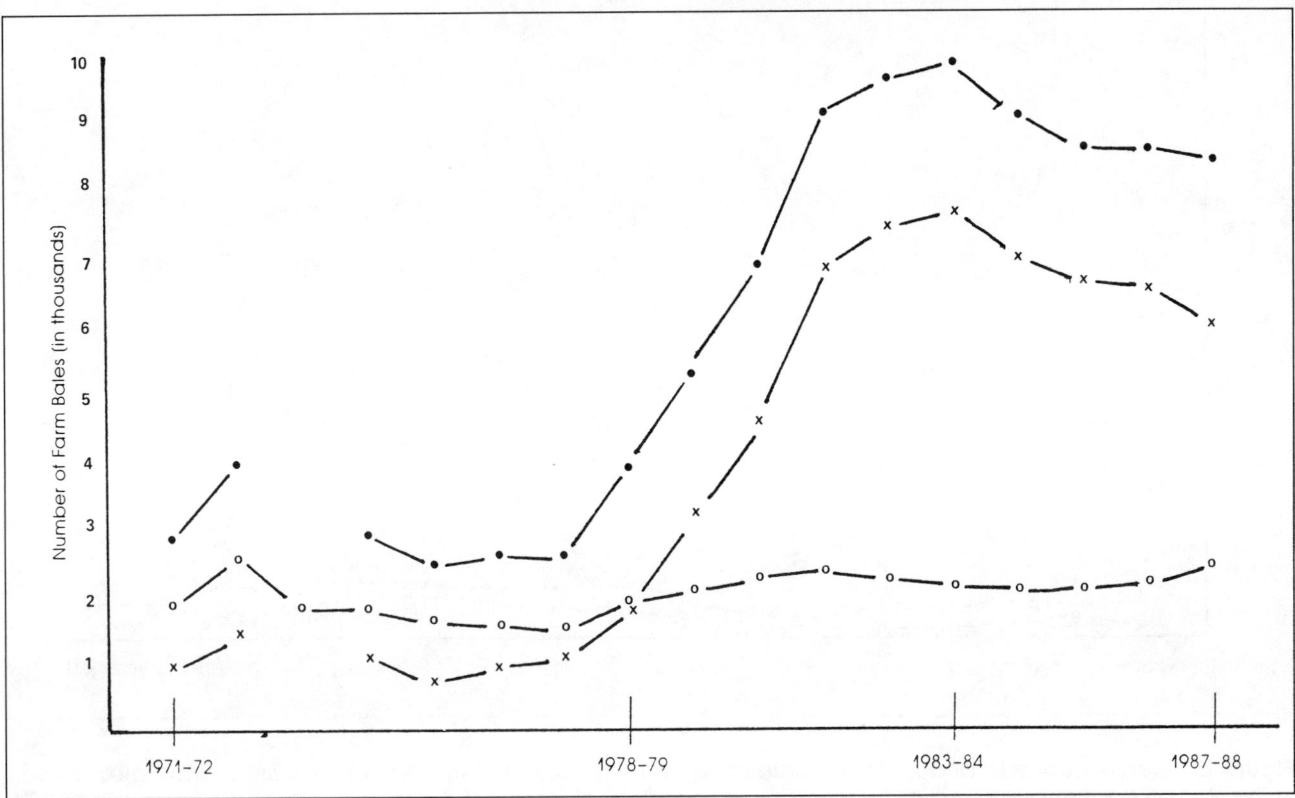

Figure 1 Greasy wool sold at auction in Australia, 1971–72 to 1987–88. Numbers of Merino and of crossbred farm bales of 'black and grey' wool. The average net bale weight in 1986–87 was 172 kg. for all 'black and grey' and white wools. o = Merino x = crossbred • = Total.

One can see from Figure 1 that crossbred wools contributed very much more to the increase in black and grey wools sold at auction than did Merino wools.

Among the crossbred wools, the increase in pieces, bellies and lambs' wool (type 609) was much greater than that of all three types of fleece wool put together (Figure 2). This is no doubt a consequence of the sale of fleece wool on the handcraft market, leaving the oddments for sale by auction.

Crossbred coloured wools are expressed as a percentage of all greasy crossbred wool auctioned in Australia in Figure 3; Merino coloured wools are presented similarly. One interprets this figure as supporting the conclusion that most coloured Merino wool auctioned has its origin in white Merino flocks—the sheep mostly being $A^a/A^a(w_3w_3)$ self colour

a. To handcraft workers, notably spinners and weavers
b. For mill spinning and weaving in specified mills
c. For processing in unspecified mills
   1) through the auction system
   2) through wool buyers

To Handcraft Workers

The best quality fleece wool is sold direct to handcraft workers, through retail shops, or at shows or similar functions from stands organised by the breeders or by the coloured sheep societies.

The types of wool produced by different sheep breeds are shown Table 2, together with comments on the end-uses considered appropriate for each type.

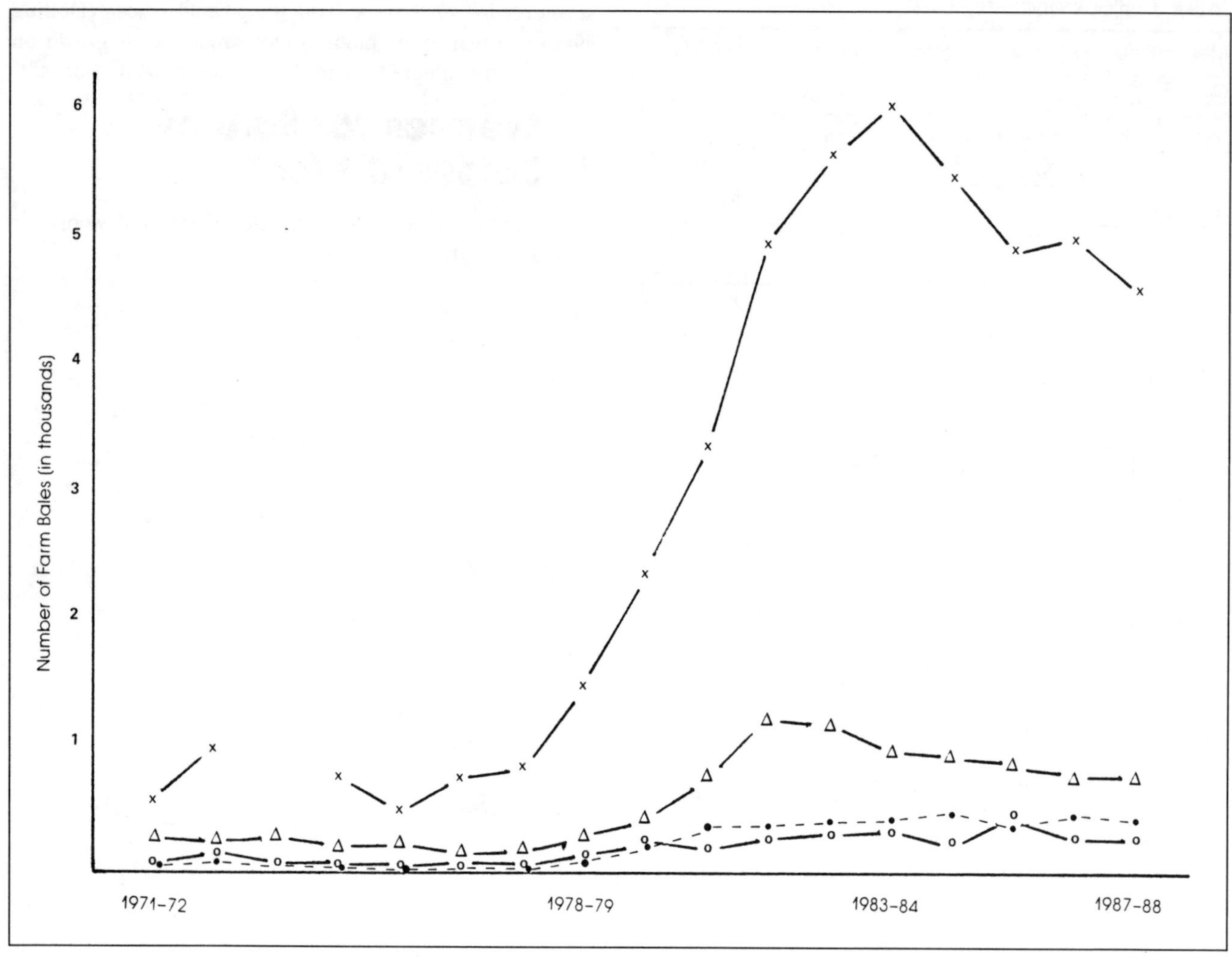

Figure 2 Greasy wool sold at auction in Australia, 1971–72 to 1987–88. Numbers of crossbred farm bales auction in the four categories described as fleece with average fibre diameter 26.5 microns or less (o), fleece 26.6 to 30.5 microns (Δ), fleece 30.6 microns or more (•), and as pieces, bellies and lambs' wool (×).

Increasing numbers of coloured sheep are rugged to keep the fleece free of dirt and vegetable matter, and without a faded, wasty tip (Dennis 1979). This highlights the change in attitude from 'growing coloured wool for spinners' to 'growing high quality wool for a very selective handcraft market.'

Some growers sell not only greasy wool but carded sliver also; they arrange for the scouring and carding of their own wools to tap a wider market.

For Mill spinning and Weaving in Specified Mills

Undyed yarns of coloured wools are mill-spun by several organisations in South Australia and Victoria. Yarn shades range from a natural black through a range of greys to a light silver; and, when adequate supply is available, of a brown. Merino wools are used for fine soft-handling yarns, crossbred wools for more bulky yarns.

The emphasis of earlier years on rugged outdoor yarns has yielded to finer and finer yarns, down to a 2 x folded yarn of 220 TEX (2 singles of 18 Yorkshire skein).

A recent significant trend is to knit a small amount of a primary colour or two into a garment of otherwise all undyed coloured wool shades. A different market is tapped.

Fabrics, also, are mill-produced. In marketing wool in this form, one needs to bear in mind the greater amounts of money tied up in each kilogram of saleable product than is the case with knitting yarn.

If processing is under the control of a breed society, a vertically integrated enterprise can be developed with the growers making their own marketing decisions (Dolling 1984b).

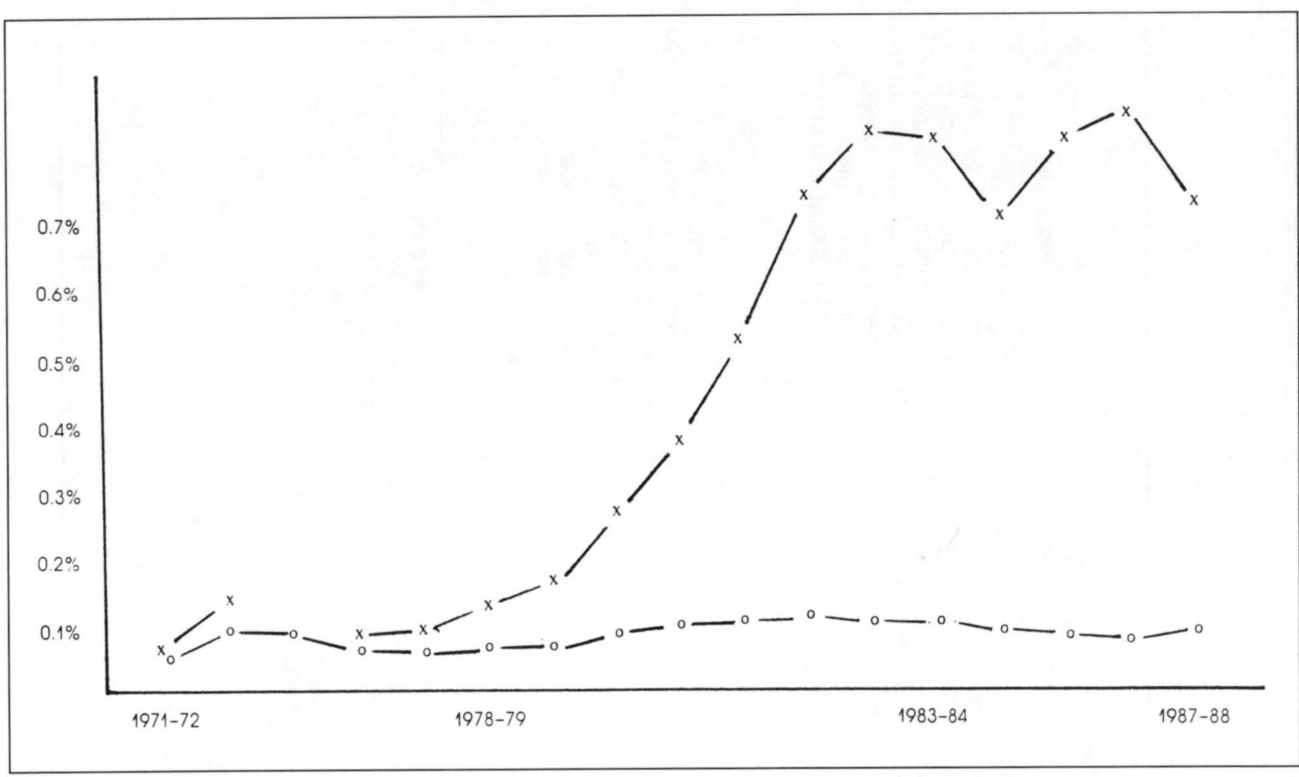

Figure 3 Greasy wool sold at auction in Australia, 1971-72 to 1987-88. 'Black and grey' Merino wool as a percent of all Merino wool sold; and 'black an grey' crossbred wool as a percentage of all crossbred wool sold. Units were farm bales. o = Merino x = Crossbred.

For Processing in Unspecified Mills

Greasy wool sold at auction goes to unspecified mills—which makes it difficult for the seller to consider the precise needs of the buyer when the wool is being prepared.

Through the Auction System. The main aim in preparation is to have at least three bales in a lot within any one of the types of 'black and grey' wools. Fibre diameter and clean scoured yield measurements are publicised on such lots, prior to sale.

The reception of wools by breed societies can greatly help in achieving this.

A society may receive the wool from each grower, class it, and sell under the name of the society; growers are reimbursed by the society.

An alternative procedure is for the society to receive and forward each grower's wool to a bulk classing store, the store doing the classing, selling the wool and paying each grower directly.

Through Wool Buyers. Growers sell to wool buyers who class and sell on the auction system. This is generally the least profitable, even if the most convenient, avenue for sale.

Prices of Merino and Crossbred Wools Sold at Auction

The low wool prices of the early 1970s and the marked increases of recent years are evident in Table 1.

The other feature is that even the best types of coloured wools reach auction prices significantly below the Australian average.

Of the avenues for sale summarised earlier, by far the most profitable is, of course, the sale of high quality wools for handcraft use.

The Coloured Sheep Breeders' Societies

The traditional pattern of development of organisations in Australia has been within each State, a pattern ingrained in each colony before federation in 1901. Coloured sheep breeders saw no reason to change the pattern so we find a coloured sheep breeders' society in each of the five mainland States—Western Australia, South Australia, Victoria, New South Wales and Queensland.

Different societies have emphasised different aspects of coloured wool production and marketing. The improved preparation of wool for sale by auction has been pioneered

TABLE 2 Crossbred Wools, British Long Wools and Downs Wools Grown in Australia—and Their Uses

Breed of sheep	*Description of wool*	*Quality or count*	*Fibre diameter (microns)*	*Crimps per 2.5 cm*	*Staple length for 12 months' growth (cm)*	*Comments on Spinning and on the yarn*	*Products*
Corriedale Perendale Romney Merino x Perendale Romney Border Leicester English Leicester Lincoln	Fine crossbred	56s to 54s	27 to 29	4 to 6	10 to 15	Easy to comb and spin. Spin finer than required but do not overspin. Good breed for looms, and elasticity of yarn overcomes tension problems. Dye well.	*Knitting and crochet:* finer qualities suitable for baby wear, lacy blouses, medium to heavy cardigans and jumpers. *Weaves:* finer qualities suitable for stoles, lightweight tweed. Stronger qualities suitable for rugs, wall hangings, carpets.
	Medium crossbred	50s to 48s	30 to 33	2 to 3	12 to 17		
British long wool breeds Romney Border Leicester English Leicester Lincoln	Strong crossbred	46s to 44s	34 to 36	1	14 to 20	Dye extemely well. Spin with firm tension.	*Knitting and crochet:* finer qualities suitable for heavy skiwear and hard wearing outer garments, pram covers, blankets, berets. *Weaving:* rugs, carpets, wall hangings, heavy tweed fabric bags.
	Extra strong crossbred	40s and lower	37 and higher	less than 1	18 to 25		
Short wool or downs breeds Dorset Suffolk Ryeland South-Suffolk Shropshire Hampshire Dorset Down Southdown Merino x Downs	Downs	64s to 50s	21 to 30	4 to 12	5 to 9	Generally spin poorly. May be too short, and may part poorly. Crimp and staples usually poorly defined. Spongy. Not generally recommended. If longer than 7.5 cm can be spun readily with practice. Useful where elasticity is required. Ideal for spinning a 'woollen' type yarn.	*Knitting and crochet:* socks, blankets, rug hooking, shawls, light jumpers and cardigans. *Weave:* uncrushable fabric e.g. tweed.

TABLE 2, continued Australian Merino and Comeback Wools—and Their Uses

Breed of sheep	*Description of wool*	*Quality or count*	*Fibre diameter (microns)*	*Crimps per 2.5 cm*	*Staple length for 12 months' growth (cm)*	*Comments on spinning on the yarn*	*Products*
Merino Poll Merino	Superfine Merino	70s and higher	19 and finer	14 to 15 and higher	5 to 7	Difficult to spin. Generally too short. Parts and dyes poorly. Felts easily.	Not recommended.
	Fine Merino	64s	21 to 22	12 to 13	7.5 to 10	Spin slowly with little tension in order to preserve softness and firmness by getting more twist. Ply firmly.	*All fine knitting crochet:* Hair-pin lace, baby wear, shawls, light jumpers and cardigans.
	Medium Merino	60s	23 to 24	10 to 11	8 to 11	Usually very good spinning. Dye evenly and well.	*Weaves:* Lace stoles and evening fabrics. Suitable for weft. Not suitable for heavy knitwear because constant washing may cause cause matting and shrinkage.
	Strong Merino	58s	25 to 26	7 to 9	9 to 12		Heavier knitted and crocheted outerwear.
	Extra-Strong Merino	56s to 50s	27 to 30	5 to 6	10 to 12		
Comeback Polwarth Cormo Zenith	Comeback	60s to 58s	23 to 26	7 to 11	10 to 15	Generally of longer staple length and easier to spin than Merino of similar quality. Shrinkage much less than Merino. Dye evenly and well.	*Knitting and crochet:* Very suitable for baby wear and adult lacy knitwear, light to medium weight jumpers and cardigans. *Weaves:* Suitable for light fabrics requiring good draping qualities, particularly lace weaves.

From Fleet (1978). Revised with Elaine Ray in 1989.

TABLE 3 Average prices of greasy wool sold at auction in Australia, 1971–72 to 1987–88 (in cents/kg of clean scoured wool). All wool sold; black and grey Merino fleece (Type 320); and black and grey crossbred fleece (Type 606). Types 320 and 606 are the best quality Merino and crossbred wools offered for auction.

	Average price in cents per kg of clean scoured wool		
		Black and Grey Wool without vegetable matter fault	
Wool Selling Season	*All wool sold—white, plus black and grey—with and without vegetable matter fault*	*Merino Fleece Good. 24.5 microns or less Type 320*	*Crossbred Fleece 26.5 microns or less Type 606*
1971–72	125.7	77.9	51.3
1972–73	313.5	232.4	none sold
1973–74	299.3	209.4	168.8
1974–75	203.3	159.6	119.0
1975–76	228.7	178.5	222.1
1976–77	302.1	216.3	212.8
1977–78	302.5	240.9	254.1
1978–79	328.5	271.1	259.8
1979–80	390.1	315.8	297.2
1980–81	410.5	312.5	277.0
1981–82	423.4	341.4	324.0
1982–83	433.9	319.1	268.0
1983–84	467.5	363.3	348.9
1984–85	498.3	416.0	337.9
1985–86	537.7	464.0	414.8
1986–87	620.0	475.9	444.1
1987–88	981.0	826.7	665.4

by Western Australia and Victoria. The preparation of wool for use by specified mills has been developed by South Australia. Two of the breed societies produce annually a Flock Book.

But in all societies, and in regions within societies, the local marketing of wool for handcraft use is of paramount importance—and, in this, liason with handspinners' and weavers' groups.

Cooperation between societies is perhaps best exemplified by the staging of the National Coloured Fleece Competition by different societies around the country. The first Competition was at the 1979 Congress; the ninth, in 1989, is organised by the Northern Regional Group of New South Wales.

The names of the five societies should be a useful reference for those interested in coloured wool production in any given area of Australia. They are:

Melanian Sheep Breeders' Society of Australia Inc. (of Western Australia).

South Australian Coloured Sheep Owners' Society Inc—S.A.C.S.O.S.

Black and Coloured Sheep Breeders' Association of Australia Inc (of Victoria)

Black and Coloured Sheep Breeders' Association (N.S.W.) Inc.

Queensland Coloured Sheep Owners' Association.

References

Australian Wool Corporation (1988). Australian Wool Sale Statistics. Analysis by wool type; Statistical Analysis 18B.

Dennis, W. S. (1979). Sheepcoats for black and coloured Polwarth sheep. *Proc. Congress on Breeding Coloured Sheep and Using Coloured Wool,* Adelaide, South Australia. Peacock Publications, pp. 100–101.

Dolling, C. H. S. (1984a). The coloured sheep industry in Australia. *Proc. Congress on Coloured Sheep and their Products,* New Zealand, pp. 21–27.

Dolling, C. H. S. (1984b). Sable wool—a growers' marketing project. *Proc. Congress on Coloured Sheep and their Products,* New Zealand, pp. 355–359.

Dolling, C. H . S. (1989). Pigmentation in the Australian Merino. *Proc. 1989 World Congress on Coloured Sheep,* Oregon, U.S.A.

Fleet, M. R. (1978). Selecting wool for hand spinning. Department of Agriculture and Fisheries, South Australia, Bulletin No 1/78.

Hayman, R. H. and Cooper, D. W. (1965). The frequency of pigmented sheep in the Australian Merino. *Wool Technology and Sheep Breeding,* 12:81.

The source of information for the figures and for Table 3 was the series of booklets published by the Australian Wool Corporation 'Australian Wool Sale Statistics', Series B from 1971–72 to 1987–88.

A Professional Guide for Marketing Handspun via Mail Order

Arlene Mintzer
2168 64th Street
Brooklyn, New York 11204 USA

The following information is what I consider to be an important check list in the setting up of a mail order hand spinning business. It is my hope that what I have written will enable you to approach the advertising and marketing of your handspun yarn in a clear and methodical manner.

Handspun colored wool is a very special product and as spinners it is our responsibility to promote those qualities as professionally as possible.

First and foremost you need to decide who your potential audience might be. Are you catering to knitters, crocheters and weavers? If so, then you have a very broad range of publications to choose from. Once you've decided who your audience is, then you can make a list of current periodicals that you wish to advertise in. Keep in mind that many magazines have new product sections in which they list products gratis. Local newspapers are always looking for human interest stories. Think about why your business is special and send out a press release, listing those qualities that make your particular products unique.

It is very important, however, that you have stationery printed up before you contact any of these publications. Even just the simplest presentation lets the world know that you take yourself seriously as a professional. You need not order stationery in huge quantities. Depending on where you live you may decide it's easier to order it from a mail order business catalog. If you don't type, then please find a good typist who can type up a press release for you and a cover letter if you wish to include one. It is also a good idea to give a follow-up phone call in about one week.

Sample Cards

Your sample card is the most important sales tool for your product. It has got to be *beautiful.* It is the best visual and tactile representation of your yarn. You want your potential customer to feel that they must have some of it. I use a business card printed on colored stock for sample card purposes. It has my business address on it, which for me is a P. O. box. This is a very cost effective way to have a separate mailing address for business purposes and again to present a highly professional image. On the back of that same card you may want to print several lines such as description, price and weight. You can then fill in this information by hand, thus allowing you to use these cards even if prices change. You can put together as many cards as you like. I have mine punched with 2 holes. One is so I can thread my yarn through one end of the card and the other is so that I can tie cards together.

Many spinners are concerned with the price that they can charge for these cards. Do not undercharge for your cards. Color cards are very time consuming to make. I have found that customers are eager to buy my sample cards. One of the special touches I add to mine is a simple mini skein of my yarn. This is just a small skein that is included with my sample cards, so that potential customers can try one of my yarns. Perhaps you will decide that your sample card cost is refundable once your potential customer places an order of a certain dollar value. Many yarn companies practice this policy. A separate price list should also be included with your sample card. Don't let your customers have to do guess work. Always be direct with them and clearly state all your policies up front. Some of the most important points to think about are as follows:

a. A brief description of your yarns, containing fiber content, yards (meters) per ounce and skein put-up.
b. Suggested gauges and needle sizes appropriate for knitting and crocheting, for example.
c. Minimum orders—How many skeins or numbers of ounces must be purchased at one time?
d. Payment—Various methods of payment includes checks, money orders, COD and credit cards. You must also think about whether you will accept foreign checks. If you do decide to accept them, then you should state that all checks be payable in U.S. funds and drawn on a U.S. bank. When I purchase goods from overseas I generally send an International Money Order, although there are businesses that allow you to send American checks with an additional amount for conversion of funds to their specific currency.
e. Do you want to sell both retail and wholesale? I feel that it is important to have a separate price list for each. As to your pricing guidelines, it is a very individual matter. I do think, however, you must not undersell your wholesale customer. If, for example, your suggested retail price for a specific yarn is $5.00 per ounce, then that's what you should charge. I've known a number of retail shop owners who stopped doing business with small cottage industry spinners, because they found out that the spinner was charging his or her retail customer the same price as it was charging the shop owner. If you do decide to sell wholesale, you must be very clear about your payment policy from the start.

   It is also a good idea to state that prices are subject to change without notice.

   Most retail shop owners would ideally like to pay you on a net 30 days basis, once they have established credit with you. Many of the cottage industry spinners I know, are not able to run their businesses profitably with such a long stretch between orders being filled and payments of them. Therefore, it is perfectly acceptable to request pre-payment on all orders and certainly considered crucial, if you do any custom spinning.
f. Returns—Do you accept returns? If so, how long can the yarn be kept before it is returned? Once again, if the order is custom spun, I would state clearly on your price list if it is returnable or not.
g. Delivery Time—This is another one of those extremely important points that I'd like to elaborate on. It is very difficult to figure out exactly how much time you'll need to fill multiple orders. The best thing I can say is allow yourself plenty of time to fill an order. You can always deliver an order earlier than originally stated. It certainly will do a great deal for your professional image, if your customer gets an order sooner than originally planned. If you are going to need additional time with an order, make sure your customer knows about it. You need not explain why you're late. Just drop a note stating the new delivery date. I'd much rather tell a customer that I'm going to be late with an order, before he or she calls me up and has to ask—where is my order?
h. Shipping—State how you will send orders and specific shipping costs. You may want to have a separate order form with shipping costs printed directly on it. Remember to include a handling charge with postage fees. It takes time to pack orders carefully. Let your customers know that you are as proud of the finished product as they will be, when they receive it.
i. Phone Orders—It is important to state your days and hours of business on your price list. If you are not available to take calls much of the time, then perhaps you can have a recorded message that asks that customers try you again, or that you will call them back. It is important that little children do not answer the phone during business hours as this does not help your professional image.

Projects

Many people love beautiful handspun yarns, but need to be encouraged to use them. Perhaps you may want to offer them a simple project such as a hat or scarf that can be made out of your yarn. You can offer the pattern as a free gift when they purchase the yarn. You may also want to sell the pattern by the dozen with wholesale orders, for a nominal fee. You could offer it free, to be copied by the shop and given out to customers for use with your yarn. Should you decide to sell an original pattern, make sure to put a copyright notice on the pattern.

Sheep are beautiful creatures and wool an excellent product of their incredibly good looks. It is through our creative talents and professional outreach that the consumer will want to use more and more of this unique and wonderful fiber. It is our job to keep the magic of handspinning alive and well. It is also important to educate the public as to just how very special it is!

Sales and Marketing of Natural Coloured Wool in New Zealand

Wendy Lynch
The Patch
RD4
Rotorua, New Zealand

Sheep were brought to New Zealand as recently as the 1820's to provide meat, spinning wool, and sheepskins for the first settlers.

Natural coloured sheep have been in New Zealand for as long as the traditional white sheep, but in such small numbers as to hardly affect wool sales. They are still regarded as the poor relations of the mainstream white flocks.

Accurate records of numbers have not been kept by The New Zealand Wool Board nor the Black and Coloured Sheep Breeders Association. However the 1988 estimates, based on the sales and use of coloured wool over a twelve month period, show that there are about 150,000 coloured sheep, producing over 4000 bales of wool per year.

The New Zealand white sheep flock is about 65 million, so natural coloured sheep represent about .23% of the total.

Types of Natural Coloured Wool

Handcraft fleeces

At the top end of the market, breeders producing handcraft fleeces generally have no trouble selling them from $6 to $20 per kilo, depending on quality and colour. Selling is by individual fleeces, not bale lines.

Colour-Sorted Commercial Wool

The second type of wool is the good colour-sorted commercial wool, for making into yarn, for knitting, weaving or blankets. At January 1989 this wool was selling at almost white wool prices in the North Island but up to $2 per kilo less in the South as end users are mostly in the North Island.

Oddments

The bottom end of the market, i.e. bellies and pieces, tender fleece and 2nd shear tends to be bought at auction by the exporters, at the lowest possible price. This wool is not normally colour sorted and is an "all-in" blend.

Present Selling of Natural Coloured Wool

Wool Sold at Auction

The New Zealand Wool board has taken a low profile in the promotion of natural coloured wool. Their argument is that black wool offered for auction is of low quality (i.e. tender wool, second shear, bellies and pieces etc.) and is therefore not worth much. This auctioned wool is mainly exported to Europe and South East Asia.

The bulk of good commercial coloured wool is sold by the present coloured wool pools and brokers by private arrangement and never hits the auction floor.

Handcraft fleeces are very rarely offered at auction. If so they tend to be lumped in with the normal commercial wool.

Individual Sales

Many breeders sell their own handcraft spinning wool, through local spinning groups, gate sales or mail order. Some breeders manufacture their own commercial yarn, operate carding companies, or have pools of knitters who spin and knit sweaters from locally grown wool. Some also buy wool in from other local growers.

The good commercial coloured wool is not sold at auction, despite efforts in 1988 to get both growers and auctioneers to insist that it be auctioned.

Company Sales

There are three main companies in New Zealand who have tried with varying success to pool natural coloured wool, and grade it into quality and colour for various end uses that are available. The main problems are the cost of freight and binning and sorting into different lines. As the quantity of wool is not large, if a customer requires a larger order of 90 bales for a container for export, it is often difficult to locate this number within one area.

Collective Sales

Many different schemes have been looked into, and are in fact still being pursued, to try to collect all the coloured wool in New Zealand into say, two places, one in the North Island and another in the South. The major problem to this proposal is that members of the Black & Coloured Sheep Breeders Association of New Zealand produce only 25% of the coloured wool. The rest comes from unregistered flocks, and from the throw-backs from white sheep farms.

How to motivate these 75% of producers to join a system is a massive problem. The Black and Coloured Sheep Breeders have been working hard over the last 10 years to breed sheep that produce a superior fleece type and market the wool, but we still only represent 25% of the coloured wool produced in the country.

The Wool Board takes 6% of the selling price, but unfortunately we have not seen any positive results directly affecting natural coloured wool. Only the promotion of wool and the Woolmark in general. However a separate levy on behalf of the Black & Coloured Sheep Breeders Association of New Zealand was operating for a short time through the approved pools to help offset the Association's marketing costs.

Market Research

Market Segmentation

The total market is segmented along the lines of fleece quality and type.

The main segments are handcraft, commercial, and oddments. Within each of these segments are both domestic and export buyers.

Market Needs

Handcraft These fleeces are sold individually and must be prepared meticulously by eliminating bellies, pieces, britch and tender patches. Buyers like to have a range of colours and qualities available to choose from. The finer wools are for knitting and the coarser wools are in demand by weavers.

Generally hand craft spinners prefer the lighter natural coloured fleeces, but these are harder to sell commercially. The darker fleeces are generally harder to sell for the spinners' markets but command a slightly higher price for the commercial yarn. There are of course other variables such as; moorit, multi-coloured or spotted fleeces.

As well as greasy fleeces there is a demand for washed and carded wool as either individual fleeces or as coiled sliver.

Domestically this wool is often bought direct from the grower who prepares the fleece.

The export trade is through firms who specialise in buying and exporting these craft fleeces mainly to the North American market. They are not interested in buying bales of unskirted wool but prefer all the preparation to be done by the grower.

Commercial This wool must be sorted for colour and if possible for fineness. Commercial buyers prefer one tonne lots as this is the minimum quantity for commercial spinning into yarn.

Oddments Oddments are used for blending with Russian Karakul coarse wool-hard twist carpet yarn, or in commercial felt and other end uses where quality and colour are not so important.

The wool trade calls all non-white wool "black", regardless of the actual colour. Most wool in this category is sold to exporters who prefer to buy to 90-bale lots, as this is enough to fill a container. The price is very dependent on the overseas market for "black wool", which is low compared with white wool. Several exporters have brought into New Zealand containers of European "black wool", mostly coarse hairy type wool and blended in the coloured oddments to make a more acceptable blend for spinning into carpet yarn for the Middle East.

Only one buyer within New Zealand has the capacity to use this type of wool in the manufacture of yarn.

The Future

One has to be stout-hearted to hang in there with natural coloured sheep. Like any business there are highs and lows, and with colour preference changing each year, we are often in a no-win situation.

There are only a handful of breeders who make a total living from the proceeds of their natural coloured sheep, the rest are mainly hobby farmers.

As the good quality commercial wool is largely sold privately and not at auction, the floor price of coloured wool as set by the Wool Board is based on the lower end of the market. To increase the overall floor price, this practice must cease.

One centralised collection point in each island would go a long way to eliminating the existing marketing problems.

Since the World Congress was held in New Zealand in 1984, there have been several new businesses founded on the use of coloured wool. Because of competition for raw material and the relatively small quantities of wool available, the coloured wool price has risen almost to the level of white wool. The general quality of wool offered for sale has improved, as has the standard of sheep.

The Black & Coloured Sheep Breeders of New Zealand now have a separate register of Pure Breeds within the flocks, all of which will add to the general value of the end products. Our Association and its members have worked hard to develop quality breed types in coloured sheep and to produce top class handcraft spinning wool.

The future for coloured wool in New Zealand is promising, provided that we work cooperatively to improve the coordinated marketing of our products.

Kahurangi: An Approach to Marketing

Sue and Ian Stewart
Kahurangi
RD4
Masterton, New Zealand

This paper has been written as a 'non professional' contribution—it really relates to how we have operated, rather than a researched 'this is how you should do it' paper. The intention is for you the reader to extract information appropriate to yourself and to compare similar ideas, hopefully, picking up some good ideas, and yet avoiding some of the mistakes we've made. It is not to be compared with the gospel according to Saint Marketing.

Marketing as Seen by Ian

Sue and I have decided to split the paper, demonstrating our own strengths and indicating that there is more than one way of selling. We both have completely differing ideals and ideas. I will talk about grasping opportunities. She will describe the methods by which she has retained and captured the buyers, many of whom I introduced to the business. I'm convinced that part of our success is the fact that being so different in attitude one of us appeals when the other one doesn't, and often some clients seem to respond to each equally.

Before anyone is successful in marketing one must have a product to market—in our case fleece and yarn. It is helpful however if that product is reliable, sound and potentially good. We were lucky at 'Kahurangi' in as much that we didn't have a lot of opposition; in both the fleece (handcraft) and yarn (commercially spun) we were early in the field of natural undyed woollen products. However to retain the market there must be adaptability to change, both by forward thinking (anticipation) and by demand, (customer preference e.g. good and bad suggestions being followed up).

If it can be said that we are successful, then the question can be asked, 'why?'

To me the most important word in marketing is 'Awareness'.

Firstly awareness from a seller's point of view:

Awareness of product and its availability.
Awareness of self and those in your team.
Awareness of clients and their demands.
Awareness of clients and their needs.
Awareness of change and an ability to do so.

Then looking from the buyer's side, 'awareness' is still an important word. How to make them aware of your product.

Very early on we chose a logo, and a name. 'Kahurangi' was chosen by a local Maori as being a meaningful name. It can be translated as 'blue skies' or 'little treasures'. In our case we didn't use a professional to design the logo, but there can be advantages in doing so.

Once we started in 'real' business I always have referred to ourselves as: Kahurangi does this, or Kahurangi does that. We always use our logo when we advertise—all our outgoing mail uses recycle stickers with the logo on—we printed our own letter head with Kahurangi Natural Wools, had boxes overprinted, and in any advertising we use our logo—in other words be proud of your company, your logo and always use every opportunity to make the public aware of your name.

Another word in my vocabulary referring to marketing is 'exposure'. Some people are extroverts and others introverts—in our case it's not difficult to see which one of us is the extrovert. To me watching for, seeing, and then seizing upon any opportunity to market is not embarrassing. I believe in the attitude of allowing every potential buyer to reach our product by touch, feel or sight. I encourage communication, especially the personal contact one. When we analyse our business, in particular our export section all contacts have been chance contacts. Two of our largest (one in Australia and one in Japan) are directly related to the World Congress held in New Zealand five years ago.

I need no convincing that one's own profile, attitude, generosity and personality are as important as the product. A successful business needs character; style and flare come later; it also needs sincerity. Compliments or complaints are usually genuine; accept and handle them with dignity, put your business and clientele ahead of self. You need them more than they need you! (Some of us may not like this reality).

Participation brings marketing opportunities. Join associations, e.g. NZBCSBA, Agricultural and Pastoral Shows, Small Farmers, and Breed Societies. Membership has many blessings: one it removes you from your own environment (change is as good as a rest), it puts you in contact with those of a similar base interest; it allows your self esteem to develop alongside others. Above all there are others whose needs are greater than yours, consequently sharing of ideas and life styles occurs. All of us have needs, it is only by teamwork, communication and participation (e.g. as we are at this Congress) we will grow. Offering to share with others has opened the door for many opportunities.

Finally, all the attitudes and principles described earlier are useless unless the product is sound, genuine, and offers value to the end user. Always be aware of standard, evenness, consistency, performance and reliability of your product. Accept the effort of others to survive and both you and they will do so. Share- care- and always be yourself. A genuine person is always a good marketer. Believe in yourself, and your products.

Marketing as Seen by Sue

I guess that if in 1974 we owned 20 coloured sheep and sold their fleeces, and in 1988 we owned 1000 coloured sheep and sold 100 handcraft fleeces, several thousand kg of processed undyed yarn, plus garments and sheepskins, then somewhere in between there has been some market development. Presenting this paper gives each of us an opportunity to look back and see just what our policy and procedure has been. During 1988 we discovered the Myers-Briggs indicator of temperament types, and this put into words and clarified some of Ian's and my personality traits, and has made us aware of our differences, and how we complement each other.

When we apply these differences to our coloured wool enterprise, I think that between us we bring a lot of skills to the task. The sheer scale of the business has meant that some division of labour has been necessary, and while I did help on the farm in the early years, the area of physical farming, production of stock and wool, decisions regarding which rams go to which ewe are now Ian's. I take responsibility for the packing and dispatch of products and organise labour to help with this.

However between the growing of the wool and the final dispatch of products it is very much a shared venture, and we approach this as two individuals.

We start in the woolshed where I'm in charge of the skirting of fleeces, and the weighing and pricing of same—my bottom-liner is 'When in doubt, throw it out'—feeling very strongly, that no-one wants to discard as unusable part of what they've paid for. Although we usually have 4 working on the fleeces to two shearers, there is a certain amount of time pressure and occasionally a larger handful than is necessary joins the skirtings. Ian keeps a watchful eye on proceedings as he moves sheep, presses bales of skirtings etc. and has some difficulty in not getting upset by the amount being discarded. However I maintain strongly that one is better to err on the heavy-handed side, and that attention to quality never goes amiss. We have certainly found in the marketing field that bad news travels faster than good news, so I go to some lengths to avoid there being any bad news. I suggest to our customers that if they don't like the product, they tell us, and if they do like it, then they should tell others!

We have grown slowly and steadily with the market, rather than by speculating in advance. We have never been in a financial position to do market research beforehand. However we are always on the lookout for new ideas. Ian tends to come up with them and I do the follow up. He's prepared to give anything a try, whereas I sound a note of caution.

We sell a small percentage of fleece and yarn to farm visitors such as spinning groups visiting from Wellington, for a day in the country, and bringing a picnic lunch. We are not on a busy highway but do get a few casual callers. We have gradually educated clients who want to come out to us to buy, to phone first. We don't always have someone at the house, and prefer not to end up in the position of paying someone to be there.

Phoning first saves them a wasted trip, and also avoids customers arriving at inconvenient times, especially just as we're getting in the car to go somewhere by a certain time.

We have tried to strike a balance between not putting all our eggs in one basket, and spreading ourselves too thinly. We have found that some time-consuming practices that were fine in the early days have had to be cut out. Choosing fleece 'on the hoof' was one, and apart from the time involved, we found that when the fleece came off it was not always what they had thought it would be. I've also said no to sending several samples of individual fleeces and waiting to hear which one a customer prefers—it is easier to send the fleece I think most likely, with emphasis on the fact that I am really willing to change it if it is not what they want.

In fact I'm always willing to change a product—you need to work on the premise that the customer is always right, even though there will be the occasional case when you know very well that they are not, and they are digging deep to try and find fault, when what has really happened is that they have actually changed their mind about what they wanted in the first place!

We sell fleeces, yarns, skins, garments—some of these are on a commission basis for clients who use a lot of our wool, but the main emphasis is on fleeces and yarns. The others add an extra dimension and interest.

New Zealand has a small and eventually limited market, so we have had to look at markets overseas. We have followed up every enquiry we have ever had, as one never knows what will lead where.

We do advertise—it is expensive, but you can't be everywhere yourself. It is vital to keep one's name in public and for people to know one is still in business. There is a story told in New Zealand of a business that spent $6 million last year on advertising and half was wasted. As it is impossible to know which half, the message is obviously—you have no choice but to carry on.

The other way we keep ourselves known is to be at events—this is Ian's speciality. As well as direct sales and personal contact, it is advertising and perhaps part of the cost of travelling to and being at Craft Fairs, Shows etc. needs to be seen as that.

Another balance needs to be between not getting too big and thus losing control and the personal touch, and yet still developing, growing and being open to new ideas. I reckon that Ian manages to come up with one of these a day, which if were all followed up, would fragment us into a million pieces. I think I bring some system to this approach and sort the wheat from the chaff. Hopefully we don't become static and boring, but remain open to new possibilities without losing sight of what we're about.

I believe attention to detail is important, and keeping to deadlines essential. Good filing and recording systems are a must—inefficiency is appreciated by no-one, and losing orders or not replying promptly is part of the bad news that spreads very quickly. One pitfall I think we all make once in New Zealand, is not keeping the envelopes that American queries come in, and not all of you repeat the address on the letter.

However, above all, I think much of our success must be attributed to the fact that we are dealing with and marketing a product that we're enthusiastic about. We both love the farm, and are interested in adding value to our own produce. I know I'm a disliker of routine and thus require a variety in my work. It is good that we have reached a level where I can afford to pay someone to do the monotonous tasks such as endless labelling and packing. At the same time we are small enough to keep a 'hands on' involvement. Our last shearing day was a good example—in between working with the fleeces all day, 8 A.M. to 5 P.M. I also fed the shearer, and at lunch time finished typing the documents for a large yarn order that was going to Canada, and phoned the shippers regarding some bales of wool that were going to Japan. They say variety is the spice of life!

People have always been important and I really enjoy meeting the large number that our business brings us into contact with. 1989 is proving to be a significant year for me, and I am about to move into another type of shepherding with another sort of flock which will no doubt include some proverbial black sheep, and which I now know from experience are just as important, if not more so, than the traditional white ones.

Past History and Current Trends in Sheep and Wool on North Haven Island, Maine

Jane Robinson Parsons
Calderwood Hall
North Haven, Maine 04853 USA

I would like to share with you how an offshore Maine island has coped with shifting consumer interests, bureaucracy, import competition, isolation from mainland markets and economic instability to maintain viable sheep and wool businesses consistent with the 200-year tradition of island living.

The island is located 12 miles out in the North Atlantic off the coast of Maine, in the once famed coastal fishing grounds of New England. At one time there were many such inhabited islands, but now due to economic decline only a handful remain as viable year round communities.

Measuring approximately 20km by 5km, the island has two sheltered harbors and a long sheltered water way between neighboring islands to the south. Accessible now only by a rundown nine car ferry, private boats and occasionally with the two seat mail plane, the ebb and flow of traffic is ruled by the weather. Travel is hindered in summer by fog, in fall by high winds and seas, in winter by wind, snow and ice, in spring by wind, mud and slush. Because of this the island has remained almost untouched by the advances of the mid and late twentieth century.

Life on the island is also ruled by the weather and the seasons. Summer is a frantic time for many of the 350 year-round residents as they strive to earn most of the year's income in two to three short months while the seasonal trade is good. It is not uncommon for islanders to hold three jobs or more in trying to make ends meet. There is no industry and now only a few full-time fishermen. Almost everyone else is involved in the support services for the affluent east coast elite who have come to make the island their summer home since the 1890's. This summer influx has now grown enough to threaten the stability of the year round community due to the lack of available land and housing at affordable prices. With the passing of July and August people begin to relax and get busy putting all the big summer estates to bed for the winter. September and October find still a few visitors coming and going from the island, but to a large degree life has resumed a pattern that has been traditional for the last two hundred years. Quiet, isolated and tuned into Mother Nature for the next six months. Winter is beautiful, possibly the most humbling time of year, where joys are from the little things in life. Lambs are born from January to April, when the ground is frozen solid to a depth of three feet and ice covers the roads and water. The wind, named appropriately "The Arctic Express" blows constantly, making all-night lambing or fishing almost life- threatening for either sheep or man. Spring brings thaw weather, mud, anxiety over the coming of summer and a sense of once again losing control of the island to the summer visitors. Lambs grow up and need less attention than in winter. Grass greens up by mid May, just in time to lighten the load on the farmers, who are most likely working other jobs by now.

Historical Background

Historically documented in the shell middens, fishing has been the major economic focus of the island for the last 5000 years. Indians and their antecedents were not keepers of domesticated animals, so this discussion will begin with the arrival of the Europeans who brought with them their food supply of domesticated animals to sustain them during extended trips off the New England coast. These animals were commonly dumped off on small islands as an easily relocatable, easily caught, fatter and healthier food cache, protected somewhat from bears, wolves and Indians. With ship's holds empty, vessels could fill up with fish, lumber or goods traded with Indians, and still have a food supply whenever it was needed. The first documented flock of sheep on the mainland was in 1661, and the first permanent settlers brought sheep with them to the island in the 1760's. Flocks were isolated, sheep were small and inbreeding was a big problem, especially on the island. Since local markets were unavailable, sheep provided meat and wool for home use only.

From 1700–1760 Indian wars decimated both settlers and livestock. Early colonists discovered quickly that islands could be used to hide sheep from Indians and the King's agents, since under colonial rule, the production of finished goods was strictly prohibited and all raw wool and spun yarn was ordered sent back to England. In some areas colonial households were required to send a predetermined quota back to England on a regular basis, so dissenters and tax evaders looked toward the convoluted coast of Maine to find places to hide. Still effective today, government agents rarely take an entire day to track down offenders, when it means braving open ocean seas on a small ferry and possibly getting stranded overnight with no lodging.

The period between 1790 to 1850 was a time of real growth for sheep and wool, as local markets expanded, textile and spinning mills opened on the Maine rivers, and there was a big effort to upgrade the mongrelized colonial flocks with both Spanish and Saxon Merinos, thus improving fleece quality and carcass size. From the 1820's to 1840's a variety of breeds were introduced on the mainland, including Leicesters, Southdowns, Suffolks, Devonshires, Cotswolds, Oxfords, Cheviots, Hampshires, Shropshires, and Dorsets, in an attempt to find just the right sheep for a particular area or purpose. Exactly which of these breeds formed the foundation for the now well known "island sheep" isn't clearly documented but many characteristics point to considerable Cheviot blood with some other admixtures.

After 1910 on the mainland there was a switch to mutton breeds which tended to wipe out the effects of the earlier upgrading for fleece quality done in the first half of the 1800's. It appears that North Haven may have been spared some of that shift, or that it had less of an impact on the island since by 1910 North Haven had a reputation for producing the best lamb and wool on the coast and it is not likely that the island farmers, who are reluctant to change anything anyway, would have been taken in by the latest techniques on the mainland. In 1850 the island sheep population peaked at 2000 sheep on 71 farms, which would have provided an adequate gene pool from which to select the next generations. Island farmers were probably made aware of the effects of inbreeding earlier than mainland farmers, since the tendency is not to bring anything unknown onto the island. Thus no new sheep, no new diseases. With 2000 sheep, it could be imagined that considerable trading of good rams occurred, and if as few as five farms participated in a unidirectional rotation of ram lambs, quite a few generations could pass with no ill effects, although increased homogeneity in appearance and other traits could have occurred. This is seen in island flocks today.

Since 1850 the sheep population on the island has continually declined due to the slow exodus of farmers and their children, and the switch from a basically subsistence agricultural economy with farming and fishing combinations, to gradually increasing dependence on the seasonal trades associated with an affluent summer colony and tourism. The current population of sheep is slightly up from the all time low in the early 1970s and now lies around 150 breeding ewes on five farms. These farms are quite typical for Maine where the average number of sheep kept is 25 per farm. Three of these five farms have an interest in colored wool for the handspinning market, and with guidance have greatly improved fleece quality in conjunction with good carcass size over the last six years. The emphasis has been on developing small manageable flocks requiring minimal capital investment, and providing meat, fleece to send to the mills in trade for spun yarn or blankets, wool for handspinning or felting, and sheepskins for tanning.

Variables Affecting the Rise and Fall of Sheep and Wool on North Haven

The first variable to be discussed will be government regulations. As mentioned earlier, the first colonists came to this country to escape a variety of governmental restrictions, but no sooner were they settled than they had to abide by tax laws and quotas from England. By 1817, as the west outmigrations began to drain population away from the coast of Maine there was a government attempt to attract people to Maine from the south by offering land for $1.00 (one US Dollar) per acre, in hopes of keeping fields and farms from being abandoned. It didn't work then and the loss of permanent residents still continues on the island. The fact remains that it is easier to make a living elsewhere.

Facilities to run a seasonal business on the island are marginal, outmoded and do not meet government standards for lighting, electricity, doors, screens, windows, floors, drains, lavatories, etc. . . Home cooking or food packaging is either done for friends or under-the-table, so to speak.

The home knitting business which in the last 10 years employed up to 30 knitters, also had to be abandoned due to government regulations and the lobbying efforts by the garment workers union. Machine knitting at home is still viable but handknitters are not interested. Lamb was traditionally sold on the hoof to escape the governmental regulations, but recently it has been shown that there is a real demand for freezer lamb in small quantities during July and August.

Up until 1937 steamers plied the coast transporting all island produce to the Boston Market overnight. Lamb was slaughtered on the island, put on board ship in the evening and sold by morning. North Haven lamb set the price standard for both the Boston and New York markets, and was specifically listed on the menu at the Ritz in Manhattan. With the change over from steam to gasoline and the opening of new roadways, the steamer routes disappeared and lambs had to be funnelled through mainland auctions and markets at greatly reduced profit. Even now if this water transportation were reestablished, it would not be possible to send slaughtered lamb, as government inspection during slaughter, packaging and weighing is all strictly controlled at USDA approved slaughterhouses. Simply the cost of the proper labels and scales inspections are enough to diminish the profit margin.

Consequently, labelling and many of the exacting government standards are not well followed, continuing in the philosophy that if there are no complaints, and the service is professionally rendered, then no one is the wiser. Some Maine islands see themselves beyond the borders of the continental USA, and have their own ways of dealing with things.

On the positive side, the federal government still allows for the application for incentive payments for shorn wool and unshorn live lambs. For small farmers the payment amount might just cover the shipping fees if wool were sent to the mill via the United States Post Office. The drawback here though is the attitude of the islanders, who abhor the idea of a government subsidy, or any kind of welfare, and wouldn't apply for the payments.

Another variable of considerable importance is consumer interest. Fortunately the seasonal residents of the island remember the reputation of North Haven lamb and are quite proud to be able to serve it to their guests. Just five years ago the only lamb available for sale on the island in summer was frozen New Zealand lamb. All of the North Haven lambs were sold in the late fall after all the seasonal residents had gone home to the cities. A survey was conducted and it was determined that if Maine lamb were available the customers would pay extra to know it was Maine lamb, how it was raised and how old it was. Flavor however, would end the discussion forever, as when the first native lamb was marketed to the summer community the word spread quickly.

Consumer interest in natural colored wool followed that of the rest of the country, and it is in continual demand. Wool in general is the fiber of choice for the elite east coast crowd as they become increasingly aware of the shortcomings of synthetics along with pollution concerns. Maine islands are cold by comparison to the mainland in summer, and wool is a part of everyday wear around the water in all seasons. Considering a 200 year history, the short intercession of synthetics between 1950 and 1980 is just a small glitch. The buying power of young Americans is so great at the moment that they can afford all the woolens they want plus all the newest synthetics.

Of constant concern to the American farmer is the question of import competition. As early as the War of 1812 the American wool market was controlled by politics. During the period of 1806 to 1815 eighty new textile businesses were started in Maine, thirty of these in 1815 alone. This was all due to the unavailability of European goods. At the end of the war came a flood of woolen imports of higher quality and lower prices than the U.S. made materials. This glut dropped the bottom out of the U.S. sheep and wool production. The response was that by 1820, with the introduction of many Merinos, the U.S. mills were competing again with finer wools from domestic sheep. This pattern has continued to the present time with only the mills surviving that are able to adapt to world trade dictates, or find special local niches. One such mill in Harmony, Maine is the Bartlett Mill, in continuous operation since 1820 and going strong. Many small island farmers have come to depend on this mill to buy their wool, or to trade wool for yarn and blankets.

After extensive analysis in 1983–84 of mainland prices for retail goods which were handcrafted from wool, it was concluded that prices were too low to make much of a profit except in the most exclusive of shops in the biggest cities. Competition from developing countries in South America and Asia was the prime concern. For machine knits prices are low on the Maine coast, but good markets can be found in the larger cities of the U.S. and possibly in Europe as well, depending on the value of the dollar. These markets are often inaccessible to the average island resident who has minimal off-island business experience, arranging for airplane travel, attending shows, marketing, purchasing contracts, shipping and transportation costs, not to mention getting around in a city and dealing with outsiders. A good business advisor or companion is needed to give support enough to brave this experience. In general, handcrafts which are labor intensive cannot thrive on the open market in competition with foreign goods unless they border on art, or hourly wage is unimportant.

For most island businesses the best market is the one at home, during the season when visitors come. Imports are not a problem when people spend a delightful vacation on their "Fantasy Island" and want to take home reminders which were made right there by someone they've met and whose sheep they've fed. In this instance isolation works in the island's favor, as the customer tries to take home a little bit of this step back in time.

Before going on to a discussion of the current island businesses, it is important to understand that the islands off the coast of Maine are much less able to adapt to economic change than the mainland, and have a long history of economic instability. There usually aren't the variety of skills and businesses found in larger communities, resources are limited, education is limited and islanders are reluctant to change their ways, even a little. Anxiety among the islanders is high, as no two years seem to be alike either in weather or economics. Due to such uncertainty over the future, islanders characteristically like to dwell on the past and on tradition.

Current Businesses and Options in Sheep and Wool

Island style laissez-faire sheep management is based on minimal care with the sheep being left alone to fend for themselves for the entire year so that only the hardiest survive. Overhead is minimal and the products of fleece and meat are adequate, but depending on the harvest season, could often be better. Lambs take much longer to reach market weight under this system, but the overhead is not a factor here. Wool can be tender, contaminated with sand or vegetation, and severely weathered, but usually not caked with mud or manure. Survival rates are based on the carrying capacity of the island, on trespassers including dogs and poachers from neighboring ports, on the severity on any particular winter, and of course on the breed of sheep selected. The selection of the proper island makes a critical difference in providing a micro-climate where sheep can find shelter on their own, can find enough seaweed and beach vegetation without getting swept into the surf and where enough moisture collects in the spring to provide lactating ewes with water. Many shepherds never concern themselves with the water question, on the assumption that the dew and fog provide an ample supply. Another concern is that at high tide an island may be considered a fenced pasture, but seasonally when especially high and low tides are normal, sheep may just walk off on a sandbar. Also in winter they may venture out onto the ice, walking off onto a neighboring island, or even to the mainland. It has been reported that sheep will also swim a mile or for about five minutes, whichever comes first. Laissez-faire management is experimented with every few years by different people with differing degrees of success. It can be noted however, that colored sheep have a greater selective advantage, in that they can't be as easily seen by poachers. Native island sheep which were born under this system are very different in personality than mainland farm sheep. They will hunker down and hide rather than run or be herded, and they prefer to browse like goats. Even their offspring born in a barnyard exhibit these tendencies. Due to this, once a year roundups and shearing can be hair-raising events. And it is obvious what the effect is on the gene pool if one ram lamb escapes capture.

Another system of sheep management is using small islands as summer pasturage only. Generally property is owned by wealthy seasonal residents who may or may not be in residence when the sheep are on the island. The sheep are used to keep the fields open. Vistas over the Atlantic are especially attractive with a few sheep contentedly grazing. This system works well for both parties involved. Land is too expensive on the coast of Maine to buy an island on a farm budget and many of the islands are too rocky to have the pastures mowed by the estate caretaker. Contracted mowing now runs from $25 to $50 (twenty-five to fifty US dollars) per hour, with no baling, so it behooves the landowner to look for a farmer with sheep. The more unusual the sheep the more the landowners seem to like it, at it seems to add status. On many islands fencing needs are minimal and can be portable, run off a boat battery, so overhead is still low. One difficulty can be transportation. Most islanders have boats and friends with boats, so many combinations can be imagined. Vessels used include a World War II landing craft, lobster boats, rowboats, fishing dories, and a Boston Whaler. Sheep are easily trained as lambs to go down on docks and into boats, but older sheep may need some encouragement. Past experience has indicated that small boat transportation works best with lambs up to five or six months old.

Sheep also provide a business with good PR when an area is well known for sheep raising such as North Haven. Customers enjoy experiencing a part of a legend and seem willing to jump back into the past for a weekend of the "good old days." The local Inn on North Haven keeps a few colored sheep descended from a flock the proprietor once kept out on some of the more remote local islands. This can make for good storytelling and provides food for the Inn, besides creating the proper mood for a good experience. The guests like to help feed the sheep and may get so caught up in the idea that they go home with a raw fleece, or a handspun item or a new wool sweater kit from the yarn shop. Besides being enjoyable, it also makes the sheep a business deduction without going through the hassle of filing as a farm for income tax purposes.

Another business on the island which started with a small flock of sheep is North Island Design, a nationwide mail order business selling wool sweater kits both wholesale and retail. In its early stage of development, 1980 to 1984 wool was collected from several island farms and taken to the Bartlett Mill once a year to trade for spun fisherman yarn, which was sold or knit into classic sweaters for the North Island Yarn shop. This employed approximately 30 island knitters and was well received by all. However, as with other small cottage industries in the Northeast they began to have difficulty with the labor regulations and were being tormented by Union pressure; this aspect of the business was discontinued. Instead the business went into nationwide marketing of island designed and packaged sweater kits. The first year of this new venture far exceeded the supply of local wool, so the shop began buying yarn outright and just assembling the kits on the island. Much of the strength of the business lies in the tradition and history of wool on the island which all knitters seem to enjoy. Recently a book of the early designs has been published, showing many of the designs which are inspired by life on the island. Many feature natural colored backgrounds. In the past year the business has used approximately 910 kg of natural colored wool, 4100 kg of natural white wool and 1800 kg of dyed wool. There was a period of time when the yarn was imported, but now yarns spun in Maine and Nebraska predominate.

The last business to be discussed will be Calderwood Hall, a seven year experiment in a vertically integrated cottage industry, run by a husband and wife team. An abandoned community building provides both history and character to a 312 square meter retail arts and crafts shop. Items from approximately 30 artisans are sold on a sliding commission depending on the home of the artist, with preference given to island residents. No sales work is required of the exhibitors, but they are encouraged to demonstrate their work in the gallery. The shop is open from June to October with July and August being the peak months. The remainder of the year is spent producing items for sale which include breeding stock, raw wool, lamb cuts, home cooked lamb entrees, sheepskins, millspun yarn and blankets, handspun yarn and handknit hats, a variety of machine-knits for all ages, sheep notecards, t-shirts, mugs, aprons, original sheep portraits and sheep photography. About 25% of the gross sales are sheep related items which also include handhooked primitive wool rugs and handwoven clothing produced by off-island cottage industries. Other popular items such as marine landscapes, nautical antiques, handmade quilts, bird prints, boating t-shirts, notecards, scenic photography, handmade lampshades and fine furniture round out the selection.

In 1983 it was said that there was no profit to be made from raising sheep on the island. It appears now six years later, that it is just a question of management. Certainly the standard farm operation producing fat market lambs does not find the necessary profit margin to make it worthwhile, if the only option is selling to an oversaturated local market in the late fall. Analyzing the seasonal island market and national market, plus defining a niche not already occupied has paid off well for several island sheep businesses. As stated by Dr. Stanley Musgrave from the University of Maine, in conclusion of the 1984 Island Institute Conference, "All we have to do is unclutter our minds, define our objective and dare to achieve it, without putting very much overhead on the sheep that they must pay for.

References

Beacom, S. E., (1985) "Pulpit Harbor, Two Hundred Years" The North Historical Society.

Beveridge, N. P., (1976) "The North Island, Early Times to Yesterday." The North Haven Bicentennial Committee.

Musgrave, S. & Barton, B., (1984) The Historical and Regional Context of Sheep Raising in New England and Maine. In Proceedings, Second Island Institute Conference. Ed. Conkling, Philip. Rockland, Maine.

Ulbrich, K., (1984) The Use of Maine Islands for Summer Pasturage. In Proceedings, Second Island Institute Conference. Ed. Conkling, Philip. Rockland, Maine.

Hagerty, P., (1984) The Use of Maine Islands for Year Round Pasturage. In Proceedings, Second Island Institute Conference. Ed. Conkling, Philip. Rockland, Maine.

Pingree, C. & Anderson, D., (1988) "Maine Island Classics" North Island Designs, Inc. North Haven, Maine.

Coloured Wool Down Under and Way Down Under, 1970–1988

Trevor D. Semmens
Dept. Agriculture
GPO Box 192B
Hobart, Tasmania Australia

Commencing around 1970, there was a revival of interest in woollen handicrafts such as spinning and weaving. As a result natural (i.e. pigmented) wool again found favour. This resulted in a large increase in coloured sheep and wool production.

Presented in this paper are some trends and characteristics of the coloured wool industry in Tasmania and Australia as a whole. It covers the period 1970/71 to 1987/88.

In addition, mention will be made of recent developments in Tasmania with regard to coloured sheep. That is to, (1) breeding of Moorit sheep and (2) the possibility of a coloured sheepskin industry based on the Swedish Gotland Sheep.

Some Background

Tasmania, the island state of Australia varies greatly from the rest of Australia in terms of:

1. Climate
2. Greater diversity of farming enterprises
3. Sheep breeds—Tasmania's main breeds are Polwarth, Corriedale, Crossbred and Comeback. It has only 13% Merinos as against 97% in Queensland.
4. Amount of coloured wool auctioned. An increase of 13.6 times as against 3.9 times for the rest of Australia.

(Interestingly, there are some similarities between Tasmania and Oregon.

1. Both count Agriculture, Forestry and Tourism as major industries, and
2. Eugene and Hobart are on similar latitudes—Eugene 44°N and Hobart 43°S)

Industry Statistics and Comments

Coloured Sheep Numbers

Using Curtis'[1] estimate that the quantity of coloured wool auctioned represented 70% of production, the estimated numbers of coloured sheep were approximately:

	1970/71	*1987/88*
Tasmania	2,500	26,000
Australia	140,000	480,000

Coloured Wool Production

The number of bales and ratios of coloured wool to white wool sold at auction for Tasmania and Australia is presented in Table 1).

TABLE 1 Numbers of Bales and Ratio of Coloured Wool to White Wool Sold at Auction

	Tasmania			*Australia*	
Year	*Bales of coloured wool*	*Ratio of coloured : white*	*Coloured as percentage of all Australian coloured wool*	*Bales of coloured wool*	*Ratio of coloured : white*
	(No.)	*(No. of bales)*	*(%)*	*(No.)*	*(No. of bales)*
1970/71	45	1 : 2932	1.8	2450	1 : 2026
1971/72	111	1 : 1158	3.9	2820	1 : 1697
1972/73	110	1 : 996	2.8	3978	1 : 945
1973/74	N.A.	N.A.	N.A.	N.A.	N.A.
1974/75	67	1 : 1832	2.3	2905	1 : 1448
1975/76	61	1 : 1968	2.5	2426	1 : 1531
1976/77	68	1 : 1486	2.6	2602	1 : 1308
1977/78	80	1 : 1306	3.0	2630	1 : 1178
1978/79	144	1 : 718	3.7	3916	1 : 839
1979/80	260	1 : 386	5.0	5245	1 : 641
1980/81	441	1 : 231	6.3	6963	1 : 497
1981/82	352	1 : 297	3.8	9176	1 : 383
1982/83	306	1 : 339	3.2	9688	1 : 360
1983/84	250	1 : 392	2.5	9855	1 : 362
1984/85	320	1 : 299	3.5	9121	1 : 431
1985/86	187	1 : 559	2.2	8661	1 : 439
1986/87	474	1 : 221	5.5	8672	1 : 461
1987/88	456	1 : 216	5.4	8449	1 : 493

N.A. = Not available.

In spite of significantly increased amounts auctioned—an increase of 13.6 times in Tasmania and 3.9 times in Australia, it's contribution to the wool industry is still very small. In 1987/88, it had risen in quantitative terms to only 0.46% in Tasmania and 0.20% in Australia. Even still, the Australian Wool Corporation and the (white wool) industry generally are concerned about the possibility of contamination. A comment was made recently by an industry representative that he felt that contamination of white wool was set to increase due to:

(i) lack of attention to pulling out "black wool" at shearing,
(ii) the increased use of wide combs and the possible missing of small "black spots", and,
(iii) the increasing number of black sheep within normal flocks.

In value terms coloured wool's approximate contribution to the wool clip was:

	1970/71		*1987/88*	
	Amount ($)	*% of Total Wool Income*	*Amount ($)*	*% of Total Wool Income*
Tasmania	1,000	0.01	230,000	0.18
Australia	65,500	0.015	5,030,000	0.11

Wool Type Auctioned

In Tasmania, it is principally Crossbred with a small amount of Merino. However, in Australia as a whole, it has gone from a predominance of Merino wool to a predominance of Crossbred wool. This has been brought about by the demand for Crossbred type wools, (longer, freer and easier to use), mainly for craft purposes. Since 1981/82, (In Australia), the proportions of Merino to Crossbred wools has shown no change. See Table 2.

TABLE 2 Percentage of Coloured Wool Auctioned by Type

	Tasmania		*Australia*	
	Merino	*Crossbred*	*Merino*	*Crossbred*
1970/71	NIL	100	66	34
1971/72	7	93	66	34
1972/73	8	92	64	36
1973/74	N.A.	N.A.	N.A.	N.A.
1974/75	12	88	63	37
1975/76	7	93	67	33
1976/77	7	93	63	37
1977/78	NIL	100	58	42
1978/79	1	99	50	50
1979/80	NIL	100	39	61
1980/81	6	94	33	67
1981/82	1	99	26	74
1982/83	3	97	23	77
1983/84	3	97	22	78
1984/85	2	98	23	77
1985/86	5	95	24	76
1986/87	NIL	100	25	75
1987/88	NIL	100	29	71

N.A. = Not available

Wool Prices

Average prices paid at auction in Australia for similar types of wool, (white vs. coloured), together with the percentage discount for coloured wool are shown in Table 3. As can be seen, the average discount for these particular types was:

Merino	33%
Crossbred	42%

Wool sold outside the auction system varies greatly in price. The very best quality wool sold to crafts people commands \$6–7/kg., while average and poorer quality wools range from about 80 cents—\$3/kg. Thus the breeding and selling of coloured wools is not a great money earner—it's really just a nice hobby!

Prices received at auction for *all* coloured wool compared with all white wool is shown in Figure 1.

Averaged over the period studied, the prices (greasy basis), for coloured wool auctioned, as a percentage of white wool were:

Tasmania	48%	(i.e., discount of 52%)
Australia	53%	(i.e., discount of 47%)

Factors affecting prices paid at auction included:

- smaller lot sizes
- lower preparation standards
- limited demand, and
- lower yields (on average, 8.1% for Tasmania and 5.7% for Australia)

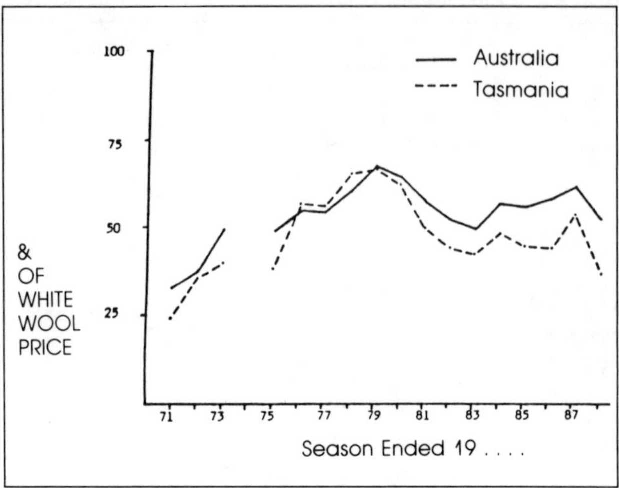

Figure 1 Prices revealed for all coloured wool as per cent of all wool auctioned (greasy basis).

Vegetable Fault

The degree of vegetable fault is categorised as:

F = Free or nearly free (to 1%)
L = Low (1–3%)
M = Medium (3–7%)
H = High (7%+)

Indications are that the percentage of vegetable fault in coloured wools auctioned in Australia has remained fairly

TABLE 3 Average Greasy Prices (cents/kg) for Similar Types of Wool (Australia)

	Wool Type (Merino)			*Wool Type (Crossbred)*		
Year	*79 (white) (a)*	*320 (coloured) (b)*	*Discount (%)*	*434 (white) (c)*	*607 (coloured) (d)*	*Discount (%)*
1970/71	83.5	32.9	61	78.3	25.5	67
1971/72	85.9	47.3	45	89.5	44.1	51
1972/73	213.5	145.2	32	248.7	142.5	43
1973/74	224.3	130.4	42	203.6	92.4	55
1974/75	159.9	101.3	37	145.5	47.0	68
1975/76	165.7	115.7	30	156.7	107.2	32
1976/77	211.6	141.8	33	213.5	139.1	35
1977/78	215.9	158.4	27	209.0	159.6	24
1978/79	226.3	177.4	22	224.1	163.8	27
1979/80	273.5	204.9	25	260.5	169.5	35
1980/81	289.6	205.9	29	267.2	169.8	36
1981/82	314.6	207.1	34	292.3	173.0	41
1982/83	331.2	212.3	36	286.3	167.9	41
1983/84	350.6	233.2	33	309.7	162.9	47
1984/85	371.9	269.5	28	290.7	174.6	40
1985/86	394.9	316.9	20	298.2	182.4	39
1986/87	463.8	308.2	34	354.8	225.8	36
1987/88	681.9	563.7	17	467.1	303.7	36
		Average	33		Average	42

Type (a) = Good topmaking, 22 micron, Merino Fleece, B length
Type (b) = Good fleece, F.N.F., Merino Black and Grey
Type (c) = Average 27 micron Crossbred Fleece, F.N.F. White
Type (d) = Good fleece, F.N.F., Medium Crossbred, Black and Grey

static up to the 1979/80 season. Since then there has been a trend to more 'free or nearly free' and 'low', and less 'medium' and 'heavy' faulted wool (Figure 2). This could be an indication of a greater proportion of fleece wool in coloured wool auctioned, and a corresponding drop in wool sold direct to crafts people.

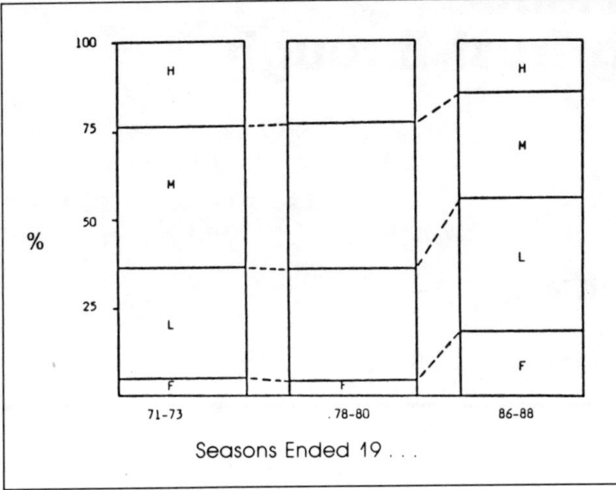

Figure 2 Vegetable fault—all Australian coloured wool auctioned

Recent Developments

Moorits in Tasmania

In 1987, two breeders imported some Moorit sheep for breeding purposes. Five came from New Zealand and three from South Australia. The idea being to breed up these sought-after sheep. Both breeders are sharing rams to reduce in-breeding and the progeny so far are promising.

Gotland Sheep Pelts in Tasmania

In 1988, a project was commenced by the Tasmanian Department of Agriculture with the following aims:

To "Investigate the objective traits of Gotland pelts that determine pelt quality and formulate recommendations on selection of animals for breeding purposes.

Specifically, to measure the traits which determine pelt quality of the complete range of Gotland Pelt types so that the objective assessment of pelt type can be made within an Australian industry.

Subsequently, to measure the objective characteristics of progeny of imported Gotland sheep and implement an objectively based breeding program".

The idea here is the possible development of a fashion (coloured) sheep pelt industry based on this breed. Obviously, it is early days, so we eagerly await results.

Conclusions

This study has shown,

1. That coloured wool production has increased markedly since 1970/71 and in terms of quantity, it reached its peak in Australia in 1983/84. In Tasmania, it is much higher but has been relatively static for the past six years or so.
2. Most coloured wool is of Crossbred type. This is because of its suitability for craft purposes.
3. Coloured wool prices showed a significant discounting due to lower yields and demand. There was a strong relationship between white and coloured wool prices. From what we have seen, the coloured wool industry is very small compared with the white wool industry. To put it in perspective, imagine total wool auctioned as a 170 kg bale. The quantity of auctioned coloured wool as a proportion of this bale would be:

	1970/71	*1987/88*
Tasmania	57 g	786 g
Australia	83 g	333g

However, it is still very important for breeders, processors, crafts people, craft shops, brokers and for the tourist industry.

References

[1]Curtis, C. (1979) "The black and coloured wool industry in Australia." Australian Wool Corporation.

[2]Australian Wool Corporation. "Australian Wool Sale Statistics, Statistical Analysis No's 1B (1970/71) to 18B (1987/88)."

Peace Fleece: Building Trust through Trade

Peter Hagerty
RFD1 Box 57
Kezar Falls, Maine 04047 USA

Moscow was cool in the early morning light and as I crossed Red Square, the clear blue sky promised another hot and dry August day. There was no dew on the grass or on the cobblestones. The corn we had seen from the air the previous evening was parched and brown. Signs of drought were everywhere.

This was my first trip to the Soviet Union. I had come with farmers from the corn, wheat and cotton growing States to meet with our Russian counterparts in agriculture, to share with them some of our experiences and to develop contacts for the future. Several of the mid-western farmers had felt first hand the pain of the Reagan Administration's embargo policies. Although they were Republican by tradition, they openly criticized our government's right to shut off the Soviet market.

I, myself, had come to the Soviet Union hoping to do some sheep shearing, meet farmers like myself and purchase some Soviet wool which I would bring home to Maine, blend with wool from mine and my neighbors sheep and spin into a yarn called Peace Fleece. This idea for Peace Fleece first arose in the spring of 1985. My neighbors and I would watch the evening news and see the leaders of the two superpowers acting like children with matches, yelling names and daring each other to drop the first flame. Although we felt in our hearts that the Soviet people were just like us, the fear of nuclear war numbed us and made us feel powerless in the face of the Administration's foreign policy.

"What can one person do?" we would ask each other. With a shrug of the shoulders, we would return to work.

But for me the sense of despair became too great. My feelings about my work on the farm began to change. What was the point of raising all these lambs, growing all this feed if all at once the sky should light up and everything I valued was gone.

With my wife's help, I began writing letters to the Soviet Union, to ministers of agriculture, doctors of veterinary medicine and to various ministers of trade. I received no reply. Then I went to the U.S. State Department, the Agricultural Department and the Commerce Department in Washington. No one knew anything of Soviet sheep farming practices. I could find no one in the domestic wool trade who had any experience with importing wool from the Soviet Union.

"They've never sold wool to the United States," one Manhatten broker told me, "and they probably never will." The only way it seemed I would get anywhere would be to go to the USSR myself. By chance there was a tour of farmers leaving the States in August. I signed aboard.

I had been told by several people here in the States that I must expect a long, drawn out process of phone calls and broken appointments in my search for Soviet wool. But thanks to the efforts of two very generous business people whom I met on my first day in Moscow, things turned out quite differently.

"It's really quite unusual," said Jesse, hanging up her office phone. "They have agreed to meet with you this afternoon and seem very interested that you are here."

Stephen and Jesse Nielsen had been living in Moscow for two years, working for an American firm that has over the years done extensive trade with the Soviets. They both spoke fluent Russian and held a high regard for the Soviets with whom they traded.

"You will find these men and women very open to a variety of ideas," they both assured me.

The Nielsen's driver took us far from the center of Moscow to the headquarters of Firm Runo, a government-owned corporation that handles the State's Five Year Plan for cotton, wool and synthetic fibres. Wool has been essential over the years for the survival of this nation. The largest producer of wool in the world, the Soviet Union's northern climate fuels an insatiable demand for wool and both Australia and New Zealand find extensive export markets here.

I felt very uncomfortable waiting in the birch panneled conference room. Caught off guard by the spontaneous meeting, I looked like a tourist and felt like a farmer. Stephen, impeccably dressed, read my mind.

"Just be yourself and let your idea stand on its own. We've been doing business with the Russians for over forty years. They will be very upfront about your prospects."

Firm Runo's Director Nikolai Emelianov was about forty years old and wore a tailored tropical suit. He smiled openly as we shook hands and sat down to read my Peace Fleece brochure, written in both Russian and English.

This was the third Russian person with whom I had talked at any length in my life. He was my age with about the same amount of gray hair. We both had gone to university, both had families. Yet most of the world would take us as enemies of the cold war. I felt very out of place. I knew how to grow lambs, make hay and help my kids with their back to school shopping. I wasn't at all sure how I would hold up in this world of international business.

"Mr. Hagerty, why do you want to do this Peace Fleece?" he asked forthrightly. I muttered something about building trust through trade.

"We have never exported wool to the United States before. Why should we do this for you?" I remembered something Gorbachev had said about the American small business person and finished up with capitalism being able to live with socialism and how I was afraid of nuclear war.

Before I knew it, we were talking about our children, his son, my daughter and parental dilemmas like television. I told him about my wife, my town, my farm. As we continued to talk about these things that really mattered to us, we both became more relaxed.

"But it is difficult, this Peace Fleece," he said. "Our country needs all the wool we can find . . . but perhaps we can do something.

And just like that, the wool samples were on the table, prices were discussed, a deal was made and we were shaking hands and smiling, agreeing to see each other again soon, perhaps in England in the spring.

"I am as close as your phone," he said. "Give me a call if you have any new ideas or run into any problems you want to discuss."

In those few short moments the world became a very small place for me. As we shook hands warmly in that Moscow office, I felt my despair give way to hope. I had felt this shifting of feeling before, sometimes in the barn when I felt the heartbeat in the chest of a newborn lamb after a difficult birthing. But it had come this time because a small bit of trust had passed between two men of warring nations, a trust that could be built on over time.

That handshake is now three years old and during this time, Peace Fleece and Soviet-American trade have become essential parts to my life. I still live on our sheep farm in Kezar Falls, Maine with my wife and children, but to my friends and neighbors I have become a world traveler and media celebrity. Our business has enjoyed prominence on the front pages of the *Wall Street Journal* and *Pravda,* has been profiled in *People* magazine and our woodstove has smoked and our sheep have been shorn for the Today Show. Our Soviet/Maine yarn is selling in over forty stores throughout the United States and our mailbox overflows with orders every day. Thanks to perestroika (restructuring of the Soviet economy) our eleven colors of yarn may be selling on Arabat Street in Moscow later this year. For the first time in my life I feel I have found my niche.

My wife Marty and I moved to southwestern Maine fourteen years ago from the city and readily embraced the work ethic of our neighbors. She started a pottery business and I joined a local logging crew, learned how to shear sheep, burned wood and made our own maple syrup. We slowly filled our old barn with sheep and fenced the rocky hillsides. Our daughter Cora was born in 1977 and our son Silas followed four years later. I shared the child care with Marty and more than one neighbor scolded me for allowing such a young girl to ride on a tractor.

One day in 1984, Cora came home and asked me to watch a television show with her that evening. "Teacher said she wants our moms or dads to be there." I agreed to sit in.

What I sat in on was "The Day After," the ABC Network depiction of nuclear war in Kansas. I vividly remember the scene that brought me to my knees that night. A rural farmwife was making her husband's bed when she looks out the window to see the ICBM leave its silo out in the back forty. At that moment she begins shaking with fear for she knows the end of the world has come. She turns back to the bed to finish its making.

Our valley where we live here in Maine is very quiet after dark. No traffic, no distant rumblings of trucks or highway sounds. Just the occasional bark of a dog. The night after that show, a sonic boom wrenched me from bed and I awoke sweating uncontrollably on the floor. The sky had turned orange, the trees were afire, the wall of the house nearest the driveway had fallen away, taking my two children to their deaths. All in the half second between the jet noise and my hitting the floor. I lay there unable to be consoled by my wife. It was several minutes before I could return to bed.

Over the next six months, I began to discover the extent of my despair. Not a day passed without a violent or depressing image robbing me of a pleasant moment. Even during the lambing and shearing seasons when there was so much activity and newborn life all around, I would find myself watching my daughter, fearing that she would never marry, never know the joy and the pain of birth. As I faced these horrible fantasies each day, I wondered how long they had been with me, lying quietly below my surface, waiting for a network docudrama to give them life. I was becoming a difficult person to be around.

My wife was the first to suggest that I travel to the Soviet Union, to meet with the "enemy" first hand, perhaps make some kind of business deal and reaffirm our ability to coexist.

I have been to the Soviet Union four times in the past four years. Every time I have had farmers with me to see the land, meet the workers, enjoy the culture and explore possibilities for trade. And each time I look forward most of all to my meetings with Mr. Emelianov.

On my latest trip to talk with Firm Runo, Emelianov was there to make me feel welcome, important, a valued customer of his firm. We talked wool, weather and the health of his wife's mother. Because Christmas was a few weeks off, I gave him gifts for his family made from our wool. We completed our business and I said goodbye.

Today business is booming and I am sleeping deep and long. My children are growing up and last week Cora had her first dance. It snowed first in November for hunting season but everyone was hoping for a fresh dusting by Christmas. Light flakes started falling early afternoon of Christmas Eve and the U.P.S. truck arrived late. But this time the driver came all the way into the house. In his hands was a package we knew right away must be special.

"From the Trade Office of the USSR," it read. Inside a bottle of Moldavian champagne, Latvian smoked fish and Russian caviar.

"Merry Christmas," wrote a familiar hand "and to all a good night."